D0323101

OXFORD MATHEMATICAL MONOGRAPHS

Series Editors

I. G. MACDONALD R. PENROSE H. MCKEAN

OXFORD MATHEMATICAL MONOGRAPHS

E. Belleni-Morante: *Applied semigroups and evolution equations*

I. G. Macdonald: *Symmetric functions and Hall polynomials*

J. W. P. Hirschfeld: *Projective geometries over finite fields*

N. Woodhouse: *Geometric quantization*

A. M. Arthurs: *Complementary variational principles* Second edition

P. L. Bhatnagar: *Nonlinear waves in one-dimensional dispersive systems*

N. Aronszajn, T. M. Creese, and L. J. Lipkin: *Polyharmonic functions*

J. A. Goldstein: *Semigroups of linear operators*

M. Rosenblum and J. Rovnyak: *Hardy classes and operator theory*

J. W. P. Hirschfeld: *Finite projective spaces of three dimensions*

Finite Projective Spaces of Three Dimensions

by

J. W. P. HIRSCHFELD
Reader in Mathematics, University of Sussex

CLARENDON PRESS · OXFORD
1985

Oxford University Press, Walton Street, Oxford OX2 6DP
Oxford New York Toronto
Delhi Bombay Calcutta Madras Karachi
Kuala Lumpur Singapore Hong Kong Tokyo
Nairobi Dar es Salaam Cape Town
Melbourne Auckland
and associated companies in
Beirut Berlin Ibadan Nicosia

Oxford is a trade mark of Oxford University Press

Published in the United States
by Oxford University Press, New York

British Library Cataloguing in Publication Data

Hirschfeld, J. W. P.

Finite projective spaces of three dimensions.—
(Oxford mathematical monographs)
1. Projective spaces 2. Finite geometrics
I. Title
516′.5 QA554
ISBN 0-19-853536-8

Library of Congress Cataloging in Publication Data
Hirschfeld, J. W. P. (James William Peter), 1940—
Finite projective spaces of three dimensions.
(Oxford mathematical monographs)
Continues: Projective geometries over finite fields.
1. Projective spaces. I. Title. II. Series.
QA471.H577 1985 516′.5 85-7300
ISBN 0-19-853536-8

Set and printed in the United Kingdom by
The Universities Press (Belfast) Ltd., Northern Ireland

To my family

Adrienne

Rachel

Benjamin

PREFACE

This is the second volume of a planned three, and represents Part IV of a treatise on projective spaces over a finite field. Parts I to III were all contained in the first volume, *Projective geometries over finite fields*. Part I gave introductory material on finite fields and projective spaces. Part II developed elementary properties of n-dimensional spaces over a finite field. Part III studied the line and the plane in detail. This volume, Part IV, is almost entirely devoted to three dimensions. Four and five dimensions necessarily appear when the lines in three dimensions are mapped to the points of a quadric in five dimensions. Part V, on n-dimensional spaces, will be contained in a further volume being written together with J. A. Thas.

The numeration of chapters in this book continues from *Projective geometries over finite fields*. So this book contains chapters 15 to 21. Any reference to *Projective geometries over finite fields* is generally given just by the section or proposition number. All references to other works are to be found in the last section of each chapter. If a work is dated before 1978 and the year is unstarred, then it is to be found in the Bibliography of *Projective geometries over finite fields*. If a work is dated before 1978 and is starred, then it is in the Bibliography of this book, as is any work from 1978 on.

I am grateful to the University of Bari for the invitation to give a course of lectures on the material of this book during the summer of 1984 and to the geometers there for their patience.

The criticisms of Aiden Bruen, Bill Kantor, and Joseph Thas led to improvements and the elimination of some errors. Aiden Bruen and Brian Wilson read the proofs. Jill Foster typed almost all of the manuscript. Sheila Collier, Phyllis Smith, Pauline Steele, and Melanie Suppel helped out in the final weeks. Janet Robertson drew the diagrams. To all of them I am deeply grateful.

University of Sussex J. W. P. H.
October 1984

CONTENTS

PART IV $PG(3, q)$

15. LINES 3
    - 15.1 Preliminaries 3
    - 15.2 Coordinates, linear complexes and polarities 4
    - 15.3 Quadrics 13
    - 15.4 The representation of lines of $PG(3, q)$ in $PG(5, q)$ 28
    - 15.5 Notes and references 31

16. OVALOIDS AND QUADRICS 33
    - 16.1 Ovaloids 33
    - 16.2 Characterization of quadrics 37
    - 16.3 Stereographic projection 40
    - 16.4 λ-polarities 41
    - 16.5 Notes and references 51

17. SPANS, SPREADS, AND PACKINGS 53
    - 17.1 Regular spreads 53
    - 17.2 Subregular spreads 62
    - 17.3 Aregular spreads 64
    - 17.4 Packings 66
    - 17.5 Packings of $PG(3, 2)$ and the geometry of $\mathscr{H}_{5,2}$ 68
    - 17.6 k-spans 76
    - 17.7 Hermitian arcs in $PG(4, q)$ 86
    - 17.8 Notes and references 90

18. k-CAPS 93
    - 18.1 Arithmetical preliminaries 93
    - 18.2 Examples of complete caps 96
    - 18.3 Caps in ovaloids for q even 99
    - 18.4 Caps in ovaloids for q odd 103
    - 18.5 Notes and references 110

19. HERMITIAN SURFACES 112
    - 19.1 Basic properties 112
    - 19.2 Lines on $\mathscr{U}$ 118
    - 19.3 Regular systems of lines on $\mathscr{U}$ 123
    - 19.4 Sets of type $(1, n, q+1)$ 138
    - 19.5 The characterization of Hermitian surfaces 156
    - 19.6 Sets of odd type in $PG(3, 4)$ 167
    - 19.7 Notes and references 178

20. CUBIC SURFACES WITH 27 LINES 182

20.1 Existence of a double-six 182
20.2 Structure of the surface 191
20.3 Surfaces over small fields 202
20.4 Mappings onto the plane 213
20.5 The representation of $\mathscr{F}_4$ in $PG(5, 2)$ 215
20.6 The classification of complex singular cubic surfaces by subsets of $PG(5, 2)$ 221
20.7 The representation in $PG(5, 2)$ of the 28 bitangents of a plane quartic curve 223
20.8 Notes and references 228

21. TWISTED CUBICS AND k-ARCS 229

21.1 Elementary properties of the twisted cubic 229
21.2 Characterization of the twisted cubic for q odd 242
21.3 $(q+1)$-arcs for q even 244
21.4 Further properties of the twisted cubic 253
21.5 Notes and references 259

APPENDIX III. ORDER OF AND ISOMORPHISMS AMONG THE SEMI-LINEAR GROUPS 260

AIII.1 Definitions 260
AIII.2 Comparative orders and isomorphisms in the same column of Table AIII.1 261
AIII.3 Simple groups 265
AIII.4 Isomorphisms between $DX(n, q)$ and $\mathbf{A}_m$ or $\mathbf{S}_m$ 265
AIII.5 Isomorphisms among $DX(n, q)$ for different characteristic p 265
AIII.6 Isomorphisms among $DX(n, q)$ for the same characteristic p 266
AIII.7 Notes and references 267

APPENDIX IV. THE NUMBER OF POINTS ON AN ALGEBRAIC VARIETY 269

AIV.1 The Weil conjectures 269
AIV.2 Curves 270
AIV.3 Notes and references 275

APPENDIX V. ERRATA FOR *Projective geometries over finite fields* (OUP, 1979) 276

BIBLIOGRAPHY 279

INDEX OF NOTATION 303

AUTHOR INDEX 309

GENERAL INDEX 311

PART IV

$PG(3, q)$

15

LINES

15.1 Preliminaries

As in § 3.1, $PG^{(r)}(n, q)$ is the set of r-spaces in $PG(n, q)$, and it will later be convenient to write

$$\Pi = PG^{(0)}(3, q), \qquad \mathscr{L} = PG^{(1)}(3, q), \qquad \Phi = PG^{(2)}(3, q).$$

Also, $\phi(r; n, q)$ is the number of r-spaces in $PG(n, q)$, $\chi(s, r; n, q)$ is the number of r-spaces through an s-space in $PG(n, q)$, and $\psi(t, s, r; n, q)$ is the number of r-spaces meeting a given s-space in some t-space. Then, from the corollary to Theorem 3.1.1,

$$\phi(0; 3, q) = \phi(2; 3, q) = \theta(3, q) = (q^2+1)(q+1),$$
$$\phi(1; 3, q) = (q^2+q+1)(q^2+1),$$
$$\chi(0, 1; 3, q) = \chi(0, 2; 3, q) = q^2+q+1,$$
$$\chi(1, 2; 3, q) = q+1,$$
$$\psi(0, 1, 1; 3, q) = q(q+1)^2.$$

In § 3.2, planes and lines are characterized as follows: a subset Π_2 of $PG(3, q)$ is a plane if and only if it has q^2+q+1 points and meets every line; a subset Π_1 of $PG(3, q)$ is a line if and only if it has $q+1$ points and meets every plane.

Lemma 15.1.1: *In $PG(3, q)$, the number of lines*
(i) *skew to a given line is q^4;*
(ii) *meeting two skew lines is $(q+1)^2$;*
(iii) *skew to each of two skew lines is $q(q^2-1)(q-1)$;*
(iv) *meeting three skew lines is $q+1$.*

Proof: (i) The number of lines through a point is $\chi(0, 1; 3, q) = q^2+q+1$. So the number of lines meeting a line l is

$$(q+1)(q^2+q) = q(q+1)^2 = \psi(0, 1, 1; 3, q)$$

as above. The number of lines skew to l is therefore $(q^2+1)(q^2+q+1) - q(q+1)^2 - 1 = q^4$.

(ii) Each point of a line l can be joined to each point of a skew line l'. Hence the number of transversals of l and l' is $|l|\,|l'| = (q+1)^2$.

(iii) The number of lines meeting l or l' is

$$2(q+1)[(q^2+q)-(q+1)] + (q+1)^2 + 2 = 2q^3+3q^2+1.$$

Hence the number skew to both l and l' is

$$(q^2+1)(q^2+q+1)-(2q^3+3q^2+1)=q(q^2-1)(q-1).$$

(iv) If the point P lies on neither of the skew lines l_1 or l_2, then $Pl_1 \cap Pl_2$ is the unique transversal of l_1 and l_2 through P. Hence, if l_3 is skew to l_1 and l_2, there is a unique transversal of l_1 and l_2 through each point of l_3. As l_1 and l_2 are skew, these $q+1$ transversals of l_1, l_2, and l_3 are mutually skew. □

The set of transversals of three skew lines is a *regulus*.

The group of projectivities of $PG(3, q)$ is $PGL(4, q)$ of order $p(4, q)= q^6(q^4-1)(q^3-1)(q^2-1)$.

A *pencil* of lines is the set of lines through a point that lie in a plane. A *star* of lines is the set of lines through a point. More generally, a *pencil* of spaces Π_r is the set of Π_r containing a given Π_{r-1} and lying in a given Π_{r+1}.

15.2 Coordinates, linear complexes and polarities

Given a line l, two of whose points are $\mathbf{P}(X)$ and $\mathbf{P}(Y)$, where $X= (x_0, x_1, x_2, x_3)$ and $Y=(y_0, y_1, y_2, y_3)$, a *coordinate vector* of l is

$$L=(l_{01}, l_{02}, l_{03}, l_{12}, l_{31}, l_{23})$$

where $l_{ij}=x_iy_j-x_jy_i$: then L is determined by l up to a factor of proportion. Write

$$l=\mathbf{l}(L).$$

If $\boldsymbol{\pi}(U)$ and $\boldsymbol{\pi}(V)$ are two planes through l, where $U=(u_0, u_1, u_2, u_3)$ and $V=(v_0, v_1, v_2, v_3)$, then a *dual coordinate vector of l* is

$$\hat{L}=(\hat{l}_{01}, \hat{l}_{02}, \hat{l}_{03}, \hat{l}_{12}, \hat{l}_{31}, \hat{l}_{23}),$$

where $\hat{l}_{ij}=u_iv_j-u_jv_i$.

Lemma 15.2.1: $\rho\hat{l}_{ij}=l_{rs}$ *where* $\{i, j, r, s\}=\{0, 1, 2, 3\}$; *that is, up to a factor of proportion, $\hat{L}$ is L with its components in reverse order.* □

Write $\varpi(l)=l_{01}l_{23}+l_{02}l_{31}+l_{03}l_{12}$. For q odd, we may write $\varpi(l)= \hat{L}L^*/2$. The *mutual invariant* of two lines $l=\mathbf{l}(L)$ and $l'=\mathbf{l}(L')$ is

$$\begin{aligned}\varpi(l, l')&=\hat{L}L'^*=L\hat{L}'^*\\ &=l_{01}l'_{23}+l_{02}l'_{31}+l_{03}l'_{12}+l_{12}l'_{03}+l_{31}l'_{02}+l_{23}l'_{01}.\end{aligned}$$

Lemma 15.2.2: (i) *For any line l,* $\varpi(l)=0$; (ii) *the two lines l and l' intersect if and only if* $\varpi(l, l')=0$. □

To a line $l = \mathbf{l}(L)$ are associated the matrices Λ and $\hat{\Lambda}$, where

$$\Lambda = \begin{bmatrix} 0 & -l_{23} & -l_{31} & -l_{12} \\ l_{23} & 0 & -l_{03} & l_{02} \\ l_{31} & l_{03} & 0 & -l_{01} \\ l_{12} & -l_{02} & l_{01} & 0 \end{bmatrix} = \begin{bmatrix} 0 & -\hat{l}_{01} & -\hat{l}_{02} & -\hat{l}_{03} \\ \hat{l}_{01} & 0 & -\hat{l}_{12} & \hat{l}_{31} \\ \hat{l}_{02} & \hat{l}_{12} & 0 & -\hat{l}_{23} \\ \hat{l}_{03} & -\hat{l}_{31} & \hat{l}_{23} & 0 \end{bmatrix}$$

and

$$\hat{\Lambda} = \begin{bmatrix} 0 & -l_{01} & -l_{02} & -l_{03} \\ l_{01} & 0 & -l_{12} & l_{31} \\ l_{02} & l_{12} & 0 & -l_{23} \\ l_{03} & -l_{31} & l_{23} & 0 \end{bmatrix} = \begin{bmatrix} 0 & -\hat{l}_{23} & -\hat{l}_{31} & -\hat{l}_{12} \\ \hat{l}_{23} & 0 & -\hat{l}_{03} & \hat{l}_{02} \\ \hat{l}_{31} & \hat{l}_{03} & 0 & -\hat{l}_{01} \\ \hat{l}_{12} & -\hat{l}_{02} & \hat{l}_{01} & 0 \end{bmatrix}$$

Both Λ and $\hat{\Lambda}$ are skew-symmetric and singular, since

$$|\Lambda| = |\hat{\Lambda}| = \varpi(l)^2.$$

Lemma 15.2.3: *$l = \mathbf{l}(L)$ contains the point $\mathbf{P}(X)$ if and only if $X\Lambda = 0$. Dually, l lies in the plane $\boldsymbol{\pi}(U)$ if and only if $U\hat{\Lambda} = 0$.* □

A *linear complex* $\mathscr{A}$ is the set of lines $l = \mathbf{l}(L)$ *satisfying*

$$AL^* = 0,$$

where $A = (a_{01}, a_{02}, a_{03}, a_{12}, a_{31}, a_{23})$. Write $\mathscr{A}(l) = AL^*$.

A convenient notation for $\mathscr{A}$ is to write $\mathscr{A} = \boldsymbol{\lambda}(A)$ or $\mathscr{A} = \boldsymbol{\lambda}(AL^*)$. Thus, if $\mathscr{A} = (1, 0, 0, 0, 0, 1)$, we write $\mathscr{A} = \boldsymbol{\lambda}(1, 0, 0, 0, 0, 1)$ or, more usually, $\mathscr{A} = \boldsymbol{\lambda}(l_{01} + l_{23})$.

Let $\varpi(A) = a_{01}a_{23} + a_{02}a_{31} + a_{03}a_{12}$. Then $\mathscr{A} = \boldsymbol{\lambda}(A)$ is *special* if $\varpi(A) = 0$, in which case it consists of all the lines meeting the line $\mathbf{l}(\hat{A})$, the *axis* of the complex; if $\varpi(A) \neq 0$, then $\mathscr{A}$ is *general*.

If $\mathscr{A} = \boldsymbol{\lambda}(A)$ is general, it consists of $(q^2+1)(q+1)$ lines. There is a pencil of $q+1$ lines of $\mathscr{A}$ through every point and in every plane of the space. To see this, it is enough to observe that the lines of $\mathscr{A}$ through a point are those in the polar plane of the point with respect to the null polarity defined by $\mathscr{A}$. So $\mathscr{A}$ contains $(q^2+1)(q+1)$ pencils and each line of $\mathscr{A}$ lies in $q+1$ of its pencils.

If $\mathscr{A} = \boldsymbol{\lambda}(A)$ is special, it consists of $q(q+1)^2+1$ lines. Through any point off the axis and in any plane not through the axis, there is a pencil of $q+1$ lines of $\mathscr{A}$. However, there are q^2+q+1 lines of $\mathscr{A}$ through any point on the axis and in any plane through the axis. It should be remarked that the axis is a line of $\mathscr{A}$.

One way to see the properties of general and special linear complexes is look ahead to the representation in § 15.4. The existence of $q+1$ or

q^2+q+1 lines of a linear complex $\mathscr{A}$ through a point or in a plane corresponds in $PG(5, q)$ to the fact that a subspace Π_4 either contains a given subspace Π_2 or meets it in a Π_1.

At the end of this section it will be shown that these properties are sufficient to characterize linear complexes.

Theorem 15.2.4: *General and special linear complexes in $PG(3, q)$ are projectively unique and have the following canonical forms:*

(i) *$\mathscr{A}$ general: $\mathscr{A}=\boldsymbol{\lambda}(l_{01}+l_{23})$;*

(ii) *$\mathscr{A}$ special: $\mathscr{A}=\boldsymbol{\lambda}(l_{01})$.*

Proof: A linear complex $\mathscr{A}$ is given by $\mathscr{A}(l)=\sum'' a_{ij}l_{ij}=\sum'' a_{ij}(x_iy_j-x_jy_i)$. The bilinear form is singular if and only if $\mathscr{A}$ is special. If $\mathscr{A}$ is general, it therefore induces a null polarity, which has the canonical form

$$(x_0y_1-x_1y_0)+(x_2y_3-x_3y_2),$$

from Theorem 5.3.2.

If $\mathscr{A}$ is special, choose its axis as $\mathbf{U}_2\mathbf{U}_3$; then $\mathscr{A}=\boldsymbol{\lambda}(l_{01})$. □

Theorem 15.2.5: *In $PG(3, q)$, the total numbers of linear complexes are as follows:*

(i) *special: $(q^2+1)(q^2+q+1)$;*

(ii) *general: $q^2(q^3-1)$.*

Proof: (i) The number of possible axes is just

$$\phi(1; 3, q)=(q^2+1)(q^2+q+1).$$

(ii) The number of general complexes is

$$\theta(5, q)-\phi(1; 3, q)=q^2(q^3-1). \quad \square$$

Corollary: *The order of $PGSp(4, q)$, the subgroup of $PGL(4, q)$ fixing a general linear complex, is $q^4(q^4-1)(q^2-1)$.*

Proof: $|PGSp(4, q)|=p(4, q)/[q^2(q^3-1)]=q^4(q^4-1)(q^2-1)$. □

A *linear congruence* is the set of lines belonging to two linear complexes. Two linear complexes $\mathscr{A}=\boldsymbol{\lambda}(A)$ and $\mathscr{B}=\boldsymbol{\lambda}(B)$ are *apolar* if $\varpi(A, B)=AB^*=0$. If a linear congruence $\mathscr{C}$ is determined by $\mathscr{A}$ and $\mathscr{B}$, it is equally well determined by any two of the linear complexes $\mathscr{A}_{st}=\boldsymbol{\lambda}(sA+tB)$, s, $t\in\gamma$. The complex $\mathscr{A}_{st}$ is special if and only if

$$s^2\varpi(A)+st\varpi(A, B)+t^2\varpi(B)=0.$$

Let this equation have N roots s/t in γ^+.

Lemma 15.2.6: *The different types of congruences $\mathscr{C}$ are as listed in Table 15.1.* □

Table 15.1 *Congruence Types of* $\mathscr{C}$

N	Name of $\mathscr{C}$	Description	$\|\mathscr{C}\|$
0	elliptic	spread	q^2+1
1	parabolic	lines meeting the axis of a special complex and lying in an apolar complex	q^2+q+1
2	hyperbolic	lines meeting two skew lines, the *axes* of $\mathscr{C}$	$(q+1)^2$
$q+1$	degenerate	lines meeting two intersecting lines; that is, lines lying in a plane π or passing through a point P in π	$2q^2+q+1$

A set of lines in $PG(3, q)$ is *linearly dependent* if their coordinate vectors are linearly dependent. Thus, the maximum number of linearly independent lines is 6. The lines $l = \mathbf{l}(L)$ linearly dependent on the k lines $l_i = \mathbf{l}(L_i)$, $i \in \mathbf{N}_k$, are given by

$$L = t_1 L_1 + \ldots + t_k L_k$$

where

$$\sum{}'' t_i t_j \varpi(l_i, l_j) = 0.$$

We now consider the configurations arising when $k<6$.

Theorem 15.2.7: *Let $\mathscr{R}$ be the set of lines linearly dependent on the linearly independent lines $l_1, l_2, \ldots, l_k$. Then, for $2 \leq k \leq 5$, the possibilities for $\mathscr{R}$ are listed in Table* 15.2. □

All these sets $\mathscr{R}$ can be determined dually as the intersection of $6-k$ linear complexes, as indeed linear congruences, for example, were defined.

Polarities

From §§ 2.1 and 5.3, Table 15.3 lists the types of polarities in $PG(3, q)$, where the name, canonical form $F(X, Y)$, variety $\mathscr{F}$ of self-conjugate points and subgroup G of $PGL(4, q)$ preserving the polarity are given in each case. If $F(X, Y) = x_0 z_0 + x_1 z_1 + x_2 z_2 + x_3 z_3$, then the polarity is $\mathbf{P}(y_0, y_1, y_2, y_3) \rightarrow \boldsymbol{\pi}(z_0, z_1, z_2, z_3)$. Also ν is a non-square in $GF(q)$ and $\bar{x} = x^{\sqrt{q}}$ for q a square.

The ordinary and Hermitian polarities are described in subsequent sections on quadrics and Hermitian varieties. Now, some properties of null and pseudo polarities will be examined.

A general linear complex $\mathscr{A}$ determines a null polarity $\mathfrak{A}$ and conversely. Every point P of $PG(3, q)$ is self-conjugate with respect to $\mathfrak{A}$ and the polar plane $P\mathfrak{A}$ of P is the plane containing the pencil of lines of $\mathscr{A}$

Table 15.2 *Possibilities for* $\mathscr{R}$ $(2 \leq k \leq 5)$

k	Conditions on l_i	$\mathscr{R}$	$\lvert\mathscr{R}\rvert$
2	l_1 skew to l_2	$\{l_1, l_2\}$	2
2	l_1 meets l_2	pencil of lines coplanar and concurrent with l_1 and l_2	$q+1$
3	l_1, l_2, l_3 concurrent at P but not coplanar	all lines through P	q^2+q+1
3	l_1, l_2, l_3 coplanar in π but not concurrent	all lines in π	q^2+q+1
3	l_3 skew to l_1 and l_2 with $l_1 \cap l_2 = P$	pencil containing l_1 and l_2 plus pencil containing l_3 and the line joining P to $l_1 l_2 \cap l_3$	$2q+1$
3	l_1, l_2, l_3 skew	regulus containing l_1, l_2 and l_3	$q+1$
4	l_1, l_2, l_3, l_4 form *skew quadrilateral* $\mathscr{Q}$	hyperbolic congruence: lines meeting both diagonals of $\mathscr{Q}$	$(q+1)^2$
4	l_1, l_2, l_3, l_4 skew such that no l_i is tangent to the hyperbolic quadric defined by the other three	elliptic congruence	q^2+1
4	l_4 meets l_1, l_2 and l_3	parabolic congruence consisting of $q+1$ pencils defined by l_4 and a line of the regulus containing l_1, l_2, and l_3	q^2+q+1
4	l_1, l_2, l_3 concurrent at P; l_2, l_3, l_4 coplanar in π	degenerate congruence: lines through P plus lines in π	$2q^2+q+1$
5	l_1, l_2, l_3, l_4 skew with unique transversal l_5	special linear complex: all lines meeting l_5	$q(q+1)^2+1$
5	l_1, l_2, l_3, l_4, l_5 skew	general linear complex: pencil of lines through every point of the space	$(q^2+1)(q+1)$

through P. Conversely, the pole $\pi\mathfrak{A}$ of a plane π is the vertex of the pencil of lines of $\mathscr{A}$ in π.

For any polarity $\mathfrak{T}$ of $PG(3, q)$, a line l has a *polar line* $l' = l\mathfrak{T}$ consisting of the poles of the planes through l or, dually, l' is the intersection of the polars of the points on l. Then l is *self-conjugate* if l' meets l, and l is *self-polar* if $l' = l$.

Table 15.3 *Polarity Types of PG*(3, *q*)

	Name	$F(X, Y)$	$\mathscr{F}$	G
q odd	hyperbolic	$x_0y_1+x_1y_0+x_2y_3+x_3y_2$	$\mathscr{H}_3$	$PGO_+(4, q)$
	elliptic	$-2\nu x_0y_0+2x_1y_1+x_2y_3+x_3y_2$	$\mathscr{E}_3$	$PGO_-(4, q)$
	null	$x_0y_1-x_1y_0+x_2y_3-x_3y_2$	$PG(3, q)$	$PGSp(4, q)$
q even	null	$x_0y_1+x_1y_0+x_2y_3+x_3y_2$	$PG(3, q)$	$PGSp(4, q)$
	pseudo	$x_0y_0+x_0y_1+x_1y_0+x_2y_3+x_3y_2$	$\mathbf{u}_0$	$PGPs^*(4, q)$
q square	Hermitian	$x_0\bar{y}_0+x_1\bar{y}_1+x_2\bar{y}_2+x_3\bar{y}_3$	$\mathscr{U}_3$	$PGU(4, q)$

Suppose that for the null polarity $\mathfrak{A}$, the line l meets its polar l' at P. Then P is conjugate to every point of l and l'. So P is conjugate to another point of l. Hence $l=l'$ and l belongs to $\mathscr{A}$. Thus the self-conjugate lines are just the self-polar lines, which are the lines of the linear complex.

Before considering pseudo polarities, it is opportune to derive the formulas for the coordinate vectors of lines under projectivities and polarities. For this, we use the kth exterior power $\Lambda^{(k)}(T)$ of a matrix T. In fact, only $\Lambda^{(2)}(T)$ is required. The components of $\Lambda^{(2)}(T)$ are the 2×2 minors of T arranged in lexicographical order. That is, if $T=(t_{ij})$ is a 4×4 matrix, $\Lambda^{(2)}(T)$ is a 6×6 matrix whose rows and columns are denoted by 01, 02, 03, 12, 13, 23 and occur in this order, where the element in row ij and column rs is $t_{ij,rs}=t_{ir}t_{js}-t_{is}t_{jr}$. To fit in with our notation for line coordinates, we use a slight variation of $\Lambda^{(2)}(T)$ and define $T^{(2)}$ to be the 6×6 matrix with rows and columns denoted by 01, 02, 03, 12, 31, 23 in this order, where the element in row ij and column rs is defined by exactly the same formula as for $\Lambda^{(2)}(T)$.

Lemma 15.2.8: (i) *Let $\mathfrak{T}$ be the projectivity in $PG(3, q)$ given by $X'=XT$. If $l'=l\mathfrak{T}$, $l=\mathbf{P}(X)\mathbf{P}(Y)$, $l'=\mathbf{P}(X')\mathbf{P}(Y')$, $l=\mathbf{l}(L)$, and $l'=\mathbf{l}(L')$, then*

$$\rho L'=LT^{(2)};$$

that is,

$$\rho l'_{ij}=\sum{}'' l_{rs}t_{rs,ij},$$

where

$$t_{rs,ij}=t_{ri}t_{sj}-t_{rj}t_{si}.$$

(ii) *Let $\mathfrak{T}$ be a polarity in $PG(3, q)$ given by $\boldsymbol{\pi}(X')=\mathbf{P}(X)\mathfrak{T}=\mathbf{P}(\tilde{X}T)$,* where $\tilde{X}=\bar{X}=X^{\sqrt{q}}$ *or $\tilde{X}=X$ according as $\mathfrak{T}$ is Hermitian or not. If $l'=l\mathfrak{T}$, $l=\mathbf{P}(X)\mathbf{P}(Y)$, $l'=\boldsymbol{\pi}(X')\cap\boldsymbol{\pi}(Y')$, $l=\mathbf{l}(L)$ and $l'=\mathbf{l}(L')$, then*

$$\rho\hat{L}'=\tilde{L}T^{(2)};$$

that is,

$$\rho l'_{ij} = \sum{}'' \tilde{l}_{rs} t_{rs,uv},$$

where

$$\{i, j, u, v\} = \{0, 1, 2, 3\}.$$

Proof: (i) $x'_i = \sum x_r t_{ri}$, $y'_i = \sum y_r t_{ri}$. So, up to a constant factor,

$$\begin{aligned} l'_{ij} &= x'_i y'_j - x'_j y'_i \\ &= \sum x_r t_{ri} \sum y_s t_{sj} - \sum x_r t_{rj} \sum y_s t_{si} \\ &= \sum x_r y_s (t_{ri} t_{sj} - t_{rj} t_{si}) \\ &= \sum{}'' (x_r y_s - x_s y_r)(t_{ri} t_{sj} - t_{rj} t_{si}) \\ &= \sum{}'' l_{rs} t_{rs,ij}. \end{aligned}$$

(ii) From Lemma 15.2.1 and (i),

$$l'_{ij} = \hat{l}'_{uv} = \sum{}'' \tilde{l}_{rs} t_{rs,uv}. \quad \square$$

Corollary: (i) *If* $\mathfrak{A}$ *is the null polarity associated with the linear complex* $\mathscr{A} = \boldsymbol{\lambda}(A)$, *then*

(a) $l' = l\mathfrak{A} \Leftrightarrow \rho l'_{ij} = \varpi(A) l_{ij} - \mathscr{A}(l) a_{rs}$
$\Leftrightarrow \rho L' = \varpi(A) L - \mathscr{A}(l) \hat{A}$;

(b) *l and m are conjugate lines* $\Leftrightarrow$

$$\varpi(A)\varpi(l, m) = \mathscr{A}(l)\mathscr{A}(m).$$

(ii) *If, in* (i), $\mathscr{A} = \boldsymbol{\lambda}(l_{01} + l_{23})$, *then* $\varpi(A) = 1$. *So*

(a) $l' = l\mathfrak{A} \Leftrightarrow \rho L' = L - \mathscr{A}(l)\hat{A}$;

(b) *l and m are conjugate* $\Leftrightarrow \varpi(l, m) = \mathscr{A}(l)\mathscr{A}(m)$.

(iii) *If* $\mathfrak{T}$ *is a pseudo polarity given by*

$$F(X, Y) = x_0 y_0 + x_0 y_1 + x_1 y_0 + x_2 y_3 + x_3 y_2,$$

then

(a) $l' = l\mathfrak{T} \Leftrightarrow \rho(l'_{01}, l'_{02}, l'_{03}, l'_{12}, l'_{31}, l'_{23})$
$= (l_{23}, l_{02}, l_{03}, l_{02} + l_{12}, l_{03} + l_{31}, l_{01})$

(b) *l and m are conjugate* $\Leftrightarrow \varpi(l, m) = \mathscr{A}(l)\mathscr{A}(m)$ *where* $\mathscr{A} = \boldsymbol{\lambda}(l_{01} + l_{23})$.

(iv) *If* $\mathfrak{T}$ *is an ordinary polarity with self-conjugate variety* $\mathscr{F} = \mathbf{V}(a_0 x_0^2 + a_1 x_1^2 + a_2 x_2^2 + a_3 x_3^2)$,
then

(a) $l' = l\mathfrak{T} \Leftrightarrow \rho l'_{ij} = a_r a_s l_{rs}$;

(b) *l and m are conjugate* $\Leftrightarrow \sum'' a_i a_j l_{ij} m_{ij} = 0$.

(v) *If, in* (iv), $\mathscr{F} = \mathbf{V}(x_0^2 + x_1^2 + x_2^2 + x_3^2)$, *then*

(a) $l' = l\mathfrak{T} \Leftrightarrow \rho L' = \hat{L}$;

(b) *l and m are conjugate* $\Leftrightarrow LM^* = 0$. $\square$

(vi) *If $\mathfrak{T}$ is a Hermitian polarity given by*

$$F(X, Y) = x_0\bar{y}_0 + x_1\bar{y}_1 + x_2\bar{y}_2 + x_3\bar{y}_3,$$

then

(a) $l' = l\mathfrak{T} \Leftrightarrow L' = \rho\hat{\bar{L}}$,

(b) *l and m are conjugate* $\Leftrightarrow L\bar{M}^* = 0$. □

Note: In (iii) (b), the canonical form chosen for a pseudo polarity may mislead when another form is considered. If $\mathfrak{T}$ is a pseudo polarity given by

$$F(X, Y) = x_0y_0 + x_1y_1 + x_2y_2 + x_3y_3,$$

then $l = \mathbf{l}(L)$ and $m = \mathbf{l}(M)$ are conjugate if and only if $LM^* = 0$. So l is self-conjugate if $LL^* = 0$; that is,

$$l_{01} + l_{02} + l_{03} + l_{12} + l_{31} + l_{23} = 0.$$

Compare the proof of Theorem 5.2.9.

Theorem 15.2.9: *Let $\mathfrak{T}$ be a pseudo polarity in $PG(3, q)$, $q = 2^h$. Then*

(i) *there are $q+1$ self-polar lines for $\mathfrak{T}$ forming a pencil in the plane π of self-conjugate points with centre $\pi\mathfrak{T}$;*

(ii) *there are $q^2(q+1)$ self-conjugate but not self-polar lines, which with the self-polar lines form a general linear complex.*

Proof: (i) If $\mathfrak{T}$ is in canonical form as above, the plane π of self-conjugate points is $\mathbf{u}_0$, whose pole $\mathbf{U}_1$ lies in $\mathbf{u}_0$. Hence the self-polar lines are just the lines in $\mathbf{u}_0$ through $\mathbf{U}_1$.

(ii) From part (iii) of the corollary to Lemma 15.2.8, the line l is self-conjugate if $\varpi(l, l) = \mathscr{A}(l)^2$ where $\mathscr{A}(l) = l_{01} + l_{23}$; so $\mathscr{A}(l) = 0$. As there are $(q^2+1)(q+1)$ lines in a general linear complex, there are $(q^2+1)(q+1) - (q+1) = q^2(q+1)$ self-conjugate but not self-polar lines. □

Corollary: *In $PG(3, q)$, $q = 2^h$, the number of pseudo polarities is $q^2(q^3-1)(q^4-1)$, each of which has a projective group $PGPs^*(4, q)$ of order $q^4(q^2-1)$.*

Proof: If a pseudo polarity $\mathfrak{T}$ is given by the bilinear form $F(X, Y) = x_0y_0 + t(x_0y_1 + x_1y_0 + x_2y_3 + x_3y_2)$, then, for each t in γ_0, the self-conjugate lines of $\mathfrak{T}$ are the complex $\boldsymbol{\lambda}(l_{01} + l_{23})$. So, for each possible plane of self-conjugate points and each possible complex of self-conjugate lines, there are $q-1$ pseudo polarities. So the number of pseudo polarities is

$$(q^2+1)(q+1)q^2(q^3-1)(q-1) = q^2(q^3-1)(q^4-1).$$

Then $|PGPs^*(4, q)| = p(4, q)/[q^2(q^3-1)(q^4-1)] = q^4(q^2-1)$. □

To characterize linear complexes we consider the property $(A_{m,n})$ for a subset $\mathscr{A}$ of $\mathscr{L}$.

$(A_{m,n})$: There exist integers m and n with $1<m\leqslant n\leqslant q^2+q+1$ such that

(a) $|\mathscr{A}\cap\mathscr{P}|=1$ or m for every pencil $\mathscr{P}$;

(b) $|\mathscr{A}\cap\mathscr{S}|=m$ or n for every star $\mathscr{S}$.

Lemma 15.2.10: *If $\mathscr{A}$ satisfies $(A_{m,n})$, then $m=q+1$.*

Proof: Let $\mathscr{S}$ be a star with centre P and $|\mathscr{A}\cap\mathscr{S}|=m$. The lines of $\mathscr{A}\cap\mathscr{S}$ meet a plane π not through P at the points of an m-set $\mathscr{K}$, which meets every line of π in 1 or m points. Therefore, if τ_i is the number of i-secants of $\mathscr{K}$ in π, then, by equations (12.1),

$$\tau_1+\tau_m=q^2+q+1,\quad \tau_1+m\tau_m=m(q+1),\quad m(m-1)\tau_m=m(m-1).$$

Hence

$$\tau_m=1,\quad \tau_1=q^2+q,\quad m=q+1.\quad \square$$

Lemma 15.2.11: *If $\mathscr{A}$ satisfies $(A_{m,n})$, then $n=q+1$ or q^2+q+1.*

Proof: Let $\mathscr{S}$ be a star with centre P and $|\mathscr{S}\cap\mathscr{A}|=n$. Let π be a plane not through P. The lines of $\mathscr{S}\cap\mathscr{A}$ meet π in an n-set which meets every line in 1 or $m=q+1$ points. Hence, again by equations (12.1),

$$\tau_1+\tau_{q+1}=q^2+q+1,\quad \tau_1+(q+1)\tau_{q+1}=n(q+1),\quad q(q+1)\tau_{q+1}=n(n-1).$$

Eliminating τ_1 and τ_{q+1} gives

$$n^2-n(q^2+2q+2)+(q+1)(q^2+q+1)=0.$$

Therefore, either (i) $\tau_1=0$, $\tau_{q+1}=q^2+q+1$, $n=q^2+q+1$ or (ii) $\tau_1=q^2+q$, $\tau_{q+1}=1$, $n=q+1$. $\square$

Corollary: *If $\mathscr{A}$ satisfies $(A_{m,n})$, then either* (i) $m=n=q+1$ *or* (ii) $m=q+1$, $n=q^2+q+1$. $\square$

Theorem 15.2.12: *If $\mathscr{A}$ satisfies (A_{q+1,q^2+q+1}) and $\mathscr{A}\neq\mathscr{L}$, then $\mathscr{A}$ is a special linear complex.*

Proof: If there were only one star $\mathscr{S}$ with $|\mathscr{S}\cap\mathscr{A}|=q^2+q+1$, then through any point Q other than the centre P of $\mathscr{S}$ there would be $q+1$ lines of $\mathscr{A}$. Since every plane meets $\mathscr{A}$ in 1 or $q+1$ lines, the $q+1$ lines of $\mathscr{A}$ through Q are coplanar. Let QR be a line of $\mathscr{A}$ other than QP. Since RQ and RP are in $\mathscr{A}$, so is the pencil with centre R in the plane PQR. Now let S be any point not in the plane PQR. Then the plane through S containing the lines of $\mathscr{A}$ meets QR in a point P'. So the lines $P'P$, $P'Q$, $P'S$ are in $\mathscr{A}$ and not coplanar. So the star with centre P' is in $\mathscr{A}$, a contradiction.

So, let P and P' be the centres of the stars $\mathscr{S}$ and $\mathscr{S}'$ contained entirely in $\mathscr{A}$. Let π be a plane through PP'. Since π contains two pencils of $\mathscr{A}$

through P and P', all lines in π lie in $\mathscr{A}$. Hence the stars through every point of PP' lie in $\mathscr{A}$.

Now, suppose that Q is a point off PP' which is the centre of a star lying in $\mathscr{A}$. Then, as above, every point on a line QP'', with P'' on PP', is the centre of a star contained in $\mathscr{A}$. Hence every point in the plane QPP' is the centre of a star in $\mathscr{A}$; that is, $\mathscr{A} = \mathscr{L}$. This contradiction means that $\mathscr{A}$ comprises all lines meeting PP'; that is, $\mathscr{A}$ is a special linear complex. □

Theorem 15.2.13: *If $\mathscr{A}$ satisfies $(A_{q+1,q+1})$, then $\mathscr{A}$ is a general linear complex.*

Proof: Through each point of $PG(3, q)$, there are $q+1$ lines, which are necessarily coplanar. Thus, associating to each point P the plane π containing the lines of $\mathscr{A}$ through it, we have a null polarity. For, if P' is any point in π, then PP' is in $\mathscr{A}$ and the polar plane π' of P' contains PP', whence P is in π'. Hence $\mathscr{A}$ is the set of self-polar lines in a null polarity; that is, $\mathscr{A}$ is a general linear complex. □

Theorem 15.2.14: *If $\mathscr{A}$ is a set of lines other than $\mathscr{L}$ satisfying $(A_{m,n})$ then $\mathscr{A}$ is either a special or a general linear complex.* □

15.3 Quadrics

A quadric surface $\mathscr{F} = \mathbf{V}(F)$ in $PG(3, q)$ is given by $F = \sum' a_{ij}x_ix_j$. The *rank* of $\mathscr{F}$ is the smallest number of indeterminates appearing in F under any change of coordinate system.

Lemma 15.3.1: *The quadrics $\mathscr{F}$ of $PG(3, q)$ fall into six orbits under $PGL(4, q)$ with typical members as listed in Table* 15.4, *where f is an irreducible binary form.*

Proof: For ranks 1, 2, and 3, the forms and description follow from those for the plane, § 7.2. For rank 4, the canonical forms are those of Theorem 5.2.4. □

We now consider the properties of the cone, elliptic quadric, and hyperbolic quadric in some detail.

I. *Cone*

The cone $\mathscr{F} = P_0\mathscr{P}_2$ consists of the points on $q+1$ lines, the *generators*, through the vertex P_0, which is $\mathbf{U}_3$ for the given form. Any plane not through P_0 meets $\mathscr{F}$ in the points of a conic. So $\mathscr{F}$ contains q^2+q+1 points.

For q odd, the cone is the dual in $PG(3, q)$ of a line conic; that is, corresponding to $q+1$ lines in a plane no three of which are concurrent are $q+1$ lines through a point no three of which are coplanar.

Table 15.4 *Quadrics $\mathscr{F}$ in $PG(3, q)$*

Rank	Symbol	Description	Reducibility	F	$\lvert\mathscr{F}\rvert$
1	$\Pi_2\mathscr{P}_0$	repeated plane	reducible	x_0^2	q^2+q+1
2	$\Pi_1\mathscr{H}_1$	pair of distinct planes	reducible	x_0x_1	$2q^2+q+1$
2	$\Pi_1\mathscr{E}_1$	line (intersection of pair of complex conjugate planes)	irreducible	$f(x_0, x_1)$	$q+1$
3	$\Pi_0\mathscr{P}_2$	(quadric) cone	absolutely irreducible	$x_0^2+x_1x_2$	q^2+q+1
4	$\mathscr{E}_3$	elliptic quadric	absolutely irreducible	$f(x_0, x_1)+x_2x_3$	q^2+1
4	$\mathscr{H}_3$	hyperbolic quadric	absolutely irreducible	$x_0x_1+x_2x_3$	$(q+1)^2$

At each point P of $\mathscr{F}$ other than P_0, there is a tangent plane containing the generator PP_0 and the tangent line at P to any conic section through P. For q even, the tangent planes are collinear in the *nuclear line* of $\mathscr{F}$, which contains the nucleus of every conic section of $\mathscr{F}$. So the cone is analogous to the conic in that the dual variety consisting of the tangent primes is of order one in both cases.

From § 7.2 we recall that there are four types of plane quadrics, $\mathscr{P}_2$, $\Pi_0\mathscr{H}_1$, $\Pi_0\mathscr{E}_1$, $\Pi_1\mathscr{P}_0$, being respectively a conic, a line pair, a point, and a line. Here we have used the general $\mathscr{P}_n$, $\mathscr{H}_n$, $\mathscr{E}_n$ symbols introduced in § 5.2 for non-singular quadrics as well as the notion of a cone from § 2.6 (xii).

For any q, there are four types of planes with respect to $\mathscr{F}$. So we write

$$\Phi=\Phi_1\cup\Phi_2^+\cup\Phi_2^-\cup\Phi_3,$$

where

$\Phi_1=\{\Pi_2\in\Phi \mid \Pi_2\cap\mathscr{F}\sim\Pi_1\mathscr{P}_0\}$ is the set of tangent planes,

$\Phi_2^+=\{\Pi_2\in\Phi \mid \Pi_2\cap\mathscr{F}\sim\Pi_0\mathscr{H}_1\}$ is the set of planes through a pair of generators,

$\Phi_2^-=\{\Pi_2\in\Phi \mid \Pi_2\cap\mathscr{F}\sim\Pi_0\mathscr{E}_1\}$ is the set of planes through P_0 containing no generator, and

$\Phi_3=\{\Pi_2\in\Phi \mid \Pi_2\cap\mathscr{F}\sim\mathscr{P}_2\}$ is the set of planes meeting $\mathscr{F}$ in a conic.

Lemma 15.3.2: $|\Phi_1|=q+1$, $|\Phi_2^+|=\frac{1}{2}q(q+1)$, $|\Phi_2^-|=\frac{1}{2}q(q-1)$, $|\Phi_3|=q^3$. □

For the remainder of the discussion on the cone, it is necessary to separate the cases of q odd and even.

(i) q odd

Each tangent plane to $\mathscr{F}$ meets the q others in distinct lines through P_0. The points other than P_0 on these lines are *external points* of $\mathscr{F}$ and form the set Ψ_0^+. The other points off $\mathscr{F}$ are *internal points* of $\mathscr{F}$ and form the set Ψ_0^-. The simple points on $\mathscr{F}$ form the set Ψ_1 and the singular point P_0 we will write as the set Ψ_1'. Then $\Pi = \Psi_0^+ \cup \Psi_0^- \cup \Psi_1 \cup \Psi_1'$.

Similarly $\mathscr{L} = \mathscr{L}_1 \cup \mathscr{L}_1^+ \cup \mathscr{L}_1^- \cup \mathscr{L}_2^+ \cup \mathscr{L}_2^- \cup \mathscr{L}_3$, where $\mathscr{L}_3$ is the set of generators; $\mathscr{L}_1$ is the set of tangents at the simple points (excluding the generators); $\mathscr{L}_1^+$ is the set of *external vertex tangents*, which are intersections of pairs of tangent planes; $\mathscr{L}_1^-$ is the set of *internal vertex tangents*, which are the remaining lines off $\mathscr{F}$ through P_0; $\mathscr{L}_2^+$ is the set of bisecants of $\mathscr{F}$; $\mathscr{L}_2^-$ is the set of lines skew to $\mathscr{F}$.

Lemma 15.3.3: (i) $|\Psi_0^+| = \frac{1}{2}q^2(q+1)$, $|\Psi_0^-| = \frac{1}{2}q^2(q-1)$, $|\Psi_1| = q(q+1)$, $|\Psi_1'| = 1$.

(ii) $|\mathscr{L}_1| = q^2(q+1)$, $|\mathscr{L}_1^+| = \frac{1}{2}q(q+1)$, $|\mathscr{L}_1^-| = \frac{1}{2}q(q-1)$, $|\mathscr{L}_2^+| = \frac{1}{2}q^3(q+1)$, $|\mathscr{L}_2^-| = \frac{1}{2}q^3(q-1)$, $|\mathscr{L}_3| = q+1$. □

Since Π, $\mathscr{L}$ and Φ have been partitioned as described, we can count the various numbers of each type of subspace on every other type of subspace and this is done below. For example, there are two tables connecting Π and $\mathscr{L}$. One shows the number of the various types of point on each type of line and the other shows the number of the various types of line through each type of point. To avoid any ambiguity, each table has a fixed row sum.

Theorem 15.3.4: *The results of the theorem are set out in Tables* 15.5(*a*)–(*g*). □

(ii) q even

Every point of $PG(3, q)$ lies on a tangent plane of $\mathscr{F}$. Hence $\Pi = \Psi_0 \cup \Psi_0' \cup \Psi_1 \cup \Psi_1'$, where, as before, Ψ_1' and Ψ_1 are respectively the vertex and the simple points of $\mathscr{F}$, but now Ψ_0 is the set of points off $\mathscr{F}$ on just one tangent plane and Ψ_0' is the set of points off $\mathscr{F}$ on more than one tangent plane, which is just the set of nuclei of conic sections of $\mathscr{F}$.

In this case, $\mathscr{L} = \mathscr{L}_1 \cup \mathscr{L}_1' \cup \mathscr{L}_1'' \cup \mathscr{L}_2^+ \cup \mathscr{L}_2^- \cup \mathscr{L}_3$, where, as before, $\mathscr{L}_3$, $\mathscr{L}_1$, $\mathscr{L}_2^+$, and $\mathscr{L}_2^-$ are respectively the set of generators, tangent lines at the simple points, bisecants, and lines skew to $\mathscr{F}$; however, $\mathscr{L}_1'$ consists of the nuclear line only and $\mathscr{L}_1''$ is the set of lines through the vertex which are neither generators nor the nuclear line.

Lemma 15.3.5: (i) $|\Psi_1| = q(q+1)$, $|\Psi_1'| = 1$, $|\Psi_0| = q(q^2-1)$, $|\Psi_0'| = q$.

(ii) $|\mathscr{L}_1| = q^2(q+1)$, $|\mathscr{L}_1'| = 1$, $|\mathscr{L}_1''| = q^2-1$, $|\mathscr{L}_2^+| = \frac{1}{2}q^3(q+1)$, $|\mathscr{L}_2^-| = \frac{1}{2}q^3(q-1)$, $|\mathscr{L}_3| = q+1$. □

Table 15.5 (*a*) *Planes Through a Point*

	Φ_1	Φ_2^+	Φ_2^-	Φ_3
Ψ_0^+	2	$(q-1)/2$	$(q-1)/2$	q^2
Ψ_0^-	0	$(q+1)/2$	$(q+1)/2$	q^2
Ψ_1	1	q	0	q^2
Ψ_1'	$q+1$	$q(q+1)/2$	$q(q-1)/2$	0

(*b*) *Points in a Plane*

	Ψ_0^+	Ψ_0^-	Ψ_1	Ψ_1'
Φ_1	q^2	0	q	1
Φ_2^+	$q(q-1)/2$	$q(q-1)/2$	$2q$	1
Φ_2^-	$q(q+1)/2$	$q(q+1)/2$	0	1
Φ_3	$q(q+1)/2$	$q(q-1)/2$	$q+1$	0

(*c*) *Points on a Line*

	Ψ_0^+	Ψ_0^-	Ψ_1	Ψ_1'
$\mathcal{L}_1$	q	0	1	0
$\mathcal{L}_1^+$	q	0	0	1
$\mathcal{L}_1^-$	0	q	0	1
$\mathcal{L}_2^+$	$(q-1)/2$	$(q-1)/2$	2	0
$\mathcal{L}_2^-$	$(q+1)/2$	$(q+1)/2$	0	0
$\mathcal{L}_3$	0	0	q	1

(*d*) *Lines Through a Point*

	$\mathcal{L}_1$	$\mathcal{L}_1^+$	$\mathcal{L}_1^-$	$\mathcal{L}_2^+$	$\mathcal{L}_2^-$	$\mathcal{L}_3$
Ψ_0^+	$2q$	1	0	$q(q-1)/2$	$q(q-1)/2$	0
Ψ_0^-	0	0	1	$q(q+1)/2$	$q(q+1)/2$	0
Ψ_1	q	0	0	q^2	0	1
Ψ_1'	0	$q(q+1)/2$	$q(q-1)/2$	0	0	$q+1$

(*e*) *Lines in a Plane*

	$\mathcal{L}_1$	$\mathcal{L}_1^+$	$\mathcal{L}_1^-$	$\mathcal{L}_2^+$	$\mathcal{L}_2^-$	$\mathcal{L}_3$
Φ_1	q^2	q	0	0	0	1
Φ_2^+	0	$(q-1)/2$	$(q-1)/2$	q^2	0	2
Φ_2^-	0	$(q+1)/2$	$(q+1)/2$	0	q^2	0
Φ_3	$q+1$	0	0	$q(q+1)/2$	$q(q-1)/2$	0

Table 15.5 (contd.)

(*f*) *Planes Through a Line*

	Φ_1	Φ_2^+	Φ_2^-	Φ_3
$\mathscr{L}_1$	1	0	0	q
$\mathscr{L}_1^+$	2	$(q-1)/2$	$(q-1)/2$	0
$\mathscr{L}_1^-$	0	$(q+1)/2$	$(q+1)/2$	0
$\mathscr{L}_2^+$	0	1	0	q
$\mathscr{L}_2^-$	0	0	1	q
$\mathscr{L}_3$	1	q	0	0

(g) *Lines Meeting a Line in a Point*

	$\mathscr{L}_1$	$\mathscr{L}_1^+$	$\mathscr{L}_1^-$	$\mathscr{L}_2^+$	$\mathscr{L}_2^-$	$\mathscr{L}_3$
$\mathscr{L}_1$	$2q^2-1$	q	0	$q^2(q+1)/2$	$q^2(q-1)/2$	1
$\mathscr{L}_1^+$	$2q^2$	$(q+2)(q-1)/2$	$q(q-1)/2$	$q^2(q-1)/2$	$q^2(q-1)/2$	$q+1$
$\mathscr{L}_1^-$	0	$q(q+1)/2$	$(q-2)(q+1)/2$	$q^2(q+1)/2$	$q^2(q+1)/2$	$q+1$
$\mathscr{L}_2^+$	$q(q+1)$	$(q-1)/2$	$(q-1)/2$	$(q-1)(q^2+4q+2)/2$	$q^2(q-1)/2$	2
$\mathscr{L}_2^-$	$q(q+1)$	$(q+1)/2$	$(q+1)/2$	$q^2(q+1)/2$	$(q+1)(q^2-2)/2$	0
$\mathscr{L}_3$	q^2	$q(q+1)/2$	$q(q-1)/2$	q^3	0	q

As for q odd, we can produce tables for numbers of incidences between points, lines, and planes. Each table has a fixed row sum.

Theorem 15.3.6: *The results of the theorem are set out in Tables* 15.6(*a*)–(g). □

II. *Elliptic quadric*

The elliptic quadric $\mathscr{E}_3$ consists of q^2+1 points. As it contains no lines and as any line can meet $\mathscr{E}_3$ in at most two points, no three of the q^2+1 points are collinear. Since $q^2+1=(q+1)(q-1)+2$, any plane through 2 points of $\mathscr{E}_3$ meets it in $q+1$ points; these form a conic. Hence through each point P of $\mathscr{E}_3$ there are $q(q+1)$ planes whose section is a conic and one, the tangent plane, meeting $\mathscr{E}_3$ only at P. Thus $\Phi=\Phi_2^-\cup\Phi_3$, where Φ_2^- is the set of tangent planes and Φ_3 the set of planes meeting $\mathscr{E}_3$ in a conic.

The lines fall into three classes: $\mathscr{L}=\mathscr{L}_1\cup\mathscr{L}_2^+\cup\mathscr{L}_2^-$, where $\mathscr{L}_1$ is the set of tangent lines, $\mathscr{L}_2^+$ the set of bisecants and $\mathscr{L}_2^-$ the set of lines skew to $\mathscr{E}_3$. For convenience, we write $\Pi=\Psi_0\cup\Psi_1$, where Ψ_1 is the set of (simple) points on $\mathscr{E}_3$ and Ψ_0 those not on $\mathscr{E}_3$.

Lemma 15.3.7: (i) $|\Psi_0|=q(q^2+1)$, $|\Psi_1|=q^2+1$;

(ii) $|\Phi_2^-|=q^2+1$, $|\Phi_3|=q(q^2+1)$;

(iii) $|\mathscr{L}_1|=(q+1)(q^2+1)$, $|\mathscr{L}_2^+|=|\mathscr{L}_2^-|=q^2(q^2+1)/2$. □

Table 15.6 (*a*) *Planes Through a Point*

	Φ_1	Φ_2^+	Φ_2^-	Φ_3
Ψ_0	1	$q/2$	$q/2$	q^2
Ψ_0'	$q+1$	0	0	q^2
Ψ_1	1	q	0	q^2
Ψ_1'	$q+1$	$q(q+1)/2$	$q(q-1)/2$	0

(*b*) *Points in a Plane*

	Ψ_0	Ψ_0'	Ψ_1	Ψ_1'
Φ_1	$q(q-1)$	q	q	1
Φ_2^+	$q(q-1)$	0	$2q$	1
Φ_2^-	$q(q+1)$	0	0	1
Φ_3	q^2-1	1	$q+1$	0

(*c*) *Points on a Line*

	Ψ_0	Ψ_0'	Ψ_1	Ψ_1'
$\mathscr{L}_1$	$q-1$	1	1	0
$\mathscr{L}_1'$	0	q	0	1
$\mathscr{L}_1''$	q	0	0	1
$\mathscr{L}_2^+$	$q-1$	0	2	0
$\mathscr{L}_2^-$	$q+1$	0	0	0
$\mathscr{L}_3$	0	0	q	1

(*d*) *Lines Through a Point*

	$\mathscr{L}_1$	$\mathscr{L}_1'$	$\mathscr{L}_1''$	$\mathscr{L}_2^+$	$\mathscr{L}_2^-$	$\mathscr{L}_3$
Ψ_0	q	0	1	$q^2/2$	$q^2/2$	0
Ψ_0'	$q(q+1)$	1	0	0	0	0
Ψ_1	q	0	0	q^2	0	1
Ψ_1'	0	1	q^2-1	0	0	$q+1$

(*e*) *Lines in a Plane*

	$\mathscr{L}_1$	$\mathscr{L}_1'$	$\mathscr{L}_1''$	$\mathscr{L}_2^+$	$\mathscr{L}_2^-$	$\mathscr{L}_3$
Φ_1	q^2	1	$q-1$	0	0	1
Φ_2^+	0	0	$q-1$	q^2	0	2
Φ_2^-	0	0	$q+1$	0	q^2	0
Φ_3	$q+1$	0	0	$q(q+1)/2$	$q(q-1)/2$	0

Table 15.6 (contd.)
(*f*) *Planes Through a Line*

	Φ_1	Φ_2^+	Φ_2^-	Φ_3
$\mathscr{L}_1$	1	0	0	q
$\mathscr{L}_1'$	$q+1$	0	0	0
$\mathscr{L}_1''$	1	$q/2$	$q/2$	0
$\mathscr{L}_2^+$	0	1	0	q
$\mathscr{L}_2^-$	0	0	1	q
$\mathscr{L}_3$	1	q	0	0

(g) *Lines Meeting a Line in a Point*

	$\mathscr{L}_1$	$\mathscr{L}_1'$	$\mathscr{L}_1''$	$\mathscr{L}_2^+$	$\mathscr{L}_2^-$	$\mathscr{L}_3$
$\mathscr{L}_1$	$2q^2-1$	1	$q-1$	$q^2(q+1)/2$	$q^2(q-1)/2$	1
$\mathscr{L}_1'$	$q^2(q+1)$	0	q^2-1	0	0	$q+1$
$\mathscr{L}_1''$	q^2	1	q^2-2	$q^3/2$	$q^3/2$	$q+1$
$\mathscr{L}_2^+$	$q(q+1)$	0	$q-1$	$(q-1)(q^2+4q+2)/2$	$q^2(q-1)/2$	2
$\mathscr{L}_2^-$	$q(q+1)$	0	$q+1$	$q^2(q+1)/2$	$(q+1)(q^2-2)/2$	0
$\mathscr{L}_3$	q^2	1	q^2-1	q^3	0	q

The subgroup of $PGL(4, q)$ fixing $\mathscr{E}_3$ is $PGO_-(4, q)$.

Theorem 15.3.8: *The orbits of* $PGO_-(4, q)$ *acting*
(i) *on* Π *are* Ψ_0 *and* Ψ_1;
(ii) *on* $\mathscr{L}$ *are* $\mathscr{L}_1$, $\mathscr{L}_2^+$, *and* $\mathscr{L}_2^-$;
(iii) *on* Φ *are* Φ_2^- *and* Φ_3. □

As for the cone, we can complete tables of incidences for the various points, lines, and planes with one another. All tables have a fixed row sum.

Theorem 15.3.9: *Incidences for* $\mathscr{E}_3$ *in* $PG(3, q)$ *are as listed in Tables* 15.7(*a*)–(g). □

The quadric $\mathscr{E}_3$ induces a polarity $\mathfrak{E}_3$, which is ordinary for q odd and null for q even.

Theorem 15.3.10: *The polarity* $\mathfrak{E}_3$ *defined by* $\mathscr{E}_3$ *induces the following bijections*:
(i) $\Psi_0 \leftrightarrow \Phi_3$;
(ii) $\Psi_1 \leftrightarrow \Phi_2^-$;
(iii) $\mathscr{L}_1 \leftrightarrow \mathscr{L}_1$;
(iv) $\mathscr{L}_2^+ \leftrightarrow \mathscr{L}_2^-$.
For q *even, the bijection* (iii) *is the identity map on* $\mathscr{L}_1$.

Table 15.7

(*a*) *Planes Through a Point*

	Φ_2^-	Φ_3
Ψ_0	$q+1$	q^2
Ψ_1	1	$q(q+1)$

(*b*) *Points in a Plane*

	Ψ_0	Ψ_1
Φ_2^-	$q(q+1)$	1
Φ_3^+	q^2	$q+1$

(*c*) *Points on a Line*

	Ψ_0	Ψ_1
$\mathcal{L}_1$	q	1
$\mathcal{L}_2^+$	$q-1$	2
$\mathcal{L}_2^-$	$q+1$	0

(*d*) *Lines Through a Point*

	$\mathcal{L}_1$	$\mathcal{L}_2^+$	$\mathcal{L}_2^-$
Ψ_0	$q+1$	$\frac{1}{2}q(q-1)$	$\frac{1}{2}q(q+1)$
Ψ_1	$q+1$	q^2	0

(*e*) *Lines in a Plane*

	$\mathcal{L}_1$	$\mathcal{L}_2^+$	$\mathcal{L}_2^-$
Φ_2^-	$q+1$	0	q^2
Φ_3^+	$q+1$	$q(q+1)/2$	$q(q-1)/2$

(*f*) *Planes Through a Line*

	Φ_2^-	Φ_3
$\mathcal{L}_1$	1	q
$\mathcal{L}_2^+$	0	$q+1$
$\mathcal{L}_2^-$	2	$q-1$

(g) *Lines Meeting a Line in a Point*

	$\mathcal{L}_1$	$\mathcal{L}_2^+$	$\mathcal{L}_2^-$
$\mathcal{L}_1$	$q(q+1)$	$q^2(q+1)/2$	$q^2(q+1)/2$
$\mathcal{L}_2^+$	$(q+1)^2$	$(q^2-1)(q+2)/2$	$q(q^2-1)/2$
$\mathcal{L}_2^-$	$(q+1)^2$	$q(q^2-1)/2$	$(q^2-1)(q+2)/2$

Proof: For P in $\Pi\backslash\mathscr{E}_3$, the points of contact of the $q+1$ tangents through P to $\mathscr{E}_3$ all lie on a conic $\mathscr{P}_2$ in a plane π, which is the polar plane of P. For q even, P lies in π and is the nucleus of $\mathscr{P}_2$. For q odd, P is not in π. So the tangents through P form a pencil or a cone as q is even or odd. Conversely, a plane π of Φ_3 meets $\mathscr{E}_3$ in a conic $\mathscr{P}_2$ at each point of which there are $q+1$ tangent lines to $\mathscr{E}_3$ giving $(q+1)^2$ lines unisecant to $\mathscr{P}_2$. There is a unique set of $q+1$ of these unisecants, one through each point of $\mathscr{P}_2$, concurrent at a point P, which is the pole of π.

For P on $\mathscr{E}_3$, its polar is the tangent plane at P and, conversely, the pole of a tangent plane is its point of contact.

If l is a bisecant, the tangent planes at the points of $l \cap \mathscr{E}_3$ meet in a line l' skew to $\mathscr{E}_3$, which is the polar of l. Conversely, through a line l' skew to $\mathscr{E}_3$, there are two tangent planes whose points of contact lie on the line l, which is the polar of l'.

For q even, $\mathfrak{E}_3$ being null, every self-conjugate line is self-polar. So each tangent to $\mathscr{E}_3$ is self-polar and $\mathscr{L}_1$ is a general linear complex. For q odd, each tangent is self-conjugate but not self-polar. If l is a tangent with point of contact P lying in the tangent plane π and π' is any other plane through l with pole P', then P' is not on l, and PP' is a tangent at P and is the polar of l. □

Let $\mathscr{S}$ be the set of sublines $PG(1, q)$ of $PG(1, q^2)$. As above, Φ_3 is the set of planes meeting $\mathscr{E}_3$ in a conic.

Theorem 15.3.11: *There exist bijections $\mathfrak{T}_1: PG(1, q^2) \to \mathscr{E}_3$ and $\mathfrak{T}_2: \mathscr{S} \to \Phi_3$ such that, if $P \in PG(1, q^2)$ and $PG(1, q) \in \mathscr{S}$ with $P \in PG(1, q)$, then $P\mathfrak{T}_1 \in PG(1, q)\mathfrak{T}_2$.*

Proof: By Lemma 6.2.1, any subline $PG(1, q)$ on $PG(1, q^2)$ is a non-singular Hermitian variety $\mathscr{U}_{1,q^2}$. So, by § 6.2,

$$PG(1, q) = \mathscr{U}_{1,q^2} = \mathbf{V}_{1,q^2}(a_0 x_0 \bar{x}_0 + a_1 x_1 \bar{x}_1 + b x_0 \bar{x}_1 + \bar{b} \bar{x}_0 x_1),$$

where $\bar{a}_0 = a_0$, $\bar{a}_1 = a_1$ and $a_0 a_1 - b\bar{b} \neq 0$, with $\bar{x} = x^q$.

Take α in $GF(q^2) \backslash GF(q)$ with $\bar{\alpha}^2 \neq \alpha^2$; this can be done in $(q-1)^2$ ways for q odd and $q^2 - q$ ways for q even. Also, let $\beta = \alpha/(\alpha^2 - \bar{\alpha}^2)$. Then $\bar{\beta} = -\bar{\alpha}/(\alpha^2 - \bar{\alpha}^2)$, and

$$\begin{aligned} \alpha\bar{\beta} + \bar{\alpha}\beta &= 0, \\ \alpha\beta + \bar{\alpha}\bar{\beta} &= 1. \end{aligned} \tag{15.1}$$

Hence

$$x_2' = \alpha x_2 + \bar{\alpha} x_3 \quad \text{and} \quad x_3' = \bar{\alpha} x_2 + \alpha x_3$$

imply

$$x_2 = \beta x_2' + \bar{\beta} x_3' \quad \text{and} \quad x_3 = \bar{\beta} x_2' + \beta x_3'.$$

Let $\mathfrak{T}$ be the projectivity of $PG(3, q^2)$ given by

$$\mathbf{P}(x_0, x_1, x_2, x_3)\mathfrak{T} = \mathbf{P}(x_0, x_1, \alpha x_2 + \bar{\alpha} x_3, \bar{\alpha} x_2 + \alpha x_3).$$

Its effect on planes is therefore given by

$$\boldsymbol{\pi}(u_0, u_1, u_2, u_3)\mathfrak{T} = \boldsymbol{\pi}(u_0, u_1, \beta u_2 + \bar{\beta} u_3, \bar{\beta} u_2 + \beta u_3).$$

Consider $\mathfrak{T}_1$ and $\mathfrak{T}_2$ defined as follows:

$$\begin{aligned} P = \mathbf{P}(x_0, x_1) &\to \mathbf{P}(x_0\bar{x}_0, x_1\bar{x}_1, x_0\bar{x}_1, \bar{x}_0x_1) \\ &= Q' \to \mathbf{P}(x_0\bar{x}_0, x_1\bar{x}_1, \alpha x_0\bar{x}_1 + \bar{\alpha}\bar{x}_0x_1, \bar{\alpha}x_0\bar{x}_1 + \alpha\bar{x}_0x_1) \\ &= Q = Q'\mathfrak{T} = P\mathfrak{T}_1; \end{aligned}$$

$$\begin{aligned} PG(1, q) \to \boldsymbol{\pi}(a_0, a_1, b, \bar{b}) = \pi' \to \boldsymbol{\pi}(a_0, a_1, \beta b + \bar{\beta}\bar{b}, \bar{\beta}b + \beta\bar{b}) \\ = \pi = \pi'\mathfrak{T} = PG(1, q)\mathfrak{T}_2. \end{aligned}$$

Now,

$$P \in PG(1, q) \Leftrightarrow a_0x_0\bar{x}_0 + a_1x_1\bar{x}_1 + bx_0\bar{x}_1 + \bar{b}\bar{x}_0x_1 = 0.$$

Also,

$$\begin{aligned} P\mathfrak{T}_1 \in PG(1, q)\mathfrak{T}_2 &\Leftrightarrow Q \in \pi \\ \Leftrightarrow a_0x_0\bar{x}_0 + a_1x_1\bar{x}_1 &+ (\beta b + \bar{\beta}\bar{b})(\alpha x_0\bar{x}_1 + \bar{\alpha}\bar{x}_0x_1) \\ &+ (\bar{\beta}b + \beta\bar{b})(\bar{\alpha}x_0\bar{x}_1 + \alpha\bar{x}_0x_1) = 0 \\ \Leftrightarrow a_0x_0\bar{x}_0 + a_1x_1\bar{x}_1 &+ [(\alpha\beta + \bar{\alpha}\bar{\beta})b + (\alpha\bar{\beta} + \bar{\alpha}\beta)\bar{b}]x_0\bar{x}_1 \\ &+ [(\bar{\alpha}\beta + \alpha\bar{\beta})b + (\alpha\beta + \bar{\alpha}\bar{\beta})\bar{b}]\bar{x}_0x_1 = 0 \\ \Leftrightarrow a_0x_0\bar{x}_0 + a_1x_1\bar{x}_1 &+ bx_0\bar{x}_1 + \bar{b}\bar{x}_0x_1 = 0, \end{aligned}$$

using (15.1). This shows that $\mathfrak{T}_1$ and $\mathfrak{T}_2$ preserve incidence.

The point Q lies over $GF(q)$ and $Q' \in \mathbf{V}_{1,q^2}(x_0x_1 - x_2x_3)$. The projectivity $\mathfrak{T}$ maps Q' to Q and $\mathbf{V}(x_0x_1 - x_2x_3)$ to

$$\mathbf{V}((\alpha^2 - \bar{\alpha}^2)^2x_0x_1 + \alpha\bar{\alpha}x_2^2 - (\alpha^2 + \bar{\alpha}^2)x_2x_3 + \alpha\bar{\alpha}x_3^2) = \mathbf{V}(F).$$

All the coefficients of F lie in $GF(q)$. Further, $\mathbf{V}_{3,q}(F) = \mathscr{E}_3$ is an elliptic quadric, since $\alpha\bar{\alpha}x_2^2 - (\alpha^2 + \bar{\alpha}^2)x_2x_3 + \alpha\bar{\alpha}x_3^2$ is irreducible over $GF(q)$ although reducible over $GF(q^2)$ to $(\alpha x_2 - \bar{\alpha}x_3)(\bar{\alpha}x_2 - \alpha x_3)$; cf. Lemma 15.3.1. Hence $\mathfrak{T}_1$ exists. But

$$|PG(1, q^2)| = q^2 + 1 = |\mathscr{E}_3|.$$

As $\mathfrak{T}_1$ is injective, it is bijective.

Since $\mathscr{U}_{1,q^2}$ is non-singular, so $a_0a_1 - b\bar{b} \neq 0$, whence π' is not a tangent plane to $\mathbf{V}(x_0x_1 - x_2x_3)$. However, $\pi'\mathfrak{T} = \pi$ and so π is not a tangent plane to $\mathscr{E}_3$. Hence $\mathfrak{T}_2$ exists. By Lemma 4.3.1, Corollary 2(ii),

$$|\mathscr{S}| = s(1, q, q^2) = q(q^2 + 1) = |\Phi_3|.$$

As $\mathfrak{T}_2$ is injective, it is also bijective. □

The isomorphism of the collineation groups of $PG(1, q^2)$ and $\mathscr{E}_{3,q}$ follows from the existence of $\mathfrak{T}_1$ and $\mathfrak{T}_2$ and is explained in Theorem 15.3.18.

III. *Hyperbolic quadric*

We begin with a classical theorem, which is perhaps the most elementary theorem, not deducible from the propositions of incidence, about lines in $PG(3, q)$.

Theorem 15.3.12: *In $PG(3, q)$, $q > 2$, let $\{l_1, l_2, l_3, l_4\}$ and $\{l'_1, l'_2, l'_3, l'_4\}$ be two sets of four lines such that any two lines of the same set are skew and such that* 15 *of the* 16 *pairs $\{l_i, l'_j\}$ meet in a point; then the last pair also does.*

Proof: If the coordinate system is chosen so that $\mathbf{U}_0 = l_1 \cap l'_1$, $\mathbf{U}_1 = l_2 \cap l'_1$, $\mathbf{U}_2 = l_1 \cap l'_2$, $\mathbf{U}_3 = l_2 \cap l'_2$ and $\mathbf{U} = l_3 \cap l'_3$, then the coordinate vectors of the points are as in Figure 15.1, where $\mathbf{U}_i = \mathbf{P}(E_i)$ and $\mathbf{U} = \mathbf{P}(E)$. As $(E_0 + sE_1) + t(E_2 + sE_3) = (E_0 + tE_2) + s(E_1 + tE_3)$, l_4 meets l'_4 at the point with this coordinate vector. □

Note: If this theorem is formulated over a division ring K, then the proof makes it clear that the theorem is then true if and only if K is a field. So this would be a suitable setting for a geometric proof, still awaited, of Wedderburn's theorem that a finite division ring is a field.

In § 15.1, a regulus was defined as the set of lines meeting three skew lines. It consists of $q+1$ skew lines and, if l_1, l_2, and l_3 are any three of them, is denoted $\mathcal{R}(l_1, l_2, l_3)$. Theorem 15.3.12 says that the lines meeting l_1, l_2, and l_3 meet all lines of $\mathcal{R}(l_1, l_2, l_3) = \mathcal{R}$ and form a regulus, called the *complementary* regulus of $\mathcal{R}$.

The hyperbolic quadric $\mathcal{H}_3$ consists of $(q+1)^2$ points, which are all the points on a pair of complementary reguli. The two reguli are the two *systems of generators* of $\mathcal{H}_3$. For P on $\mathcal{H}_3$, the tangent plane at P is the plane containing the two generators through P. Thus, through P, there are two generators, $q-1$ tangent lines lying in the tangent plane, and q^2 bisecants, each meeting $\mathcal{H}_3$ in a further point. Hence any plane through P

	l_1	l_2	l_3	l_4
l_1'	E_0	E_1	E_0+E_1	E_0+sE_1
l_2'	E_2	E_3	E_2+E_3	E_2+sE_3
l_3'	E_0+E_2	E_1+E_3	E	$E_0+E_2+s(E_1+E_3)$
l_4'	E_0+tE_2	E_1+tE_3	$E_0+E_1+t(E_2+E_3)$	

Fig. 15.1

other than the tangent plane meets $\mathscr{H}_3$ in a conic. Hence $\Pi = \Psi_0 \cup \Psi_1$, where Ψ_1 is the set of points on $\mathscr{H}_3$ and Ψ_0 the set of points not on $\mathscr{H}_3$; $\mathscr{L} = \mathscr{L}_1 \cup \mathscr{L}_2^+ \cup \mathscr{L}_2^- \cup \mathscr{L}_3$, where $\mathscr{L}_3$, $\mathscr{L}_1$, $\mathscr{L}_2^+$, and $\mathscr{L}_2^-$ are respectively the sets of generators, tangents, bisecants, and lines skew to $\mathscr{H}_3$; $\Phi = \Phi_2^+ \cup \Phi_3$, where Φ_2^+ and Φ_3 are respectively the tangent planes and those meeting $\mathscr{H}_3$ in a conic.

Lemma 15.3.13: (i) $|\Psi_0| = q(q^2-1)$, $|\Psi_1| = (q+1)^2$; (ii) $|\Phi_2^+| = (q+1)^2$, $|\Phi_3| = q(q^2-1)$; (iii) $|\mathscr{L}_1| = (q+1)(q^2-1)$, $|\mathscr{L}_2^+| = q^2(q+1)^2/2$, $|\mathscr{L}_2^-| = q^2(q-1)^2/2$, $|\mathscr{L}_3| = 2(q+1)$. □

The subgroup of $PGL(4, q)$ fixing $\mathscr{H}_3$ is $PGO_+(4, q)$.

Theorem 15.3.14: *The orbits of $PGO_+(4, q)$ acting*
(i) *on Π are Ψ_0 and Ψ_1;*
(ii) *on $\mathscr{L}$ are $\mathscr{L}_1$, $\mathscr{L}_2^+$, $\mathscr{L}_2^-$, and $\mathscr{L}_3$;*
(iii) *on Φ are Φ_2^+ and Φ_3.* □

Now we can give tables for the number of incidences of the various types of points, lines, and planes. All tables have a fixed row sum.

Theorem 15.3.15: *Incidences for $\mathscr{H}_3$ in $PG(3, q)$ are as listed in Tables* 15.8(*a*)–(g). □

As in the case of $\mathscr{E}_3$, the quadric $\mathscr{H}_3$ induces a polarity $\mathfrak{H}_3$ which is ordinary for q odd and null for q even.

Theorem 15.3.16: *The polarity $\mathfrak{H}_3$ defined by $\mathscr{H}_3$ induces the following bijections:*
(i) $\Psi_0 \leftrightarrow \Phi_3$; (ii) $\Psi_1 \leftrightarrow \Phi_2^+$; (iii) $\mathscr{L}_1 \leftrightarrow \mathscr{L}_1$;
(iv) $\mathscr{L}_2^+ \leftrightarrow \mathscr{L}_2^+$; (v) $\mathscr{L}_2^- \leftrightarrow \mathscr{L}_2^-$; (vi) $\mathscr{L}_3 \leftrightarrow \mathscr{L}_3$.
For q odd, only the bijection (vi) *is the identity. For q even, both* (iii) *and* (vi) *are the identity map.*

Proof: The proof is similar to that for Theorem 15.3.10.

For P on $\mathscr{H}_3$, its polar plane is the tangent plane containing the two generators through P and, conversely, the pole of a tangent plane π is the intersection of the two generators in π.

For P in $\Pi \backslash \mathscr{H}_3$, the points of contact of the $q+1$ tangents through P to $\mathscr{H}_3$ all lie on a conic $\mathscr{P}_2$ in a plane π, which is the polar plane of P. For q even, P lies in π and is the nucleus of $\mathscr{P}_2$. For q odd, P is not in π. So the tangents through P form a pencil or a cone as q is even or odd. Conversely, a plane π of Φ_3 meets $\mathscr{H}_3$ in a conic $\mathscr{P}_2$ at each point of which there are $q+1$ tangent lines to $\mathscr{H}_3$ giving $(q+1)^2$ lines unisecant to $\mathscr{P}_2$. There is a unique set of $q+1$ of these unisecants, one through each point of $\mathscr{P}_2$, concurrent at a point P, which is the pole of π.

If l is a bisecant, then $l = (g_1 \cap g_1')(g_2 \cap g_2')$ where g_1, g_2 are generators

Table 15.8

(a) Planes Through a Point

	Φ_2^+	Φ_3
Ψ_0	$q+1$	q^2
Ψ_1	$2q+1$	$q(q-1)$

(b) Points in a Plane

	Ψ_0	Ψ_1
Φ_2^+	$q(q-1)$	$2q+1$
Φ_3	q^2	$q+1$

(c) Points on a Line

	Ψ_0	Ψ_1
$\mathcal{L}_1$	q	1
$\mathcal{L}_2^+$	$q-1$	2
$\mathcal{L}_2^-$	$q+1$	0
$\mathcal{L}_3$	0	$q+1$

(d) Lines Through a Point

	$\mathcal{L}_1$	$\mathcal{L}_2^+$	$\mathcal{L}_2^-$	$\mathcal{L}_3$
Ψ_0	$q+1$	$\frac{1}{2}q(q+1)$	$\frac{1}{2}q(q-1)$	0
Ψ_1	$q-1$	q^2	0	2

(e) Lines in a Plane

	$\mathcal{L}_1$	$\mathcal{L}_2^+$	$\mathcal{L}_2^-$	$\mathcal{L}_3$
Φ_2^+	$q-1$	q^2	0	2
Φ_3	$q+1$	$\frac{1}{2}q(q+1)$	$\frac{1}{2}q(q-1)$	0

(f) Planes Through a Line

	Φ_2^+	Φ_3
$\mathcal{L}_1$	1	q
$\mathcal{L}_2^+$	2	$q-1$
$\mathcal{L}_2^-$	0	$q+1$
$\mathcal{L}_3$	$q+1$	0

(g) Lines Meeting a Line in a Point

	$\mathcal{L}_1$	$\mathcal{L}_2^+$	$\mathcal{L}_2^-$	$\mathcal{L}_3$
$\mathcal{L}_1$	$(q-1)(q+2)$	$\frac{1}{2}q^2(q+3)$	$\frac{1}{2}q^2(q-1)$	2
$\mathcal{L}_2^+$	$(q-1)(q+3)$	$\frac{1}{2}(q-1)(q^2+5q+2)$	$\frac{1}{2}q(q-1)^2$	4
$\mathcal{L}_2^-$	$(q+1)^2$	$\frac{1}{2}q(q+1)^2$	$\frac{1}{2}(q-2)(q+1)^2$	0
$\mathcal{L}_3$	$(q-1)^2$	$q^2(q+1)$	0	$q+1$

of one system and g_1', g_2' of the other. Then the polar l' of l is $l' = (g_1 \cap g_2')(g_2 \cap g_1')$ as in Fig. 15.2. Each generator is self-polar and each tangent is self-conjugate. So, for q odd, the polar of each tangent is another tangent through the same point of $\mathcal{H}_3$. In fact, at each point P of $\mathcal{H}_3$, the polarity induces an involution (see § 6.3) on the lines through P of the tangent plane at P. For q odd, this involution is hyperbolic (elliptic for the quadric $\mathcal{E}_3$) with the generators fixed; for q even, it is the identity.

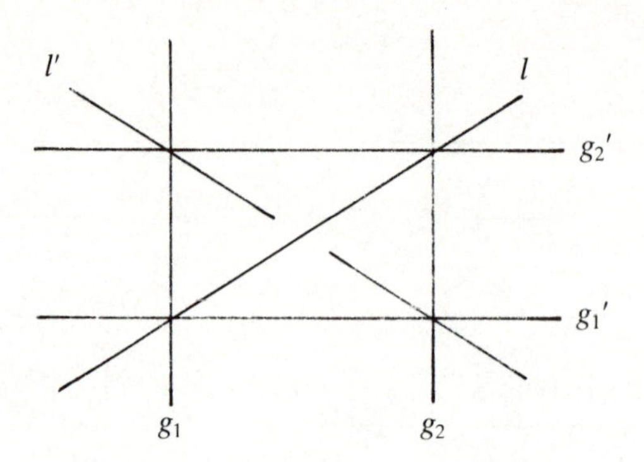

Fig. 15.2

For q even, the self-conjugate lines are self-polar so that the generators and tangents form a general linear complex.

If l is skew to $\mathcal{H}_3$, the $q+1$ planes through l all meet $\mathcal{H}_3$ in a conic. So the poles of these planes lie on a line l' skew to $\mathcal{H}_3$. □

We now know enough about quadrics to complete Lemma 15.3.1 and calculate the number of quadrics of each type in $PG(3, q)$.

Theorem 15.3.17: *The $\theta(9)$ quadrics of $PG(3, q)$ have the numbers of quadrics $\mathcal{F}$ in the six orbits under $PGL(4, q)$ as listed in Table* 15.9.

Proof: $n(\Pi_2\mathcal{P}_0)=\theta(3)$ and $n(\Pi_1\mathcal{H}_1)=\mathbf{c}(\theta(3), 2)$. Each line can be represented in $N(2, q)=q(q-1)/2$ ways; see § 1.6. So $n(\Pi_1\mathcal{E}_1)=\phi(1; 3, q)$ $N(2, q)$. As there are q^5-q^2 conics in a plane, each point of $PG(3, q)$ can be the vertex of q^5-q^2 cones, whence $n(\Pi_0\mathcal{P}_2)=(q^5-q^2)\ \theta(3, q)$. Using Lemma 15.1.1, the number of reguli in $PG(3, q)$ is $(q^2+q+1)\times$ $(q^2+1)q^4\cdot q(q^2-1)(q-1)/[(q+1)q(q-1)]=q^4(q^3-1)(q^2+1)$. As each hyperbolic quadric contains two reguli, $n(\mathcal{H}_3)=q^4(q^3-1)(q^2+1)/2$. Finally, $n(\mathcal{E}_3)$ is found by subtracting the other numbers from $\theta(9, q)=$ $(q^{10}-1)/(q-1)$. □

Table 15.9

Symbol for $\mathcal{F}$	Description of $\mathcal{F}$	Number $n(\mathcal{F})$
$\Pi_2\mathcal{P}_0$	repeated plane	$(q^2+1)(q+1)$
$\Pi_1\mathcal{H}_1$	pair of planes	$q(q^2+q+1)(q^2+1)(q+1)/2$
$\Pi_1\mathcal{E}_1$	line	$q(q^3-1)(q^2+1)/2$
$\Pi_0\mathcal{P}_2$	cone	$q^2(q^3-1)(q^2+1)(q+1)$
$\mathcal{E}_3$	elliptic quadric	$q^4(q^3-1)(q^2-1)/2$
$\mathcal{H}_3$	hyperbolic quadric	$q^4(q^3-1)(q^2+1)/2$

Corollary: (i) $|PGO_-(4,q)|=2q^2(q^4-1)$;
(ii) $|PGO_+(4,q)|=2q^2(q^2-1)^2$.

Proof: (i) $|PGO_-(4,q)|=p(4,q)/n(\mathscr{E}_3)$;
(ii) $|PGO_+(4,q)|=p(4,q)/n(\mathscr{H}_3)$. □

The next theorem is actually a corollary of the correspondence established in Theorem 15.3.11. For the definitions of the groups, see Appendix III.

Theorem 15.3.18:
(i) $P\Gamma L(2,q^2)\cong P\Gamma O_-(4,q)$; (ii) $P\gamma L(2,q^2)\cong PGO_-(4,q)$;
(iii) $P\gamma SL(2,q^2)\cong PO_-(4,q)$; (iv) $PSL(2,q^2)\cong PS'O_-(4,q)$.

Proof: (i) Every collineation of $PG(1,q^2)$ gives a collineation of $\mathscr{E}_{3,q}$ using the mappings $\mathfrak{T}_1$ and $\mathfrak{T}_2$ of Theorem 15.3.11. For, with $X=(x_0,x_1)$, then $\mathbf{P}(X)\mathfrak{S}=\mathbf{P}(XS)$ induces $\mathbf{V}(XA\bar{X}^*)\mathfrak{S}=\mathbf{V}(XSA\bar{S}^*X^*)$; here

$$A=\begin{pmatrix} a_0 & b \\ \bar{b} & a_1 \end{pmatrix}$$

and $S=(s_{ij})$. If we now write $A'=(a_0,a_1,b,\bar{b})$, then $A\to SA\bar{S}$ can be rewritten as $A'\to A'R$, where $R=(r_{ij})$. This defines a projectivity $\mathbf{P}(A')\mathfrak{R}=\mathbf{P}(A'R)$ of $PG(3,q^2)$ which fixes $\mathbf{V}(x_0x_1-x_2x_3)$ and hence $\mathscr{E}_{3,q}$. If $\mathfrak{R}$ is the identity, so is $\mathfrak{S}$. This means that there is a monomorphism from $P\Gamma L(2,q^2)$ into $P\Gamma O_-(4,q)$. As these groups have the same order $2hq^2(q^4-1)$, they are isomorphic.

(ii) $P\gamma L(2,q^2)$ is the subgroup of $P\Gamma L(2,q^2)$ containing the involutory automorphism $\mathbf{P}(x_0,x_1)\to\mathbf{P}(\bar{x}_0,\bar{x}_1)$ but none of higher order. This effects the projectivity $\mathbf{P}(x_0,x_1,x_2,x_3)\to\mathbf{P}(x_0,x_1,x_3,x_2)$ fixing $\mathscr{E}_3$.

(iii) For q even, these groups are the same as those in (ii). For q odd, $G_0=PSL(2,q^2)$ is of index four in $G=P\gamma L(2,q^2)$. There are three groups, G_1, G_2, G_3 of index two in G obtained by adding to G_0 each of its three cosets in G. With $\Delta=ad-bc$ and ν a specific non-square,

$$G_1=\{x\to(ax+b)/(cx+d)\mid\Delta=1 \text{ or } \nu\}=PGL(2,q^2),$$

$$G_2=\{x\to(a\tilde{x}+b)/(c\tilde{x}+d)\mid\Delta=1,\ \tilde{x}=x \text{ or } x^q\}=P\gamma SL(2,q^2),$$

$$G_3=\{x\to(a\tilde{x}+b)/(c\tilde{x}+d)\mid\Delta=1 \text{ and } \tilde{x}=x, \text{ or } \Delta=\nu \text{ and } \tilde{x}=x^q\}.$$

Since $PO_-(4,q)$ is of index two in $PGO_-(4,q)$ it is isomorphic to one of G_1, G_2, G_3. The collineation given in (ii) is in G_2 and does give an element of $PO_-(4,q)$, which is therefore isomorphic to G_2.

(iv) These groups are the simple subgroups of index two of the groups in (iii). □

It should be noted that $PS'O(4,q)$ is often written $P\Omega_-(4,q)$ and is the commutator subgroup of $PSO_-(4,q)$. For q odd, $PS'O_-(4,q)$ is the same as $PSO_-(4, q)$; for q even, the former is of index two in the latter. For q even, "special" is usually differently defined for orthogonal groups: see Appendix III. However, even with this other definition, (iv) holds.

Theorem 15.3.19: (i) *$PGO_-(4, q)$ is triply transitive on the points of $\mathscr{E}_3$, and transitive on the points of $\Pi\backslash\mathscr{E}_3$.*

(ii) *$PGO_+(4, q)$ is transitive on ordered triples of points on $\mathscr{H}_3$ no two of which lie on the same generator, and transitive on the points of $\Pi\backslash\mathscr{H}_3$.*

Proof: (i) Since $PGL(2, q^2)$ acts triply transitively on $PG(1, q^2)$ (see § 6.1) so also does $P\gamma L(2, q^2)$. Hence the bijection $\mathfrak{T}_1$ of Theorem 15.3.10 and the isomorphism of Theorem 15.3.18(ii) imply that $PGO_-(4, q)$ acts triply transitively on $\mathscr{E}_3$. Since $\mathfrak{T}_1$ and $\mathfrak{T}_2$ of Theorem 15.3.11 are incidence preserving, this means that $PGO_-(4, q)$ acts transitively on Φ_3 and, by the polarity of $\mathscr{E}_3$, also transitively on Ψ_0.

(ii) Since a regulus $\mathscr{R}$ is determined by three of its lines, $PGL(4, q)$ induces on $\mathscr{R}$ a group isomorphic to $PGL(2, q)$. Thus $PGO_+(4, q)$ has a subgroup G of index two isomorphic to $PGL(2, q)\times PGL(2, q)$ fixing both its reguli. Since $PGL(2, q)$ acts triply transitively on $PG(1, q)$, we have the first part of the result. The second part now follows, since any point off $\mathscr{H}_3$ has polar plane meeting $\mathscr{H}_3$ in a conic no two of whose points lie on the same generator of $\mathscr{H}_3$.

One may note that, with $\mathscr{H}_3=\mathbf{V}(x_0x_3-x_1x_2)=\{\mathbf{P}(ts, t, s, 1) \mid t, s\in\gamma^+\}$, the elements of G are $\mathbf{M}(T)$, where $t\to(at+b)/(ct+d)$ and $s\to(a's+b')/(c's+d')$ gives

$$T=\begin{bmatrix} aa' & ac' & ca' & cc' \\ ab' & ad' & cb' & cd' \\ ba' & bc' & da' & dc' \\ bb' & bd' & db' & dd' \end{bmatrix}. \quad \square$$

15.4 The representation of lines of $PG(3, q)$ in $PG(5, q)$

The mapping $\mathfrak{G}:\mathscr{L}\to\mathscr{H}_5$ from the set of lines in $PG(3, q)$ to the set of points of the hyperbolic quadric $\mathscr{H}_5=\mathbf{V}(x_0x_5+x_1x_4+x_2x_3)$ in $PG(5, q)$ is given by

$$\mathbf{l}(L)\mathfrak{G}=\mathbf{P}(X)$$

where $x_0=l_{01}$, $x_1=l_{02}$, $x_2=l_{03}$, $x_3=l_{12}$, $x_4=l_{31}$, $x_5=l_{23}$; that is, a line with vector L in $PG(3, q)$ becomes a point in $PG(5, q)$ with vector L and lying on $\mathscr{H}_5$. Linear dependence of lines in $PG(3, q)$ is represented by linear dependence of the corresponding points on $\mathscr{H}_5$.

$\mathfrak{G}$ is bijective and its image is the simplest, non-trivial example of a Grassmannian $\mathscr{G}_{r,n}$, the variety whose points represent the r-spaces in n-space. Thus $\mathscr{H}_5=\mathscr{G}_{1,3}$.

Table 15.10 *Properties of* $\mathfrak{E}$

$\Pi = PG(3, q)$	q	$PG(5, q)$
$(q^2+1)(q^2+q+1)$ lines in Π	–	$(q^2+1)(q^2+q+1)$ points on $\mathscr{H}_5$
Two skew lines	–	The two points of $\mathscr{H}_5$ on a bisecant
Two intersecting lines	–	Two points whose join is on $\mathscr{H}_5$
A pencil of lines	–	A line on $\mathscr{H}_5$
The three sides of a triangle	–	The three vertices of a triangle whose sides lie on $\mathscr{H}_5$
A plane (of lines)	–	A Greek plane on $\mathscr{H}_5$
(The lines through) a point	–	A Latin plane on $\mathscr{H}_5$
q^3+q^2+q+1 planes	–	q^3+q^2+q+1 Greek planes on $\mathscr{H}_5$
q^3+q^2+q+1 points	–	q^3+q^2+q+1 Latin planes on $\mathscr{H}_5$
$q+1$ points on a line	–	$q+1$ Latin planes through a point on $\mathscr{H}_5$
$q+1$ planes through a line	–	$q+1$ Greek planes through a point on $\mathscr{H}_5$
There is a unique line through two points	–	Two Latin planes meet in a point
There is a unique line on two planes	–	Two Greek planes meet in a point
A pencil of lines lies in just one plane and passes through just one point	–	A line on $\mathscr{H}_5$ lies in just one Greek plane and just one Latin plane
The set of lines through a point and in a plane form a pencil or is empty as the point does or does not lie in the plane	–	A Latin plane and a Greek plane either meet in a line or have empty intersection
The edges of a trihedral angle and the pencils generated by pairs of the sides	–	The vertices and sides of a triangle in a Latin plane
The sides of a triangle and the pencils generated by pairs of the sides	–	The vertices and sides of a triangle in a Greek plane
The three face planes of a trihedral angle	–	The three Greek planes through the sides of a triangle in a Latin plane
The three vertices of a triangle	–	The three Latin planes through the sides of a triangle in a Greek plane
The 3-dimensional Principle of Duality	–	The consistent interchange of Latin and Greek planes
A dual conic	–	A conic in a Greek plane
A cone $\Pi_0\mathscr{P}_2$	–	A conic in a Latin plane
Two non-coplanar pencils with a common line and distinct centres	–	A pair of intersecting lines: the section of $\mathscr{H}_5$ by a tangent plane
A regulus consisting of the $q+1$ transversals to three skew lines	–	A conic: the section of $\mathscr{H}_5$ by the plane of a triangle with vertices but not sides on $\mathscr{H}_5$
A hyperbolic congruence: the $(q+1)^2$ transversals of two skew lines l, l'	–	A hyperbolic quadric $\mathscr{H}_3$: the section of $\mathscr{H}_5$ by a solid Π_3 whose polar line meets $\mathscr{H}_5$ in two points
Two families of $q+1$ pencils each joining a point on one of l,l' to all the points on the other	–	The two reguli on $\mathscr{H}_3$

Table 15.10 (contd.)

$\Pi = PG(3, q)$	q	$PG(5, q)$
An elliptic congruence: q^2+1 lines forming a regular spread	–	An elliptic quadric $\mathscr{E}_3$: the section of $\mathscr{H}_5$ by a solid whose polar line is skew to $\mathscr{H}_5$. No two points of $\mathscr{E}_3$ are conjugate with respect to $\mathscr{H}_5$
The regulus containing three lines of an elliptic congruence lies in the congruence	–	The conic section of $\mathscr{H}_5$ by a plane through 3 points on a solid section $\mathscr{E}_3$ lies on $\mathscr{E}_3$
An arbitrary spread of q^2+1 lines	–	q^2+1 points of $\mathscr{H}_5$, one in each Latin and in each Greek plane.
A packing	–	A partition of $\mathscr{H}_5$ into q^2+q+1 sets of q^2+1 points as above
A packing by regular spreads	–	A partition of $\mathscr{H}_5$ into elliptic quadrics $\mathscr{E}_3$
A parabolic congruence: q^2+q+1 lines meeting the axis l, no two of which meet off l	–	A cone $\Pi_0\mathscr{P}_2$: the section of $\mathscr{H}_5$ by a tangent solid Π_3, whose polar line is therefore a tangent line with contact Π_0
The $q+1$ pencils of lines in the parabolic congruence	–	The $q+1$ generators of $\Pi_0\mathscr{P}_2$
The q^3 reguli in the parabolic congruence	–	The q^3 conic sections of $\Pi_0\mathscr{P}_2$ by planes of Π_3 not through Π_0
There is a unique line of a hyperbolic, elliptic, or parabolic congruence through a point P unless P lies on an axis	–	The solid Π_3 containing $\mathscr{F} = \mathscr{H}_3$, $\mathscr{E}_3$, or $\Pi_0\mathscr{P}_2$ meets a Latin plane in a unique point unless the plane contains a point of intersection of the polar line of Π_3 and $\mathscr{F}$
A special linear complex: the $q(q+1)^2+1$ lines meeting or equal to the axis	–	A cone $\Pi_0\mathscr{H}_3$: the section of $\mathscr{H}_5$ by a tangent prime
A general linear complex: $(q^2+1)(q+1)$ lines	–	A quadric $\mathscr{P}_4$: the section of $\mathscr{H}_5$ by a non-tangent prime Π_4
The $(q^2+1)(q+1)$ pencils in a general linear complex, one through each point and one in each plane	–	The $(q^2+1)(q+1)$ lines on $\mathscr{P}_4$; one in every Latin plane and one in every Greek plane
A pair of apolar linear complexes	–	A pair of conjugate primes
The null polarity whose self-polar lines form a general linear complex	odd	The harmonic homology whose axial prime and centre are the representing Π_4 and its pole
The null polarity whose self-polar lines form a general linear complex	even	The elation whose axial prime and centre are the representing Π_4 and its pole, which lies in Π_4 and is the nucleus of $\mathscr{P}_4$
The pseudo polarity with bilinear form $x_0y_0+x_0y_1+x_1y_0+x_2y_3+x_3y_2$	even	The projectivity $\mathbf{P}(x_0, x_1, x_2, x_3, x_4, x_5) \to \mathbf{P}(x_5, x_1, x_2, x_1+x_3, x_2+x_4, x_0)$ with fixed points the plane $\mathbf{V}(x_1, x_2, x_0+x_5)$

Table 15.10 (contd.)

$\Pi = PG(3, q)$	q	$PG(5, q)$
The ordinary polarity with self-conjugate quadric $\mathscr{E}_3 = \mathbf{V}(-\nu x_0^2 + x_1^2 + x_2 x_3)$	odd	The projectivity $\mathbf{P}(x_0, x_1, x_2, x_3, x_4, x_5) \rightarrow \mathbf{P}(x_5, 2x_3, 2x_4, 2\nu x_1, 2\nu x_2, 4\nu x_0)$ with no fixed points
The ordinary polarity with self-conjugate quadric $\mathscr{H}_3 = \mathbf{V}(x_0 x_1 + x_2 x_3)$	odd	The projectivity $\mathbf{P}(x_0, x_1, x_2, x_3, x_4, x_5) \rightarrow \mathbf{P}(x_5, x_1, -x_2, -x_3, x_4, x_0)$ with fixed points the planes $\mathbf{V}(x_0 - x_5, x_2, x_3)$, $\mathbf{V}(x_0 + x_5, x_1, x_4)$
A pair of complementary reguli on $\mathscr{H}_3$	–	A pair of conics on $\mathscr{H}_5$ in polar planes Π_2, Π_2'
$\mathscr{H}_3$ does not define a null polarity	odd	$\Pi_2 \cap \Pi_2' = \Pi_{-1}$
$\mathscr{H}_3$ does define a null polarity	even	$\Pi_2 \cap \Pi_2' = \Pi_0$
Four skew lines with 0, 1, or 2 transversals	–	Four points generating a solid whose polar line is skew, tangent, or bisecant to $\mathscr{H}_5$
A double-six	–	A pair of simplexes inscribed and circumscribed to $\mathscr{H}_5$ and to one another
The double-six has a null polarity	even	The representing simplexes are in perspective from the nucleus of $\mathscr{P}_4$

The properties of $\mathfrak{G}$ are listed in Table 15.10. They vary little from the classical properties over the complex field. However, one should again be wary of the fields of even order, since then the polarity induced by $\mathscr{H}_5$ is null. In particular, for q even, every point of $PG(5, q)$ lies in its polar prime; whereas, for q odd, the only points lying in their polar primes are those of $\mathscr{H}_5$.

$\mathscr{H}_5$ has two systems of generating planes called *Greek* and *Latin* planes for convenience. Some statements in Table 15.10 are made with reference to those immediately preceding it. For the definitions of *spread*, *packing*, and *double-six*, see §§ 17.1, 17.4, and 20.1 respectively.

15.5 Notes and references

§ 15.1. For calculations via a cohomology theory of the number of lines satisfying certain incidence conditions, see Mielants and Leemans (1983), Morikawa (1983).

§ 15.2. For more details on line coordinates, see for example Todd (1946). Theorem 15.2.14 and its preliminaries follow de Resmini (1984a).

§ 15.3. For results on quadrics over finite fields, see Segre (1959a), Primrose (1951), Ray Chaudhuri (1962a), Zeitler (1980, 1981), as well as Volume 3.

An approach to a geometric proof of Wedderburn's theorem is given by Segre (1958a, 1960a, p. 333). This is formulated as in Theorem 15.3.12. Another approach is via the configurations of five associated lines or planes in four dimensions, Segre (1960a, p. 362). See also Al-Dhahir (1972).

A related result is given by Buekenhout (1966b); if, in a projective plane Π of order q, a $(q+1)$-arc $\mathscr{C}$ contains a hexagon H which satisfies Pascal's theorem, then $\mathscr{C}$ is a conic and $\Pi = PG(2, q)$. Variations of this theorem have been given by Artzy (1968), G. Conti (1975), Karzel and Sörenson (1971), Rigby (1969), Amici and Casciaro (1983), Korchmáros (1981b).

§ 15.4. The style of Table 15.10 follows Coxeter (1962), who did it for the field of real numbers.

16

OVALOIDS AND QUADRICS

16.1 Ovaloids

In $PG(3, q)$, a set $\mathcal{K}$ of k points no three of which are collinear is a *k-cap*. A k-cap is *complete* if it is not contained in a $(k+1)$-cap.

Lemma 16.1.1: *In $PG(3, q)$, q odd, a k-cap $\mathcal{K}$ satisfies $k \leqslant q^2+1$.*

Proof: If $P_1, P_2 \in \mathcal{K}$, then each plane through P_1P_2 meets $\mathcal{K}$ in a k'-arc and so $k' \leqslant q+1$, by Theorem 8.1.3. So $k \leqslant 2+(q+1)(q-1)=q^2+1$. □

To prove a similar result for q even, some further definitions will be needed. A line of $\Pi_3 = PG(3, q)$ is a *bisecant*, a *tangent*, or an *external* line to a k-cap $\mathcal{K}$ as it meets $\mathcal{K}$ in 2, 1, or 0 points. Let t be the number of tangents through a point P of $\mathcal{K}$. For any point Q of $\Pi_3 \backslash \mathcal{K}$, let $\sigma_2(Q)$ and $\sigma_1(Q)$ be the respective numbers of bisecants and tangents through Q.

Lemma 16.1.2:

$$\text{(i) } t+k=q^2+q+2; \qquad \text{(ii) } \sigma_1(Q)+2\sigma_2(Q)=k. \quad \square$$

Theorem 16.1.3: *A k-cap $\mathcal{K}$ in $PG(3, q)$, q even, has no tangents if and only if $q=2$, $k=8$ and $\mathcal{K}$ is the complement of a plane.*

Proof: If $t=0$, then Lemma 16.1.2 implies that $k=q^2+q+2$. So the number of bisecants of $\mathcal{K}$ is $\frac{1}{2}(q^2+q+2)(q^2+q+1)$, which is less than $(q^2+1)(q^2+q+1)$, the total number of lines of $PG(3, q)$. Hence there exists an external line l to $\mathcal{K}$. Since $t=0$, a plane through l is either skew to $\mathcal{K}$ or meets $\mathcal{K}$ in a $(q+2)$-arc. So $q+2$ divides q^2+q+2 and therefore $q^2=(q-2)(q+2)+4$, whence $q+2$ divides 4. Therefore $q=2$ and $k=8$. Thus $\Pi_3 \backslash \mathcal{K}$ is a set of seven points such that, if a line has two points in $\Pi_3 \backslash \mathcal{K}$, it has three. So $\Pi_3 \backslash \mathcal{K}$ is a plane. The converse is immediate. □

Lemma 16.1.4: *If $\mathcal{K}$ is a complete k-cap in $PG(3, q)$, q even and $q>2$, then $k \leqslant q^2+1$.*

Proof: From Theorem 16.1.3, we may suppose that $q^2+1<k<q^2+q+2$.

Let $P \in \mathcal{K}$ and let l be a tangent to $\mathcal{K}$ through P. Then any plane through l meets $\mathcal{K}$ in at most $q+1$ points. If all planes through l meet $\mathcal{K}$ in at most q points, then $k \leqslant 1+(q+1)(q-1)=q^2$. So there exists a plane π through l such that $\pi \cap \mathcal{K}$ is a $(q+1)$-arc; it has a nucleus Q (Lemma

8.1.4). If every line through Q were an external line or a tangent to $\mathcal{K}$, then $\mathcal{K}\cup\{Q\}$ would be a $(k+1)$-cap. So there is a bisecant b of $\mathcal{K}$ through Q and b cannot lie in π. So the planes through b all contain one of the tangents in π through Q. So no plane through b meets $\mathcal{K}$ in a $(q+2)$-arc. Counting the points of $\mathcal{K}$ on the planes through b gives $k \leqslant 2+(q+1)(q-1)=q^2+1$: contradiction. □

As in § 3.3, the maximum value of k for which a k-cap exists in $PG(3, q)$, that is the maximum number of points no three of which are collinear, is denoted by $m_2(3, q)$.

Theorem 16.1.5:

$$\text{(i) } m_2(3, q)=q^2+1 \quad \textit{for} \quad q>2; \qquad \text{(ii) } m_2(3, 2)=8.$$

Proof: Theorem 16.1.3 gives (ii). Lemma 16.1.1 means that $m_2(3, q) \leqslant q^2+1$ for q odd and Theorem 16.1.4 gives the same result for q even with $q>2$. However, as in Theorem 15.3.1, an elliptic quadric is a (q^2+1)-cap. □

For $q>2$, a (q^2+1)-cap is an *ovaloid.*

Lemma 16.1.6: *If $\mathcal{K}$ is an ovaloid in $PG(3, q)$, then*

(i) *at each point P of $\mathcal{K}$, there is a unique tangent plane π such that $\pi \cap \mathcal{K}=\{P\}$;*

(ii) *apart from the q^2+1 tangent planes, every plane meets $\mathcal{K}$ in a $(q+1)$-arc;*

(iii) *the lines through P consist of q^2 bisecants and $q+1$ tangents, the latter of which all lie in π;*

(iv) *through a point off $\mathcal{K}$, there are $\frac{1}{2}q(q-1)$ bisecants, $\frac{1}{2}q(q+1)$ external lines, and $q+1$ tangents, the last of which are coplanar when q is even.*

Proof: (a) For q odd, any plane through two points of $\mathcal{K}$ meets $\mathcal{K}$ in a $(q+1)$-arc since $q^2+1=2+(q+1)(q-1)$. So the number of planes through P in $\mathcal{K}$ meeting $\mathcal{K}$ in a $(q+1)$-arc is $q^2(q+1)/q=q(q+1)$, leaving one plane π meeting $\mathcal{K}$ only in P. The $q+1$ lines through P are tangents to $\mathcal{K}$ and are necessarily all the tangents through P. If Q is on a tangent l through P, then each plane through l other than π meets $\mathcal{K}$ in a conic and so this plane contains exactly one other tangent through Q. Hence (i)–(iv) are proved.

(b) For q even, take P in $\mathcal{K}$ and l a tangent through P. Then, exactly as in Lemma 16.1.4, there is a plane π' through l meeting $\mathcal{K}$ in a $(q+1)$-arc with nucleus Q. So there are $q+1$ tangents through Q in π'.

Since $\mathcal{K}$ is complete, there exists a bisecant b to $\mathcal{K}$ through Q. Each plane β through b contains a tangent through Q in π' and so meets $\mathcal{K}$ in at most $q+1$ points. A count of the points of $\mathcal{K}$ on all the planes through b gives that $\beta \cap \mathcal{K}$ is a $(q+1)$-arc.

Every tangent to $\mathcal{K}$ through Q is in π'. For, if l' were another, then each plane through l' contains a tangent to $\mathcal{K}$ in π' and so meets $\mathcal{K}$ in at most q points. But $l'b$ must be one of these planes, and it has just been shown that $l'b \cap \mathcal{K}$ comprises $q+1$ points: a contradiction. This also means that if a plane α through l other than π' contains a point P' of $\mathcal{K}$ other than P, then QP' is a bisecant and so $\alpha \cap \mathcal{K}$ is a $(q+1)$-arc.

Thus each plane through l meets $\mathcal{K}$ in one point or in a $(q+1)$-arc. Counting the points of $\mathcal{K}$ on the planes through l gives that q of these planes meet $\mathcal{K}$ in a $(q+1)$-arc and one, π, in P alone. Hence the $q+1$ tangents to $\mathcal{K}$ through P all lie in π.

The number of planes through a tangent other than a tangent plane is $(q^2+1)(q+1)q/(q+1)=q^3+q$, which is all the planes other than the tangent planes. So each non-tangent plane α' meets $\mathcal{K}$ in a $(q+1)$-arc and through the nucleus of $\alpha' \cap \mathcal{K}$ there are no tangents other than those in α'. □

Corollary: *For $q=4$ and q odd, every plane which is not a tangent plane meets $\mathcal{K}$ in a conic.*

Proof: For $q=4$, every $(q+1)$-arc is a conic. Likewise, by Theorem 8.2.4, every $(q+1)$-arc is a conic for q odd. □

Theorem 16.1.7: *In $PG(3, q)$, q odd or $q=4$, an ovaloid $\mathcal{K}$ is an elliptic quadric.*

Proof: Suppose firstly that $q \geqslant 4$. Let π_1 be a plane meeting $\mathcal{K}$ in a conic $\mathcal{C}_1$ and let $P_1, \ldots, P_5$ be five points on the conic. Take another plane π_2 through P_1 and P_2 and let Q_1, Q_2, Q_3 be three other points of the conic $\mathcal{C}_2 = \mathcal{K} \cap \pi_2$. The tangent plane at P_i, $i=1, 2$, to $\mathcal{K}$ meets π_2 in a line which is tangent to $\mathcal{C}_2$. Take a third plane π_3 through P_1 and P_2, containing a point R of $\mathcal{K}$ not in π_1 or π_2. Likewise π_3 meets the tangent plane at P_i, $i=1, 2$, in a line tangent to $\mathcal{C}_3 = \pi_3 \cap \mathcal{K}$.

The quadric $\mathcal{F}$ through the nine points P_1, P_2, P_3, P_4, P_5, Q_1, Q_2, Q_3, and R contains five points of both $\mathcal{C}_1$ and $\mathcal{C}_2$, and so the conics themselves. The tangents at P_1 to $\mathcal{C}_1$ and $\mathcal{C}_2$ are therefore tangent lines to $\mathcal{F}$; so the tangent plane to $\mathcal{K}$ at P_1 is tangent to $\mathcal{F}$. Hence the section $\mathcal{C}$ of $\mathcal{F}$ by π_3 will have as tangent at P_1 the meet of the tangent plane at P_1 with π_3; similarly for P_2. So $\mathcal{C}_3$ and $\mathcal{C}$ have three points in common as well as tangents at two of them and hence coincide.

Now take a point $P \neq P_1, P_2$ on $\mathcal{K}$ and a plane π through PP_1 which contains neither P_2 nor any of the tangents at P_1 to $\mathcal{C}_1$, $\mathcal{C}_2$, or $\mathcal{C}_3$. This plane meets $\mathcal{C}_1$, $\mathcal{C}_2$, and $\mathcal{C}_3$ in three distinct points apart from P_1 and so its intersection with $\mathcal{F}$ is a conic which contains these four points and is tangent at P_1 to the meet of π and the tangent plane to $\mathcal{K}$ at P_1. But this conic consists of all the points of $\pi \cap \mathcal{K}$ and so P belongs to $\mathcal{F}$. Hence $\mathcal{K} \subset \mathcal{F}$. As $\mathcal{F}$ can only be an elliptic quadric, the theorem is proved.

Now let $q=3$. Let l be an external line to $\mathscr{K}$. Of the four planes through l, two, π_1 and π_2, meet $\mathscr{K}$ in a conic while the other two, π_3 and π_4, are tangent planes. Let $\mathscr{F}$ be the quadric through the nine points of $\mathscr{K}$ on π_1, π_2, and π_3. Since no three of these points are collinear, $\mathscr{F}$ is an elliptic quadric. It must be shown that the tenth point P of $\mathscr{F}$ coincides with the point P_4 of $\mathscr{K}$ in π_4.

Let P_3 be the point of $\mathscr{K}$ in π_3. Then P_4 does not lie in the tangent plane to $\mathscr{F}$ at P_3, since otherwise this plane containing two points of $\mathscr{K}$ would contain two more, which would also be on $\mathscr{F}$. So P_3P_4 is not tangent to $\mathscr{F}$ and, since it does not lie in π_3, it is one of the nine lines through P_3 not in π_3. Since it cannot be any of the eight lines joining P_3 to the points of $\mathscr{K}$ in π_1 and π_2, it meets $\mathscr{F}$ in a point, apart from P_3, which is not in π_1, π_2 or π_3 and so is in π_4. This point is therefore P_4 and so $P=P_4$. Hence $\mathscr{F}=\mathscr{K}$. □

This theorem, on the one hand, characterizes ovaloids for q odd and $q=4$ and on the other hand characterizes elliptic quadrics. It is therefore natural to ask if, for q even with $q>4$, there exist ovaloids which are not elliptic quadrics.

For any ovaloid, the tables of incidences for elliptic quadrics (see § 15.3) hold, with the sole proviso that the planes of Φ_3 meet the ovaloid in $(q+1)$-arcs which are not necessarily conics.

Let $\mathscr{K}$ be an ovaloid in $PG(3, q)$, q even and $q>4$. Associate to each tangent plane its point of contact. Any other plane π meets K in a $(q+1)$-arc with nucleus Q: associate π to Q. A point and its associated plane are respectively called *pole* and *polar*.

Theorem 16.1.8: *If $\mathscr{K}$ is an ovaloid in $PG(3, q)$, q even, then the correspondence which interchanges pole and polar is a null polarity.*

Proof: Suppose the point P corresponds to the plane π. It must be shown that the poles of all the planes through P are all the points of π. We note that, if P is on $\mathscr{K}$, the tangent lines through P to $\mathscr{K}$ form the pencil through P in the tangent plane at P; if P is not on $\mathscr{K}$, the tangent lines through P to $\mathscr{K}$ form the pencil of tangents to the $(q+1)$-arc in the polar plane of P.

If π' is some other plane through P, the line $\pi' \cap \pi$ is tangent to $\mathscr{K}$. But the pole of π' is the intersection of all the tangents lying in π', so it lies on the line $\pi' \cap \pi$ and therefore in π. □

Corollary 1:

(i) *A tangent line to $\mathscr{K}$ is its own polar;*

(ii) *the polar line of a bisecant l is the external line which is the intersection of the tangent planes at the two points of $l \cap \mathscr{K}$.* □

Corollary 2: *The tangent lines to an ovaloid form a linear complex.* □

16.2 Characterization of quadrics

In § 16.1, elliptic quadrics for q odd were characterized by ovaloids. In this section, we complete the characterization of quadrics by considering cones and hyperbolic quadrics.

A subset $\mathcal{K}$ of $PG(3, q)$ is *ruled* if any line with three points in $\mathcal{K}$ lies entirely in $\mathcal{K}$. In the terminology of Chapter 19, this is equivalent to $\mathcal{K}$ being a set of type $(0, 1, 2, q+1)$.

Theorem 16.2.1: *If $\mathcal{K}$ is a ruled set of $(q+1)^2$ points in $PG(3, q)$, q odd, then $\mathcal{K}$ is a hyperbolic quadric or consists of a plane and a line.*

Proof: If a plane π contains no lines belonging to $\mathcal{K}$, then $|\pi \cap \mathcal{K}| \leq q+1$. If π contains only one line belonging to $\mathcal{K}$, it can contain only one point of $\mathcal{K}$ off the line and $|\pi \cap \mathcal{K}| \leq q+2$. If π contains two lines of $\mathcal{K}$ and some other point P of $\mathcal{K}$, then q of the lines through P in π are in $\mathcal{K}$ and it follows that all lines of π belong to $\mathcal{K}$. Hence when $|\pi \cap \mathcal{K}| > 2q+1$, all lines of π belong to $\mathcal{K}$. Thus if $\mathcal{K}$ contains all the points of π, its remaining q points can only consist of the points of a line l not in π.

Now suppose that no plane contains more than $2q+1$ points of $\mathcal{K}$. Since $(q+1)^2 > q^2+1$, $\mathcal{K}$ contains three collinear points and therefore some line l. Then each plane through l can contain at most q points outside l; as $|\mathcal{K}| = (q+1)^2 = q+1+q(q+1)$, each plane through l meets $\mathcal{K}$ in exactly q points off l. However, if such a set of q points were not on one line, the existence of further points on $\mathcal{K}$ would be implied. So $\mathcal{K}$ consists of l and $q+1$ lines l_i, $i \in \mathbf{N}_{q+1}$, with the planes ll_i distinct; hence no two lines l_i intersect except perhaps on l. If all the lines l_i are concurrent at a point P of l, then, since the $q+2$ lines l_i, l cannot meet a plane not through P in a $(q+2)$-arc, three of the $q+2$ lines lie in a plane and $\mathcal{K}$ contains all of this plane. If at least two lines l_1 and l_2 are concurrent at a point P of l and $P \notin l_3$, then l_3 meets the plane $\pi = l_1 l_2$ at a point off l_1 and l_2; hence π contains $2q+2$ points of $\mathcal{K}$ and therefore lies in $\mathcal{K}$. So the $q+1$ lines l_i meet l at separate points and are therefore mutually skew. Now the regulus of transversals of l_1, l_2, and l_3 lies on $\mathcal{K}$ and $\mathcal{K}$ is the hyperbolic quadric containing l_1, l_2, and l_3. □

Theorem 16.2.2: *If $\mathcal{K}$ is a ruled set of $(q+1)^2$ points in $PG(3, q)$, q even, then $\mathcal{K}$ is a hyperbolic quadric or consists of a plane and a line or consists of lines joining an oval to a vertex.*

Proof: The proof proceeds exactly as for q odd, except that the possibility of the $q+1$ lines l_i being concurrent at a point of l is no longer discarded. □

Theorem 16.2.3: *If $\mathcal{K}$ is a ruled set of q^2+q+1 points in $PG(3, q)$, q odd, then $\mathcal{K}$ is a cone or a plane.*

Proof: If any plane π contains more than $2q+1$ points of $\mathcal{K}$, then π lies entirely in $\mathcal{K}$ as in Theorem 16.2.1, and in this case coincides with $\mathcal{K}$.

Suppose no plane meets $\mathcal{K}$ in more than $2q+1$ points. As $q^2+q+1>q^2+1$, so $\mathcal{K}$ contains three collinear points and therefore some line l. Then, q of the $q+1$ planes through l contain a further line of $\mathcal{K}$; these q lines with l fill up $\mathcal{K}$. If two of these lines l_1 and l_2 met at some point off l, then the plane ll_1l_2 would belong to $\mathcal{K}$. If l_1 and l_2 were skew, the transversal from a point P of $\mathcal{K}\backslash(l \cup l_1 \cup l_2)$ to l_1 and l_2 would lie entirely in $\mathcal{K}$ and be skew to l. So the q lines of $\mathcal{K}$ meeting l are concurrent at a point P of l. As no three of these $q+1$ lines are coplanar, they meet a plane not through P in a $(q+1)$-arc. Hence $\mathcal{K}$ is a cone. □

Theorem 16.2.4: *If $\mathcal{K}$ is a ruled set of q^2+q+1 points in $PG(3, q)$, q even, then $\mathcal{K}$ consists of lines joining a plane $(q+1)$-arc to a vertex or is a plane.* □

Corollary: *If $\mathcal{K}$ is a ruled set of q^2+q+1 points in $PG(3, q)$, $q=2$ or 4, then $\mathcal{K}$ is a cone or a plane.* □

So far, the characterization of quadrics has depended on being given the size of the set $\mathcal{K}$. The treatment is now generalized by removing this restriction. Unfortunately this allows other examples.

We recall that the notation $\Pi_r\mathcal{S}$ means the set of all points on the lines PQ, where $P \in \Pi_r$, $Q \in \mathcal{S}$ and $\mathcal{S}$ is contained in a subspace Π_s skew to Π_r; generally $\Pi_r\mathcal{S}$ is called a *cone* with *vertex* Π_r and *base* $\mathcal{S}$. In the earlier part of this section, the word cone was only used for a quadric cone $\Pi_0\mathcal{P}_2$ with vertex a point and base a conic.

For $q=2$, every subset of $PG(n, q)$ is of type $(0, 1, 2, q+1)$. So this case is excluded from the following classification.

Lemma 16.2.5: *A set of type $(0, 1, 2, q+1)$ in $PG(2, q)$, $q>2$, is one of the following:*

(1) *a k-arc;*

(2) $\Pi_1 \cup \Pi'_i$ $(i=-1, 0, 1)$;

(3) $PG(2, q)$. □

Theorem 16.2.6: *A set $\mathcal{K}$ of type $(0, 1, 2, q+1)$ in $PG(3, q)$ not contained in a plane is one of the following list.*

(1) A *k-cap*;

(2) (i) $l_1 \cup l_2$;

(ii) $l_1 \cup l_2 \cup m_1$;

(iii) $l_1 \cup l_2 \cup m_1 \cup m_2$;

here $l_1m_1l_2m_2$ is a skew quadrilateral (Fig. 16.1).

(3) $\mathcal{H}_3$;

(4) $\Pi_0\mathcal{P}_2$;

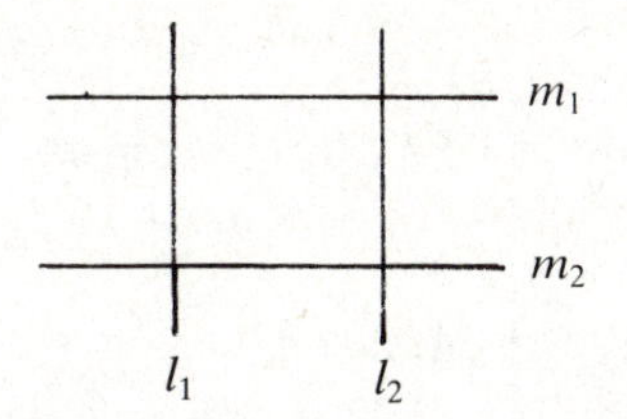

Fig. 16.1

(5) $\Pi_0\mathcal{K}_1 \cup \mathcal{K}_2$, where $\mathcal{K}_1$ is a plane k_1-arc and $\mathcal{K}_2$ is a k_2-cap (possibly $k_2 = 0$) such that
   (α) the plane containing two lines of $\Pi_0\mathcal{K}_1$ contains no point of $\mathcal{K}_2$;
   (β) no plane contains a line of $\Pi_0\mathcal{K}_1$ and two points of $\mathcal{K}_2$;
(6) $\Pi_2 \cup \Pi'_i \quad (i = 0, 1, 2)$;
(7) $PG(3, q)$.

Proof: If $\mathcal{K}$ contains no lines, then it is of type (1). There are three other cases.

(a) *$\mathcal{K}$ contains two skew lines l_1 and l_2 but no plane.*

Suppose $P_1 \in \mathcal{K} \backslash (l_1 \cup l_2)$. Then the unique line m_1 through P_1 meeting l_1 and l_2 is in $\mathcal{K}$. If $\mathcal{K}$ contains a further point P_2, then then the unique line m_2 through P_2 meeting l_1 and l_2 is also in $\mathcal{K}$. Now, m_2 cannot meet m_1, as otherwise the plane $m_1 m_2 l_1$ would be in $\mathcal{K}$. This gives types (2)(i), (ii), (iii).

Let $\mathcal{K}$ contain a skew quadrilateral as above and a further point P_3. Through P_3 there is a transversal l_3 to m_1 and m_2 as well as a transversal m_3 to l_1 and l_2 (Fig. 16.2). So $\mathcal{K}$ contains the reguli $\mathcal{R}(l_1, l_2, l_3)$ and $\mathcal{R}(m_1, m_2, m_3)$. Thus $\mathcal{K}$ contains $\mathcal{H}_3$, which gives type (3). If $\mathcal{K}$ contains a point P off $\mathcal{H}_3$, it contains all bisecants of $\mathcal{K}$ through P and hence $\mathcal{K}$ is $PG(3, q)$.

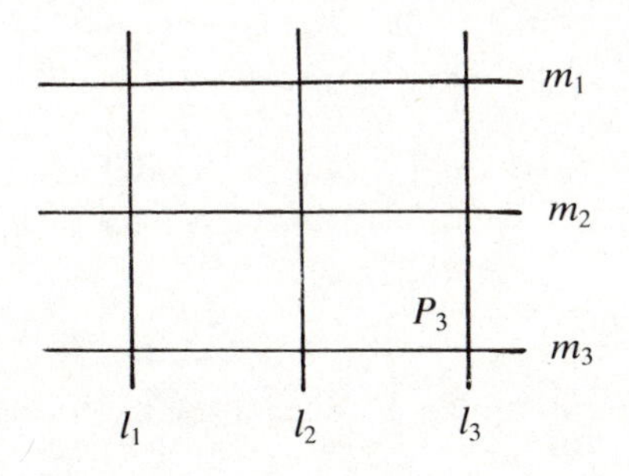

Fig. 16.2

(b) *All lines of $\mathcal{K}$ meet in a point P.*

The lines of $\mathcal{K}$ meet a plane π in a k_1-arc $\mathcal{K}_1$. So $\mathcal{K}$ contains the cone $P\mathcal{K}_1$. Any points of $\mathcal{K}$ not on $P\mathcal{K}_1$ form a cap $\mathcal{K}_2$. This gives case (5); the conditions (α) and (β) follow from Lemma 16.2.5.

(c) *$\mathcal{K}$ contains a plane π.*

Since $\mathcal{K}$ contains a point P_1 not in π, the set $\pi \cup \{P_1\}$ is a possible set. If $\mathcal{K}$ contains a further point P_2, then $\mathcal{K}$ contains the line P_1P_2, which gives the set $\pi \cup P_1P_2$. If $\mathcal{K}$ contains a further point P_3, let $\pi' = P_1P_2P_3$. The plane π' meets $\mathcal{K}$ in a set containing the line $\pi \cap \pi'$ and the points P_1, P_2, P_3. By Lemma 16.2.5, the plane π' is wholly contained in $\mathcal{K}$. This gives the set $\pi \cup \pi'$. Now, let P_4 be in $\mathcal{K}\backslash(\pi \cup \pi')$; so all lines through P_4 meeting π and π' in separate points lie in $\mathcal{K}$, whence $\mathcal{K} = PG(3, q)$. □

Note: The only set in the theorem whose size is not precisely specified is that of type (5).

16.3 Stereographic projection

Occasionally, problems in $PG(3, q)$ can be more easily investigated in $PG(2, q)$ and so we consider the projection from a point of itself of a hyperbolic quadric, an elliptic quadric, and, for q even, an ovaloid.

I. *Hyperbolic quadric*

Let l and l' be the generators through P_0 on $\mathcal{H}_3$ and let π be a plane not containing P_0. Let $\pi \cap ll' = l_0 = QQ'$, where $Q = l \cap \pi$ and $Q' = l' \cap \pi$. Then $\mathfrak{S}: \mathcal{H}_3\backslash\{P_0\} \to \pi$ is given by $P\mathfrak{S} = PP_0 \cap \pi$. So each point on $l\backslash\{P_0\}$ maps to Q and each point on $l'\backslash\{P_0\}$ to Q'; the q^2 points of $\mathcal{H}_3$ not on l or l' map to the q^2 points of $\pi\backslash l_0$. The q generators of $\mathcal{H}_3$ other than l' meeting l map to the q lines of π other than l_0 through Q and, similarly, the q generators other than l meeting l' map to the q lines of π other than l_0 through Q'.

From Theorem 15.3.15, there are $2q+1$ tangent planes to $\mathcal{H}_3$ through P_0 and q^2-q planes meeting $\mathcal{H}_3$ in a conic. These q^2-q conics map to the q^2-q lines in π containing neither Q nor Q'. The q^3-q^2 conics on $\mathcal{H}_3$ not through P_0 map to the q^3-q^2 conics in π through Q and Q'.

If $\mathcal{H}_3$ is partitioned by $q+1$ conics, then this configuration maps to a line l_1 not through Q or Q' and q conics through both Q and Q'. Thus the q^2 points of $\pi\backslash l_0$ are partitioned into $q-1$ on each of q conics and q on l_1.

II. *Elliptic quadric*

Let π be a plane not containing P_0 on $\mathcal{E}_3$, let π_0 be the tangent plane to $\mathcal{E}_3$ at P_0 and let $l_0 = \pi \cap \pi_0$. Then $\mathfrak{S}: \mathcal{E}_3\backslash\{P_0\} \to \pi$ is given by

$P\mathfrak{S} = PP_0 \cap \pi$, as for the hyperbolic quadric. The q^2 points of $\mathscr{E}_3 \backslash \{P_0\}$ map to the q^2 points of $\pi \backslash l_0$. The q^2+q conics on $\mathscr{E}_3$ through P_0 map to the q^2+q lines of $\pi \backslash l_0$. The q^3-q^2 conics on $\mathscr{E}_3$ not through P_0 map to the q^3-q^2 conics in π containing a fixed pair of conjugate complex points on l_0.

In particular, if $\mathscr{E}_3$ is partitioned into P_0, P_1 and $q-1$ conics, then π is partitioned into l_0, the point $P_1\mathfrak{S}$ and $q-1$ conics.

III. *Ovaloids*

If $\mathscr{K}$ is an ovaloid, then exactly the same process can be carried out as for the elliptic quadric. We merely replace $\mathscr{E}_3$ by $\mathscr{K}$ and 'conic' by '$(q+1)$-arc' throughout II.

So we have $\mathfrak{S}: \mathscr{K} \backslash \{P_0\} \to \pi$ is given by $P\mathfrak{S} = PP_0 \cap \pi$. Then the q^2 points of $\mathscr{K} \backslash \{P_0\}$ map to the q^2 points of $\pi \backslash l_0$. The q^2+q plane $(q+1)$-arcs on $\mathscr{K}$ through P_0 map to the q^2+q lines of $\pi \backslash l_0$ and the q^3-q^2 plane $(q+1)$-arcs on $\mathscr{K}$ not through P_0 map to the same number of $(q+1)$-arcs in π none of which meet l_0. This is the only essential difference from $\mathscr{E}_3$: we cannot say that such a $(q+1)$-arc meets l_0 in a pair of complex conjugate points.

16.4 λ-polarities

The main aim of this section is to give an example of an ovaloid, which is not a quadric, over $GF(2^h)$ with h odd and $h>1$. This is done by considering appropriate bijections between $\Pi = PG(3, q)$ and a general linear complex $\mathscr{A}$. These bijections resemble polarities of the space and are in fact polarities of the tactical configuration defined by Π and $\mathscr{A}$. We remark that $|\Pi| = |\mathscr{A}| = (q^2+1)(q+1)$, that $q+1$ points of Π lie on a line of $\mathscr{A}$ and that $q+1$ lines of $\mathscr{A}$ pass through a point of Π.

Until almost the end of the section, $q = 2^h$ with no further restriction on h for the moment. Let $\mathscr{L}$ be the set of lines of Π and let $\mathfrak{T}$ be the null polarity defined by $\mathscr{A}$. For $P \in \Pi$, $\pi = P\mathfrak{T}$ is the polar plane; then, if $Q \in \Pi$, $PQ \in \mathscr{A}$ if and only if $Q \in \pi$. If $l \in \mathscr{A}$ and $P \notin l$, then $P' = l \cap \pi$ is the only point of l conjugate to P. Also, PP' is the only line of $\mathscr{A}$ containing P and meeting l.

The section $\mathscr{P}_4$ of $\mathscr{H}_5$ representing $\mathscr{A}$ in $PG(5, q)$ has the property that all its tangent solids have a point N in common, § 5.3; the lines through N in the 4-space Π_4 containing $\mathscr{P}_4$ are all tangent lines to $\mathscr{P}_4$ and each meets $\mathscr{P}_4$ in a single point. So, the projection of $\mathscr{P}_4$ from N onto a solid Π' of Π_4 not through N is a bijection. Hence the function $\mathfrak{T}_1: \mathscr{A} \to \Pi'$ which maps a line l of $\mathscr{A}$ to a point of $\mathscr{P}_4$ and thence by projection from N to a point P' of Π' is a bijection.

In coordinates, let $\mathscr{A} = \boldsymbol{\lambda}(l_{01} + l_{23})$. Then $N = \mathbf{P}(1, 0, 0, 0, 0, 1)$ and $\Pi_4 = \boldsymbol{\pi}(x_0 + x_5)$. Let $\Pi' = \mathbf{V}(x_0, x_5)$. If

$$l = \mathbf{l}(l_{01}, l_{02}, l_{03}, l_{12}, l_{31}, l_{23})$$

with

$$l_{01} = l_{23} = \surd(l_{02}l_{31} + l_{03}l_{12}),$$

then $P' = \mathbf{P}(0, l_{02}, l_{03}, l_{12}, l_{31}, 0)$. More conveniently, we can take homogeneous coordinates x_0', x_1', x_2', x_3', in Π' so that $l\mathfrak{T}_1 = P' = \mathbf{P}(x_0', x_1', x_2', x_3')$ with

$$x_0' = l_{03}, \qquad x_1' = l_{12}, \qquad x_2' = l_{31}, \qquad x_3' = l_{02}.$$

Also, if $\pi = \boldsymbol{\pi}(u_0, u_1, u_2, u_3)$ is a plane of Π' containing P', then $\pi\mathfrak{T}_1^{-1}$ is the set of lines $l = \mathbf{l}(L)$ of $\mathscr{A}$ such that

$$u_0l_{03} + u_1l_{12} + u_2l_{31} + u_3l_{02} = 0.$$

Hence $\pi\mathfrak{T}_1^{-1}$ is the set of lines of $\mathscr{A}$ meeting the line

$$\mathbf{l}(\surd(u_0u_1 + u_2u_3), u_3, u_0, u_1, u_2, \surd(u_0u_1 + u_2u_3)).$$

The lines of $\mathscr{A}$ through a point P in Π are represented by points of a line on $\mathscr{P}_4$ and thence projected from N to points of a line l' in Π'. If $P = \mathbf{P}(x_0, x_1, x_2, x_3)$, then

$$l' = \mathbf{l}(l_{01}', l_{02}', l_{03}', l_{12}', l_{31}', l_{23}'),$$

where

$$l_{01}' = l_{23}' = x_0x_1 + x_2x_3, \qquad l_{02}' = x_3^2, \qquad l_{03}' = x_0^2, \qquad l_{12}' = x_1^2, \qquad l_{31}' = x_2^2;$$

here the line coordinates l_{ij}' are defined as usual with $l_{ij}' = x_i'y_j' + x_j'y_i'$, l' being the join of $\mathbf{P}(x_0', x_1', x_2', x_3')$ and $\mathbf{P}(y_0', y_1', y_2', y_3')$. Thus $\mathfrak{T}_1$ induces the bijection $\mathfrak{T}_2 : \Pi \to \mathscr{A}'$ where $\mathscr{A}' = \boldsymbol{\lambda}(l_{01}' + l_{23}')$ is a linear complex of Π'. So we write $l' = P\mathfrak{T}_2$.

Thus we can describe the effect of $\mathfrak{T}_1$ as follows:

line of $\mathscr{A}$ $\mapsto$ point of Π'

lines of $\mathscr{A}$ through a point of Π $\mapsto$ points of a line in $\mathscr{A}'$

lines of $\mathscr{A}$ meeting a fixed line of $\mathscr{A}$ $\mapsto$ points of a plane in Π'.

In what follows, the pair $(\Pi', \mathscr{A}')$ is often identified by $(\mathfrak{T}_1, \mathfrak{T}_2)$ with the pair $(\mathscr{A}, \Pi)$, the context always indicating whether an element of $\mathscr{A} = \Pi'$ is being considered as a point or a line.

Let $\mathfrak{S} : \Pi \to \Pi'$ be an incidence-preserving bijection mapping the linear complex $\mathscr{A}$ to the linear complex $\mathscr{A}'$. As $\mathfrak{S}$ maps every plane of Π to a plane of Π', it is a collineation given by

$$\rho X' = X^{\sigma}S,$$

where $X' = (x_0', x_1', x_2', x_3')$, $X^\sigma = (x_0^\sigma, x_1^\sigma, x_2^\sigma, x_3^\sigma)$, x^σ is the image of x under the automorphism σ of γ, and S is a 4×4 non-singular matrix, § 2.1. If $\mathfrak{S}$ is involutory, that is, if the mapping $\mathscr{A} \to \mathscr{A}'$ induced by $\mathfrak{S}$ coincides with $\mathfrak{S}^{-1}$, then $\mathfrak{S}$ is a λ-*polarity*. More precisely, this means that the following diagram commutes:

$$\begin{array}{ccc} \Pi' & \xrightarrow{\mathfrak{S}^{-1}} & \Pi \\ {\scriptstyle \mathfrak{T}_1^{-1}}\big\downarrow & & \big\downarrow{\scriptstyle \mathfrak{T}_2} \\ \mathscr{A} & \xrightarrow{\mathfrak{S}} & \mathscr{A}' \end{array}$$

λ-polarities are similar in their behaviour to polarities and an analogous terminology is used. For a point P in Π, the line $P\mathfrak{S}$ is the λ-*polar* of P and P is the λ-*pole* of $P\mathfrak{S}$. Two points P and P' in Π are λ-*conjugate* if P' lies on $P\mathfrak{S}$ or, equivalently, P lies on $P'\mathfrak{S}$; two lines l and l' of $\mathscr{A}$ are λ-*conjugate* if l' contains the point $l\mathfrak{S}^{-1}$ or, equivalently, l contains the point $l'\mathfrak{S}^{-1}$. The term *conjugate* is still used with respect to the null polarity $\mathfrak{T}$. However, throughout this section, the term *self-conjugate* will be used in relation to the λ-polarity $\mathfrak{S}$. So a point P is self-conjugate if it lies on its λ-polar $P\mathfrak{S}$ and a line l of $\mathscr{A}$ is self-conjugate if it contains its λ-pole $l\mathfrak{S}^{-1}$.

Lemma 16.4.1: *Let P be a point of Π which is not self-conjugate and let $l = P\mathfrak{S}$. The unique point P' of l conjugate to P has as its λ-polar the line $l' = PP'$. Hence both P' and l' are self-conjugate.*

Proof: As was remarked at the beginning of the section, the line l of $\mathscr{A}$ meets just one of the pencil of lines of $\mathscr{A}$ through P. Let $l' = P'\mathfrak{S}$; then l' contains P. However, the λ-pole of PP' is $l \cap l'$. As the only line of $\mathscr{A}$ through P meeting l is PP', so $l' = PP'$. Thus P' and l' are self-conjugate. □

Lemma 16.4.2: (i) *Every line l of $\mathscr{A}$ contains exactly one self-conjugate point.*

(ii) *Every point P of Π belongs to exactly one self-conjugate line.*

Proof: It suffices to prove (i). If $l\mathfrak{S}^{-1} \in l$, then $l\mathfrak{S}^{-1}$ is self-conjugate. If Q is another self-conjugate point of l, then Q is λ-conjugate to both itself and $l\mathfrak{S}^{-1}$. So $Q\mathfrak{S} = l$ and $Q = l\mathfrak{S}^{-1}$.

If $l\mathfrak{S}^{-1} \notin l$, then the point P of l conjugate to $l\mathfrak{S}^{-1}$ is self-conjugate by the previous lemma. If P' on l is self-conjugate, then P' is λ-conjugate to both itself and $l\mathfrak{S}^{-1}$. So $P'\mathfrak{S}$ is the join of $l\mathfrak{S}^{-1}$ and P'. But the only line of $\mathscr{A}$ through $l\mathfrak{S}^{-1}$ meeting l is the join of $l\mathfrak{S}^{-1}$ to P. So $P = P'$. □

Theorem 16.4.3: (i) *In $PG(3, q)$, $q = 2^h$, a coordinate system may be chosen so that, in a λ-polarity $\mathfrak{S}$, the points $\mathbf{U}_0$ and $\mathbf{U}_1$ are self-conjugate,*

$\mathbf{U}_2$ *is the point of* $\mathbf{U}_1\mathfrak{S}$ *conjugate to* $\mathbf{U}_0$, *and* $\mathbf{U}_3$ *is the point of* $\mathbf{U}_0\mathfrak{S}$ *conjugate to* $\mathbf{U}_1$.

(ii) $\mathfrak{S}$ *has canonical equation*

$$X' = X^\sigma,$$

where σ *is an automorphism of* γ *satisfying* $x^{\sigma^2} = x^2$ *for all* x *in* γ.

(iii) *A* λ*-polarity exists over* γ *if and only if* γ *has order* 2^h *with* h *odd: in this case,* $x^\sigma = x^{\sqrt{(2q)}}$.

Proof: Let P_0 and P_1 be self-conjugate points for $\mathfrak{S}$. As the λ-polar $P_1\mathfrak{S}$ of P_1 does not contain P_0, there is a unique point P_2 of $P_1\mathfrak{S}$ conjugate to P_0, by Lemma 16.4.1, and similarly a unique point P_3 of $P_0\mathfrak{S}$ conjugate to P_1. So

$$P_0\mathfrak{S} = P_0P_3, \qquad P_1\mathfrak{S} = P_1P_2.$$

From the definition of P_2, its λ-polar contains P_1 and meets $P_0\mathfrak{S}$. But, from the definition of P_3, the only line of $\mathscr{A}$ through P_1 meeting $P_0\mathfrak{S}$ is P_1P_3. Similarly, the λ-polar of P_3 contains P_0 and meets $P_1\mathfrak{S}$. Hence

$$P_2\mathfrak{S} = P_1P_3, \qquad P_3\mathfrak{S} = P_0P_2.$$

As the lines $P_0P_3, P_1P_2, P_1P_3, P_0P_2$ belong to $\mathscr{A}$, the selection of P_i as $\mathbf{U}_i$, $i \in \bar{\mathbf{N}}_3$, as well as a suitable unit point gives that $\mathscr{A} = \boldsymbol{\lambda}(l_{01} + l_{23})$. Now, $\mathfrak{S}: \Pi \to \Pi'$ is defined by the equation

$$x_i' = a_i x_i^\sigma$$

where $\mathbf{P}(X') = \mathbf{P}(X)\mathfrak{S}$. To simplify these equations, replace x_0, x_1, x_2, x_3 by $x_0, a_0^{\sigma+1}a_3x_1, a_0^\sigma a_3 x_2, a_0x_3$ respectively. Then $l_{01}, l_{02}, l_{03}, l_{12}, l_{31}, l_{23}$ are replaced by $a_0^{\sigma+1}a_3l_{01}$, $a_0^\sigma a_3 l_{02}$, $a_0 l_{03}$, $a_0^{2\sigma+1}a_3^2 l_{12}$, $a_0^{\sigma+2}a_3l_{31}$, $a_0^{\sigma+1}a_3l_{23}$. Thus $\mathscr{A}$ is left fixed and x_0', x_1', x_2', x_3' are replaced by a_0x_0', $a_0^{2\sigma+1}a_3^2x_1'$, $a_0^{\sigma+2}a_3x_2'$, $a_0^\sigma a_3 x_3'$. The equations of $\mathfrak{S}$ are now

$$a_0x_0' = a_0x_0^\sigma, \qquad a_0^{2\sigma+1}a_3^2x_1' = a_1a_0^{\sigma^2+\sigma}a_3^\sigma x_1^\sigma,$$
$$a_0^{\sigma+2}a_3x_2' = a_2a_0^{\sigma^2}a_3^\sigma x_2^\sigma, \qquad a_0^\sigma a_3 x_3' = a_3a_0^\sigma x_3^\sigma.$$

So the equations of $\mathfrak{S}$ can be written

$$x_0' = x_0^\sigma, \qquad x_1' = a_1x_1^\sigma, \qquad x_2' = a_2x_2^\sigma, \qquad x_3' = x_3^\sigma.$$

It remains to apply the conditions that $\mathscr{A}\mathfrak{S} = \mathscr{A}'$ and that $\mathfrak{S}$ is a λ-polarity. Firstly $\mathfrak{S}: \mathscr{L} \to \mathscr{L}'$ is given by

$$\mathbf{l}(l_{01}, l_{02}, l_{03}, l_{12}, l_{31}, l_{23})\mathfrak{S} = \mathbf{l}(a_1l_{01}^\sigma, a_2l_{02}^\sigma, l_{03}^\sigma, a_1a_2l_{12}^\sigma, a_1l_{31}^\sigma, a_2l_{23}^\sigma).$$

Hence $a_1 = a_2$. So the equations of $\mathfrak{S}$ are

$$x_0' = x_0^\sigma, \qquad x_1' = cx_1^\sigma, \qquad x_2' = cx_2^\sigma, \qquad x_3' = x_3^\sigma.$$

It remains to consider the commutative diagram; that is, for a point $\mathbf{P}(Y)$ of Π',

$$\mathbf{P}(Y)\mathfrak{T}_1^{-1}\mathfrak{S} = \mathbf{P}(Y)\mathfrak{S}^{-1}\mathfrak{T}_2.$$

$$\mathbf{P}(y_0, y_1, y_2, y_3)\mathfrak{T}_1^{-1}\mathfrak{S} = \mathbf{l}(\sqrt{(y_0y_1 + y_2y_3)}, y_3, y_0, y_1, y_2, \sqrt{(y_0y_1 + y_2y_3)})\mathfrak{S}$$
$$= \mathbf{l}(c(y_0y_1 + y_2y_3)^{\sigma/2}, cy_3^\sigma, y_0^\sigma, c^2y_1^\sigma, cy_2^\sigma, c(y_0y_1 + y_2y_3)^{\sigma/2});$$
$$\mathbf{P}(y_0, y_1, y_2, y_3)\mathfrak{S}^{-1}\mathfrak{T}_2 = \mathbf{P}(y_0^{\sigma^{-1}}, c^{-\sigma^{-1}}y_1^{\sigma^{-1}}, c^{-\sigma^{-1}}y_2^{\sigma^{-1}}, y_3^{\sigma^{-1}})\mathfrak{T}_2$$
$$= \mathbf{l}(c^{-\sigma^{-1}}(y_0y_1 + y_2y_3)^{\sigma^{-1}}, y_3^{2\sigma^{-1}}, y_0^{2\sigma^{-1}}, c^{-2\sigma^{-1}}y_1^{2\sigma^{-1}}, c^{-2\sigma^{-1}}y_2^{2\sigma^{-1}},$$
$$c^{-\sigma^{-1}}(y_0y_1 + y_2y_3)^{\sigma^{-1}}).$$

Hence $c = 1$ and $x^{\sigma^2} = x^2$ for all x in γ.

Since σ is an automorphism of γ, so $x^\sigma = x^{2^r}$ for some r. Since $x^{\sigma^2} = x^2$, so $x^{2^{2r}} = x^2$. Therefore $x^{2^{2r-1}} = x$ and $h = 2r - 1$. Hence $r = (h+1)/2$ and $2^r = 2^{(h+1)/2} = \sqrt{(2q)}$. □

The set of self-conjugate points of a λ-polarity $\mathfrak{S}$ is a *λ-quadric* Ω. The lines of Π through P in Ω meeting Ω in no further point form the *tangent space* to Ω at P.

Theorem 16.4.4:

(i) *A λ-quadric Ω is an ovaloid for $q > 2$.*

(ii) *The tangent space at P in Ω is the polar plane of P (with respect to the null polarity $\mathfrak{T}$).*

(iii) *Every line of $\mathscr{L}\backslash\mathscr{A}$ meets Ω in 0 or 2 points and every line of $\mathscr{A}$ meets Ω in one point.*

Proof: It suffices to prove (iii) and, by Lemma 16.4.2, that a line l of $\mathscr{L}\backslash\mathscr{A}$ with $l \cap \Omega \neq \varnothing$ meets Ω in two points. For, then (ii) follows and, as there are then $q+1$ coplanar lines of $\mathscr{A}$ through each point of Ω, so $|\Omega| = |\mathscr{A}|/(q+1) = q^2 + 1$.

Take $P \in l \cap \Omega$ and let $l' = l\mathfrak{T}$. Now, $P\mathfrak{T}$ is the plane containing the lines of $\mathscr{A}$ through P. So both l' and $P\mathfrak{S}$ lie in $P\mathfrak{T}$. So let $Q = P\mathfrak{S} \cap l'$. Since $Q \in P\mathfrak{S}$, so $P \in Q\mathfrak{S}$; whence $Q\mathfrak{S}$ lies in $P\mathfrak{T}$. Therefore, let $R = Q\mathfrak{S} \cap l'$. Then both $R\mathfrak{S}$ and l belong to $Q\mathfrak{T}$. As every point of l is conjugate to every point of l', so $R\mathfrak{S} \cap l$ is conjugate to R and, by Lemma 16.4.1, is the only point of $R\mathfrak{S}$ conjugate to R as well as being a self-conjugate point of l.

Now, let $P' \in l\backslash\{P\}$. Since $Q \in l'$, the points P' and Q are conjugate; so $P'Q \in \mathscr{A}$. Hence let $R' = (P'Q)\mathfrak{S}^{-1} = P'\mathfrak{S} \cap Q\mathfrak{S}$; therefore $P' = R'\mathfrak{S} \cap l$. If $P' \in \Omega$, then $P'\mathfrak{S} = P'R' \in \mathscr{A}$, whence $R' = P'\mathfrak{T} \cap Q\mathfrak{S} = P'\mathfrak{T} \cap P\mathfrak{T} \cap Q\mathfrak{S} = l' \cap Q\mathfrak{S} = R$. So $P' = R'\mathfrak{S} \cap l = R\mathfrak{S} \cap l$. □

For convenience, we will continue to write x^σ instead of $x^{\sqrt{(2q)}}$. It is useful to note that exponentiation by $\sigma+1$, $\sigma-1$, $\sigma+2$, or $\sigma-2$ is a permutation of γ_0.

Theorem 16.4.5: *The canonical form of a* λ*-quadric* Ω *is*

$$\Omega = \mathbf{V}(x_1^\sigma(x_0x_1+x_2x_3)+x_2^{\sigma+2}+x_3^\sigma x_1^2,\ x_0^\sigma(x_0x_1+x_2x_3)+x_2^\sigma x_0^2+x_3^{\sigma+2})$$

or, more usefully,

$$\Omega = \{\mathbf{U}_1\}\cup\{\mathbf{P}(1, z, y, x)\mid z = xy+x^{\sigma+2}+y^\sigma\}.$$

Proof: From § 15.1, the condition that the point $\mathbf{P}(X)$ lies on the line $\mathbf{l}(L)$ is $X\Lambda = 0$. Taking the λ-polarity $\mathfrak{S}$ to have canonical form as in Theorem 16.4.3,

$$\mathbf{P}(x_0, x_1, x_2, x_3)\mathfrak{S} = \mathbf{l}((x_0x_1+x_2x_3)^{\sigma/2}, x_3^\sigma, x_0^\sigma, x_1^\sigma, x_2^\sigma, (x_0x_1+x_2x_3)^{\sigma/2}).$$

Hence $\mathbf{P}(X)$ lies on $\mathbf{P}(X)\mathfrak{S}$ if and only if

$$\begin{aligned}
x_1^\sigma(x_0x_1+x_2x_3)+x_2^{\sigma+2}+x_3^\sigma x_1^2 &= 0,\\
x_0^\sigma(x_0x_1+x_2x_3)+x_2^\sigma x_0^2+x_3^{\sigma+2} &= 0,\\
x_0^\sigma x_2^2+x_1^\sigma x_0^2+x_3^\sigma(x_0x_1+x_2x_3) &= 0,\\
x_0^\sigma x_1^2+x_1^\sigma x_3^2+x_2^\sigma(x_0x_1+x_2x_3) &= 0.
\end{aligned}$$

Since there are two conditions for a point to lie on a line, only two of these four conditions are independent. Substituting $x_0 = 0$, we find that $\mathbf{u}_0\cap\Omega = \{\mathbf{U}_1\}$. For points off $\mathbf{u}_0$, the point $\mathbf{P}(1, z, y, x)$ lies on the variety defined by the first two equations if

$$\begin{aligned}
z^{\sigma+1}+z^\sigma xy+y^{\sigma+2}+z^2x^\sigma &= 0\\
z+xy+y^\sigma+x^{\sigma+2} &= 0.
\end{aligned}$$

The second equation implies the first. It is perhaps worth noting that $\mathbf{u}_0\cap\Omega = \{\mathbf{U}_1\}$ does not follow from any two of the above four equations, but does follow from the first two. Hence Ω can be thought of either as the intersection of the primals defined by the first two equations, or as $(\Omega\cap\mathbf{u}_0)\cup(\Omega\backslash\mathbf{u}_0)$, whence the second form for Ω. □

Corollary: *For* $q = 2^h$, *h odd and* $h > 1$, *there exists an ovaloid which is not a quadric.* □

Let $Sz(q)$ be the group of projectivities of Π fixing the linear complex $\mathscr{A}$ and the λ-polarity $\mathfrak{S}$. Hence $Sz(q)$ also fixes Ω. Let $G(\Omega)$ be the group of projectivities fixing Ω; so $Sz(q) < G(\Omega)$.

Lemma 16.4.6: (i) *$Sz(q)$ is doubly transitive on* Ω;
(ii) *$Sz(q)$ is transitive on pairs of* λ*-conjugate points of* $\Pi\backslash\Omega$;
(iii) *$Sz(q)$ is transitive on* $\Pi\backslash\Omega$.

Proof: (i) follows from Theorem 16.4.3. If P and P' in $\Pi\backslash\Omega$ are a pair of λ-conjugate points and if $\mathbf{U}_0$ and $\mathbf{U}_1$ are taken as the respective poles of the self-conjugate lines containing P' and P, then $\mathbf{U}_2$ and $\mathbf{U}_3$ are respectively P and P'. □

Let G_1 be the stabilizer in $Sz(q)$ of $\mathbf{U}_1$ and let G_2 be the stabilizer in G_1 of $\mathbf{U}_0$. Then, by Theorem 16.4.3, G_2 also fixes $\mathbf{U}_2$ and $\mathbf{U}_3$. Let G_2' be the subgroup of $G(\Omega)$ fixing both $\mathbf{U}_0$ and $\mathbf{U}_1$.

Lemma 16.4.7: (i) $G_2 = \{\mathfrak{S}_t \mid t \in \gamma_0\}$, *where* $\mathfrak{S}_t$ *is given in non-homogeneous coordinates by*

$$x' = tx, \qquad y' = t^{\sigma+1}y, \qquad z' = t^{\sigma+2}z.$$

(ii) *For* $q > 2$, $G_2' = G_2$.

Proof: (i) Since an element of G_2 fixes $\mathbf{U}_0$, $\mathbf{U}_1$, $\mathbf{U}_2$, $\mathbf{U}_3$, it is given by $\rho x_i' = a_i x_i$, $i \in \bar{\mathbf{N}}_3$, and hence by $z' = a_1 z$, $y' = a_2 y$, $x' = a_3 x$. Substitution in the equation for Ω gives the result.

(ii) A projectivity fixing $\mathbf{U}_1$ and Ω fixes $\mathbf{u}_0$. So, if $\mathfrak{T}$ fixes Ω, $\mathbf{U}_1$, and $\mathbf{U}_0$, it is given by $(x', y', z') = (x, y, z)T$ for some matrix T. Hence, for $q > 2$, the terms $z, xy, x^{\sigma+2}, y^{\sigma}$ are fixed up to a scalar multiple. Hence T is diagonal and $\mathfrak{T}$ has the same form as an element of G_2. □

Corollary:

(i) *For* $q > 2$, $G(\Omega) = Sz(q)$; (ii) $|Sz(q)| = q^2(q^2+1)(q-1)$.

Proof: From the lemma, $|G_2| = q-1$ and, for $q > 2$, also $|G_2'| = q-1$. Since, by Lemma 16.4.6, $Sz(q)$ is doubly transitive on Ω, so $|Sz(q)| = (q^2+1)q^2|G_2|$ and $|G(\Omega)| = (q^2+1)q^2|G_2'|$. As $Sz(q) < G(\Omega)$, the result follows. □

Lemma 16.4.8:

(i) $G_1 = G_2 H$ *where* $H \triangleleft G_1$; (ii) $|G_1| = q^2(q-1)$.

Proof: For all b, c in γ, let $\mathfrak{T}_{b,c}$ be the projectivity with equations, in non-homogeneous coordinates,

$$x' = x + b$$
$$y' = y + b^{\sigma}x + c$$
$$z' = z + by + (b^{\sigma+1} + c)x + bc + b^{\sigma+2} + c^{\sigma}.$$

Then it follows that

$$\mathfrak{S}_t^{-1}\mathfrak{T}_{b,c}\mathfrak{S}_t = \mathfrak{T}_{b',c'} \quad \text{with} \quad b' = bt, \qquad c' = ct^{\sigma+1}; \tag{16.1}$$

$$\mathfrak{T}_{b,c}\mathfrak{T}_{b',c'} = \mathfrak{T}_{b'',c''} \quad \text{with} \quad b'' = b + b', \qquad c'' = c + c' + bb'^{\sigma}; \tag{16.2}$$

$$\mathfrak{T}_{b,c}^{-1} = \mathfrak{T}_{b,c'} \quad \text{with} \quad c' = c + b^{\sigma+1}; \tag{16.3}$$

$$\mathfrak{T}_{b,c}^{-1}\mathfrak{T}_{b',c'}^{-1}\mathfrak{T}_{b,c}\mathfrak{T}_{b',c'} = \mathfrak{T}_{0,c''} \quad \text{with} \quad c'' = bb'^{\sigma} + b'b^{\sigma}. \tag{16.4}$$

Now, let $H=\{\mathfrak{T}_{b,c}\}$ and let $H_0=\{\mathfrak{T}_{0,c}\}$. Firstly, it follows by substitution in the equation for Ω that $\mathfrak{T}_{b,c}\in G_1$. From (16.1), (16.2) and (16.3), H and G_2H are groups with H normal in G_2H. Also G_1 is transitive on $\Omega\backslash\{\mathbf{U}_1\}$, by Lemma 16.4.6. So $|G_1|/|G_2|=q^2$, whence $|G_1|=q^2(q-1)$. However, $|G_2\cap H|=1$; so $|G_2H|=|G_2|\,|H|=q^2(q-1)$. As $G_2H\subset G_1$, so $G_2H=G_1$. □

Lemma 16.4.9: *If $\iota\neq 1$, then $\mathfrak{T}_{b,c}\mathfrak{S}_t$ is conjugate to $\mathfrak{S}_t$ in G_1.*

Proof: $\mathfrak{T}_{b,c}\mathfrak{S}_t$ has a fixed point in $\Pi\backslash\mathbf{u}_0$ and so a fixed point P in $\Omega\backslash\{\mathbf{u}_1\}$. There is at least one element of G_1 taking P to $\mathbf{U}_0$. So $\mathfrak{T}_{b,c}\mathfrak{S}_t$ is conjugate to some element of G_2, which a little manipulation shows to be $\mathfrak{S}_t$ itself. □

Corollary: *The only non-trivial normal subgroups of G_1 are H_0 and subgroups KH, where K is any subgroup of G_2.*

Proof: This follows from the lemma and (16.4). □

Lemma 16.4.10: *If $\mathfrak{R}$ in $Sz(q)$ transforms $\mathbf{U}_0$ to $\mathbf{U}_1$ then $\mathfrak{R}^{-1}\mathfrak{S}_t\mathfrak{R}=\mathfrak{S}_{t^{-1}}\mathfrak{T}_{b,c}$ for some $\mathfrak{T}_{b,c}$ in H.*

Proof: If $\mathfrak{R}$ is an element of $Sz(q)$ given by $X'=XR$, where $X'=(x_0', x_1', x_2', x_3')$ and $X=(x_0, x_1, x_2, x_3)$, such that $\mathbf{U}_0\mathfrak{R}=\mathbf{U}_1$, then $\mathbf{u}_1\mathfrak{R}=\mathbf{u}_0$ since $\mathscr{A}=\boldsymbol{\lambda}(l_{01}+l_{23})$ is preserved. So we can write

$$R=\begin{bmatrix}0 & a_{01} & 0 & 0\\ 1 & a_{11} & a_{12} & a_{13}\\ 0 & a_{21} & a_{22} & a_{23}\\ 0 & a_{31} & a_{32} & a_{33}\end{bmatrix}.$$

Then the corresponding transformation on Π' is given by $Y'=X'R'$ where

$$R'=\begin{bmatrix}0 & a_{01}a_{32} & a_{01}a_{33} & 0\\ a_{23} & a_{11}a_{22}+a_{12}a_{21} & a_{11}a_{23}+a_{21}a_{13} & a_{22}\\ a_{33} & a_{11}a_{32}+a_{31}a_{12} & a_{11}a_{33}+a_{13}a_{31} & a_{32}\\ 0 & a_{01}a_{22} & a_{01}a_{23} & 0\end{bmatrix}.$$

The conditions that $\mathscr{A}$ is preserved are

$$a_{01}=a_{22}a_{33}+a_{23}a_{32},\qquad a_{21}=a_{12}a_{23}+a_{13}a_{22},\qquad a_{31}=a_{12}a_{33}+a_{13}a_{32}.$$

The condition that $\mathfrak{S}$ of Theorem 16.4.3(ii) is preserved is

$$\rho R'=R^{\sigma}.$$

A little manipulation gives the result that

$$R=\begin{bmatrix} 0 & 1 & 0 & 0 \\ 1 & de+d^{\sigma+2}+e^{\sigma} & e & d \\ 0 & d^{\sigma+1}+e & d^{\sigma} & 1 \\ 0 & d & 1 & 0 \end{bmatrix}$$

for some d and e in γ. The matrix of $\mathfrak{T}_{b,c}$ is

$$\begin{bmatrix} 1 & bc+b^{\sigma+2}+c^{\sigma} & c & b \\ 0 & 1 & 0 & 0 \\ 0 & b & 1 & 0 \\ 0 & b^{\sigma+1}+c & b^{\sigma} & 1 \end{bmatrix}.$$

A little more manipulation gives that

$$\mathfrak{R}^{-1}\mathfrak{S}_t\mathfrak{R}=\mathfrak{S}_{t^{-1}}\mathfrak{T}_{b,c}$$

$$bt=d(t+1), \qquad ct^{\sigma+1}=d^{\sigma+1}(t^{\sigma}+1)+e(t^{\sigma+1}+1). \quad \square$$

Theorem 16.4.11: *For $q=2^h$, h odd, $Sz(q)$ has order $q^2(q^2+1)(q-1)$ and, for $h>1$, $Sz(q)$ is simple.*

Proof: The order of $Sz(q)$ was obtained in the corollary to Lemma 16.4.7.

Suppose that G is a non-trivial normal subgroup of $Sz(q)$. Since $Sz(q)$ is doubly transitive on Ω, so G is transitive on Ω. Let $\mathfrak{R}$ in G be a projectivity such that $\mathbf{U}_0\mathfrak{R}=\mathbf{U}_1$. Then $\mathfrak{R}^{-1}\mathfrak{S}_t\mathfrak{R}\in G_1$ and, by Lemma 16.4.10, is of the form $\mathfrak{S}_{t^{-1}}\mathfrak{T}_{b,c}$. So $\mathfrak{S}_t^{-1}\mathfrak{R}^{-1}\mathfrak{S}_t\mathfrak{R}=\mathfrak{S}_{t^{-2}}\mathfrak{T}_{b,c}$ is in G. So, by Lemma 16.4.9, $\mathfrak{S}_{t^{-2}}\in G$ for all t in γ_0. Thus $G_2\subset G$. If $q>2, |G_2|>1$. So, by the corollary to Lemma 16.4.9, $G\cap G_1=G_1$. Hence $G_1\subset G$ and, since G is transitive on Ω, so $G=Sz(q)$.

For $q=2$, $Sz(q)$ is of order 20 and therefore has a normal subgroup of order 5. $\square$

For $q=2$, the set Ω is a 5-cap and hence $G(\Omega)\cong\mathbf{S}_5$ with order 120.

For q even, it is possible to relate ovaloids and certain spreads of lines in $PG(3,q)$ using the map $\mathfrak{T}_1$.

Theorem 16.4.12: *There is a one-to-one correspondence between the ovaloids of $PG(3,q)$, q even, and the spreads whose lines lie in a linear complex.*

Proof: Let $\mathcal{K}$ be an ovaloid. Then, by Theorem 16.1.9, Corollary 2, its tangent lines form a linear complex $\mathcal{A}$. If P is a point of $\mathcal{K}$, the $q+1$ tangent lines to $\mathcal{K}$ at P are mapped by $\mathfrak{T}_1$ to the points of a line in $\mathcal{A}'$. So the points of $\mathcal{K}$ are mapped to a set of q^2+1 lines of $\mathcal{A}'$. As the tangents at two points of $\mathcal{K}$ are all different, the q^2+1 lines of $\mathcal{A}'$ are mutually skew and hence form a spread of Π'.

Conversely, suppose $\mathcal{F} \subset \mathcal{A}'$ is a spread of lines of Π'. The q^2+1 lines of $\mathcal{F}$ are mapped by $\mathfrak{T}_1^{-1}$ onto q^2+1 pencils of lines $\mathcal{L}_0, \ldots, \mathcal{L}_{q^2}$ belonging to $\mathcal{A}$. As no two of the pencils have a line in common, $\mathcal{L}_0 \cup \ldots \cup \mathcal{L}_{q^2} = \mathcal{A}$. If P_i is the centre of $\mathcal{L}_i$ and $\mathcal{K} = \{P_i \mid i \in \bar{\mathbf{N}}_{q^2}\}$, then $|\mathcal{K}| = q^2+1$. It remains to show that $\mathcal{K}$ is an ovaloid.

Firstly, every line of $\mathcal{A}$ contains just one point of $\mathcal{K}$; for, if $P_iP_j = l \in \mathcal{A}$, then $\mathcal{L}_i \cap \mathcal{L}_j = \{l\}$. Now take a plane π in Π and consider its intersection with $\mathcal{K}$. The lines of $\mathcal{A}$ which belong to π form a pencil with centre P. If $P \in \mathcal{K}$, then $\mathcal{K} \cap \pi = \{P\}$; if $P \notin \mathcal{K}$, then $|\mathcal{K} \cap \pi| = q+1$. So every plane meets $\mathcal{K}$ in 1 or $q+1$ points. Therefore, if l is a line such that $|l \cap \mathcal{K}| = n > 1$, then counting the points of $\mathcal{K}$ on the planes of the pencil through l gives

$$q^2+1 = (q+1)(q+1-n)+n;$$

whence $n = 2$. So $\mathcal{K}$ is an ovaloid and the pencils $\mathcal{L}_i$ consitute the tangent lines at its points. □

The existence of the non-quadric ovaloids arose from the existence of λ-polarities or, alternatively described, from the existence of 'polarities' of the tactical configuration defined by the null system $(\Pi, \mathcal{A})$. It is natural to ask if a linear complex gives rise to any comparable properties for q odd.

Let us return to the definitions of $\mathfrak{T}_1$ and $\mathfrak{T}_2$. If Π' is identified with Π and so $\mathcal{A}'$ with $\mathcal{A}$ by $\mathbf{P}(X') \to \mathbf{P}(X)$, then $\mathfrak{T}_1$ and $\mathfrak{T}_2$ are inverses and their existence shows that the null system $(\Pi, \mathcal{A})$ is self-dual; that is, there exist $\mathfrak{T}_1 : \mathcal{A} \to \Pi$ and $\mathfrak{T}_2 : \Pi \to \mathcal{A}$ (and, since $\mathfrak{T}_2$ was induced by $\mathfrak{T}_1$, we may unambiguously write $\mathfrak{T}_2 = \mathfrak{T}_1$) such that, for $P \in \Pi$ and $l \in \mathcal{A}$, we have that $P \in l \Rightarrow l\mathfrak{T}_1 \in P\mathfrak{T}_1$. That there is no comparable situation for q odd is shown by the following result.

Theorem 16.4.13: *The null system* $(\Pi, \mathcal{A})$ *is self-dual for q even but not self-dual for q odd.*

Proof: For q even, the self-duality of $(\Pi, \mathcal{A})$ was shown above.

For q odd, suppose there exists a duality $\mathfrak{T}_0$ of $(\Pi, \mathcal{A})$; that is, a bijection interchanging $\mathcal{A}$ and Π, but preserving incidence.

Consider a line l not in $\mathcal{A}$ and two distinct points P_1 and P_2 of l. If $P_1\mathfrak{T}_0$ and $P_2\mathfrak{T}_0$ had a point P in common, then $P\mathfrak{T}_0^{-1}$ would be in $\mathcal{A}$ and would contain both P_1 and P_2; so we would have $P_1P_2 = l \in \mathcal{A}$, a contradiction. Let $l' = l\mathfrak{T}$, the polar of l in the null polarity defined by $\mathcal{A}$. If $Q_0, \ldots, Q_q$ are the points of l', then $P_1Q_i \in \mathcal{A}$ and $P_2Q_i \in \mathcal{A}$ for $i \in \bar{\mathbf{N}}_q$.

Since $(P_1Q_i)\mathfrak{T}_0 \in P_1\mathfrak{T}_0$ and $(P_2Q_i)\mathfrak{T}_0 \in P_2\mathfrak{T}_0$, the line $Q_i\mathfrak{T}_0$ meets both $P_1\mathfrak{T}_0$ and $P_2\mathfrak{T}_0$. So, to the $q+1$ points of l' correspond the $q+1$ lines of $\mathscr{A}$ meeting both $P_1\mathfrak{T}_0$ and $P_2\mathfrak{T}_0$; these $q+1$ lines form a regulus $\mathscr{R}$.

If P is an arbitrary point of l, then $(PQ_i)\mathfrak{T}_0 \in Q_i\mathfrak{T}_0$ and so the line $P\mathfrak{T}_0$ is in the complementary regulus $\mathscr{R}'$ of $\mathscr{R}$. So, to the $q+1$ points of l correspond the $q+1$ lines of $\mathscr{R}'$. Hence both $\mathscr{R}$ and $\mathscr{R}'$ are subsets of $\mathscr{A}$.

Now we consider an arbitrary point Q of the quadric $\mathscr{H}_3$ containing $\mathscr{R}$ and $\mathscr{R}'$. Since the lines of $\mathscr{R}$ and $\mathscr{R}'$ through Q belong to $\mathscr{A}$, it follows that the tangent plane to $\mathscr{H}_3$ at Q is the polar plane of Q in the null polarity $\mathfrak{T}$. If now π is a plane meeting $\mathscr{H}_3$ in a conic $\mathscr{P}_2$, then the $q+1$ tangents to $\mathscr{P}_2$ all belong to $\mathscr{A}$ and so meet in a point, which only occurs for q even. □

16.5 Notes and references

§ 16.1. Theorem 16.1.5 is due to Bose (1947) for q odd and to Qvist (1952) for q even; the latter has been followed. For another treatment of the even case, see Segre (1967). Theorem 16.1.7 is due to Barlotti (1955) and Panella (1955). Here we follow Barlotti (1965), which also contains an account of the rest of § 16.1. Theorem 16.1.8 is due to Segre (1959b).

§ 16.2. Theorem 16.2.6 comes from Lefèvre (-Percsy) (1975). Another characterization of quadrics is given by Buekenhout (1969a) as part of a more general theorem. The set $\mathscr{K}$ in $PG(3, q)$ is *quadratic* if it is ruled and if at any point P of $\mathscr{K}$ the tangents and the lines of $\mathscr{K}$ through P form a pencil. Then a quadratic set is either $\mathscr{H}_3$ or an ovaloid. See also Buekenhout (1976a), Schröder (1976), Tallini (1976).

A further point of view is to consider generalized quadrangles, § 2.2. See Payne and Thas (1984), Buekenhout and Lefèvre (1974), Buekenhout and Shult (1974), Olanda (1972*), (1973*), (1977*), Lefèvre-Percsy (1977a*), (1977b*), and the works of Payne and Thas.

§ 16.4. This is based on Tits (1962a). For another treatment, see Lüneburg (1965, 1980). The groups $Sz(q)$ were discovered by Suzuki (1962*). The identification of $G(\Omega)$ and $Sz(q)$ of Lemma 16.4.7 follows Tits (1960).

An ovaloid other than an elliptic quadric was also found for $q=8$ by Segre (1959b), who gave sufficient conditions for such an ovaloid to [illegible]ist. He also showed that these conditions were not fulfilled wher [illegible] Häring and Heise (1979) showed that the conditions cannot [illegible]illed for $q>8$. Fellegara (1962) showed that Segre's example for $q=8$ is a λ-quadric; that is, Tits' example is the same as Segre's. Fellegara further showed, by computer, that every ovaloid in $PG(3, 8)$ is either the elliptic quadric $\mathscr{E}_3$ or the λ-quadric Ω.

A characterization of λ-quadrics is given by Tits (1966). As in

$PG(3, q)$, a *k-cap* in $PG(n, q)$ is a set of k points no three of which are collinear; an *ovoid* in $PG(n, q)$ is a k-cap such that the tangent lines at each point form a prime. If the group $G(\mathscr{K})$ of projectivities of an ovoid $\mathscr{K}$ is doubly transitive on $\mathscr{K}$, then either (1) $n=2$ and $\mathscr{K}$ is a conic; (2) $n=3$ and $\mathscr{K}$ is an elliptic quadric; or (3) $n=3$ and $\mathscr{K}$ is a λ-quadric. For other results, see Tits (1962b).

Tits (1962a) also shows that, if $\mathscr{O}$ is an oval obtained as a section of a λ-quadric plus its nucleus, then $\mathscr{O}$ is projectively equivalent to $\mathscr{D}(\sigma)$ or $\mathscr{D}(\sigma+2)$, in the notation of § 8.4; that is,

$$\mathscr{O}=\{\mathbf{P}(t^m, t, 1) \mid t \in \gamma^+\} \cup \{\mathbf{U}_1\}$$

where $m=\sigma$ or $m=\sigma+2$.

Theorems 16.4.12 and 16.4.13 come from Thas (1972).

17

SPANS, SPREADS, AND PACKINGS

17.1 Regular spreads

From § 4.1, a spread of lines of $PG(3, q)$ is a set of q^2+1 mutually skew lines, which necessarily partition the space. Let us first give an example.

Lemma 17.1.1: (i) *The pencil* $\mathscr{F}$ *of quadrics* $\mathscr{F}_t$, $t \in \gamma^+$, *where*

$$\mathscr{F}_t = \mathbf{V}(f(x_0, x_1) + tf(x_2, x_3))$$

and f is an irreducible binary quadratic, consists of $\mathscr{F}_0 = \mathbf{U}_2\mathbf{U}_3$, $\mathscr{F}_\infty = \mathbf{U}_0\mathbf{U}_1$ *and* $q-1$ *hyperbolic quadrics* $\mathscr{F}_t$, $t \in \gamma_0$.

(ii) $\mathscr{F}_0$ *and* $\mathscr{F}_\infty$ *are polar lines with respect to each* $\mathscr{F}_t$, $t \in \gamma_0$.

(iii) $\mathscr{F}_0$ *and* $\mathscr{F}_\infty$ *together with one of the reguli on each* $\mathscr{F}_t$, $t \in \gamma_0$, *form a spread.*

(iv) 2^{q-1} *spreads can be formed in this way.*

Proof: Since f is irreducible, no two of the $q+1$ quadrics $\mathscr{F}_t$ intersect. Each point of the space lies on some $\mathscr{F}_t$. As $\mathscr{F}_0$ and $\mathscr{F}_\infty$ each consist of a line, the other $q-1$ quadrics contain $(q^2+1)(q+1)-2(q+1)=(q-1)(q+1)^2$ points altogether. As each $\mathscr{F}_t$, $t \in \gamma_0$, is of rank 4 (Lemma 15.3.1) it is elliptic or hyperbolic and contains q^2+1 or $(q+1)^2$ points respectively. Hence each $\mathscr{F}_t$, $t \in \gamma_0$, contains $(q+1)^2$ points and is hyperbolic. Parts (ii), (iii), and (iv) follow immediately. □

A spread $\mathscr{S}$ is *regular* if, when l_1, l_2, l_3 are in $\mathscr{S}$, then the whole regulus $\mathscr{R}(l_1, l_2, l_3)$ is contained in $\mathscr{S}$.

Lemma 17.1.2: *The following are equivalent:*

(i) $\mathscr{S}$ *is an elliptic congruence;*

(ii) $\mathscr{S}$ *is a regular spread;*

(iii) $\mathscr{S}$ *is a spread such that the lines of* $\mathscr{S}$ *meeting any line not in* $\mathscr{S}$ *form a regulus.*

Proof: (i) ⇒ (ii). $\mathscr{S}$ is represented by the section $\mathscr{E}_3$ of a solid Π_3 with $\mathscr{H}_5$. If two points of $\mathscr{E}_3$ were conjugate their join would lie in Π_3 and therefore on $\mathscr{E}_3$, contradicting that no three points of $\mathscr{E}_3$ are collinear. So no two points of $\mathscr{E}_3$ are conjugate and hence no two lines of $\mathscr{S}$ intersect. Therefore $\mathscr{S}$ is a spread.

The plane containing three points of $\mathscr{E}_3$ meets $\mathscr{H}_5$ in a conic lying entirely on $\mathscr{E}_3$; this conic represents the regulus containing the three lines represented by the three points. So $\mathscr{S}$ is regular.

To see this argument solely in $PG(3, q)$, consider $\mathcal{S}$ as the set of lines meeting two skew conjugate lines m and $\bar{m}$. If l is a line of $PG(3, q)$ meeting m and $\bar{m}$, it is the join of a point P of m and the conjugate point $\bar{P}$ of $\bar{m}$. Since m has q^2+1 points in $PG(3, q^2)$, there are q^2+1 such lines l and, since m and $\bar{m}$ are skew, so are any two lines of $\mathcal{S}$. If l_1, l_2, l_3 are in $\mathcal{S}$, then the quadric they determine also contains m and $\bar{m}$; hence $\mathcal{S}$ is regular.

(ii) $\Rightarrow$ (i). By definition, a regular spread is the set of lines $\mathcal{S}$ linearly dependent on n skew independent lines for some n. For $n = 3$, $\mathcal{S}$ is a regulus and, for $n = 4$, $\mathcal{S}$ is an elliptic congruence, by Theorem 15.2.7.

(ii) $\Rightarrow$ (iii). If $\mathcal{S}$ is a regular spread and l is not in $\mathcal{S}$, then through each point P_i of l there is a unique line l_i of $\mathcal{S}$, $i \in \mathbf{N}_{q+1}$. Let $\mathcal{R} = \mathcal{R}(l_1, l_2, l_3)$. Then $\mathcal{R} \subset \mathcal{S}$. If l_i is not in $\mathcal{R}$, then there will be two lines of $\mathcal{S}$ through P_i. So each l_i is in $\mathcal{R}$. Conversely, if l_1, l_2, and l_3 are in $\mathcal{S}$ and l is one of their transversals, then the $q+1$ lines of $\mathcal{S}$ meeting l form $\mathcal{R}(l_1, l_2, l_3)$. □

Corollary: *A regular spread contains* $q(q^2+1)$ *reguli.*

Proof: Under the map $\mathfrak{G}$ of § 15.4, a regulus of a regular spread corresponds to a conic on $\mathscr{E}_3$. Hence Lemma 15.3.7 gives the result. □

The general construction of spreads in § 4.1 gives a regular spread in the case of lines in $PG(3, q)$.

A *subregular sequence of spreads of length* k is a sequence $\mathcal{S}_0, \mathcal{S}_1, \ldots, \mathcal{S}_k$ of spreads such that (i) $\mathcal{S}_0$ is regular, (ii) $\mathcal{S}_{i+1} = (\mathcal{S}_i \backslash \mathcal{R}) \cup \mathcal{R}'$, where $\mathcal{R}'$ is the complementary regulus of a regulus $\mathcal{R}$ of $\mathcal{S}_i$.

A spread $\mathcal{S}$ is *subregular of index* k if (i) there exists a subregular sequence $\mathcal{S}_0, \mathcal{S}_1, \ldots, \mathcal{S}_k$ with $\mathcal{S} = \mathcal{S}_k$, (ii) there is no shorter sequence beginning with a regular spread and ending with $\mathcal{S}$. It follows that a regular spread is subregular of index 0.

Lemma 17.1.3: *Exactly two of the spreads formed in Lemma* 17.1.1 *from the pencil* $\mathscr{F}$ *of quadrics* $\mathscr{F}_t = \mathbf{V}(f(x_0, x_1) + tf(x_2, x_3))$ *with* f *irreducible are regular.*

Proof: An elliptic congruence is represented on $\mathscr{H}_5$ by an elliptic quadric $\mathscr{E}_3$ in a solid Π_3 whose polar line Π_1 is skew or *external* to $\mathscr{H}_5$. The line Π_1 may also be considered as meeting $\mathscr{H}_5$ in a pair of conjugate complex points and similarly an elliptic congruence is the set of lines meeting a pair of conjugate complex axes.

Over $\gamma' = GF(q^2)$, $f(x, y) = (x - \alpha y)(x - \beta y)$ with $\beta = \alpha^q$. Let $\bar{\mathscr{F}}_t = \mathbf{V}(f(x_0, x_1) + tf(x_2, x_3))$ over γ'. If we write

$$y_0 = x_0 - \alpha x_1, \qquad y_1 = x_0 - \beta x_1, \qquad y_2 = x_2 - \alpha x_3, \qquad y_3 = x_2 - \beta x_3,$$

then $\bar{\mathscr{F}}_t = \mathbf{V}(y_0 y_1 + t y_2 y_3)$. Let $\mathcal{R}_t$, $\mathcal{R}'_t$ be the reguli of $\mathscr{F}_t$ and $\bar{\mathcal{R}}_t$, $\bar{\mathcal{R}}'_t$ those of $\bar{\mathcal{F}}_t$. Then

$$\bar{\mathcal{R}}_t = \{\mathbf{V}(y_0 - \lambda y_2, \lambda y_1 + t y_3) \mid \lambda \in \gamma'^+\},$$
$$\bar{\mathcal{R}}'_t = \{\mathbf{V}(y_0 - \mu y_3, \mu y_1 + t y_2) \mid \mu \in \gamma'^+\}.$$

Each line of $\bar{\mathcal{R}}_t$ meets $\mathbf{V}(y_0, y_2)$ and $\mathbf{V}(y_1, y_3)$ for each t in γ_0; similarly, every line of $\bar{\mathcal{R}}'_t$ meets $\mathbf{V}(y_0, y_3)$ and $\mathbf{V}(y_1, y_2)$ for each t in γ_0. Hence the q^2+1 lines comprising $\mathbf{V}(x_0, x_1)$, $\mathbf{V}(x_2, x_3)$ and the q^2-1 lines in all the reguli $\mathcal{R}_t$, t in γ_0, form an elliptic congruence $\mathcal{C}$ with complex conjugate axes $\mathbf{V}(x_0-\alpha x_1, x_2-\alpha x_3)$ and $\mathbf{V}(x_0-\beta x_1, x_2-\beta x_3)$. Similarly, with the reguli $\mathcal{R}'_t$ instead of $\mathcal{R}_t$, we have an elliptic congruence $\mathcal{C}'$ with complex conjugate axes $\mathbf{V}(x_0-\alpha x_1, x_2-\beta x_3)$ and $\mathbf{V}(x_0-\beta x_1, x_2-\alpha x_3)$.

As two elliptic congruences can have 0, 1, 2, or $q+1$ lines in common, $\mathcal{C}$ and $\mathcal{C}'$ are the only ones that can be formed from $\mathcal{F}_0$, $\mathcal{F}_\infty$, and either $\mathcal{R}_t$ or $\mathcal{R}'_t$ for each t in γ_0. □

Corollary 1: *For $q=2$, every spread is regular.*

Proof: A spread consists of five lines. If we start with any two, the remaining three form a regulus $\mathcal{R}$. So the two lines together with $\mathcal{R}$ or its complement $\mathcal{R}'$ form an elliptic congruence as in the lemma.

Alternatively, any four skew lines define an elliptic congruence whose fifth line can consist only of the remaining three points of $PG(3, 2)$. □

In the case of $PG(3, 2)$ we may begin from first principles. Given three skew lines l_1, l_2, l_3 and a point P on none of them, there are three lines through P meeting pairs of the l_i and hence three lines through P meeting just one of the l_i. Thus the seventh line l_4 through P is skew to l_1, l_2, l_3. If Q is a point on none of the four lines, there are six lines through Q to pairs of them and no lines through Q meeting just one. So the seventh line through Q is skew to l_1, l_2, l_3, l_4. Hence every regulus lies in a unique spread and every pair of skew lines lies in two spreads. There are 280 pairs of skew lines, 560 reguli, and 56 spreads.

Corollary 2: *For $q>2$, there exists a spread which is not regular.* □

Corollary 3: *All 2^{q-1} spreads obtained in Lemma 17.1.1 are subregular.* □

Corollary 4: *In $PG(3, 3)$, if a spread $\mathcal{S}$ contains a regulus, then it is subregular of index 0 or 1.*

Proof: Given a regulus $\mathcal{R}$, then the number of lines skew to the quadric $\mathcal{H}$ containing $\mathcal{R}$ is $\frac{1}{2}q^2(q-1)^2=18$, by Theorem 15.3.13(iii). These 18 lines fall into nine polar pairs. If l is one of the 18, then, by Theorem 15.3.15(g), there are $\frac{1}{2}(q-2)(q+1)^2=8$ of the 18 lines that meet l in a point. This leaves nine, which can therefore only comprise the polar l' of l

and eight lines forming two complementary reguli of a quadric $\mathcal{H}'$, by Lemma 17.1.1. The four quadrics $\mathcal{H}$, $\mathcal{H}'$, l, l' form a pencil as in that lemma, and, by Lemma 17.1.3, $\mathcal{R} \cup \{l, l'\}$ together with one regulus of $\mathcal{H}'$ forms a regular spread; if the other regulus of $\mathcal{H}'$ is used, the spread is subregular of index one. □

Theorem 17.1.4: *Every spread in $PG(3, 3)$ is either regular or obtained from a regular spread by reversing one regulus.*

Proof: By the preceding corollary, it suffices to show that a spread $\mathcal{S}$ contains a regulus.

Suppose therefore that $\mathcal{S}$ contains no regulus. Then each set of four lines in $\mathcal{S}$ has 0, 1, or 2 transversals. As a line contains only four points, any two distinct sets of four lines cannot share a transversal. Thus, if every such set has at least one transversal, there would be at least $\mathbf{c}(10, 4) = 210$ such transversals, which is more than the 130 lines of $PG(3, 3)$. So we may assume that $\mathcal{S}$ contains a set of four lines l_1, l_2, l_3, l_4 with no transversals.

The number of lines meeting none of the l_i is, by the Principle of Inclusion and Exclusion,

$$\begin{aligned}&(q^2+1)(q^2+q+1)-4[q(q+1)^2+1]+6(q+1)^2-4(q+1)\\&=q^4-3q^3+5q-1\\&=14.\end{aligned}$$

Among these fourteen, there are four lines m_1, m_2, m_3, m_4 such that m_i is the fourth line of the regulus containing $\{l_1, l_2, l_3, l_4\}\backslash\{l_i\}$.

The lines l_1, l_2, l_3, l_4 may be chosen arbitrarily. So, let

$$l_1 = \mathbf{V}(x_0, x_2), \qquad l_2 = \mathbf{V}(x_1, x_3), \qquad l_3 = \mathbf{V}(x_0 + x_3, x_1 + x_2),$$
$$l_4 = \mathbf{V}(x_0 + x_1 + x_3, x_1 - x_2 - x_3).$$

Then

$$m_1 = \mathbf{V}(x_2 - x_3, x_0 - x_1 + x_2), \qquad m_2 = \mathbf{V}(x_2 + x_3, x_0 + x_1 + x_2),$$
$$m_3 = \mathbf{V}(x_0 + x_2 + x_3, x_0 + x_1 - x_2), \qquad m_4 = \mathbf{V}(x_0 - x_3, x_1 - x_2).$$

The ten lines skew to l_1, l_2, l_3, l_4 other than m_1, m_2, m_3, m_4 are

$$\begin{array}{ll}\mathbf{V}(x_0+x_1-x_2, x_0-x_3), & \mathbf{V}(x_0-x_1, x_1+x_2+x_3),\\ \mathbf{V}(x_0+x_1, x_0-x_2-x_3), & \mathbf{V}(x_0+x_1, x_2-x_3),\\ \mathbf{V}(x_0+x_1+x_2, x_0-x_1+x_3), & \mathbf{V}(x_0+x_1, x_0-x_2+x_3),\\ \mathbf{V}(x_0-x_1-x_3, x_1-x_2), & \mathbf{V}(x_0-x_1+x_2, x_1+x_2+x_3),\\ \mathbf{V}(x_0-x_1, x_2+x_3), & \mathbf{V}(x_0-x_1, x_1-x_2+x_3).\end{array}$$

If $\mathcal{S}$ contains no regulus, then there must be six lines among these ten partitioning the 24 points on no l_i. However, the point $\mathbf{P}(1, 0, 1, 1)$ only lies on $\mathbf{V}(x_0+x_1-x_2, x_0-x_3)$ of these ten lines, and $\mathbf{P}(1, 1, 1, -1)$ only lies on $\mathbf{V}(x_0-x_1, x_2+x_3)$. These two lines meet at $\mathbf{P}(1, 1, -1, 1)$, a contradiction. So $\mathcal{S}$ contains a regulus. □

We now consider a result on the partitioning of a regular spread into reguli. The problem may be given four alternative settings: $PG^{(1)}(3, q)$, $\mathscr{E}_3$, $PG(2, q)$, and $PG(1, q^2)$. These four settings are explained in the following lemma.

Lemma 17.1.5: *The following are equivalent:*

(i) *If an elliptic linear congruence $\mathscr{C}$ is partitioned into the lines l_1 and l_2, and $q-1$ reguli, then the quadrics containing the reguli lie in a pencil $\mathscr{F}$ such that l_1 and l_2 are members of $\mathscr{F}$ as quadrics of rank 2 and such that l_1 and l_2 are polars with respect to each of the $q-1$ hyperbolic quadrics of $\mathscr{F}$.*

(ii) *If an elliptic quadric surface $\mathscr{E}_3$ is partitioned into the points P_1 and P_2, and $q-1$ conics, then the planes containing the conics lie in a pencil such that the planes of the pencil through P_1 and P_2 are the tangent planes to $\mathscr{E}_3$ at these points.*

(iii) *If a plane is partitioned into a point P, a line l, and $q-1$ conics with a pair of complex conjugate points lying on l in common, then the conics lie in a pencil of quadric curves such that P and l are members of rank 2 and 1 respectively. For q odd, P is the pole of l for each conic of the pencil.*

(iv) *If $PG(1, q^2)$ is partitioned into two points P_1 and P_2, and $q-1$ sublines, then the sublines lie (as Hermitian varieties) in a pencil containing P_1 and P_2 as singular varieties.*

Proof: Statements (i) and (ii) are equivalent by the map $\mathfrak{G}: \mathscr{L} \to \mathscr{H}_5$: elliptic congruences go to elliptic quadrics and reguli to conics. We pass from (ii) to (iii) via stereographic projection from P_1 to a plane not containing it, as in § 16.3. Finally, (i) and (iv) are equivalent by the mappings $\mathfrak{T}_1$ and $\mathfrak{T}_2$ of Theorem 15.3.11. □

To show that (i)–(iv) of Lemma 17.1.5 are all true, we distinguish the cases of q odd and q even. We start with the latter case, where a more general result is established.

Theorem 17.1.6: *If, in $PG(3, q)$ with q even, an ovaloid $\mathscr{K}$ is partitioned into points P_1 and P_2, and $(q+1)$-arcs $\mathscr{C}_1, \ldots, \mathscr{C}_{q-1}$ lying in respective planes $\pi_1, \ldots, \pi_{q-1}$, then the planes π_i have a line in common, which is the intersection of the tangent planes to $\mathscr{K}$ at P_1 and P_2.*

Proof: Non-tangent planes meet $\mathscr{K}$ in $(q+1)$-arcs by Lemma 16.1.6. Let the nucleus of $\mathscr{C}_i$ be N_i, $i \in \mathbf{N}_{q-1}$, and let $l = \{P_1, P_2, N_1, N_2, \ldots, N_{q-1}\}$.

Then it will be shown that every plane π has at least one point in common with l. It follows by Theorem 3.2.2 that l is a line.

(i) If π contains P_1 or P_2, then $\pi \cap l \neq \varnothing$.

(ii) If π is the tangent plane to $\mathscr{K}$ at P, where $P \neq P_1$ or P_2, then P lies on one of the $(q+1)$-arcs, $\mathscr{C}_1$ say. Then the tangent l_1 to $\mathscr{C}_1$ at P lies in π. As N_1 is on l_1, so N_1 is in π.

(iii) If π is π_i, then N_i lies in π.

(iv) There remains the case that $\pi \cap \mathscr{K}$ is a $(q+1)$-arc $\mathscr{C}$, which contains neither P_1 nor P_2 and which is not some $\mathscr{C}_i$. Then $\mathscr{C}$ meets each $\mathscr{C}_i$ in 0, 1, or 2 points. As $\mathscr{C}$ contains $q+1$ points, which is odd, there exists a $(q+1)$-arc, $\mathscr{C}_1$ say, such that $|\mathscr{C} \cap \mathscr{C}_1| = 1$. Hence $\mathscr{C}$ and $\mathscr{C}_1$ have a common tangent l_1 at this common point. As N_1 is on l_1, so N_1 is in π.

Hence l is a line.

In Theorem 16.1.8, it was shown that $\mathscr{K}$ defines a null polarity in which a point on $\mathscr{K}$ is polar to the tangent plane at that point and a point off $\mathscr{K}$ is polar to the unique plane π for which it is the nucleus of $\pi \cap \mathscr{K}$. So the polar planes of P_1 and P_2 are the tangent planes there and the polar plane of N_i is π_i, $i \in \mathbf{N}_{q-1}$. Hence these $q+1$ planes all pass through the polar line of l. □

Corollary: *For q even, the equivalent statements of Lemma* 17.1.5 *are true.*

Proof: This is just the case that, in the theorem, $\mathscr{K}$ is an elliptic quadric. □

It will now be shown that (ii) of Lemma 17.1.5 is true for q odd.

Let $\mathscr{F}$ be a non-singular quadric in $PG(3, q)$, q odd, so that, for the moment, $\mathscr{F} = \mathscr{E}_3$ or $\mathscr{H}_3$. Let $\mathscr{F} = \mathbf{V}(F)$ with $F(X) = XTX^*$. Let $P = \mathbf{P}(A)$ and $Q = \mathbf{P}(B)$. Then

$$\mathbf{P}(A+tB) \in \mathscr{F} \Leftrightarrow F(A+tB) = 0$$

and

$$F(A+tB) = ATA^* + 2tATB^* + t^2BTB^*.$$

The discriminant

$$\Delta_{\mathscr{F}}(P, Q) = (ATB^*)^2 - (ATA^*)(BTB^*).$$

Then the relation of PQ to $\mathscr{F}$ is as follows:

$\lvert PQ \cap \mathscr{F} \rvert$	0	1	2
$\Delta_{\mathscr{F}}(P, Q)$	non-square	zero	non-zero square

Let π_P be the polar plane of P and, for P not on $\mathscr{F}$, let $\mathscr{C}_P = \pi_P \cap \mathscr{F}$, the conic of points on $\mathscr{F}$ conjugate to P.

Lemma 17.1.7: *The number of points in which PQ meets $\mathscr{F}$ determines the intersection of the polar conics $\mathscr{C}_P \cap \mathscr{C}_Q$ as follows:*

	$\lvert PQ \cap \mathscr{F} \rvert$	0	1	2
$\mathscr{F} = \mathscr{E}_3$:	$\lvert \mathscr{C}_P \cap \mathscr{C}_Q \rvert$	2	1	0
$\mathscr{F} = \mathscr{H}_3$:	$\lvert \mathscr{C}_P \cap \mathscr{C}_Q \rvert$	0	1	2

Proof: When $\mathscr{F} = \mathscr{E}_3$, the sets of bisecants and external lines are interchanged by the polarity, from Theorem 15.3.10(iv); when $\mathscr{F} = \mathscr{H}_3$, they are fixed by the polarity, from Theorem 15.3.15(*d*), (*e*). □

$\mathscr{F}$ defines an equivalence relation with exactly two classes on the set Ψ_0 of points not on $\mathscr{F}$. Let $P = \mathbf{P}(A)$, $Q = \mathbf{P}(B)$. Then P is equivalent to Q, written $P \underset{\mathscr{F}}{\sim} Q$, if $ATA^*/BTB^* = t^2$ for some t in γ_0. This gives an equivalence relation whose classes are

$$\{P \in \Psi_0 \mid P = \mathbf{P}(A),\ ATA^* \text{ is a square}\},$$
$$\{P \in \Psi_0 \mid P = \mathbf{P}(A),\ ATA^* \text{ is a non-square}\}.$$

The relation $P \underset{\mathscr{F}}{\sim} Q$ defines the relation $\mathscr{C}_P \underset{\mathscr{F}}{\sim} \mathscr{C}_Q$, whose geometry is explained by the following lemma.

Lemma 17.1.8: *$\mathscr{C}_P \underset{\mathscr{F}}{\sim} \mathscr{C}_Q$ if and only if there is a point R such that $\mathscr{C}_R$ is tangent to both $\mathscr{C}_P$ and $\mathscr{C}_Q$.*

Proof: Let $\mathscr{C}_P \underset{\mathscr{F}}{\sim} \mathscr{C}_Q$ so that, with $P = \mathbf{P}(A)$ and $Q = \mathbf{P}(B)$, we have $ATA^*/BTB^* = t^2$. Let $R = \mathbf{P}(C)$ be the intersection of a tangent line to $\mathscr{F}$ through P with the polar plane of $\mathbf{P}(A - tB)$. Then

$$(A - tB)TC^* = 0 \tag{17.1}$$

$$\Delta_{\mathscr{F}}(A, C) = 0. \tag{17.2}$$

So

$$ATA^*/BTB^* = t^2 = (ATC^*)^2/(BTC^*)^2 \tag{17.3}$$

$$(ATC^*)^2 = (ATA^*)(CTC^*). \tag{17.4}$$

Thus

$$(BTC^*)^2 = (BTB^*)(CTC^*); \tag{17.5}$$

that is

$$\Delta_{\mathscr{F}}(B, C) = 0. \tag{17.6}$$

So $\mathscr{C}_R$ is tangent to $\mathscr{C}_P$ and $\mathscr{C}_Q$ by Lemma 17.1.7. Conversely, (17.5) and (17.4) imply (17.3). □

With regard to $\mathscr{F}$ there are three types of pencils of planes according as two, one, or no planes in the pencil are tangent planes; the pencils are correspondingly called *hyperbolic*, *parabolic*, or *elliptic*.

Lemma 17.1.9: *Half the conics in a hyperbolic or elliptic pencil are in each class. All the conics in a parabolic pencil are in the same class.*

Proof: Write $F(A+tB)=a+2bt+ct^2$. If $b^2-ac\neq 0$, then $a+2bt+ct^2$ has half its non-zero values square and half non-square, since $\mathbf{V}(x_2-tx_1)$ is correspondingly a bisecant or external line of the conic $\mathbf{V}(ax_2^2+2bx_1x_2+cx_1^2-dx_0^2)$ for an equal number of values of t, as in Table 8.2. If $b^2-ac=0$, then $a+2bt+ct^2=c(t+b/c)^2$, whence either all its non-zero values are square or all non-square. □

If P and Q are conjugate, then the conics $\mathscr{C}_P$ and $\mathscr{C}_Q$ are also *conjugate*. In this case, still with $P=\mathbf{P}(A)$ and $Q=\mathbf{P}(B)$, we have $ATB^*=0$ and

$$\Delta_{\mathscr{F}}(P, Q)=-(ATA^*)(BTB^*). \tag{17.7}$$

Lemma 17.1.10:

(i) $\mathscr{F}=\mathscr{E}_3$.

(a) $q\equiv 1 \pmod 4$. *Conjugate conics intersect if and only if they belong to different classes.*

(b) $q\equiv -1 \pmod 4$. *Conjugate conics intersect if and only if they belong to the same class.*

(ii) $\mathscr{F}=\mathscr{H}_3$.

(a) $q\equiv 1 \pmod 4$. *Conjugate conics intersect if and only if they belong to the same class.*

(b) $q\equiv -1 \pmod 4$. *Conjugate conics intersect if and only if they belong to different classes.*

Proof: $|\mathscr{C}_P\cap\mathscr{C}_Q|=0$ or 2. Since -1 is a square for $q\equiv 1 \pmod 4$ and a non-square for $q\equiv -1 \pmod 4$, the result follows from (17.7) and Lemma 17.1.7. □

A *flock* $\mathscr{G}$ is a maximal set of disjoint conic sections of $\mathscr{F}$. When $\mathscr{F}=\mathscr{E}_3$ and $|\mathscr{G}|=q-1$, the two points of $\mathscr{F}$ on no conic of $\mathscr{G}$ are the *carriers* of $\mathscr{G}$. Let $\mathscr{M}$ be the set of conic sections of $\mathscr{E}_3$ through the point P_1 of $\mathscr{E}_3$.

Lemma 17.1.11: *If $\mathscr{G}$ is any flock of $\mathscr{E}_3$ with carriers P_1 and P_2, then any conic of $\mathscr{M}$ either contains P_2 and is tangent to no conic of $\mathscr{G}$ or it is tangent to exactly one conic in $\mathscr{G}$.*

Proof: Each of the q^2+q conics in $\mathscr{M}$ has an odd number q of points other than P_1. If $\mathscr{C}$ is not one of the $q+1$ conics through P_2, it must be tangent to an odd number of conics of $\mathscr{G}$. So each of the q^2-1 conics $\mathscr{C}_i$ of $\mathscr{M}$ not through P_2 is tangent to $n_i\geqslant 1$ conics of $\mathscr{G}$.

Through any point P of $\mathscr{E}_3\backslash\{P_1, P_2\}$ there is a unique conic $\mathscr{C}$ of $\mathscr{G}$. Exactly one of the conics of $\mathscr{M}$ is tangent at P to $\mathscr{C}$, since q planes through PP_1 meet $\mathscr{C}$ again. So $\sum n_i \leqslant q^2-1$, whence $n_i = 1$ and no conic through P_1 and P_2 is tangent to a conic in $\mathscr{G}$. □

Consider the pencil of planes containing the tangent planes to $\mathscr{E}_3$ at P_1 and P_2, and let $\mathscr{G}_0$ be the set of $q-1$ conic sections of $\mathscr{E}_3$ by the other $q-1$ planes in the pencil. Let $\mathscr{M}_0$ be the pencil of conics through P_1 and P_2.

Lemma 17.1.12: *If $\mathscr{C}$ is any conic in the flock $\mathscr{G}$, then the conics in $\mathscr{M}_0$ that meet $\mathscr{C}$ all belong to the same equivalence class.*

Proof: Consider a point P of $\mathscr{C}$. There are $q+1$ conics of $\mathscr{M}$ through P. One, let it be $\mathscr{C}'$, contains P_2; another, say $\mathscr{C}''$, is tangent to $\mathscr{C}$ and distinct from $\mathscr{C}'$, by Lemma 17.1.11. The other $q-1$ are tangent in pairs to the other conics of $\mathscr{G}$, since through P and P_1 there are two or zero conics tangent to a given conic and each of these $q-1$ conics is tangent to exactly one conic of $\mathscr{G}$. Since, by Lemma 17.1.9, tangent conics are in the same class, an even number of the $q-1$ conics are in each class. Again by Lemma 17.1.9, exactly half the conics through P and P_1 are in each class. Thus $\mathscr{C}'$ and $\mathscr{C}''$ are in the same class or not according as $\frac{1}{2}(q+1)$ is even or odd; and similarly, by Lemma 17.1.9, for $\mathscr{C}$ and $\mathscr{C}'$. In either case all conics of $\mathscr{M}_0$ that meet $\mathscr{C}$ are in the same class. □

Lemma 17.1.13: *If $\mathscr{C}$ is a conic section of $\mathscr{E}_3$ for which all the conics of $\mathscr{M}_0$ that contain some point of $\mathscr{C}$ are in the same class, then $\mathscr{C} \in \mathscr{G}_0$.*

Proof: All conics in $\mathscr{G}_0$ are conjugate to all conics in $\mathscr{M}_0$. For $i = 1, 2$, let Q_i be a point of $\mathscr{C}$ through which passes a conic $\mathscr{C}_i$ of $\mathscr{G}_0$ and a conic $\mathscr{D}_i$ of $\mathscr{M}_0$ (Fig. 17.1). By hypothesis, $\mathscr{D}_1 \sim \mathscr{D}_2$. Lemma 17.1.10 gives two

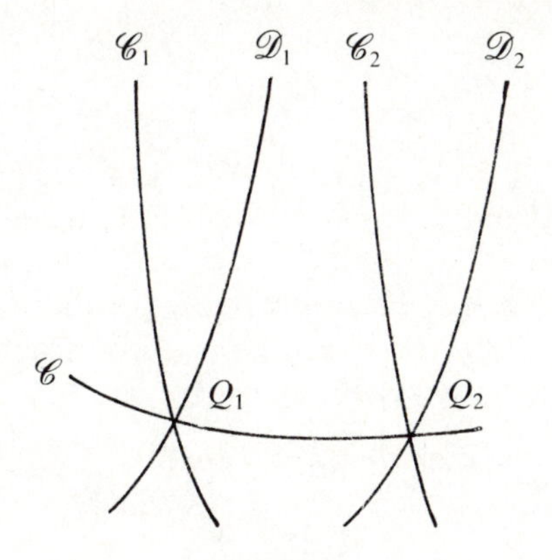

Fig. 17.1

cases. When $q \equiv 1 \pmod 4$, $\mathscr{C}_1 \nsim \mathscr{D}_1$ and $\mathscr{C}_2 \nsim \mathscr{D}_2$, whence $\mathscr{C}_1 \sim \mathscr{C}_2$. On the other hand, when $q \equiv -1 \pmod 4$, $\mathscr{C}_1 \sim \mathscr{D}_1$ and $\mathscr{C}_2 \sim \mathscr{D}_2$, whence $\mathscr{C}_1 \sim \mathscr{C}_2$. So, in both cases, all conics in $\mathscr{G}_0$ meeting $\mathscr{C}$ are in the same class. □

Theorem 17.1.14: *Any partition of $\mathscr{E}_3$ into conics $\mathscr{C}_1, \ldots, \mathscr{C}_{q-1}$ and points P_1 and P_2, is cut out by the pencil of planes through the polar of P_1P_2.*

Proof: By Lemmas 17.1.12 and 17.1.13, if $\mathscr{C}_i$ is in the flock $\mathscr{G}$, it lies in the linear flock $\mathscr{G}_0$, which is precisely the theorem. □

Corollary: *For q odd, the equivalent statements of Lemma* 17.1.5 *are true.* □

17.2 Subregular spreads

If in a spread $\mathscr{S}$ a regulus $\mathscr{R}$ is replaced by its complementary regulus $\mathscr{R}'$, we say that $\mathscr{R}$ has been *reversed*: the spread obtained is $\mathscr{S}' = (\mathscr{S}\backslash\mathscr{R}) \cup \mathscr{R}'$. If $\mathscr{S}$ is regular and $\mathscr{K} = \{\mathscr{R}_1, \ldots, \mathscr{R}_k\}$ is a set of pairwise disjoint reguli in $\mathscr{S}$, then $\mathscr{S}^{\mathscr{K}}$ is the subregular spread obtained by reversing the reguli of $\mathscr{K}$. The set $\mathscr{K}$ is *linear* if the quadrics containing $\mathscr{R}_1, \mathscr{R}_2, \ldots, \mathscr{R}_k$ lie in a pencil.

Theorem 17.2.1: *If $\mathscr{S}$ is a regular spread and $\mathscr{K}$ is not linear, then every regulus in $\mathscr{S}^{\mathscr{K}}$ is either a regulus of $\mathscr{S}$ or the complementary regulus $\mathscr{R}_i'$ of $\mathscr{R}_i$ for some i in $\mathbf{N}_k$.*

Proof: Suppose that $\mathscr{R}$ is a regulus in $\mathscr{S}^{\mathscr{K}}$ such that $\mathscr{R} \not\subset \mathscr{S}$ and $\mathscr{R} \neq \mathscr{R}_i'$, $i \in \mathbf{N}_k$. So $|\mathscr{R} \cap \mathscr{S}| \leq 2$ and $|\mathscr{R} \cap \mathscr{R}_i'| \leq 2$, for all i. Hence there exists m in $\mathbf{N}_k$ such that $|\mathscr{R} \cap \mathscr{R}_i'| \geq 1$ for $i = 1, \ldots, m$ and $|\mathscr{R} \cap \mathscr{R}_i'| = 0$ for $i = m+1, \ldots, k$. Then each line of $\mathscr{R}$ is either in $\mathscr{S}$ or in some $\mathscr{R}_i'$ for i in $\mathbf{N}_m$. Since $|\mathscr{R}| = q+1$,

$$q+1 \leq 2m+2 \tag{17.8}$$

whence

$$q-1 \leq 2m \leq 2k. \tag{17.9}$$

It will be shown that $\mathscr{J}_m = \{\mathscr{R}_i \mid i \in \mathbf{N}_m\}$ is not linear. Suppose the contrary. Then $m < k$ and $\mathscr{J}_m$ can be extended to a complete linear set $\mathscr{J}$ in $\mathscr{S}$ with common polar lines l_1 and l_2. Since $\mathscr{K}$ is not linear, there exists j in $\mathbf{N}_k \backslash \mathbf{N}_m$ such that $\mathscr{R}_j \notin \mathscr{J}$. Each line of $\mathscr{R}_j$ is either l_1 or l_2 or a line of one of the $q-1-m$ reguli in $\mathscr{J} \backslash \mathscr{J}_m$. Since $\mathscr{R}_j$ has at most two lines in common with each of these reguli,

$$2(q-1-m)+2 \geq q+1,$$

whence

$$q-1 \geq 2m. \tag{17.10}$$

Thus equality holds in (17.8) and (17.10), and $\mathcal{R}_j \supset \{l_1, l_2\}$. Hence $|\mathcal{R} \cap \mathcal{R}'_i| = 2$ for i in $\mathbf{N}_m$ and $|\mathcal{R} \cap \mathcal{S}| = 2$. Since, now $2m+2 = q+1$, we have that $\mathcal{R} \subset \mathcal{S}^{\mathcal{I}}$ and $\mathcal{S} \cap \mathcal{S}^{\mathcal{I}} = \{l_1, l_2\}$. Thus $\mathcal{R}$ contains l_1 and l_2, and is therefore not disjoint from $\mathcal{R}_j$, a contradiction. Hence $\mathcal{I}_m = \{\mathcal{R}_1, \ldots, \mathcal{R}_m\}$ is not linear. This also means that $m \geqslant 2$.

Now it will be shown that $|\mathcal{R} \cap \mathcal{R}'_i| = 1$ for i in $\mathbf{N}_m$. Suppose that $|\mathcal{R} \cap \mathcal{R}'_1| = 2$. Then the pair of reguli $\{\mathcal{R}_1, \mathcal{R}_2\}$ extends to a complete linear set $\mathcal{H}$ of reguli in $\mathcal{S}$ with common polar lines l'_1 and l'_2. As in Lemma 17.1.3, $\mathcal{S}^{\mathcal{H}}$ is regular and $\mathcal{S} \cap \mathcal{S}^{\mathcal{H}} = \{l'_1, l'_2\}$. Since $|\mathcal{R} \cap \mathcal{R}'_1| = 2$ and $|\mathcal{R} \cap \mathcal{R}'_2| \geqslant 1$, the regulus $\mathcal{R}$ has three lines in $\mathcal{S}^{\mathcal{H}}$ and so is contained in $\mathcal{H}$. Hence each line l of $\mathcal{R}$ not in $\mathcal{S}$ is in the complementary regulus of one in $\mathcal{H}$. So the regulus $\mathcal{R}(l)$ of lines of $\mathcal{S}$ meeting l is in $\mathcal{H}$. But each $\mathcal{R}'_i$, $i \in \mathbf{N}_m$, contains some l. So each $\mathcal{R}_i$, $i \in \mathbf{N}_m$, is some $\mathcal{R}(l)$. Thus $\mathcal{I}_m \subset \mathcal{H}$, whence $\mathcal{I}_m$ is linear, a contradiction.

It now follows that $|\mathcal{R} \cap \mathcal{R}'_i| = 1$ for i in $\mathbf{N}_m$, and since also $|\mathcal{R} \cap \mathcal{S}| \leqslant 2$,

$$m+2 \geqslant q+1,$$

that is,

$$m \geqslant q-1. \tag{17.11}$$

But $m \leqslant k \leqslant q-1$. So $m = k = q-1$ and $|\mathcal{R} \cap \mathcal{S}| = 2$. Thus Theorems 17.1.6 and 17.1.14 expressed in the form of part (i) of Lemma 17.1.5 say that $\mathcal{K}$ is linear, a contradiction. □

Now it can be shown that however complicated a sequence of reversing reguli when one begins from a regular spread, the result always has a very simple form.

Theorem 17.2.2: *Every subregular spread in $PG(3, q)$ is obtainable by reversing a set of disjoint reguli in some regular spread $\mathcal{S}$.*

Proof: Suppose $\mathcal{S}_0, \mathcal{S}_1, \ldots, \mathcal{S}_k$ is a subregular sequence of spreads of length k, so that $\mathcal{S}_0$ is regular and $\mathcal{S}_{i+1} = (\mathcal{S}_i \backslash \mathcal{R}_i) \cap \mathcal{R}'_i$ for $i \in \bar{\mathbf{N}}_{k-1}$.

The proof now proceeds by induction. When $k=1$, the spread $\mathcal{S}_1$ is obtained by reversing the single spread $\mathcal{R}_0$ of $\mathcal{S}_0$.

For i in $\mathbf{N}_{k-1}$, suppose that $\mathcal{S}_i = \mathcal{S}^{\mathcal{K}_i}$ for some set $\mathcal{K}_i$ of disjoint reguli in the regular spread $\mathcal{S}$. Then $\mathcal{S}_{i+1}$ is obtained from $\mathcal{S}_i$ by reversing $\mathcal{R}_i$. If $\mathcal{K}_i$ is not linear, then by the previous theorem, $\mathcal{R}_i$ is either a regulus of $\mathcal{S}$ or the complementary regulus of one in $\mathcal{K}_i$. In the first case, $\mathcal{S}_{i+1} = \mathcal{S}^{\mathcal{K}_i \cup \{\mathcal{R}_i\}}$; in the second case $\mathcal{S}_{i+1} = \mathcal{S}^{\mathcal{K}_i \backslash \{\mathcal{R}'_i\}}$.

If $\mathcal{K}_i$ is linear, it is contained in a complete linear set $\mathcal{K}$. If $\mathcal{R}_i$ is not in $\mathcal{S}$, it contains at most two lines of $\mathcal{S}$ and so its remaining lines lie in $\mathcal{S}^{\mathcal{K}}$. For $q>3$, the regulus $\mathcal{R}_i$ has at least three lines in the regular spread $\mathcal{S}^{\mathcal{K}}$ and so $\mathcal{R}_i \subset \mathcal{S}^{\mathcal{K}}$. Then for each regulus $\mathcal{R}$ in $\mathcal{K} \backslash \mathcal{K}_i$, its complementary regulus $\mathcal{R}'$ is disjoint from $\mathcal{R}_i$. The set $\mathcal{J}=\{\mathcal{R}_i\}\cup\{\mathcal{R}' \mid \mathcal{R}\in\mathcal{K}\backslash\mathcal{K}_i\}$ comprises disjoint reguli of $\mathcal{S}^{\mathcal{K}}$, and $\mathcal{S}_{i+1}=(\mathcal{S}^{\mathcal{K}})^{\mathcal{J}}$. Thus, for $q>3$, the result holds for $\mathcal{S}_{i+1}$ and by induction for $\mathcal{S}_k$.

For $q=2$, all spreads are regular, by Lemma 17.1.3, Corollary 1.

When $q=3$, a spread is subregular of index 0 or 1, by Theorem 17.1.4. □

17.3 Aregular spreads

So far all the spreads encountered have been subregular and hence have contained a regulus. A spread is *aregular* if it contains no regulus.

Now we give an example of an aregular spread $\mathcal{S}$ for every field not of prime order in the following way. Let $l_{01}=\mathbf{U}_0\mathbf{U}_1$; then q^2 lines are formed by joining the points of $\mathbf{u}_2\backslash l_{01}$ to the points of $\mathbf{u}_3\backslash l_{01}$ in such a way that these q^2 lines are mutually skew. Then $\mathcal{S}$ will consist of these q^2 lines plus l_{01}.

Lemma 17.3.1: *Let $\gamma=GF(q)$, $q=p^h$ and $h>1$. Then for each b in γ_0, there exists c in γ_0 such that $x^{p+1}+bx-c$ has no roots in γ.*

Proof: As $b\neq 0$, the map $\sigma:\gamma\to\gamma$ given by $x\sigma=x^{p+1}+bx$ is not injective, since both 0 and $-b^{1/p}$ map to 0. So σ is not surjective. So there exists c in γ and hence in γ_0 which is not in the image of σ; that is, $x^{p+1}+bx-c$ has no roots in γ. □

Lemma 17.3.2: *Over $\gamma=GF(q)$, $q=p^h$ with $h>1$, let $\mathfrak{T}':\mathbf{u}_3\backslash l_{01}\to\mathbf{u}_2\backslash l_{01}$ be the map given by*

$$\mathbf{P}(z, y, 1, 0)\mathfrak{T}'=\mathbf{P}(cy^p, z^p+by^p, 0, 1),$$

where $x^{p+1}+bx-c$ has no roots in γ. Then

(i) *$\mathfrak{T}'$ is induced by the collineation $\mathfrak{T}$ of $PG(3, q)$ given by*

$$\mathbf{P}(x_0, x_1, x_2, x_3)\mathfrak{T}=\mathbf{P}(cx_1^p, x_0^p+bx_1^p, x_3^p, x_2^p);$$

(ii) *the joins of corresponding points under $\mathfrak{T}'$ are mutually skew.*

Proof: (i) follows by substituting $x_0=z$, $x_1=y$, $x_2=1$, $x_3=0$. For (ii), the join of $\mathbf{P}(z, y, 1, 0)$ and $\mathbf{P}(cy^p, z^p+by^p, 0, 1)$ is

$$\mathbf{l}(z^{p+1}+bzy^p-cy^p, -cy^p, z, -(z^p+by^p), -y, 1).$$

If l_0 is the join of $P_0=\mathbf{P}(z_0, y_0, 1, 0)$ with $P_0\mathfrak{T}'$ and l_1 the join of $P_1=\mathbf{P}(z_1, y_1, 1, 0)$ with $P_1\mathfrak{T}'$, then

$$\begin{aligned}\varpi(l_0, l_1)&=z_0^{p+1}+bz_0y_0^p-cy_0^p+cy_0^py_1-z_0(z_1^p+by_1^p)\\&\quad+z_1^{p+1}+bz_1y_1^p-cy_1^p+cy_1^py_0-z_1(z_0^p+by_0^p)\\&=(z_0-z_1)^{p+1}+b(z_0-z_1)(y_0-y_1)^p-c(y_0-y_1)^{p+1}.\end{aligned}$$

As $x^{p+1}+bx-c$ has no roots in γ, so $\varpi(l_0, l_1)=0$ if and only if $z_0=z_1$ and $y_0=y_1$; that is $l_0=l_1$. □

Theorem 17.3.3: *Over $\gamma=GF(q)$, $q=p^h$ with $h>1$, the set $\mathscr{S}$ formed by l_{01} and the q^2 lines $\mathbf{P}(z, y, 1, 0)\ \mathbf{P}(cy^p, z^p+by^p, 0, 1)$, where $x^{p+1}+bx-c$ has no roots in γ, is an aregular spread.*

Proof: It was shown in the previous lemma that the q^2 lines indicated are mutually skew. As they are all skew to l_{01}, together with l_{01} they form a spread.

Let $\mathscr{C}$ be any conic in $\mathbf{u}_3$. Then the map $\mathfrak{T}$ of the lemma induces a bijection $\mathfrak{T}_0$ between $\mathscr{C}$ and a conic $\mathscr{C}'$ in $\mathbf{u}_2$. As $\mathfrak{T}$ is not a projectivity, so also is $\mathfrak{T}_0$ not a projectivity. The same conclusion holds if $\mathfrak{T}_0$ maps a line in $\mathbf{u}_3$ to a line in $\mathbf{u}_2$.

Suppose $\mathscr{S}$ contains a regulus $\mathscr{R}$. If l_{01} is in $\mathscr{R}$, each of the planes $\mathbf{u}_2$ and $\mathbf{u}_3$ meets the other lines of $\mathscr{R}$ in the points of a line. If l_{01} is not in $\mathscr{R}$, each of $\mathbf{u}_2$ and $\mathbf{u}_3$ meets the lines of $\mathscr{R}$ in a conic. But such a correspondence $\mathfrak{T}_0$ between lines in $\mathbf{u}_2$ and $\mathbf{u}_3$ or conics in $\mathbf{u}_2$ and $\mathbf{u}_3$ is a projectivity induced by a projectivity on the complementary regulus $\mathscr{R}'$. This contradicts the previous observation that $\mathfrak{T}_0$ is not a projectivity. □

For $q=2^h$ with h odd and $h\geqslant 3$, we can give another example.

Let $\Pi=PG(3, q)$, q even, and $\mathscr{A}$ the general linear complex $\boldsymbol{\lambda}(l_{01}+l_{23})$. In § 16.4, we considered the map $\mathfrak{T}_1:\Pi\to\mathscr{A}$ given by

$$\mathbf{P}(x_0, x_1, x_2, x_3)\mathfrak{T}_1=\mathbf{l}(\sqrt{(x_0x_1+x_2x_3)}, x_3, x_0, x_1, x_2, \sqrt{(x_0x_1+x_2x_3)}).$$

If $\mathfrak{T}_1$ is restricted to an ovaloid $\mathscr{K}$, its image $\mathscr{K}\mathfrak{T}_1$ is a spread, by Theorem 16.4.12. If $\mathscr{K}$ is an elliptic quadric $\mathscr{E}_3$, the spread $\mathscr{E}_3\mathfrak{T}_1$ turns out to be regular. This is unsurprising, since we have already observed in § 15.4 that an elliptic congruence or regular spread is represented in $PG(5, q)$ by a solid section of $\mathscr{H}_5$ which is in fact an elliptic quadric $\mathscr{E}_3$.

Theorem 17.3.4: *If $q=2^h$, h odd and $h\geqslant 3$, the spread $\mathscr{S}=\Omega\mathfrak{T}_1$, where Ω is a λ-quadric, is aregular.*

Proof: To recall from Theorem 16.4.5, the canonical form for Ω is

$$\Omega=\{\mathbf{U}_1\}\cup\{\mathbf{P}(1, z, y, x)\mid z=xy+y^\sigma+x^{\sigma+2}\}$$

where σ is an automorphism of γ such that $t^{\sigma^2}=t^2$; that is, $t^\sigma=t^{\sqrt{(2q)}}$.

$\mathbf{U}_0$ and $\mathbf{U}_1$ are points of Ω; their images are

$$\mathbf{U}_0\mathfrak{T}_1=\mathbf{l}(0, 0, 1, 0, 0, 0) \quad\text{and}\quad \mathbf{U}_1\mathfrak{T}_1=\mathbf{l}(0, 0, 0, 1, 0, 0).$$

Let $l=\mathbf{l}(\sqrt{(bc+d)}, b, 1, d, c, \sqrt{(bc+d)})$ where $d=bc+c^\sigma+b^{\sigma+2}$. Since $Sz(q)$ is doubly transitive on Ω (Lemma 16.4.6), it suffices to show that the regulus $\mathcal{R} = \mathcal{R}(\mathbf{U}_0\mathfrak{T}_1, \mathbf{U}_1\mathfrak{T}_1, l)$ is not in $\mathcal{S}$. If

$$l' = \mathbf{l}(\sqrt{(b'c'+d')}, b', 1, d', c', \sqrt{(b'c'+d')}),$$

where $d' = b'c' + c'^{\sigma} + b'^{\sigma+2}$, is in $\mathcal{R}$, then, for some ρ, λ, μ in γ,

$$l' = \mathbf{l}(\mu\sqrt{(bc+d)}, \mu b, \rho + \mu, \lambda + \mu d, \mu c, \mu\sqrt{(bc+d)}).$$

So $\sqrt{(b'c'+d')} = \mu\sqrt{(bc+d)}$, $b' = \mu b$, $c' = \mu c$, $\rho = 1 + \mu$, $\lambda = \mu d + d'$. For $\mu = 0$, there are two lines l' in $\mathcal{R}$, namely $\mathbf{U}_0\mathfrak{T}_1$ and $\mathbf{U}_1\mathfrak{T}_1$. From the equations, $d' = \mu^2 d$. Now

$$d' = b'c' + c'^{\sigma} + b'^{\sigma+2} = \mu^2 bc + \mu^{\sigma}c^{\sigma} + \mu^{\sigma+2}b^{\sigma+2}$$

and

$$\mu^2 d = \mu^2 bc + \mu^2 c^{\sigma} + \mu^2 b^{\sigma+2},$$

whence

$$\mu^{\sigma+2}b^{\sigma+2} + \mu^{\sigma}c^{\sigma} + \mu^2(c^{\sigma} + b^{\sigma+2}) = 0.$$

As for each value of $\mu \neq 0$ there can only be one line l' in $\mathcal{R}$, the maximum number of lines of $\mathcal{R}$ in $\mathcal{S}$ is $\sqrt{(2q)}+2$. As this is less than $q+1$, so $\mathcal{S}$ contains neither $\mathcal{R}$ nor any other regulus. □

For $q \equiv -1 \pmod 3$, a spread is constructed in Theorem 21.4.2 based on the twisted cubic. It is aregular when q is odd and contains exactly one regulus when q is even with $q>2$. In particular, for every odd prime $p \equiv -1 \pmod 3$, there exists an aregular spread. It also means that, for $q = 2^{2m+1}$ with $m \geq 1$, there is a spread containing a regulus without being subregular. For other examples of this phenomenon, see § 17.8.

17.4 Packings

A packing $\mathcal{P}$ of $PG(3, q)$ is a partition of the lines of the space into spreads. So $\mathcal{P}$ comprises q^2+q+1 spreads no two of which have a line in common.

In this section, the existence of packings is proved for all $q>2$ and, in the next section, packings are completely classified for $q=2$. The latter case corresponds to a very old problem.

Lemma 17.4.1: *Let $\mathcal{S}$ be a regular spread in $PG(3, q)$, $q>2$, and let l be a line of $\mathcal{S}$. Then*

(i) *there are q^2+q reguli $\mathcal{R}_{ij}$ in $\mathcal{S}$ containing l, $i \in \bar{\mathbf{N}}_q$, $j \in \mathbf{N}_q$;*

(ii) *there exist q^2+q regular spreads $\mathcal{S}_{ij}$, $i \in \bar{\mathbf{N}}_q$, $j \in \mathbf{N}_q$, such that*

  (a) $\mathcal{S} \cap \mathcal{S}_{ij} = \mathcal{R}_{ij}$;

  (b) *every line skew to l and not in $\mathcal{S}$ lies in exactly one $\mathcal{S}_{ij}$.*

Proof: We shall use the properties of the representation $\mathfrak{G}: \mathcal{L} \to \mathcal{H}_5$ described in § 15.4. Let the polarity induced by $\mathcal{H}_5$ be $\mathfrak{H}_5$ and let $l\mathfrak{G} = P$.

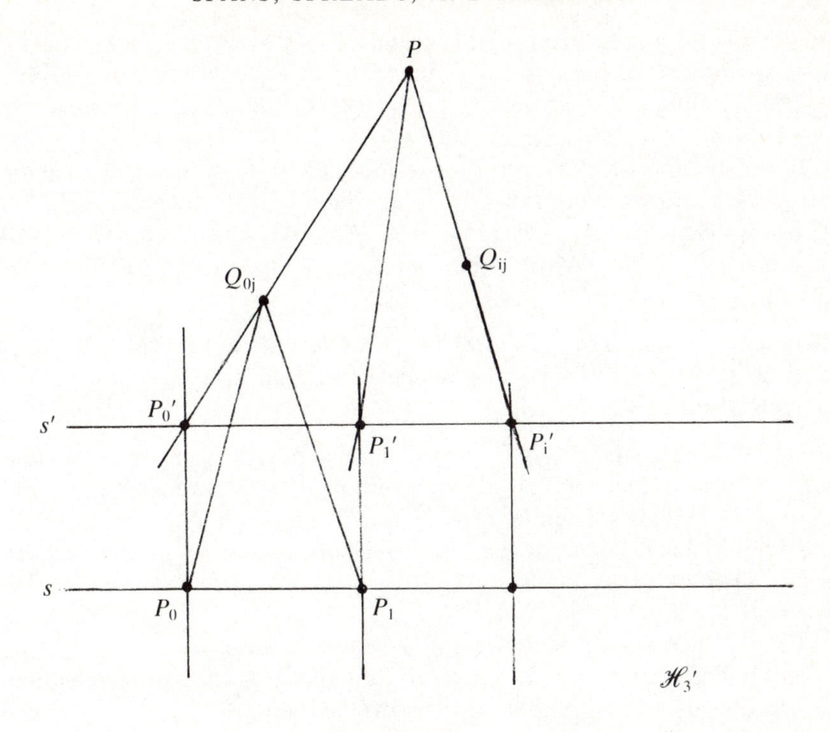

Fig. 17.2

Then the tangent prime Π_4 at P to $\mathcal{H}_5$ meets $\mathcal{H}_5$ in a cone $P\mathcal{H}_3$. Let Π_3 be a solid in Π_4 not containing P; then Π_3 meets $P\mathcal{H}_3$ in a hyperbolic quadric surface, which may be taken as $\mathcal{H}_3$. Let $\mathcal{H}_3'$ be another hyperbolic quadric in Π_3 disjoint from $\mathcal{H}_3$; Lemma 17.1.1 shows that this can be done. The cone $P\mathcal{H}_3'$ meets $\mathcal{H}_5$ only at P. Let s and s' be two generators of the same system on $\mathcal{H}_3'$ (Fig. 17.2).

Let the points of s be $P_0, \ldots, P_q$ and the points of s' be $P_0', \ldots, P_q'$, labelled so that P_iP_i' is a generator of $\mathcal{H}_3'$ for all i in $\bar{\mathbf{N}}_q$. Let the points of the line PP_i' other than P be Q_{ij}, $j \in \mathbf{N}_q$. We consider the lines P_iQ_{ij} and the planes sQ_{ij}. In fact P_iQ_{ij} lies on $P\mathcal{H}_3'$ as it lies in a generating plane PP_iP_i' of the cone; so P_iQ_{ij} does not meet $\mathcal{H}_5$. The plane sQ_{ij} lies in Π_4 and so $sQ_{ij} \cap \mathcal{H}_5$ is $sQ_{ij} \cap P\mathcal{H}_3$. Hence sQ_{ij} meets $\mathcal{H}_5$ in a repeated line, a line pair, or a conic. However, if $sQ_{ij} \cap \mathcal{H}_5$ contains a line, this line would meet s, which is impossible as $P\mathcal{H}_3' \cap P\mathcal{H}_3$ is P. So $sQ_{ij} \cap \mathcal{H}_5$ is a conic.

Let $\mathcal{S}_{ij}$ and $\mathcal{R}_{ij}$ be the sets of lines in $PG(3, q)$ represented by sections of $\mathcal{H}_5$ with the polar spaces of P_iQ_{ij} and sQ_{ij} respectively. As P_iQ_{ij} is skew to $\mathcal{H}_5$, so $\mathcal{S}_{ij}$ is a regular spread. As $sQ_{ij} \cap \mathcal{H}_5$ is a conic, so $\mathcal{R}_{ij}$ is a regulus. As P_iQ_{ij} lies in sQ_{ij}, which in turn lies in Π_4, so $\mathcal{S}_{ij}$ contains $\mathcal{R}_{ij}$, which in turn contains l. As the line s is skew to $\mathcal{H}_5$, the section of $\mathcal{H}_5$ by its polar solid represents a regular spread $\mathcal{S}$. As each plane sQ_{ij} contains s, so $\mathcal{S}$ contains each regulus $\mathcal{R}_{ij}$. Hence the q^2+q reguli $\mathcal{R}_{ij}$ are all those in $\mathcal{S}$ containing l. We already have that $\mathcal{S}\cap\mathcal{S}_{ij}$ contains $\mathcal{R}_{ij}$. However, as the join of s and P_iQ_{ij} is the plane sQ_{ij}, so $\mathcal{S}\cap\mathcal{S}_{ij}=\mathcal{R}_{ij}$.

It remains to prove (b). Let l' be a line skew to l and not in $\mathcal{S}$. Then l' is represented by a point $l'\mathfrak{G}$ with polar prime Π_4', which contains neither P nor s. Hence $s\cap\Pi_4'=P_0$, say. Also $PP_0'\cap\Pi_4'=Q_{01}$, say. Hence Π_4' contains P_0Q_{01} but no other line P_iQ_{ij}. So l' lies in $\mathcal{S}_{01}$ but no other $\mathcal{S}_{ij}$. □

Theorem 17.4.2: *If $q>2$, then $PG(3,q)$ has at least two projectively distinct packings consisting of one regular spread and q^2+q subregular spreads of index* 1.

Proof: Consider the regular spreads $\mathcal{S}$ and $\mathcal{S}_{ij}$ from the previous lemma. The spread $\mathcal{S}'_{ij}$ is obtained from $\mathcal{S}_{ij}$ by reversing $\mathcal{R}_{ij}$; thus $\mathcal{S}'_{ij}=(\mathcal{S}_{ij}\backslash\mathcal{R}_{ij})\cup\mathcal{R}'_{ij}$. Then let $\mathcal{P}=\{\mathcal{S}\}\cup\{\mathcal{S}'_{ij}\mid i\in\bar{\mathbf{N}}_q, j=\mathbf{N}_q\}$.

To show that $\mathcal{P}$ is a packing, it is sufficient to show that every line l' occurs in $\mathcal{S}$ or some $\mathcal{S}'_{ij}$.

(i) If $l'=l$, then l' is in $\mathcal{S}$.

(ii) If l' lies in one of the reguli $\mathcal{R}_{ij}$, then l' lies in $\mathcal{S}$.

(iii) If l' is skew to l and in none of the reguli $\mathcal{R}_{ij}$, then, by the lemma, l' is in exactly one $\mathcal{S}_{ij}$ and so in the corresponding $\mathcal{S}'_{ij}$.

(iv) If l' meets l, then, by Lemma 17.1.2, the $q+1$ lines of $\mathcal{S}$ meeting l' form a regulus, which is some $\mathcal{R}_{ij}$. Hence l' lies in the corresponding $\mathcal{R}'_{ij}$ and $\mathcal{S}'_{ij}$.

Therefore $\mathcal{P}$ is a packing.

We now obtain another packing $\mathcal{P}'$. The regulus $\mathcal{R}_{01}$ lies in $\mathcal{S}$ and $\mathcal{S}_{01}$; also $\mathcal{R}'_{01}$ lies in $\mathcal{S}'_{01}$. Let $\mathcal{S}'$ be the spread $(\mathcal{S}\backslash\mathcal{R}_{01})\cup\mathcal{R}'_{01}$. Let $\mathcal{P}'$ be the packing $(\mathcal{P}\backslash\{\mathcal{S},\mathcal{S}'_{01}\})\cup\{\mathcal{S}',\mathcal{S}_{01}\}$. Then both $\mathcal{P}$ and $\mathcal{P}'$ consist of one regular spread and q^2+q subregular spreads of index 1. However, the regular spreads $\mathcal{S}$ and $\mathcal{S}_{01}$ are differently related to the q^2+q-1 subregular spreads common to $\mathcal{P}$ and $\mathcal{P}'$. For, the reguli $\mathcal{R}_{ij}$ of $\mathcal{S}$ other than $\mathcal{R}_{01}$ are complementary to the respective reguli $\mathcal{R}'_{ij}$ of $\mathcal{S}'_{ij}$, whereas no regulus of $\mathcal{S}_{01}$ is complementary to a regulus of any $\mathcal{S}'_{ij}$ other than $\mathcal{S}'_{01}$. Therefore $\mathcal{P}$ and $\mathcal{P}'$ are projectively distinct. □

17.5 Packings of $PG(3,2)$ and the geometry of $\mathcal{H}_{5,2}$

The main purpose of this section is to describe fully the packings of $PG(3,2)$. A subsidiary purpose is to prove a classical isomorphism concerning $PGL(4,2)$.

A problem first posed in 1850 turns out to be pertinent. Kirkman's Fifteen Schoolgirls problem is to find an arrangement whereby 15 schoolgirls go walking each day in five rows of three so that in a week each girl has walked in the same row as every other girl.

The setting for most of the section is $PG(5,2)$ and the hyperbolic quadric $\mathscr{H}_5$. As usual, a line l is a *generator*, a *bisecant*, a *tangent*, or an *external line* to $\mathscr{H}_5$ as it has 3, 2, 1, or 0 points in common with $\mathscr{H}_5$.

Lemma 17.5.1: (i) *A packing of $PG(3,2)$ gives a solution of the fifteen schoolgirls problem.*

(ii) *The following are equivalent:*

(a) *a packing of $PG(3,2)$;*

(b) *a set of seven mutually skew external lines of $\mathscr{H}_5$ such that the polar line of the solid spanned by any two is also external to $\mathscr{H}_5$.*

Proof: (i) If each girl corresponds to a point of $PG(3,2)$, then each row corresponds to a line, each day to a spread, and each week to a packing.

(ii) To prove the equivalence of (a) and (b), we use the representation $\mathfrak{G}:\mathscr{L}\to\mathscr{H}_5$, as in § 15.4. Since each spread of $PG(3,2)$ is regular (Lemma 17.1.3, Corollary 1), and hence an elliptic congruence, a spread is represented by an elliptic quadric $\mathscr{E}_3$ on $\mathscr{H}_5$, where $\mathscr{E}_3$ lies in a solid Π_3 with polar line Π_1 external to $\mathscr{H}_5$. So a spread of $PG(3,2)$ corresponds to an external line of $\mathscr{H}_5$. If two spreads $\mathscr{S}$, $\mathscr{S}'$ have no line in common, their representing quadrics $\mathscr{E}_3$ and $\mathscr{E}_3'$ have no point in common, whence the solids Π_3 and Π_3' containing the quadrics meet in an external line l. So the polar lines Π_1 and Π_1' of Π_3 and Π_3' span a solid $\Pi_1\Pi_1'$ whose polar line is l. Thus the mapping $\mathfrak{G}\mathfrak{H}_5$, consisting of the representation of lines of $PG(3,q)$ by points of $\mathscr{H}_5$ followed by the polarity $\mathfrak{H}_5$ induced by $\mathscr{H}_5$ takes (a) to (b). □

At this stage we can observe that the existence of a packing of $PG(3,q)$ gives a generalization of Kirkman's problem as follows: if $(q^2+1)(q+1)$ schoolgirls go walking each day in q^2+1 rows of $q+1$, they can walk for q^2+q+1 days so that each girl has walked in the same row as has every other girl and hence with no girl twice.

Next we need to recall some numbers from § 3.1 and § 5.2. As usual, $\theta(n)$ denotes the number of points in $PG(n,q)$. For $q=2$,

$$\theta(1)=3,\qquad \theta(2)=7,\qquad \theta(3)=15,\qquad \theta(4)=31,\qquad \theta(5)=63.$$

Also $\phi(r;n)=\phi(r;n,q)$ is the number of r-spaces in $PG(n,q)$. Finally, $\psi_+(n)$, $\psi_-(n)$, and $\psi(n)$ are the respective numbers of points on $\mathscr{H}_n$, $\mathscr{E}_n$, and $\mathscr{P}_n$.

For $q=2$,

$$\phi(1;5)=\phi(3;5)=651,\qquad \phi(2;5)=1395,$$

$$\phi(1;3)=\psi_+(5)=35,\qquad \theta(5)-\psi_+(5)=28,$$

$$\phi(1;4)=105,\qquad \psi(4)=15.$$

Let $\mathscr{L}_3$, $\mathscr{L}_2^+$, $\mathscr{L}_1$, $\mathscr{L}_2^-$ denote the respective sets of generators, bisecants, tangents, and external lines to $\mathscr{H}_5$.

Lemma 17.5.2: (i) *If Q_1 and Q_2 are points off $\mathscr{H}_5$ in $PG(5, 2)$, then Q_1Q_2 is a tangent or an external line according as Q_1 and Q_2 are conjugate or not.*

(ii) $|\mathscr{L}_3| = 105$, $|\mathscr{L}_2^+| = 280$, $|\mathscr{L}_2^-| = 56$, $|\mathscr{L}_1| = 210$.

Proof: (i) Let $\mathscr{H}_5 = \mathbf{V}(F)$ with $F(X) = x_0x_5 + x_1x_4 + x_2x_3$. So $\mathbf{P}(X)$ is on $\mathscr{H}_5$ or not as $F(X) = 0$ or 1. Also

$$F(X+Y) = F(X) + F(Y) + \hat{X}Y^*.$$

If $Q_1 = \mathbf{P}(X)$ and $Q_2 = \mathbf{P}(Y)$ are off $\mathscr{H}_5$, then $F(X) = F(Y) = 1$. So $F(X+Y) = 0$ or 1 as $\hat{X}Y^* = 0$ or 1.

(ii) $|\mathscr{L}_3|$ is the same as the number of pencils of lines in $PG(3, 2)$ and hence is $7 \times 15 = 105$. Similarly, as there are nine pencils containing any line in $PG(3, 2)$, there are nine lines on $\mathscr{H}_5$ through any point P of $\mathscr{H}_5$. As these nine lines comprise nineteen points, there are $35 - 19 = 16$ bisecants of $\mathscr{H}_5$ through P and hence $35 \times 16/2 = 280$ bisecants altogether. Also there are $31 - 16 - 9 = 6$ tangents through P and so $35 \times 6 = 210$ altogether.

The polar prime π of Q, a point off $\mathscr{H}_5$, meets $\mathscr{H}_5$ in a quadric $\mathscr{P}_4$ comprising fifteen points. So $\pi \backslash \mathscr{P}_4$ consists of sixteen points. So there are $28 - 16 = 12$ points off $\mathscr{H}_5$ not conjugate to Q and, by (i), they lie in pairs on six external lines through Q. So $|\mathscr{L}_2^-| = 28 \times 6/3 = 56$. This number could also have been obtained by subtracting the others from $\phi(1; 5) = 651$. □

There are five different types of planes. Let Φ_4, Φ_3, Φ_2^+, Φ_1, and Φ_2^- be the sets of planes meeting $\mathscr{H}_5$ in respectively a plane, a conic, two distinct lines, a single line, and a single point.

Lemma 17.5.3: $|\Phi_4| = 30$, $|\Phi_3| = 560$, $|\Phi_2^+| = 630$, $|\Phi_2^-| = 70$, $|\Phi_1| = 105$.

Proof: $|\Phi_4|$ is the number of points plus the number of planes in $PG(3, 2)$ and so equals 30. In a spread there are ten reguli, but each regulus lies in a unique spread; so $|\Phi_3|$, which is the number of reguli in $PG(3, 2)$, is $|\mathscr{L}_2^-| = 560$. If l_1, l_2, l_3 form a pencil in $PG(3, 2)$ with centre P, then each of the six points other than P on the lines is the centre of three pencils, one of which lies in the plane $\pi = l_1l_2l_3$. So there are twelve pencils having a line in common with $\{l_1, l_2, l_3\}$ but with a centre other than P and lying in a plane other than π. So $|\Phi_2^+| = 105 \times 12/2 = 630$. If a plane meets $\mathscr{H}_5$ in a line, its other four points are mutually conjugate, by Lemma 17.5.2(i). For Q off $\mathscr{H}_5$, its polar prime meets $\mathscr{H}_5$ in $\mathscr{P}_4$ containing fifteen points. So there are sixteen points off $\mathscr{H}_5$ conjugate to Q, among which is

Q itself. If Q' is one of these, QQ' is a tangent whose polar solid meets $\mathcal{H}_5$ in a cone $\Pi_0\mathcal{P}_2$ containing seven points. So there are eight points off $\mathcal{H}_5$ conjugate to both Q and Q', and both these are among the eight. So $|\Phi_1| = 28\times 15\times 6/(4\times 3\times 2) = 105$. A plane in Φ_2^- contains four external lines and three tangents. If Q is off $\mathcal{H}_5$ the number of points off $\mathcal{H}_5$ not conjugate to Q is $28-16=12$. If Q' is one of these, then the polar solid of QQ' is an elliptic quadric meeting $\mathcal{H}_5$ in five points. So the number of points off $\mathcal{H}_5$ conjugate to both Q and Q' is 10. Hence the number of points conjugate to Q but not to Q' is $16-10=6$ and the number conjugate to neither Q nor Q' is $28-10-6-6=6$, one of which is on QQ'. So $|\Phi_2^-| = 28\times 12\times 5/(6\times 4) = 70$. The five numbers add up to 1395 as required. □

Nearly all the incidences among subspaces are summarized at the end of the section.

Lemma 17.5.4: (i) *Given two non-conjugate points Q_1 and Q_2 off $\mathcal{H}_5$, there are five points off $\mathcal{H}_5$ and not on Q_1Q_2 which are conjugate to neither Q_1 nor Q_2.*

(ii) *Together with Q_1 and Q_2 these five points form a heptad of mutually non-conjugate points determined by any pair.*

(iii) *The 28 points off $\mathcal{H}_5$ form eight heptads any two of which have a point in common.*

Proof: Part (i) was actually proved in the previous lemma. Let us look at it again from another angle. The only planes containing two external lines of $\mathcal{H}_5$ are those of Φ_2^-. There are 70 of these planes each containing four external lines and so $70\times 4/56=5$ planes of Φ_2^- through an external line. As in the proof of Lemma 17.5.2(ii), there are six external lines through a point off $\mathcal{H}_5$ and no three of these are coplanar, as otherwise their plane would contain no point of $\mathcal{H}_5$.

So through each of Q_1 and Q_2 there are five external lines other than their join. One of these lines through Q_1 and one of these lines through Q_2 lie in each of the planes of Φ_2^- through Q_1Q_2. So the intersections of these five pairs of lines are all the points off Q_1Q_2 conjugate to neither Q_1 nor Q_2. Call them Q_3, Q_4, Q_5, Q_6, Q_7.

To prove (ii), it must be shown that Q_iQ_j is an external line for $i, j \neq 1, 2$. Consider the plane $\pi_1 = Q_1Q_3Q_4$. As Q_1Q_3 and Q_1Q_4 are external lines, π_1 contains five points off $\mathcal{H}_5$ and therefore six. Let $\pi_1 \cap \mathcal{H}_5 = Q$. If Q is on Q_3Q_4, then $\pi_2 = Q_2Q_3Q_4$ is a second plane containing the tangent line QQ_3Q_4 and such that $\pi_2 \cap \mathcal{H}_5 = Q$. However, there are three tangent lines in every plane of Φ_2^-. As $|\Phi_2^-| = 70$ and $|\mathcal{L}_1| = 210$, so the number of planes of Φ_2^- through a tangent line is

$3 \times 70/210 = 1$. Hence Q is not on Q_3Q_4, and Q_3Q_4 is an external line. This proves (ii).

Since a heptad is determined by a pair of non-conjugate points, their number is $28 \times 12/(7 \times 6) = 8$. As there are twelve points off $\mathscr{H}_5$ not conjugate to a given point Q off $\mathscr{H}_5$, the number of heptads through Q is $12/6 = 2$. We note that each of the six external lines through Q has another point in both the heptads containing Q. □

The existence of the heptads allows an eight letter notation for the subspaces of $PG(5, 2)$. Let the heptads be H_i, $i \in \mathbf{N}_8$, and, for $i \neq j$, let $Q_{ij} = H_i \cap H_j$. We can then deduce the following:

(i) As the three points of an external line are mutually non-conjugate, any two belong to the same heptad. So the three points of an external line are Q_{ij}, Q_{jk}, Q_{ki}: label this line l_{ijk}.

(ii) Four mutually conjugate points off $\mathscr{H}_5$ have suffixes which comprise all eight digits: the points lie in a plane of Φ_1.

(iii) As $|\Phi_2^-| = 70$, there are two planes of Φ_2^- through every point of $\mathscr{H}_5$. The four external lines in such a plane are, say, l_{123}, l_{124}, l_{134}, l_{234}. So this plane is denoted π_{1234} and the others similarly. In confirmation, $\mathbf{c}(8, 4) = 70$. The other plane of Φ_2^- through the same point of $\mathscr{H}_5$ as π_{1234} is π_{5678}: the point of contact is denoted P_{1234} or P_{5678}.

(iv) Let S_{ijk} be the section of $\mathscr{H}_5$ by a solid with polar line l_{ijk}; then $S_{ijk} = \{P_{ijkm} \mid m \neq i, j, k\}$.

(v) Two solid sections S_{ijk} and S_{rst} have no point in common if and only if $|\{i, j, k\} \cap \{r, s, t\}| = 1$.

(vi) The polar line of the solid spanned by l_{123} and l_{145} is l_{678}.

(vii) The typical sets of three points on the different types of lines are as follows:

$$\mathscr{L}_2^-: \quad Q_{12}, Q_{13}, Q_{23}$$

$$\mathscr{L}_1: \quad Q_{12}, Q_{34}, P_{1234} = P_{5678}$$

$$\mathscr{L}_2^+: \quad Q_{12}, P_{1345} = P_{2678}, P_{1678} = P_{2345}$$

$$\mathscr{L}_3: \quad P_{1234} = P_{5678}, P_{1256} = P_{3478}, P_{1278} = P_{3456}.$$

The preliminaries have now been completed and the two theorems emerge.

Theorem 17.5.5: (i) $PGL(4, 2) \cong \mathbf{A}_8$; (ii) $PGSp(4, 2) \cong \mathbf{S}_6$.

Proof: (i) The order of $PGL(4, 2)$ is $8!/2$. A heptad corresponds in $PG(3, 2)$ to a set of seven mutually non-apolar general linear complexes. So the action of $PGL(4, 2)$ on $PG(3, 2)$ induces an action on the set of the eight heptads of $PG(5, 2)$. So there is a homomorphism $\sigma: PGL(4, 2) \to \mathbf{S}_8$. If $\mathfrak{T}\sigma$ is the identity in $\mathbf{S}_8$, then, as each point off $\mathscr{H}_5$ is the intersection of a pair of heptads, $\mathfrak{T}\sigma$ fixes all such points, hence all points on $\mathcal{H}_5$ and hence all generating planes. So $\mathfrak{T}$ fixes all general linear complexes, all lines and all points of $PG(3, 2)$; that is, $\mathfrak{T}$ is the identity. So σ is injective and hence its image has order $8!/2$. So $PGL(4, 2) \cong \mathbf{A}_8$.

(ii) Since $\mathbf{A}_8$ acts doubly transitively on the heptads, it acts transitively on the 28 external points. Therefore the stabilizer S_Q in $\mathbf{A}_8$ of an external point Q has order $8!/(2\times 28) = 6!$. However, S_Q acts on the external lines through Q inducing a homomorphism $\phi: S_Q \to \mathbf{S}_6$. If $Q = Q_{12}$, then the six external lines through Q are l_{12i}, $i = 3, \ldots, 8$: we note that l_{ijk} is a bisecant of the heptads H_i, H_j, H_k and that Q_{12} belongs to H_1 and H_2. So, if $\mathfrak{T}\phi$ is the identity, the six external lines through Q are fixed as are the six heptads not containing Q. As a transposition of heptads is impossible, the remaining two are also fixed. So $\mathfrak{T}$ is the identity and ϕ is an isomorphism. As $S_Q\sigma^{-1} = PGSp(4, 2)$, so $PGSp(4, 2) \cong \mathbf{S}_6$. □

Corollary 1: *$PGO_+(6, 2) \cong \mathbf{S}_8 \cong$ the group of all projectivities and correlations of $PG(3, 2)$.* □

Corollary 2: *The projective group G of a spread has order 360 and is isomorphic to M, where $M < \mathbf{A}_8$ and $\mathbf{A}_5 \times \mathbf{A}_3 < M < \mathbf{S}_5 \times \mathbf{S}_3$.*

Proof: A spread corresponds to an external line l_{ijk} of $\mathcal{H}_5$. So G has order $\frac{1}{2}(8!)/56 = 360$ and is isomorphic to that subgroup of $\mathbf{A}_8$ fixing both $\{1, 2, 3, 4, 5\}$ and $\{6, 7, 8\}$. Hence M is the third group between $\mathbf{A}_5 \times \mathbf{A}_3$ and $\mathbf{S}_5 \times \mathbf{S}_3$ other than $\mathbf{S}_5 \times \mathbf{A}_3$ and $\mathbf{A}_5 \times \mathbf{S}_3$. □

From the terminology of the problem's origin, a *week* is a set of seven external lines of $\mathcal{H}_5$ which defines a packing of $PG(3, 2)$ as in Lemma 17.5.1.

Theorem 17.5.6: (i) *Every week $\mathcal{W}$ is associated with a particular heptad H and a particular generating plane Π as follows:*

(a) *there are 21 distinct external lines which are polars of the solids spanned by pairs of lines of $\mathcal{W}$, and these 21 lines are all the bisecants of H;*

(b) *given l in $\mathcal{W}$, then l is not a bisecant of H and there exists a unique plane of Φ_2^- containing l and three bisecants of H; the seven points of contact of the seven planes of Φ_2^- thus determined by the seven lines of $\mathcal{W}$ are the points of Π.*

(ii) *There are $8\times 30 = 240$ weeks. They fall into two conjugacy classes of 120 each under the action of $\mathbf{A}_8$. Each conjugacy class is associated to a system of generators. Equivalently there are 240 packings of $PG(3, 2)$ which fall into two conjugacy classes of 120 under the action of $PGL(4, 2)$: a correlation interchanges these two classes.*

(iii) *The seven lines of $\mathcal{W}$ and the seven heptads besides H form an incidence structure which is a $PG(2, 2)$.*

(iv) *The projective group G of a packing is the stabilizer in* $\mathbf{A}_8$ *of both the associated heptad and the associated generating plane and so G has order* $8!/(8\times 15)=168$. *In fact, by* (iii), $G\cong PGL(3,2)$. *Also the group of a packing of* $PG(3,2)$ *fixes a point or a plane.*

Proof: An external line l_{ijk} is a bisecant of the three heptads H_i, H_j, H_k. If two external lines belong to a week, then they have one index in common, by (v) and (vi) above. A week $\mathcal{W}$ determines an incidence structure $\mathcal{W}_H=(\mathcal{P},\mathcal{B},I)$ of points and blocks, where $\mathcal{P}$ is the set of heptads other than H, $\mathcal{B}=\mathcal{W}$ and (H_i,l) in $(\mathcal{P},\mathcal{B})$ is in I if l is a bisecant of H_i. Then $\mathcal{W}_H$ satisfies the following: (a) $|\mathcal{P}|=|\mathcal{B}|=7$; (b) every block is incident with three points; (c) every two blocks are incident with one point; (d) two points are incident with at most one line. So $\mathcal{W}_H$ is a $PG(2,2)$, which proves (iii).

Suppose H_8 is the heptad which is not a 'point' of $\mathcal{W}_H$. Then the seven lines l_{ijk} of $\mathcal{W}$ have indices such that if $1,2,\ldots,7$ are the points of a $PG(2,2)$, the triples ijk are the lines; for example

$$\begin{matrix}1&2&3&4&5&6&7\\2&3&4&5&6&7&1\\4&5&6&7&1&2&3.\end{matrix}$$

So, 21 pairs of lines in the $PG(2,2)$ have the 21 pairs of points as residuals in the plane. So the 21 external lines which are polars of the 21 solids spanned by a pair of lines of $\mathcal{W}$ are the 21 lines l_{ij8}, which are all the bisecants of H_8; see (i) and (vi) on p. 72. This proves (i)(a).

To prove (i)(b), take the above scheme for $\mathcal{W}_H$ and consider l_{124}. The unique plane of Φ_2^- containing it and three bisecants of H_8 is π_{1248} with point of contact P_{1248}. So the indices of the seven such points on $\mathcal{H}_5$ are obtained by adding an 8 to each of the above columns. That they lie in a generating plane becomes clear when we see that, for example, P_{1248} and P_{2358} are collinear with P_{2678}. That is, three lines of a week which are bisecants of the same heptad correspond to three collinear points on $\mathcal{H}_5$. This proves (i)(b). Parts (ii) and (iv) now follow. □

In the above proof, it should be noted that the generating plane of $\mathcal{H}_5$ obtained from the scheme for $\mathcal{W}_H$ has points as follows:

$$\left.\begin{matrix}P_{1248}\\P_{3567}\end{matrix}\right\},\quad\left.\begin{matrix}P_{2358}\\P_{1467}\end{matrix}\right\},\quad\left.\begin{matrix}P_{3468}\\P_{1257}\end{matrix}\right\},\quad\left.\begin{matrix}P_{4578}\\P_{1236}\end{matrix}\right\},\quad\left.\begin{matrix}P_{1568}\\P_{2347}\end{matrix}\right\},\quad\left.\begin{matrix}P_{2678}\\P_{1345}\end{matrix}\right\},\quad\left.\begin{matrix}P_{1378}\\P_{2456}\end{matrix}\right\}.$$

So no heptad is particularly related to this plane as might have first appeared. In fact, if any one digit is deleted, the remaining triads must give the scheme of a $PG(2,2)$.

In Table 17.1, we list a week in $PG(5,2)$ by giving the required set of seven external lines to $\mathcal{H}_5$ and their corresponding polar solids. The five

Table 17.1 *A Week in PG*(5, 2)

$l_{124} \to S_{124}$:	P_{1234},	P_{1245},	P_{1246},	P_{1247},	P_{1248}
$l_{235} \to S_{235}$:	P_{1235},	P_{2345},	P_{2356},	P_{2357},	P_{2358}
$l_{346} \to S_{346}$:	P_{1346},	P_{2346},	P_{3456},	P_{3467},	P_{3468}
$l_{457} \to S_{457}$:	P_{1457},	P_{2457},	P_{3457},	P_{4567},	P_{4578}
$l_{561} \to S_{561}$:	P_{1256},	P_{1356},	P_{1456},	P_{1567},	P_{1568}
$l_{672} \to S_{672}$:	P_{1267},	P_{2367},	P_{2467},	P_{2567},	P_{2678}
$l_{713} \to S_{713}$:	P_{1237},	P_{1347},	P_{1357},	P_{1367},	P_{1378}

points of each solid correspond to the five lines of a spread and so to one day of the schoolgirl problem. It should be recalled that $P_{ijkl} = P_{i'j'k'l'}$, where

$$\{i, j, k, l, i', j', k', l'\} = \{1, 2, 3, 4, 5, 6, 7, 8\}.$$

In Table 17.2 we give two projectively inequivalent packings of $PG(3, 2)$ and so two solutions of the schoolgirl problem in terms of the points $P(0), P(1), \ldots, P(14)$ of $PG(3, 2)$ and its lines as listed in Table 4.2. In this table, the point $P(i)$ is simply written as i. In terms of the coordinates of § 4.2, the polarity $\mathbf{P}(a_0, a_1, a_2, a_3) \to \boldsymbol{\pi}(a_0, a_1, a_2, a_3)$ interchanges the two packings.

Finally, Table 17.3 gives most of the incidences of the different types of subspaces of $PG(5, 2)$ with respect to $\mathcal{H}_5$. Here $\Pi_r(\mathcal{V})$ is an r-space Π_r such that $\Pi_r \cap \mathcal{H}_5 = \mathcal{V}$. The notation used above for the set of all subspaces $\Pi_r(\mathcal{V})$ is indicated at the head of the table. There are no incidences given for $\Pi_r(\mathcal{V})$ and $\Pi_r(\mathcal{V}')$ with $\mathcal{V} \neq \mathcal{V}'$. For $\Pi_r(\mathcal{V})$ and $\Pi_s(\mathcal{V}')$ with $r < s$, the part of the table

	$\Pi_r(\mathcal{V})$	$\Pi_s(\mathcal{V}')$
$\Pi_r(\mathcal{V})$	a_{rr}	a_{rs}
$\Pi_s(\mathcal{V}')$	a_{sr}	a_{ss}

Table 17.2 *Two Projectively Distinct Packings of PG*(3, 2) *or Two Solutions of the Fifteen-Schoolgirls Problem*

0 1 4	1 2 5	4 5 8	1 12 13	0 3 14	5 6 9	0 5 10
2 3 6	3 4 7	6 7 10	0 7 9	2 9 11	0 2 8	1 6 11
7 8 11	8 9 12	0 11 12	2 4 10	4 6 12	3 10 12	2 7 12
9 10 13	10 11 14	2 13 14	3 5 11	5 7 13	4 11 13	3 8 13
5 12 14	0 6 13	1 3 9	6 8 14	1 8 10	1 7 14	4 9 14
2 3 6	0 3 14	3 10 12	6 8 14	1 2 5	0 11 12	3 5 11
0 1 4	2 4 10	4 11 13	2 9 11	0 7 9	1 3 9	0 6 13
9 10 13	8 9 12	5 6 9	3 4 7	4 6 12	4 5 8	4 9 14
7 8 11	5 7 13	1 7 14	0 5 10	10 11 14	6 7 10	1 8 10
5 12 14	1 6 11	0 2 8	1 12 13	3 8 13	2 13 14	2 7 12

Table 17.3 *Incidence of Subspaces*

			$\mathcal{L}_1$	$\mathcal{L}_2^-$	$\mathcal{L}_2^+$	$\mathcal{L}_3$	Φ_1	Φ_2^-
	$\Pi_0(\Pi_{-1}\mathcal{P}_0)$	$\Pi_0(\Pi_0)$	$\Pi_1(\Pi_0\mathcal{P}_0)$	$\Pi_1(\mathcal{E}_1)$	$\Pi_1(\mathcal{H}_1)$	$\Pi_1(\Pi_1)$	$\Pi_2(\Pi_1\mathcal{P}_0)$	$\Pi_2(\Pi_0\mathcal{E}_1)$
$\Pi_0(\Pi_{-1}\mathcal{P}_0)$	28	–	2	3	1	0	4	6
$\Pi_0(\Pi_0)$	–	35	1	0	2	3	3	1
$\Pi_1(\Pi_0\mathcal{P}_0)$	15	6	210	–	–	–	6	3
$\Pi_1(\mathcal{E}_1)$	6	0	–	56	–	–	0	4
$\Pi_1(\mathcal{H}_1)$	10	16	–	–	280	–	0	0
$\Pi_1(\Pi_1)$	0	9	–	–	–	105	1	0
$\Pi_2(\Pi_1\mathcal{P}_0)$	15	9	3	0	0	1	105	–
$\Pi_2(\Pi_0\mathcal{E}_1)$	15	2	1	5	0	0	–	70
$\Pi_2(\Pi_0\mathcal{H}_1)$	45	90	3	0	9	12	–	–
$\Pi_2(\mathcal{P}_2)$	80	48	8	10	6	0	–	–
$\Pi_2(\Pi_2)$	0	6	0	0	0	2	–	–
$\Pi_3(\Pi_1\mathcal{H}_1)$	15	33	3	0	6	13	1	0
$\Pi_3(\mathcal{H}_3)$	60	72	12	10	18	16	0	0
$\Pi_3(\mathcal{E}_3)$	20	8	4	10	2	0	0	4
$\Pi_3(\Pi_0\mathcal{P}_2)$	45+15	36+6	16	15	9	6	6	3
$\Pi_4(\Pi_0\mathcal{H}_3)$	15	19	7	5	9	11	3	1
$\Pi_4(\mathcal{P}_4)$	15+1	12	7+1	10	6	4	4	6

gives the parameters of a tactical configuration: a_{rr} is the total number of $\Pi_r(\mathcal{V})$, a_{ss} of $\Pi_s(\mathcal{V}')$, a_{rs} is the number of $\Pi_r(\mathcal{V})$ in a $\Pi_s(\mathcal{V}')$, and a_{sr} is the number of $\Pi_s(\mathcal{V}')$ containing a $\Pi_r(\mathcal{V})$. Some numbers in the table are given as sums to indicate peculiarities. For example, 45+15 indicates that there are 60 tangents in a non-tangent prime; however 15 of these are concurrent at the nucleus of the section.

The table is arranged to show the polarity of $\mathcal{H}_{5,2}$. The polar subspaces are as follows:

$$\Pi_0(\Pi_{-1}\mathcal{P}_0) \leftrightarrow \Pi_4(\mathcal{P}_4), \qquad \Pi_0(\Pi_0) \leftrightarrow \Pi_4(\Pi_0\mathcal{H}_3),$$
$$\Pi_1(\Pi_0\mathcal{P}_0) \leftrightarrow \Pi_3(\Pi_0\mathcal{P}_2), \qquad \Pi_1(\mathcal{E}_1) \leftrightarrow \Pi_3(\mathcal{E}_3),$$
$$\Pi_1(\mathcal{H}_1) \leftrightarrow \Pi_3(\mathcal{H}_3), \qquad \Pi_1(\Pi_1) \leftrightarrow \Pi_3(\Pi_1\mathcal{H}_3),$$
$$\Pi_2(\Pi_1\mathcal{P}_0) \leftrightarrow \Pi_2(\Pi_1\mathcal{P}_0), \qquad \Pi_2(\Pi_0\mathcal{E}_1) \leftrightarrow \Pi_2'(\Pi_0\mathcal{E}_1),$$
$$\Pi_2(\Pi_0\mathcal{H}_1) \leftrightarrow \Pi_2'(\Pi_0\mathcal{H}_1), \qquad \Pi_2(\mathcal{P}_2) \leftrightarrow \Pi_2'(\mathcal{P}_2),$$
$$\Pi_2(\Pi_2) \leftrightarrow \Pi_2(\Pi_2).$$

The dash indicates that although the polar plane is of the same type, it is not the same plane. See § 20.6 for a similar table for $\mathcal{E}_{5,2}$ and § 20.7 for further properties of $\mathcal{H}_{5,2}$.

17.6 *k*-spans

A *k-span* $\mathcal{S}$ in $PG(3, q)$ is a set of k lines no two of which intersect. A k-span $\mathcal{S}$ is *complete* if it is not contained in a $(k+1)$-span; that is, every line of the space meets some line of $\mathcal{S}$. Elsewhere, a k-span is called a *of PG*(5, 2) *with Respect to* $\mathcal{H}_5$

	Φ_2^+ $\Pi_2(\Pi_0\mathcal{H}_1)$	Φ_3 $\Pi_2(\mathcal{P}_2)$	Φ_4 $\Pi_2(\Pi_2)$	$\Pi_3(\Pi_1\mathcal{H}_1)$	$\Pi_3(\mathcal{H}_3)$	$\Pi_3(\mathcal{E}_3)$	$\Pi_3(\Pi_0\mathcal{P}_2)$	$\Pi_4(\Pi_0\mathcal{H}_3)$	$\Pi_4(\mathcal{P}_4)$
$\Pi_0(\Pi_{-1}\mathcal{P}_0)$	2	4	0	4	6	10	7+1	12	15+1
$\Pi_0(\Pi_0)$	5	3	7	11	9	5	7	19	15
$\Pi_1(\Pi_0\mathcal{P}_0)$	1	3	0	6	9	15	16	36+6	45+15
$\Pi_1(\mathcal{E}_1)$	0	1	0	0	2	10	4	8	20
$\Pi_1(\mathcal{H}_1)$	4	3	0	16	18	10	12	72	60
$\Pi_1(\Pi_1)$	2	0	7	13	6	0	3	33	15
$\Pi_2(\Pi_1\mathcal{P}_0)$	–	–	–	1	0	0	3	9	15
$\Pi_2(\Pi_0\mathcal{E}_1)$	–	–	–	0	0	5	1	2	15
$\Pi_2(\Pi_0\mathcal{H}_1)$	630	–	–	12	9	0	3	90	45
$\Pi_2(\mathcal{P}_2)$	–	560	–	0	6	10	8	48	80
$\Pi_2(\Pi_2)$	–	–	30	2	0	0	0	6	0
$\Pi_3(\Pi_1\mathcal{H}_1)$	2	0	7	105	–	–	–	9	0
$\Pi_3(\mathcal{H}_3)$	4	3	0	–	280	–	–	16	10
$\Pi_3(\mathcal{E}_3)$	0	1	0	–	–	56	–	0	6
$\Pi_3(\Pi_0\mathcal{P}_2)$	1	3	0	–	–	–	210	6	15
$\Pi_4(\Pi_0\mathcal{H}_3)$	5	3	7	3	2	0	1	35	–
$\Pi_4(\mathcal{P}_4)$	2	4	0	0	1	3	2	–	28

partial spread and a complete k-span which is not a spread is called a *maximal partial spread.* We have already had some discussion of (q^2+1)-spans or spreads. To develop a theory of k-spans, two related questions must be considered.

I. If $\mathcal{S}$ is a complete k-span with $k<q^2+1$, what values can k take? In particular, what are its maximum and minimum values?

II. If $\mathcal{S}$ is any k-span, how large must k be so that it has a unique completion to a spread?

If $\mathcal{S}$ is a k-span with $k=q^2+1-d$, then d is the *deficiency* of $\mathcal{S}$. When $d>0$, we define $\Pi(\mathcal{S})$, the *set of points residual to* $\mathcal{S}$, to be those points on no line of $\mathcal{S}$; dually, $\Phi(\mathcal{S})$, the *set of planes residual to* $\mathcal{S}$, is the set of planes containing no line of $\mathcal{S}$. Throughout, $\mathcal{L}$ denotes the set of lines of $PG(3, q)$.

Lemma 17.6.1: *Let $\mathcal{S}$ be a k-span of deficiency $d>0$ with residual sets $\Pi(\mathcal{S})$, $\Phi(\mathcal{S})$. Then*

(i) $|\Pi(\mathcal{S})|=|\Phi(\mathcal{S})|=d(q+1)$;

(ii) *If l in $\mathcal{L}\backslash\mathcal{S}$ meets r lines of $\mathcal{S}$, then l contains $q+1-r$ points of $\Pi(\mathcal{S})$ and lies in $q+1-r$ planes of $\Phi(\mathcal{S})$;*

(iii) *a point P is contained in $q+d$ or d planes of $\Phi(\mathcal{S})$ according as P is or is not in $\Pi(\mathcal{S})$;*

(iv) *a plane π contains $q+d$ or d points of $\Pi(\mathcal{S})$ according as π is or is not in $\Phi(\mathcal{S})$.*

Proof: Since $k=q^2+1-d$, the lines of $\mathcal{S}$ contain $(q+1)(q^2+1-d)$ points. So $|\Pi(\mathcal{S})|=d(q+1)$ and by duality $|\Phi(\mathcal{S})|=|\Pi(\mathcal{S})|$.

If l meets r lines of $\mathcal{S}$ and is not in $\mathcal{S}$, then r of its points lie on lines of $\mathcal{S}$ and $q+1-r$ do not; the rest of part (ii) follows by duality.

A plane π containing e lines of $\mathcal{S}$ with $e=0$ or 1 meets the remaining lines of $\mathcal{S}$ in $q^2+1-d-e$ distinct points. So there are

$$e(q+1)+q^2+1-d-e=q^2+eq+1-d$$

points of π on lines of $\mathcal{S}$. Hence π contains $(1-e)q+d$ points of $\Pi(\mathcal{S})$, which is $q+d$ or d as π is or is not in $\Phi(\mathcal{S})$. Part (iii) is the dual of (iv). □

Lemma 17.6.2: *If a k-span $\mathcal{S}$ with deficiency $d \leq q$ is contained in a spread $\mathcal{S}'$, then $\mathcal{S}'$ is the only spread containing $\mathcal{S}$.*

Proof: Any line l of $\mathcal{L}\backslash\mathcal{S}'$ meets the d lines of $\mathcal{S}'\backslash\mathcal{S}$ in at most d points. As $d \leq q$, the line l meets some line of $\mathcal{S}$. So the d lines of $\mathcal{S}'\backslash\mathcal{S}$ are the only lines skew to all the lines of $\mathcal{S}$. □

Before proving any substantial results, we define the parameters $\lambda_0, \lambda_1, \ldots, \lambda_{q+1}$ for a k-span $\mathcal{S}$. Let λ_i be the number of lines not in $\mathcal{S}$ meeting exactly i lines of $\mathcal{S}$. We observe that $\mathcal{S}$ is complete if and only if $\lambda_0=0$.

Lemma 17.6.3: *For a k-span $\mathcal{S}$ in $PG(3, q)$, the constants $\lambda_0, \lambda_1, \ldots, \lambda_{q+1}$ satisfy the following equations:*

$$\text{(i)} \quad \sum_{i=0}^{q+1} \lambda_i = (q^2+1)(q^2+q+1)-k; \tag{17.12}$$

$$\text{(ii)} \quad \sum_{i=1}^{q+1} i\lambda_i = q(q+1)^2 k; \tag{17.13}$$

$$\text{(iii)} \quad \sum_{i=2}^{q+1} \tfrac{1}{2}i(i-1)\lambda_i = (q+1)^2\tfrac{1}{2}k(k-1); \tag{17.14}$$

$$\text{(iv)} \quad \sum_{i=3}^{q+1} \tfrac{1}{6}i(i-1)(i-2)\lambda_i = (q+1)\tfrac{1}{6}k(k-1)(k-2). \tag{17.15}$$

Proof: Each equation counts a set in two different ways. The sets are the following:

(i) $\mathcal{L}\backslash\mathcal{S}$;

(ii) $\{(l, m) \mid l \in \mathcal{S},\ m \in \mathcal{L}\backslash\mathcal{S},\ l \cap m = \Pi_0\}$;

(iii)

$$\{(\{l_1, l_2\}, m) \mid l_1, l_2 \text{ distinct lines of } \mathcal{S},\ m \in \mathcal{L}\backslash\mathcal{S},\ m \text{ meets } l_1 \text{ and } l_2\};$$

(iv)

$$\{(\{l_1, l_2, l_3\}, m) \mid l_1, l_2, l_3 \text{ distinct lines of } \mathcal{S},\ m \notin \mathcal{S},\ m \text{ meets each } l_i\}.$$

□

Now, some bounds for a complete k-span are given.

Theorem 17.6.4: *If $\mathcal{S}$ is a complete k-span with $k<q^2+1$, then $q+\sqrt{q}+1\leq k\leq q^2-\sqrt{q}$.*

Proof: Let π be a plane of $\Phi(\mathcal{S})$, which exists since $d=q^2+1-k>0$. Let $\mathcal{B}$ be the set of points in which π meets the lines of S; so $|\mathcal{B}|=k$. Since $\mathcal{S}$ is complete, every line of π contains some point of $\mathcal{B}$. Suppose l is a line of π lying entirely in $\mathcal{B}$. Then the joins of l with the $q+1$ lines of $\mathcal{S}$ meeting it are the $q+1$ planes of the space through l. But one of these planes must be π, contradicting that π contains no line of $\mathcal{S}$. Hence for every line l in π, we have $1\leq|\mathcal{B}\cap l|\leq q$. So $\mathcal{B}$ is a blocking k-set in π, and, by Theorem 13.3.3, $q+\sqrt{q}+1\leq k\leq q^2-\sqrt{q}$. □

The upper bound is now considered in more detail. In Theorem 17.6.12, the lower bound is improved. For an improvement to the upper bound, see § 17.8.

Corollary 1: *A k-span $\mathcal{S}$ with $k>q^2-\sqrt{q}$ can be completed uniquely to a spread.*

Proof: By the theorem, a complete span containing $\mathcal{S}$ is a spread. As $d=q^2+1-k<\sqrt{q}+1$, so $d\leq q$. Then, by Lemma 17.6.2, there is only one spread containing $\mathcal{S}$. □

Corollary 2: *In $PG(3,2)$, the only complete k-spans are spreads.*

Proof: This follows from the theorem, since, for $q=2$, $q+\sqrt{q}+1>q^2-\sqrt{q}$. It was also proved directly in the remarks following Lemma 17.1.3, Corollary 1: any three skew lines form a regulus, which lies in a unique spread. □

In Theorem 13.2.2, it was shown that, for q square, a blocking $(q^2-\sqrt{q})$-set in $PG(2,q)$ is the complement of a subgeometry $PG(2,\sqrt{q})$. Now we prove a comparable result for spreads.

Theorem 17.6.5: *If $\mathcal{S}$ is a $(q^2-\sqrt{q})$-span in $PG(3,q)$, q square, then $\Pi(\mathcal{S})$ is a subgeometry $PG(3,\sqrt{q})$.*

Proof: The theorem is consistent with the characterization of a blocking $(q^2-\sqrt{q})$-set, since it implies that every plane in $\Phi(\mathcal{S})$ meets the lines of $\mathcal{S}$ in the complement of a $PG(2,\sqrt{q})$.

To prove the theorem, the planes of the proposed $PG(3,\sqrt{q})$ must be defined. In fact, $\Phi'(\mathcal{S})=\{\pi\cap\Pi(\mathcal{S})\mid\pi\in\Phi(\mathcal{S})\}$ is the set of planes required. The incidence of points of $\Pi(\mathcal{S})$ with planes of $\Phi'(\mathcal{S})$ is that of the space.

Since $d=\sqrt{q}+1$, $|\Pi(\mathcal{S})|=|\Phi'(\mathcal{S})|=(q+1)(\sqrt{q}+1)$ by Lemma 17.6.1(i). By part (iii) of the same lemma, each point of $\Pi(\mathcal{S})$ lies in $q+\sqrt{q}+1$

planes of $\Phi'(\mathscr{S})$ and, by part (iv), each plane of $\Phi'(\mathscr{S})$ contains $q+\sqrt{q}+1$ points of $\Pi(\mathscr{S})$. So the numbers are as expected. The sets $\Pi(\mathscr{S})$ and $\Phi'(\mathscr{S})$ satisfy the axioms for points and planes of a projective space $PG(3, \sqrt{q})$ as in § 2.1. □

Conjecture: If q is square, the points of $PG(3, q)\backslash PG(3, \sqrt{q})$ lie on the lines of a complete $(q^2-\sqrt{q})$-span.

We proceed to prove this conjecture for $q=4$.

Lemma 17.6.6: *If $\mathscr{C}_1$ and $\mathscr{C}_2$ are elliptic linear congruences in $PG(3, q)$ with no line in common, then the complex conjugate axes of $\mathscr{C}_1$ and $\mathscr{C}_2$ are mutually skew.*

Proof: Let the axes of $\mathscr{C}_i$ be l_i and l_i'. Then l_1, l_1', l_2, l_2' lie over $\gamma' = GF(q^2)$ but not over $\gamma = GF(q)$. In the representation $\mathfrak{G}: \mathscr{L} \to \mathscr{H}_5$ in $PG(5, q)$, each $\mathscr{C}_i$ is represented by an elliptic quadric $\mathscr{E}_3^{(i)}$ lying in a solid $\Pi_3^{(i)}$. As $\mathscr{C}_1$ and $\mathscr{C}_2$ have no line in common, $\Pi_3^{(1)}$ and $\Pi_3^{(2)}$ meet in an external line Π_1; the polar lines $\Pi_1^{(1)}$ and $\Pi_1^{(2)}$ of $\Pi_3^{(1)}$ and $\Pi_3^{(2)}$ span the polar solid Π_3 of Π_1. Also Π_3 meets $\mathscr{H}_5$ in a quadric $\mathscr{E}_3$.

Now our imagination must extend itself from γ to γ'. Each variety is given by the same equations but considered over γ'. So $\mathscr{E}_3$ becomes $\bar{\mathscr{H}}_3$ and $\mathscr{H}_5$ becomes $\bar{\mathscr{H}}_5$. Also $\Pi_1^{(i)}$ meets $\bar{\mathscr{H}}_5$ in P_i and P_i'.

Let us see how $\mathscr{E}_3$ lies on $\bar{\mathscr{H}}_3$; the former has q^2+1 points and the latter $(q^2+1)^2$. No two points of $\mathscr{E}_3$ can lie on the same generator of $\bar{\mathscr{H}}_3$, as such a generator would then have an equation over γ and so would be a line on $\mathscr{E}_3$. So the points of $\mathscr{E}_3$ lie on $\bar{\mathscr{H}}_3$ in such a way that each generator of $\bar{\mathscr{H}}_3$ contains exactly one point of $\mathscr{E}_3$.

Now, P_1, P_1', P_2, P_2' are all points of $\bar{\mathscr{H}}_3\backslash\mathscr{E}_3$. The situation we wish to exclude is that l_1 or l_1' meets l_2 or l_2'. If P_i and P_i' represent l_i and l_i' respectively, then l_1 meets l_2, say, if P_1P_2 lies on $\bar{\mathscr{H}}_5$. If this happens then P_1P_2 lies on $\bar{\mathscr{H}}_3$ as does its complex conjugate $P_1'P_2'$. But then P_1P_2 and $P_1'P_2'$ are in complementary reguli of $\bar{\mathscr{H}}_3$ and meet in a point P of $\mathscr{E}_3$, as in Fig. 17.3. So P is conjugate to both P_1 and P_1' and so lies in $\mathscr{E}_3^{(1)}$;

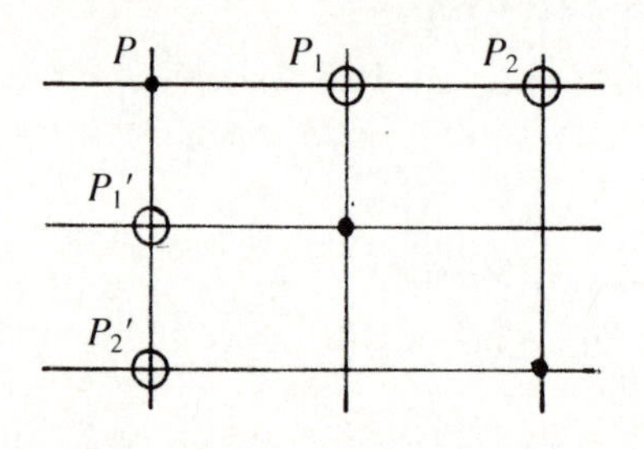

Fig. 17.3

likewise P lies in $\mathscr{E}_3^{(2)}$. So P lies in $\mathscr{E}_3^{(1)}\cap\mathscr{E}_3^{(2)}$ and represents a line common to $\mathscr{C}_1$ and $\mathscr{C}_2$, contrary to the hypothesis. □

Corollary 1: *If q is a square and there exist r mutually disjoint regular spreads in $PG(3, \sqrt{q})$, then there exists a $(2r)$-span in $PG(3, q)$.*

Proof: As each regular spread is an elliptic congruence in $PG(3, \sqrt{q})$, its two axes lie in $PG(3, q)\backslash PG(3, \sqrt{q})$. Then the $2r$ axes of the r congruences form the required span. □

Corollary 2: *The 70 points of $PG(3, 4)\backslash PG(3, 2)$ lie on the lines of a complete 14-span of $PG(3, 4)$.*

Proof: By Theorem 17.5.6, there exists a packing of $PG(3, 2)$ by regular spreads. Embed $PG(3, 2)$ in $PG(3, 4)$. Then the axes of the seven spreads of a packing of $PG(3, 2)$ form a 14-span $\mathscr{S}$ in $PG(3, 4)$. As each line of each spread in the packing of $PG(3, 2)$ meets the appropriate axes in points of $PG(3, 4)\backslash PG(3, 2)$, so $\mathscr{S}$ lies entirely in $PG(3, 4)\backslash PG(3, 2)$. As the points on no line of $\mathscr{S}$ are the 15 points of $PG(3, 2)$, of which at most three are collinear, so $\mathscr{S}$ is complete. □

This proves the conjecture for $q = 4$.

Now we give examples of complete k-spans which are not spreads for each $q > 2$.

Lemma 17.6.7: *In $PG(3, q)$, $q > 2$, there exists a complete k-span $\mathscr{S}$ with $k = q^2 - q + 1$ or $k = q^2 - q + 2$.*

Proof: By Lemma 17.1.3, Corollary 3, a spread $\mathscr{S}'$ which is not regular exists for all $q > 2$. By Lemma 17.1.2, there is some line l_0 not in $\mathscr{S}'$ such that the set of lines $\mathscr{L}'$ of $\mathscr{S}'$ meeting l_0 do not form a regulus. Let $\mathscr{S}_0 = (\mathscr{S}'\backslash\mathscr{L}')\cup\{l_0\}$. If l is any line such that $\mathscr{S}_0\cup\{l\}$ is a span, then the points of l all lie on the lines of $\mathscr{L}'$. As $|\mathscr{L}'| = q + 1 \geq 4$ and $\mathscr{L}'$ is not a regulus, there are at most two transversals to $\mathscr{L}'$ of which l_0 is one. So either $\mathscr{S} = \mathscr{S}_0$ is a complete $(q^2 - q + 1)$-span if $\mathscr{L}'$ has just the transversal l_0 or, if $\mathscr{L}'$ has transversals l_0 and l_1, then $\mathscr{S} = \mathscr{S}_0\cup\{l_1\}$ is a complete $(q^2 - q + 2)$-span. □

In the next two theorems the construction indicated by the lemma is carried out.

Theorem 17.6.8: *In $PG(3, q)$, $q > 2$, there exists a complete $(q^2 - q + 1)$-span.*

Proof: From the previous lemma, it suffices to find a spread $\mathscr{S}'$ and a line l' not in $\mathscr{S}'$ such that the lines of $\mathscr{S}'$ meeting l' have just one transversal. Let $\mathscr{S}_0$ be a regular spread containing the reguli $\mathscr{R}_1$ and $\mathscr{R}_2$, whose

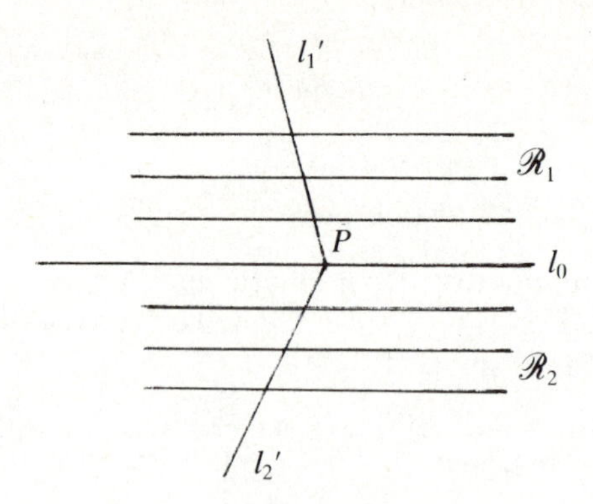

Fig. 17.4

respective complementary reguli are $\mathcal{R}_1'$ and $\mathcal{R}_2'$. Suppose that $\mathcal{R}_1 \cap \mathcal{R}_2 = \{l_0\}$. Then a line of $\mathcal{R}_1'$ can meet a line of $\mathcal{R}_2'$ only at some point of l_0.

Let l_1' and l_2' be the unique lines of $\mathcal{R}_1$ and $\mathcal{R}_2$ respectively through a point P of l_0 (Fig. 17.4). Let $\mathcal{S}_1 = (\mathcal{S}_0 \backslash \mathcal{R}_2) \cup \mathcal{R}_2'$: this spread is subregular of index one. Since l_2' is in $\mathcal{S}_1$, the line l_1' is not. Let $\mathcal{L}'$ be the set of lines of $\mathcal{S}_1$ meeting l_1'; in fact $\mathcal{L}' = (\mathcal{R}_1 \backslash \{l_0\}) \cup \{l_2'\}$. Then, as in the lemma, let $\mathcal{S} = (\mathcal{S}_1 \backslash \mathcal{L}') \cup \{l_1'\}$. If l is a line such that $\mathcal{S} \cup \{l\}$ is a span, then l must be a transversal of $\mathcal{L}'$. Since $q \geq 3$, $\mathcal{R}_1 \backslash \{l_0\}$ contains at least three lines. So l meets three lines of $\mathcal{R}_1$ and therefore lies in $\mathcal{R}_1'$. But, from above, l meets l_2' on l_0 and therefore at P. So l meets l_1' and hence $\mathcal{S} \cup \{l\}$ is not a span. Therefore $\mathcal{S}$ is a complete $(q^2 - q + 1)$-span. □

Corollary: *In $PG(3, 3)$, the only complete spans are spreads and 7-spans.*

Proof: By Theorem 17.6.4, a complete k-span which is not a spread has $k = 6$ or 7. From the theorem just proved, a complete 7-span exists. So it remains to show that a complete 6-span cannot exist.

First we note that if $\mathcal{S}'$ is any 5-span and π is a plane in $\Phi(\mathcal{S}')$, then π contains either one or two lines skew to all the lines of $\mathcal{S}'$. For, if π contains no such lines, it would meet the lines of $\mathcal{S}'$ in a blocking 5-set, contradicting Theorem 13.3.3; if π contains three such lines, they occupy at least nine points, whereas only $13 - 5 = 8$ are available.

Now let $\mathcal{S} = \mathcal{S}' \cup \{l\}$ be a complete 6-span. By the dual of the previous paragraph, through each point of l there are either one or two lines skew to all the lines of $\mathcal{S}'$. Therefore we have two possibilities:

(i) there is some point P on l such that l is the only line through P skew to the lines of $\mathcal{S}'$;

(ii) through every point of l there is one other line skew to the lines of $\mathcal{S}'$.

In case (i), the joins of P with the lines of $\mathcal{S}'$ are five distinct planes. As there are four planes through l, there are $13-5-4=4$ planes through P containing no line of $\mathcal{S}$. Let π be one of these. Then, by the above, there is a line l' in π skew to the lines of $\mathcal{S}'$. The definition of π implies that $l' \neq l$. Also l' does not contain P as l is the only line through P skew to the lines of $\mathcal{S}'$. So l' is skew to the lines of $\mathcal{S}'$ and to l. Hence $\mathcal{S} \cup \{l'\}$ is a 7-span, contradicting the completeness of $\mathcal{S}$.

In case (ii), let $\mathcal{L}'$ be the set of four lines meeting l which are skew to the lines of $\mathcal{S}'$. The lines of $\mathcal{S}'$ account for 20 points, while l and the lines of $\mathcal{L}'$ account for at most 16 points. So there exists a point Q on no line of $\mathcal{S}$ or $\mathcal{L}'$. Therefore there is a line l' through Q skew to the lines of $\mathcal{S}'$. Also l' cannot meet l or it would be in $\mathcal{L}'$. Hence $\mathcal{S} \cup \{l'\}$ is a 7-span, contradicting the completeness of $\mathcal{S}$. □

Theorem 17.6.9: *In $PG(3, q)$, q odd and $q>3$, there exists a complete (q^2-q+2)-span.*

Proof: From Lemma 17.6.7, we need to find a spread $\mathcal{S}$ and a line l not in $\mathcal{S}$ such that the lines of $\mathcal{S}$ meeting l have exactly two transversals.

Let $\mathcal{R}$ and $\mathcal{R}'$ be the two reguli on the quadric $\mathcal{H}_3$. Let $\mathcal{R}=\{r_i \mid i \in \bar{\mathbf{N}}_q\}$ and $\mathcal{R}'=\{r_i' \mid i \in \bar{\mathbf{N}}_q\}$; also let $P_{ij}=r_i \cap r_j'$, $l=P_{00}P_{11}$ and $l'=P_{01}P_{10}$. Then l and l' are mutually skew bisecants of $\mathcal{H}_3$. Now let P be a point of $l \backslash \mathcal{H}_3$ and let l_i be the unique line through P meeting l' and r_i, $i=2, 3, \ldots, q$. As PP_{01} lies in the tangent plane r_0r_1' at P_{01} and PP_{10} lies in the tangent plane r_1r_0' at P_{10}, so PP_{01} and PP_{10} are tangents to $\mathcal{H}_3$. As l and l' are skew, so each l_i is skew to r_0 and r_1 (Fig. 17.5).

Suppose that the $q-1$ lines l_i are distinct. Then l_i meets just one line r_j, namely r_i, and is therefore tangent to $\mathcal{H}_3$. As the plane Pl' is not a tangent plane, it meets $\mathcal{H}_3$ in a conic $\mathcal{P}_2$. As l_i lies in Pl', it is tangent to $\mathcal{P}_2$. So there are $q-1$ tangents to $\mathcal{P}_2$ concurrent at P. As q is odd, $q-1 \leqslant 2$: a contradiction. So some two of the l_i coincide.

The $q+1$ lines through P meeting l' and some line of $\mathcal{R}$ are PP_{01}, PP_{10}, and the $q-1$ lines l_i. As two of these coincide, there is a line l_0 through P meeting l' and skew to all the lines of $\mathcal{R}$. So the linear congruence containing l_0 and $\mathcal{R}$ is elliptic and thus a regular spread $\mathcal{S}_0$. The regulus $\mathcal{R}_0=\mathcal{R}(r_0, r_1, l_0)$ therefore lies in $\mathcal{S}_0$. As l and l' meet r_0, r_1, and l_0, they lie in the complementary regulus $\mathcal{R}_0'$.

Let $\mathcal{S}=(\mathcal{S}_0 \backslash \mathcal{R}) \cup \mathcal{R}'$. The line l is not in $\mathcal{S}$. If $\mathcal{L}'$ is the set of lines of $\mathcal{S}$ meeting l, then $\mathcal{L}' \supset \mathcal{R}_0 \backslash \{r_0, r_1\}$. As $q \geqslant 5$, $\mathcal{L}'$ contains at least four lines of $\mathcal{R}_0$. So, if $\mathcal{L}'$ is a regulus, it is $\mathcal{R}_0$. However, r_0 and r_1 are not in $\mathcal{S}$. So $\mathcal{L}'$ is not a regulus. In fact $\mathcal{L}'=(\mathcal{R}_0 \backslash \{r_0, r_1\}) \cup \{r_0', r_1'\}$ and has just the two transversals l and l'. Hence $(\mathcal{S} \backslash \mathcal{L}') \cup \{l, l'\}$ is a complete (q^2-q+2)-span. □

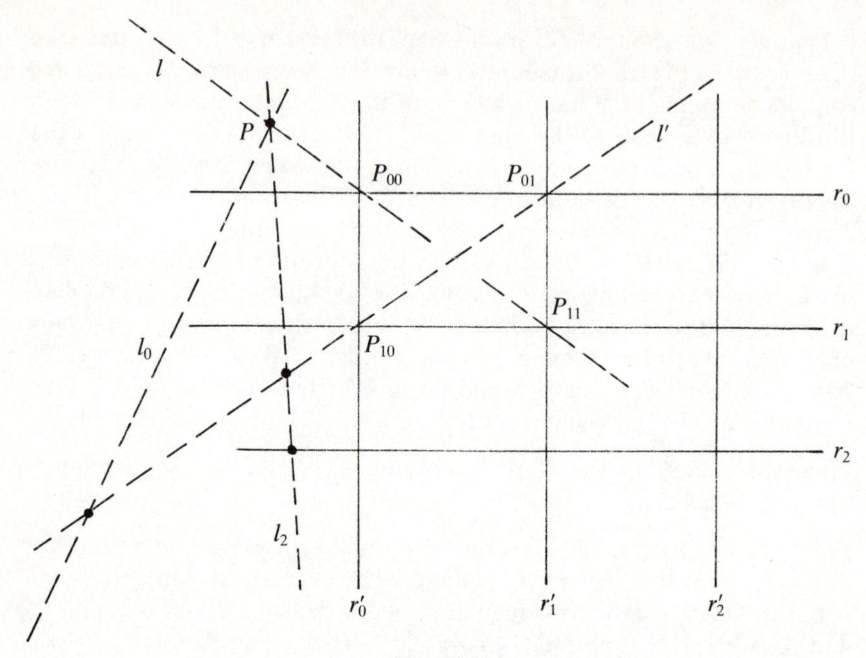

Fig. 17.5

The construction of Theorem 17.6.8 can be generalized to give smaller complete spans.

Theorem 17.6.10: *In* $PG(3, q)$ *there exists a complete* k*-span* $\mathcal{S}$ *with* $k = q^2+1-nq$ *for every integer* n *such that* $0 \leq n \leq \frac{1}{2}q - 1$. *Further, there is a line* m *covered by* $\mathcal{S}$ *and* $n+1$ *hyperbolic quadrics* $\mathcal{H}^{(0)}, \ldots, \mathcal{H}^{(n)}$ *such that*

(a) $\mathcal{H}^{(i)} \cap \mathcal{H}^{(j)} = m$ *for* $i \neq j$;

(b) *a line not on any* $\mathcal{H}^{(i)}$ *meets at least* $q-1-2n$ *lines of* $\mathcal{S}$.

Proof: Let $\mathcal{S}_0$ be a regular spread and $\mathcal{R}_0, \ldots, \mathcal{R}_n$ be $n+1$ distinct reguli in $\mathcal{S}_0$ such that $\mathcal{R}_i \cap \mathcal{R}_j = \{m\}$ for $i \neq j$ and some line m of $\mathcal{S}$. Let $\mathcal{H}^{(i)}$ be the quadric containing the complementary reguli $\mathcal{R}_i$ and $\mathcal{R}'_i$.

Choose a set $\mathcal{M} = \{l_0, \ldots, l_q\}$ of $q+1$ lines in $\mathcal{R}'_0 \cup \ldots \cup \mathcal{R}'_n$ such that

(i) through any point of m there is exactly one line of $\mathcal{M}$;

(ii) $\mathcal{M} \cap \mathcal{R}'_i \neq \varnothing$ for all i.

Let

$$\mathcal{S} = (\mathcal{S}_0 \backslash \bigcup \mathcal{R}_i) \cup \mathcal{M}.$$

Then $\mathcal{S}$ is a k-span with

$$\begin{aligned} k &= (q^2+1) - [(n+1)q+1] + (q+1) \\ &= q^2+1-nq. \end{aligned}$$

Let l be a line not in $\mathcal{S}$. If $l \in \mathcal{R}_i$, then (ii) implies that l meets a line of $\mathcal{R}'_i$ in $\mathcal{S}$; if $l \in \mathcal{R}'_i$, then l meets m and so a line of $\mathcal{M}$. So, in both these cases, l meets at least one line of $\mathcal{S}$. If l is not on $\mathcal{H}^{(i)}$, it meets $\mathcal{H}^{(i)}$ in at most two points. By construction, any point of l not on $\mathcal{H}^{(i)}$ is on a line of $\mathcal{S}$. So l meets at least $q+1-2(n+1)$ lines of $\mathcal{S}$. Thus, if $q+1-2(n+1) \geqslant 1$, that is $n \leqslant \frac{1}{2}q-1$, the spread $\mathcal{S}$ is complete. □

Lemma 17.6.11: *In $PG(3,4)$, a complete k-span has $k \geqslant 8$.*

Proof: By Theorem 17.6.4, $k \geqslant 7$. So suppose a complete 7-span $\mathcal{S}$ exists. Then $\lambda_0 = 0$, and equations (17.12) to (17.15) of Lemma 17.6.3 become

$$\begin{aligned} \lambda_1 + \lambda_2 + \lambda_3 + \lambda_4 + \lambda_5 &= 350 \\ \lambda_1 + 2\lambda_2 + 3\lambda_3 + 4\lambda_4 + 5\lambda_5 &= 700 \\ \lambda_2 + 3\lambda_3 + 6\lambda_4 + 10\lambda_5 &= 525 \\ \lambda_3 + 4\lambda_4 + 10\lambda_5 &= 175. \end{aligned}$$

Eliminating λ_1 and λ_2 gives

$$\lambda_3 + 3\lambda_4 + 6\lambda_5 = 175.$$

Hence $\lambda_2 = \lambda_4 = \lambda_5 = 0$ and $\lambda_1 = \lambda_3 = 175$.

Let $\mathcal{S} = \{l_i \mid i \in \mathbf{N}_7\}$ and let $\mathcal{R}$ be the regulus containing l_1, l_2, l_3. If m is another line of $\mathcal{R}$, it meets one of the lines of $\mathcal{S}$, say l_4, since $\mathcal{S}$ is complete. Through $P = l_4 \cap m$, there is a line m' of the regulus $\mathcal{R}'$ complementary to $\mathcal{R}$. So m' meets l_1, l_2, l_3, l_4, which is impossible since $\lambda_4 = \lambda_5 = 0$. Hence a complete 7-span does not exist. □

Theorem 17.6.12: *If $\mathcal{S}$ is a complete k-span, then $k \geqslant 2q$.*

Proof: We use equations (17.12) to (17.15) of Lemma 17.6.3. Since $\mathcal{S}$ is complete, $\lambda_0 = 0$. The right-hand sides of these equations are respectively

$$F_1 = (q^2+1)(q^2+q+1) - k, \qquad F_2 = q(q+1)^2 k,$$
$$F_3 = (q+1)^2 \tfrac{1}{2}k(k-1), \qquad F_4 = (q+1)\tfrac{1}{6}k(k-1)(k-2).$$

Now,

$$\begin{aligned} -12 + 12i - 10 \cdot \tfrac{1}{2}i(i-1) + 6 \cdot \tfrac{1}{6}i(i-1)(i-2) \\ = (i-1)(i-3)(i-4). \end{aligned}$$

As $\sum_{i=1}^{q+1} (i-1)(i-3)(i-4)\lambda_i \geqslant 0$, so

$$E = -12F_1 + 12F_2 - 10F_3 + 6F_4 \geqslant 0.$$

Arranging the polynomial E in powers of k gives

$$\begin{aligned}E &= k^3(q+1)-k^2(q+1)(5q+8)+k(12q^3+29q^2+24q+19)\\&\quad -12(q^4+q^3+2q^2+q+1)\\&=(k-2q+1)[k^2(q+1)-k(q+1)(3q+9)\\&\quad +(6q^3+8q^2+18q+28)]-(2q^3-4q^2-26q+40)\\&=(k-2q+1)\{(q+1)[k^2-k(3q+9)+2(3q^2+q+8)]+12\}\\&\quad -2(q-4)(q^2+2q-5).\end{aligned}$$

Now, suppose $q>4$. Then, if $k\leqslant 2q-1$, we have that $E<0$, a contradiction. Hence $k\geqslant 2q$.

For $q=4$, the previous lemma shows that $k\geqslant 8$. For $q=3$, $k\geqslant 7$ by the corollary to Theorem 17.6.8. For $q=2$, $k=5$ by Theorem 17.6.4, Corollary 2. □

17.7 Hermitian arcs in $PG(4, q)$

In this section, we take up the construction of $PG(2, q^2)$ in $PG(4, q)$ used in § 4.1, and show how to construct a Hermitian arc which is not a Hermitian curve in any $PG(2, q^2)$ with $q>2$. This is the counterexample promised at the end of § 12.3.

Consider a prime Π_3 in $PG(4, q)$ and a regular spread $\mathcal{S}$ of Π_3. The q^4 points off Π_3 are the points of an affine plane $AG(2, q^2)$. The lines of $AG(2, q^2)$ are the planes of $PG(4, q)$ meeting Π_3 in a line of $\mathcal{S}$: Two such planes Π_2 and Π_2' through different lines l and l' of $\mathcal{S}$ cannot meet in a line Π_1; for then Π_1 would meet l and l', and so would lie in Π_3 as would Π_2 and Π_2'. Thus Π_2 and Π_2' meet exactly in a point of $PG(4, q)\backslash\Pi_3$. Given any two points P and P' off Π_3, the line Π_1 joining them meets Π_3 in one point and hence meets exactly one line l of $\mathcal{S}$. The plane $\Pi_1 l$ is the line of $AG(2, q^2)$ through P and P'. The ideal points of $AG(2, q^2)$ are the lines of $\mathcal{S}$, and the ideal line is $\mathcal{S}$ itself. So, a $PG(2, q^2)$ has been formed as follows, where $AG(4, q)=PG(4, q)\backslash\Pi_3$:

$$PG^{(0)}(2, q^2)=AG(4, q)\cup\mathcal{S},$$

$$PG^{(1)}(2, q^2)=\{\Pi_2\backslash\Pi_3 \mid \Pi_2\in PG(4, q), \Pi_2\cap\Pi_3\in\mathcal{S}\}\cup\{\mathcal{S}\}.$$

Each line of $PG(2, q^2)$ which corresponds to a plane π of $AG(4, q)$ comprises q^2 points in π plus one point consisting of the line of $\mathcal{S}$ in π.

Now let us start from $PG(2, q^2)$ and consider a Hermitian curve $\mathcal{U}$ in the plane. Let $\mathcal{A}$ be the affine plane $PG(2, q^2)\backslash l_0$ and let $\mathcal{V}=\mathcal{U}\backslash l_0$. Then $\mathcal{V}$ is *parabolic* or *hyperbolic* in $\mathcal{A}$ according as $|\mathcal{U}\cap l_0|=1$ or $q+1$; that is, according as $|\mathcal{V}|=q^3$ or q^3-q.

Consider $\mathcal{V}$ in the $AG(4, q)$ associated with $\mathcal{A}$, and in $PG(4, q)$. Let $\bar{\mathcal{V}}$

be the union of the points corresponding to $\mathcal{V}$ in $AG(4, q)$ and of the points on the lines of $\mathcal{S}$ corresponding to points of l_0. If $\mathcal{V}$ is parabolic, then $|\bar{\mathcal{V}}| = q^3 + q + 1$; if $\mathcal{V}$ is hyperbolic, then $|\bar{\mathcal{V}}| = q^3 - q + (q+1)^2 = q^3 + q^2 + q + 1$.

Theorem 17.7.1: *The set $\bar{\mathcal{V}}$ is quadric in $PG(4, q)$. If $\mathcal{V}$ is parabolic, then $\bar{\mathcal{V}} = \Pi_0 \mathscr{E}_3$; if $\mathcal{V}$ is hyperbolic, then $\bar{\mathcal{V}} = \mathscr{P}_4$.*

Proof: Let α in $\gamma' \backslash \gamma$, $\gamma' = GF(q^2)$, satisfy $\alpha^2 - u\alpha - v = 0$, where $t^2 - ut - v$ is irreducible over γ. If $x \in \gamma'$, then $x = x_1 + \alpha x_2$ with x_1, x_2 in γ. With $\bar{x} = x^q$, we have $\alpha + \bar{\alpha} = u$ and $\alpha\bar{\alpha} = -v$.

We follow as above the sequence of mappings

$$PG(2, q^2) \rightarrow AG(2, q^2) \rightarrow AG(4, q) \rightarrow PG(4, q)$$

given by

$$\mathbf{P}(x, y, 1) \rightarrow (x, y) \rightarrow (x_1, x_2, y_1, y_2) \rightarrow \mathbf{P}(x_1, x_2, y_1, y_2, 1),$$

where z will be used for the last coordinate in both $PG(2, q^2)$ and $PG(4, q)$.

A line of $AG(2, q^2)$ of equation $x = c$ or $y = ax + b$ is represented in $AG(4, q)$ by

$$x_1 = c_1, \qquad x_2 = c_2$$

or

$$y_1 = a_1 x_1 + v a_2 x_2 + b_1, \qquad y_2 = a_2 x_1 + (a_1 + u a_2) x_2 + b_2.$$

The lines of the spread associated with $AG(2, q^2)$ are

$$\mathbf{V}(x_1, x_2, z)$$

or

$$\mathbf{V}(y_1 - a_1 x_1 - v a_2 x_2,\ y_2 - a_2 x_1 - (a_1 + u a_2) x_2,\ z).$$

With $\mathcal{U} = \mathbf{V}(x\bar{z} + y\bar{z} + \bar{y}z)$ a Hermitian curve and $l_0 = \mathbf{V}(z)$, we have $l_0 \cap \mathcal{U} = \mathbf{U}_1$ and $\mathcal{V} = \mathcal{U} \backslash l_0$ is parabolic.

$$\begin{aligned}\bar{\mathcal{V}} &= \mathbf{V}((x_1 + \alpha x_2)(x_1 + \bar{\alpha} x_2) + (y_1 + \alpha y_2 + y_1 + \bar{\alpha} y_2) z) \\ &= \mathbf{V}(x_1^2 + u x_1 x_2 - v x_2^2 + (2y_1 + u y_2) z).\end{aligned}$$

$\Pi_3 \cap \bar{\mathcal{V}} = \mathbf{V}(x_1, x_2, z)$ and the only singular point of $\bar{\mathcal{V}}$ is $\Pi_0 = \mathbf{P}(0, 0, u, -2, 0)$. As $\mathbf{V}(y_1 + u y_2)$ meets $\bar{\mathcal{V}}$ in $\mathbf{V}(y_1 + u y_2,\ x_1^2 + u x_1 x_2 - v x_2^2 + yz)$ and the latter form is the canonical one for $\mathscr{E}_3$, so $\bar{\mathcal{V}} = \Pi_0 \mathscr{E}_3$. We note that u and 2 are not both zero.

Now, let $\mathcal{U} = \mathbf{V}(x\bar{x} + y\bar{y} + z\bar{z})$ and $l_0 = \mathbf{V}(z)$ so that $|l_0 \cap \mathcal{U}| = q + 1$ and $\mathcal{V} = \mathcal{U} \backslash l_0$ is hyperbolic. Then

$$\begin{aligned}\bar{\mathcal{V}} &= \mathbf{V}((x_1 + \alpha x_2)(x_1 + \bar{\alpha} x_2) + (y_1 + \alpha y_2)(y_1 + \bar{\alpha} y_2) + z^2) \\ &= \mathbf{V}(x_1^2 + u x_1 x_2 - v x_2^2 + y_1^2 + u y_1 y_2 - v y_2^2 + z^2)\end{aligned}$$

which is non-singular and hence $\mathscr{P}_4$. $\square$

Theorem 17.7.2: *Let $\mathcal{A}$ and $\mathcal{A}'$ be affine planes $AG(2, q^2)$ respectively determined by the spreads $\mathcal{S}$ and $\mathcal{S}'$ of Π_3 in $PG(4, q)$. Let l be a line common to $\mathcal{S}$ and $\mathcal{S}'$, and let $\mathcal{V}$ be a parabolic Hermitian curve in $\mathcal{A}$ so that l corresponds to the ideal point of $\mathcal{V}$. Then the set $\bar{\mathcal{V}}$ in $PG(4, q)$ corresponds in $\mathcal{A}'$ to a set $\mathcal{V}'$ which is a parabolic Hermitian arc with ideal point corresponding to l; that is, $\mathcal{V}'$ together with its ideal point is a Hermitian arc $\mathcal{U}'$ in $PG(2, q^2)$.*

Proof: Since $\mathcal{V}$ is parabolic, $\bar{\mathcal{V}}$ meets $\mathcal{S}$ in the points of l. A plane of $PG(4, q)$ through l corresponds to a line in both $\mathcal{A}$ and $\mathcal{A}'$ and so meets $\mathcal{V}'$ in $q+1$ points. The ideal line of $\mathcal{A}'$ is tangent to $\mathcal{V}'$.

Let l' be a line of $\mathcal{A}'$ through two points P_1 and P_2 of $\mathcal{V}'$ such that it corresponds to a plane π' of $PG(4, q)$ skew to l. It must be shown that $|\pi' \cap \bar{\mathcal{V}}| = q+1$. Since $\bar{\mathcal{V}}$ is a quadric $\Pi_0\mathcal{E}_3$ and π' contains two points of $\bar{\mathcal{V}}$, so $\pi' \cap \bar{\mathcal{V}}$ is a conic $\mathcal{P}_2$ or it contains a line m. In the latter case, m meets Π_3 in a point P which lies on l, as $\bar{\mathcal{V}} \cap \Pi_3 = l$. So $P \in \pi' \cap l$, a contradiction. So $\pi' \cap \bar{\mathcal{V}}$ is $\mathcal{P}_2$ and comprises $q+1$ points. Thus $\mathcal{V}'$ is a parabolic Hermitian arc in $\mathcal{A}'$. Equivalently, if $\mathcal{A}' \cup l_0'$ is $PG(2, q^2)$, then $\mathcal{U}'$ is a Hermitian arc in $PG(2, q^2)$, where $\mathcal{V}' = \mathcal{U}' \backslash l_0'$ and l_0' is unisecant to $\mathcal{U}'$. □

The Hermitian arc $\mathcal{U}'$ in this theorem may be constructed so as not to be projectively equivalent to the Hermitian curve $\mathcal{U}_{2,q^2}$.

Two properties of subgeometries $PG(1, q)$ of $PG(1, q^2)$ are used:

(i) if l is a line in $PG(2, q^2)$ and $PG(2, q)$ is any subgeometry such that $|l \cap PG(2, q)| = q+1$, then $l \cap PG(2, q)$ is a subline of l;

(ii) if l is a line in $PG(2, q^2)$ and $\mathcal{U}_{2,q^2}$ is some Hermitian curve, then $|l \cap \mathcal{U}_{2,q^2}| = 1$ or $l \cap \mathcal{U}_{2,q^2}$ is a subline. See Lemma 6.2.1.

Lemma 17.7.3: *Let π be a plane of $PG(3, q)$, $\mathcal{C}$ a conic in π and l a line of π external to $\mathcal{C}$. Let π' be another plane through l and P a point of $\pi' \backslash l$. Then there exists an elliptic quadric $\mathcal{E}_3$ containing $\mathcal{C}$ and such that π' is the tangent plane to $\mathcal{E}_3$ at P.*

Proof: Let $\pi = \mathbf{V}(x_3)$, $\pi' = \mathbf{V}(x_2)$, $\mathcal{C} = \mathbf{V}(f(x_0, x_1) + x_2^2, x_3)$ with f irreducible. Then $l = \mathbf{V}(x_2, x_3)$. So $P = \mathbf{P}(a_0, a_1, 0, 1)$. The projectivity $\mathfrak{T}$ given by

$$\mathbf{P}(x_0, x_1, x_2, x_3)\mathfrak{T} = \mathbf{P}(x_0 - a_0x_3, x_1 - a_1x_3, x_2, x_3)$$

fixes π, π' and $\mathcal{C}$ but transforms P to $\mathbf{U}_3$. Then

$$\mathcal{E}_3 = \mathbf{V}(f(x_0, x_1) + x_2^2 + x_2x_3)$$

contains $\mathcal{C}$ and π' is the tangent plane at $\mathbf{U}_3$. □

Lemma 17.7.4: *The number of conics external to a fixed line l in $PG(2, q)$ is greater than the number of subgeometries $PG(1, q)$ of $PG(1, q^2)$ when $q > 2$.*

Proof: The number of lines external to a conic $\mathscr{C}$ is $\frac{1}{2}q(q-1)$, Lemma 8.1.1, Corollary 1. The number of conics in $PG(2,q)$ is q^5-q^2, § 7.2, and the number of lines is q^2+q+1. So, if N is the number of conics external to l, it is given by

$$\tfrac{1}{2}q(q-1)(q^5-q^2)=(q^2+q+1)N,$$

whence $N=\frac{1}{2}q^3(q-1)^2$. The number of $PG(1,q)$ in $PG(1,q^2)$ is $s(1,q,q^2)=q(q^2+1)$, by Lemma 4.3.1, Corollary 2. Thus, for $q>2$, $N>s(1,q,q^2)$. □

Theorem 17.7.5: *For any $q>2$, there exists a Hermitian arc in $PG(2,q^2)$ which is not a Hermitian curve.*

Proof: Suppose $PG(2,q^2)$ is determined by the spread $\mathscr{S}$ of the solid Π_3 in $PG(4,q)$. Let π be a plane of $PG(4,q)$ representing a line of $AG(2,q^2)$. Then, by Lemma 17.7.4, $\pi\backslash\Pi_3$ contains a conic $\mathscr{C}$ which does not represent a subline of $PG(2,q^2)$.

Let Π_3' be a solid containing π and meeting Π_3 in the plane π'. Let P_0 be a point of $\Pi_3\backslash\pi'$ and let the unique line of $\mathscr{S}$ through P_0 meet π' in P. Then, by Lemma 17.7.3, there exists a quadric $\mathscr{E}_3$ in Π_3' containing $\mathscr{C}$ and such that π' is the tangent plane at P to $\mathscr{E}_3$; so $\mathscr{E}_3\cap\Pi_3=\{P\}$.

By Theorems 17.7.1 and 17.7.2, the quadric cone $\bar{\mathscr{V}}=P_0\mathscr{E}_3$ represents a parabolic Hermitian arc $\mathscr{V}$ in $AG(2,q^2)$ which with its ideal point is a Hermitian arc $\mathscr{K}$ in $PG(2,q^2)$. By construction the line l of $PG(2,q^2)$ represented by π meets $\mathscr{K}$ in a set of $q+1$ points which is not a $PG(1,q)$. Thus $\mathscr{K}$ is not a Hermitian curve. □

It is now possible to write down the specific equation of a Hermitian arc $\mathscr{K}$ which is not a Hermitian curve. Such an arc is elsewhere called a non-classical unital.

Let $f(x_1,x_2)$ and $g(x_1,x_2)=x_1^2-ux_1x_2-vx_2^2$ be irreducible over $\gamma=GF(q)$ and let $g(\alpha,1)=0$; thus $\alpha\in\gamma'\backslash\gamma$, where $\gamma'=GF(q^2)$.

In the notation of Theorem 17.7.5 and with point coordinates $\mathbf{P}(x_1,x_2,y_1,y_2,z)$ in $PG(4,q)$, let

$$\Pi_3=\mathbf{V}(z),\qquad \Pi_3'=\mathbf{V}(y_1),\qquad \pi'=\mathbf{V}(y_1,z),$$
$$\mathscr{E}_3=\mathbf{V}(f(x_1,x_2)+y_2z,\,y_1),$$
$$P=\mathbf{U}_3=\Pi_3\cap\mathscr{E}_3,\qquad P_0=\mathbf{U}_2.$$

Then

$$P_0\mathscr{E}_3=\mathbf{V}(f(x_1,x_2)+y_2z). \tag{17.16}$$

So, passing back to coordinates in $AG(2,q^2)$ as in the proof of Theorem 17.7.1, we have, from

$$x=x_1+\alpha x_2,\qquad \bar{x}=x_1+\bar{\alpha}x_2,$$
$$y=y_1+\alpha y_2,\qquad \bar{y}=y_1+\bar{\alpha}y_2,$$

that

$$(\bar{\alpha}-\alpha)x_1=\bar{\alpha}x-\alpha\bar{x}, \qquad (\bar{\alpha}-\alpha)x_2=\bar{x}-x,$$
$$(\bar{\alpha}-\alpha)y_1=\bar{\alpha}y-\alpha\bar{y}, \qquad (\bar{\alpha}-\alpha)y_2=\bar{y}-y.$$

Hence $\mathcal{K}\backslash\{Q\}$, where $Q=\mathbf{P}(0,1,0)$, has the following equation in $AG(2,q^2)$:

$$f(\bar{\alpha}x-\alpha\bar{x},\bar{x}-x)+(\bar{\alpha}-\alpha)(\bar{y}-y)=0. \tag{17.17}$$

Thus, in $PG(2,q^2)$ with $f(x_1,x_2)\neq g(x_1,-x_2)$,

$$\mathcal{K}=\mathbf{V}(f(\bar{\alpha}xz^{q-1}-\alpha x^q, x^q-xz^{q-1})+(\bar{\alpha}-\alpha)(y^qz^q-yz^{2q-1})); \tag{17.18}$$

that is, $\mathcal{K}$ is a curve of degree $2q$.

If $f(x_1,x_2)=g(x_1,-x_2)$, then $\mathcal{K}$ is a Hermitian curve $\mathcal{U}_{2,q^2}$ exactly as in the proof of Theorem 17.7.1 and has degree $q+1$. This occurs because the terms in x^2 and $\bar{x}^2$ of (17.17) disappear.

Let us specify two small examples of $\mathcal{K}\neq\mathcal{U}_{2,q^2}$. First, let $q=3$, $\bar{x}=x^3$,

$$f(x_1,x_2)=x_1^2+x_2^2, \qquad g(x_1,x_2)=x_1^2-x_1x_2-x_2^2.$$

In the notation of § 1.7, $\alpha=\sigma$. Then (17.17) becomes

$$(\sigma^3x-\sigma\bar{x})^2+(\bar{x}-x)^2+\sigma^2(\bar{y}-y)=0,$$

whence

$$\sigma^3x^2-\sigma\bar{x}^2+\bar{y}-y=0; \tag{17.19}$$

that is, in $PG(2,9)$,

$$\mathcal{K}=\mathbf{V}(\sigma^3x^2z^4-\sigma x^6+y^3z^3-yz^5). \tag{17.20}$$

Now, let $q=4$, $\bar{x}=x^4$,

$$f(x_1,x_2)=x_1^2+x_1x_2+\eta^{10}x_2^2,$$
$$g(x_1,x_2)=x_1^2+x_1x_2+\eta^5x_2^2,$$

where in the notation of § 1.7, $\eta^4=\eta+1$. Thus $\alpha=\eta$, and (17.17) becomes

$$x^2+x\bar{x}+\bar{x}^2+\bar{y}+y=0; \tag{17.21}$$

that is, in $PG(2,16)$,

$$\mathcal{K}=\mathbf{V}(x^2z^6+x^5z^3+x^8+y^4z^4+yz^7). \tag{17.22}$$

17.8 Notes and references

Every spread determines a translation plane and vice versa as in § 17.7, where the plane is Desarguesian if and only if $\mathcal{S}$ is regular; see André

(1954), Bruck and Bose (1964, 1966), Lüneburg (1980), Ostrom (1968), Segre (1964b).

§ 17.1. The spread of r-spaces in $PG(n, q)$ given in § 4.1 is regular when $r = 1$ and $n = 3$. Theorem 17.1.4 is stated both by Bruck (1969, p. 442), and Denniston (1973c). Theorem 17.1.6 is due to Thas (1973f). Theorem 17.1.14 is originally due to Orr in his Ph.D. thesis; however, the treatment here including Lemmas 17.1.7–17.1.12 follows Fisher and Thas (1979).

§ 17.2. The theory of subregular spreads was developed in detail by Bruck (1969) and continued in (1973c). Theorems 17.2.1 and 17.2.2 follow Orr (1976). Bruen (1975b) gives a variation of Orr's treatment.

§ 17.3. Theorem 17.3.3 follows Denniston (1973a). Theorem 17.3.4 is given by Bruck (1969). Bruen (1972b) also gives an example of an aregular spread for all $q = p^{2m}$ with p odd. Denniston (1976) gives an example, for all $q = 2^{2m+1}$ with $m \geqslant 1$, of a spread which contains exactly $q-1$ reguli but is not subregular. This is done by a method similar to Lemma 17.1.1; but, in this case, the family $\mathscr{F}$ comprises the quadrics $\mathscr{F}_t$, $t \in \gamma^+$, where

$$\mathscr{F}_t = \mathbf{V}(t^6x_0^2 + t^4x_0x_1 + t^2x_1^2 + x_2^2 + x_2x_3 + x_3^2).$$

For $t \neq 0, \infty$, each $\mathscr{F}_t$ is hyperbolic and no two have a point in common.

§ 17.4. This section follows Denniston (1972). An alternative treatment is given by Beutelspacher (1974), who also generalizes the result to packings of lines in $PG(n, q)$ with $n = 2^m - 1$ for any $m \geqslant 2$. A packing is *cyclic* if it can be obtained by acting on one spread with a collineation of order $q^2 + q + 1$. Denniston (1973a) finds that no packing in $PG(3, 3)$ or $PG(3, 4)$ is cyclic, even though that of $PG(3, 2)$ is. In (1973b), Denniston finds a cyclic packing in $PG(3, 8)$. For examples of packings in higher dimensions, see Fuji-Hara and Vanstone (1984).

§ 17.5. For some history of the fifteen-schoolgirls problem, see Biggs (1981). The treatment here mainly follows Conwell (1910). Other geometric accounts of the isomorphism between $\mathbf{A}_8$ and $PGL(4, 2)$ are given by Edge (1954) and McDonough (1980). There are solutions to the schoolgirls problem which cannot be embedded in $PG(3, 2)$. Apart from the two in $PG(3, 2)$ as in Table 17.2, there are five other solutions. The full set was first given by Woolhouse in 1862 and 1863. For an account of the seven solutions, see Cole (1922). Goudarzi (1967a) gives more than two solutions in $PG(3, 2)$, but uses a smaller group than $\mathbf{A}_8$ and seems to have misread Pickert (1955, p. 298).

§ 17.6. The initial part of this section as far as Lemma 17.6.2 follows Mesner (1967). Lemma 17.6.3 is due to Glynn (1982). Theorem 17.6.4 follows Bruen (1971b), although the upper bound is due to Mesner (1967) as is Theorem 17.6.5 and Lemma 17.6.6, Corollary 2.

For $q=p^r$ with r odd, Bruen (1975a) gives an improvement on the upper bound of a complete k-span. He showed that if $k=q^2+1-d$, then $f(d)\geq q^2$, where $f(d)=\frac{1}{2}(d-1)(d^3-d^2+d+2)$. For large q, this gives $d\geq 2^{\frac{1}{4}}\sqrt{q}$ as opposed to $d\geq\sqrt{q}$ in Theorem 17.6.4. This result exploits the connection between k-spans and nets; see also Bruen (1971b, 1972a), Jungnickel (1984a). Lemma 17.6.7 and Theorems 17.6.8 and 17.6.9 follow Bruen (1971b). Theorem 17.6.10 comes from Beutelspacher (1980). In Theorem 17.6.9, a complete (q^2-q+2)-span is constructed for q odd with $q>3$. Bruen and Thas (1976) construct a complete (q^2-q+2)-span for $q=2^{2m+1}$ with $m\geq 1$ and Freeman (1980) constructs such a span for $q=2^{2m}$ with $m\geq 1$. Hence a complete (q^2-q+2)-span exists for all $q>3$; see also Jungnickel (1984a). Lemma 17.6.11 is due to Bruen and Thas (1976), and Theorem 17.6.12 follows Glynn (1982). Ebert (1978a) gives other examples of complete k-spans, while Ebert (1979) showed a weaker result than Glynn's.

§ 17.7. Hermitian arcs in $PG(2,q^2)$ other than Hermitian curves (non-classical unitals) were found by Buekenhout (1976b) for $q=2^{2m+1}$ with $m\geq 1$ using the λ-quadrics of § 16.4. The variation and extension to all $q>2$ given here follows Metz (1979). Hermitian arcs constructed in this way are called Buekenhout–Metz unitals and, as explained, include Hermitian curves. Lefèvre-Percsy (1981) gives the following characterization of these unitals, where, as in § 6.2 and § 17.7, a subline in $PG(2,q^2)$ is a $PG(1,q)$ on a line.

Theorem 17.8.1: *Let $\mathscr{K}$ be a Hermitian arc in $PG(2,q^2)$, $q>2$, and let l be a unisecant to $\mathscr{K}$. If every subline with a point on l meets $\mathscr{K}$ in 0, 1, 2, or $q+1$ points, then K is a Buekenhout–Metz unital.* □

This leads to the following, pleasing characterization of Hermitian curves by Lefèvre-Percsy (1982a).

Theorem 17.8.2: *Let $\mathscr{K}$ be a Hermitian arc in $PG(2,q^2)$. If every $(q+1)$-secant of $\mathscr{K}$ meets $\mathscr{K}$ in a subline, then $\mathscr{K}$ is a Hermitian curve.* □

For another proof of the latter theorem, see Faina and Korchmáros (1983), and for another characterization see Wilbrink (1983).

Other papers relevant to this chapter are Bruen (1972a, 1975a, 1977), Bruen and Fisher (1969), Bruen and Freeman (1982), Bruen and Silverman (1974), Cameron and Liebler (1982), Drake and Freeman (1979), Ebert (1978c), Edge (1953, 1955a), Herzer and Lunardon (1980), Kantor (1982a–e), Lunardon (1976*, 1977*, 1978), Metz (1981), Panella (1967, 1970), Pickert (1982), Sherk (1979), Walker (1976, 1979).

18

k-CAPS

18.1 Arithmetical preliminaries

k-caps were defined and some of their properties presented in § 16.1. As for plane k-arcs, § 9.1, and plane $(k; n)$-arcs, § 12.1, there are many diophantine equations associated to a k-cap $\mathcal{K}$. Some of these are presented below.

Let τ_i be the number of i-secant lines of the k-cap $\mathcal{K}$. As usual, 2-secants, 1-secants, and 0-secants are respectively called bisecants, tangents, and external lines.

Lemma 18.1.1:

$$\tau_2 = k(k-1)/2, \qquad \tau_1 = k(q^2+q+2-k),$$
$$\tau_0 = (q^2+1-k)(q^2+q+1)+k(k-1)/2. \quad \square$$

Let κ_i be the number of planes meeting $\mathcal{K}$ in an i-arc.

Lemma 18.1.2:

$$\kappa_0+\kappa_1+\ldots+\kappa_{q+2} = (q+1)(q^2+1) \tag{18.1$_1$}$$
$$\kappa_1+2\kappa_2+\ldots+(q+2)\kappa_{q+2} = k(q^2+q+1) \tag{18.1$_2$}$$
$$\kappa_2+3\kappa_3+\ldots+(q+2)(q+1)\kappa_{q+2}/2 = k(k-1)(q+1)/2. \tag{18.1$_3$}$$

Proof: The three equations count the respective sets

$$\{\pi \mid \pi \text{ a plane}\}, \quad \{(P, \pi) \mid P \in \mathcal{K} \cap \pi\}, \quad \{(\{P, P'\}, \pi) \mid P, P' \in \mathcal{K} \cap \pi\}. \quad \square$$

Let $c_i = c_i^P$ be the number of planes through a point P of $\mathcal{K}$ meeting $\mathcal{K}$ in an i-arc.

Lemma 18.1.3:

$$c_1+c_2+\ldots+c_{q+2} = q^2+q+1, \tag{18.2$_1$}$$
$$c_2+2c_3+\ldots+(q+1)c_{q+2} = (k-1)(q+1), \tag{18.2$_2$}$$
$$i\kappa_i = \sum_P c_i^P. \quad \square \tag{18.2$_3$}$$

Let r_i be the number of points off $\mathcal{K}$ on exactly i bisecants and let $m = [k/2]$, the integral part of $k/2$.

Lemma 18.1.4:

$$r_0+r_1+\ldots+r_m=(q+1)(q^2+1)-k, \quad (18.3_1)$$

$$r_1+2r_2+\ldots+mr_m=k(k-1)(q-1)/2. \quad \square \quad (18.3_2)$$

In § 16.1, it was shown that, if $\mathcal{K}$ is a k-cap, then

$$k \leq q^2+1 \quad \text{for} \quad q>2,$$

$$k \leq 8 \quad \text{for} \quad q=2.$$

It was also shown that an elliptic quadric is a (q^2+1)-cap for all q, and that, for q odd or $q=4$, the converse holds.

The main problem in this chapter is to find how large a complete cap can be when it is less than the maximum value.

Lemma 18.1.5: *If a k-cap $\mathcal{K}$ meets a plane π in an n-arc and*

$$k(k-2n-1)<2(q^2+q+1)+n(q-4)-n^2q,$$

then $\mathcal{K}$ is incomplete.

Proof: The bisecants of $\mathcal{K}$ in π contain $n(n-1)(q-1)/2+n$ points. The bisecants of $\mathcal{K}\backslash\pi$ meet π in at most $(k-n)(k-n-1)/2$ points. So, if

$$n(n-1)(q-1)/2+n+(k-n)(k-n-1)/2<q^2+q+1,$$

then there is a point of π on no bisecant of $\mathcal{K}$, whence $\mathcal{K}$ is incomplete. Manipulation gives the required inequality. $\square$

Lemma 18.1.6: *In $PG(3, q)$ with q odd, let $\mathcal{K}$ be a (q^2+1-d)-cap with $d \leq q+1$ and let l be a bisecant of $\mathcal{K}$. Then*

(i) *any plane through l contains at least $q+1-d$ points of $\mathcal{K}$;*

(ii) *at least $q+1-d$ of the planes through l meet $\mathcal{K}$ in a conic.*

Proof: (i) $(q^2+1-d)-q(q-1)=q+1-d$. So if q planes through l meet $\mathcal{K}$ in a conic, the remaining plane meets $\mathcal{K}$ in a $(q+1-d)$-arc.

(ii) $(q^2+1-d)-d(q-d)=(q+1-d)(q-1)+2$. So, even if d planes through l meet $\mathcal{K}$ in a q-arc, the remaining $q+1-d$ must meet $\mathcal{K}$ in $(q+1)$-arcs, which are conics. $\square$

Lemma 18.1.7: *In $PG(3, q)$, the maximum values of k for which a k-cap $\mathcal{K}$ can lie on the different types of quadric surfaces are as listed in Table 18.1.*

Proof: The list of quadrics is as in Theorem 15.3.1. For $\Pi_1\mathscr{E}_1$, a line, the result is by definition. For $\Pi_2\mathscr{P}_0$ and $\Pi_1\mathscr{H}_1$, respectively a plane and a repeated plane, the result follows from the number of points on an oval. Each of $\Pi_0\mathscr{P}_2$ and $\mathscr{H}_3$ can have at most two points on each generator, and so $k \leq 2(q+1)$. If $\mathscr{Q}$ is either of these quadrics and l is a line external to

Table 18.1

	q even	q odd
$\Pi_2\mathcal{P}_0$	$q+2$	$q+1$
$\Pi_1\mathcal{E}_1$	2	2
$\Pi_1\mathcal{H}_1$	$2(q+2)$	$2(q+1)$
$\Pi_0\mathcal{P}_2$	$2(q+1)$	$2(q+1)$
$\mathcal{H}_3$	$2(q+1)$	$2(q+1)$
$\mathcal{E}_3$	q^2+1	q^2+1

$\mathcal{Q}$, let π and π' be planes through l meeting $\mathcal{Q}$ in the respective conics $\mathcal{C}$ and $\mathcal{C}'$; then $\mathcal{C} \cup \mathcal{C}'$ is a $2(q+1)$-cap on $\mathcal{Q}$. □

Lemma 18.1.8: *In $PG(3, q)$ with $q > 2$, let $\mathcal{K}$ be a k-cap contained in an ovaloid $\mathcal{O}$. Then, if $k \geq (q^2+q+4)/2$, any cap $\mathcal{K}'$ containing $\mathcal{K}$ lies entirely in $\mathcal{O}$.*

Proof: Suppose on the contrary that there exists some point Q in $\mathcal{K}'\backslash\mathcal{O}$. Each of the $q+1$ tangents from Q to $\mathcal{O}$ can meet $\mathcal{K}$ in at most one point. Hence there are at least $k-(q+1) \geq (q^2-q+2)/2$ lines through Q which meet $\mathcal{K}$ and are bisecants of $\mathcal{O}$. So $\mathcal{O}$ contains at least $q+1+2(q^2-q+2)/2 = q^2+3$ points, contradicting that $\mathcal{O}$ has only q^2+1 points. □

Corollary: *In $PG(3, q)$, let $\mathcal{K}$ be a k-cap contained in a quadric $\mathcal{F}$. Then, if*

$$k \geq (q^2+q+4)/2 \quad \text{for} \quad q \geq 5,$$

$$k \geq (q^2+q+6)/2 \quad \text{for} \quad q = 3, 4,$$

$\mathcal{F}$ is an elliptic quadric $\mathcal{E}_3$ and every cap containing $\mathcal{K}$ lies on $\mathcal{F}$.

Proof: For $q \geq 5$, we have that $2(q+2) < (q^2+q+4)/2$; for $q = 4$, we have $2(q+2) < 13 = (q^2+q+6)/2$; for $q = 3$, we have $2(q+1) < 9 = (q^2+q+6)/2$. So, by Lemma 18.1.7, the quadric $\mathcal{F}$ is an $\mathcal{E}_3$. Now, by Lemma 18.1.8, a cap containing $\mathcal{K}$ lies on $\mathcal{F}$. □

We now give a lower bound for the number of points on a complete cap, which can surely be improved.

Theorem 18.1.9: *If $\mathcal{K}$ is a complete k-cap in $PG(3, q)$, then $k > \sqrt{2q}+1$.*

Proof: Since $\mathcal{K}$ is complete, the parameter r_0 of Lemma 18.1.4 is zero. Equations (18.3) then become

$$r_1 + r_2 + \ldots + r_m = (q+1)(q^2+1) - k,$$

$$r_1 + 2r_2 + \ldots + mr_m = k(k-1)(q-1)/2.$$

As $\sum_1^m (i-1)r_i \geqslant 0$, so

$$k(k-1)(q-1)/2-(q+1)(q^2+1)+k \geqslant 0,$$
$$k^2(q-1)-k(q-3)-2(q+1)(q^2+1) \geqslant 0,$$

whence

$$k \geqslant \{q-3+\sqrt{[(q-3)^2+8(q^4-1)]}\}/[2(q-1)]=M.$$

A little manipulation shows that $M>\sqrt{2q}+1$. □

18.2 Examples of complete caps

In this section we give examples for all q of complete k-caps which are not ovaloids. In the subsequent sections we find how large k must be in order that the only complete cap containing a given k-cap is an ovaloid. For $q=2$, all complete caps are readily identified.

Theorem 18.2.1: *In $PG(3, 2)$, a complete k-cap $\mathcal{K}$ is one of the following:*
(i) *$k=8$ and $\mathcal{K}$ is the complement of a plane;*
(ii) *$k=5$ and $\mathcal{K}$ is an elliptic quadric.*

Proof: By Lemma 16.1.2, $k \leqslant 8$. By Theorem 16.1.3, $k=8$ if and only if $\mathcal{K}$ is the complement of a plane.

If $\mathcal{K}$ consists of four non-coplanar points, its bisecants contain ten points; as this is less than 15, so $\mathcal{K}$ is incomplete.

If $\mathcal{K}$ consists of five points no four of which are coplanar, then each of the ten bisecants of $\mathcal{K}$ contains a point not in $\mathcal{K}$; so $\mathcal{K}$ is complete. If $\mathcal{K}=\{\mathbf{U}_0, \mathbf{U}_1, \mathbf{U}_2, \mathbf{U}_3, \mathbf{U}\}$, then any quadric $\mathcal{F}=\mathbf{V}(F)$ containing $\mathcal{K}$ has the form $F=\sum'' a_{ij}x_ix_j$ with $\sum'' a_{ij}=0$. So of the six a_{ij}, there are (a) two non-zero, (b) four non-zero, or (c) six non-zero. In case (a) there are twelve $\Pi_2\mathcal{H}_1$ like $\mathbf{V}(x_0x_1+x_0x_2)$ and three $\mathcal{H}_3$ like $\mathbf{V}(x_0x_1+x_2x_3)$. In case (b) there are three $\Pi_2\mathcal{H}_1$ like $\mathbf{V}((x_0+x_1)(x_2+x_3))$ and twelve $\mathcal{H}_3$ like $\mathbf{V}((x_0+x_1+x_2)x_3+x_1x_2)$. In case (c), $\mathbf{V}(\sum'' x_ix_j)$ is an $\mathcal{E}_3$. So there is just one elliptic quadric $\mathcal{E}_3$ containing $\mathcal{K}$ and so $\mathcal{K}$ is $\mathcal{E}_3$. By definition $\mathcal{K}$ is also a 5-arc.

If $\mathcal{K}$ contains four points in a plane π, then the three points of $\pi\backslash\mathcal{K}$ form a line l. So, if $\mathcal{K}$ consists of these four points and one or two off π, then there is a plane π' through l containing no point of $\mathcal{K}$. As $k<8$ and $\mathcal{K}$ is contained in the complement of a plane, $\mathcal{K}$ is incomplete. If $k=7$ and is not in the complement of a plane, then $\mathcal{K}$ consists of four points in π, say, and one in each of the three planes through l. But then the join of any two of these points of $\mathcal{K}\backslash\pi$ cannot meet l and so contains a point of $\pi\cap\mathcal{K}$. So $\mathcal{K}$ is not a cap. □

Corollary: *In $PG(3, 2)$, any k-cap with four coplanar points, and in particular a 6-cap or a 7-cap, is contained in the complement of a plane.* □

Lemma 18.2.2: *In* $PG(3, 3)$, *an example of a complete* 8-*cap is*

$$\mathcal{K}_3 = \{\mathbf{U}_0, \mathbf{U}_1, \mathbf{U}_2, \mathbf{U}_3, \mathbf{U}, \mathbf{P}(1, 1, -1, -1), \mathbf{P}(1, -1, 1, -1), \mathbf{P}(1, -1, -1, 1)\}.$$

Proof: $PG(3, 3)$ has 40 points. On the edges of the tetrahedron of reference there are 16 points. This leaves four points on each face, which lie on the joins of the opposite vertex to the four points of $\mathcal{K}_3$ other than the vertices. So there are $40-16-16=8$ points on no face of the tetrahedron. Four of these are in $\mathcal{K}_3$ and the remaining four are the points not on a face on the joins of any vertex to the points of $\mathcal{K}_3$ other than the vertices. □

Lemma 18.2.3: *In* $PG(3, 4)$, *there exists* (i) *a complete* 12-*cap* $\mathcal{K}_4$; (ii) *a complete* 14-*cap* $\mathcal{K}'_4$.

Proof: (i) Firstly we show that in $PG(2, 4)$ every line meets one or other of two ovals (6-arcs) $\mathcal{K}$ and $\mathcal{K}'$ which have no points in common. As in § 14.3, an oval has fifteen bisecants and six external lines. The five lines through P in $\mathcal{K}$ are all bisecants of $\mathcal{K}$ and so comprise three bisecants and two external lines of $\mathcal{K}'$. Hence there are $6\times 3/2=9$ lines bisecant to both $\mathcal{K}$ and $\mathcal{K}'$, and $6\times 2/2=6$ lines bisecant to $\mathcal{K}$ and external to $\mathcal{K}'$; likewise there are six lines bisecant to $\mathcal{K}'$ and external to $\mathcal{K}$. Therefore there are $21-9-6-6=0$ lines external to both $\mathcal{K}$ and $\mathcal{K}'$.

Now take two planes π and π' in $PG(3, 4)$ containing ovals $\mathcal{K}$ and $\mathcal{K}'$ respectively such that, if $l=\pi\cap\pi'$, then l is external to both $\mathcal{K}$ and $\mathcal{K}'$. Let $\mathcal{K}_4=\mathcal{K}\cup\mathcal{K}'$; then $\mathcal{K}_4$ is a 12-cap. Every point of π is on a bisecant of $\mathcal{K}$ and every point of π' on a bisecant of $\mathcal{K}'$. If Q is some point of $PG(3, 4)$ not on π or π', then Q projects $\mathcal{K}$ to an oval $\mathcal{K}''$ in π' such that $l\cap\mathcal{K}''=\emptyset$. Either $\mathcal{K}''$ has a point in common with $\mathcal{K}'$, in which case Q lies on a bisecant of $\mathcal{K}_4$, or $\mathcal{K}''$ and $\mathcal{K}'$ have no common points, in which case l meets $\mathcal{K}'$ or $\mathcal{K}''$ and this has already been excluded.

(ii) Let $E=(1, 1, 1, 1)$, $E_0=(1, 0, 0, 0), \ldots, E_3=(0, 0, 0, 1)$. Define

$$P_i=\mathbf{P}(\omega E_i+E), \qquad P_{ij}=\mathbf{P}(\omega E_i+\omega E_j+E), \qquad Q_i=\mathbf{P}(E_i+\omega E).$$

Then

$$\mathcal{K}'_4=\{P_0, P_1, P_2, P_3, P_{01}, P_{02}, P_{03}, P_{12}, P_{13}, P_{23}, Q_0, Q_1, Q_2, Q_3\}. \quad \square$$

Theorem 18.2.4: *In* $PG(3, q)$, *there exists a complete* k-*cap* $\mathcal{K}$ *with*

$$(q^2+q+4)/2\leqslant k\leqslant(q^2+3q+6)/2.$$

Proof: Let Q be a point off the elliptic quadric $\mathcal{E}_3$. From Theorem 15.3.9(d), there are $q(q+1)/2$ external lines, $q+1$ tangents and $q(q-1)/2$ bisecants of $\mathcal{E}_3$ through Q. The points of contact of the $q+1$ tangents lie on a conic $\mathcal{C}$ in the polar plane π of Q. Let $\mathcal{K}_1$ be the set of points consisting of Q, the points of $\mathcal{C}$ and one point of $\mathcal{E}_3$ on each of the bisecants through Q. Then $\mathcal{K}_1$ is a k_1-cap with

$$k_1 = 1 + (q+1) + q(q-1)/2 = (q^2+q+4)/2.$$

Then either $\mathcal{K}_1$ is complete or is contained in a complete k-cap $\mathcal{K}$ where $k = k_1 + r$ and $\mathcal{K}\backslash\mathcal{K}_1 = \{Q_1, \ldots, Q_r\}$. Each line QQ_i is external to $\mathcal{E}_3$ and the polar plane π_i of Q_i meets $\mathcal{E}_3$ in a conic $\mathcal{C}_i$, which meets $\mathcal{C}$ in the distinct points P_i and P_i'; the polar of QQ_i is P_iP_i'. As Q_i is not on any bisecant of $\mathcal{K}_1$, the $k_1 - 1$ joins of Q_i to the points of $\mathcal{K}_1\backslash\{Q\}$ are all distinct and therefore consist of the $q+1$ tangents and $q(q-1)/2$ bisecants through Q_i. Hence the $q-1$ points of $\mathcal{C}_i\backslash\{P_i, P_i'\}$ all lie on the bisecants through Q and are in $\mathcal{K}_1$.

If, for $i \neq j$, the conics $\mathcal{C}_i$ and $\mathcal{C}_j$ do not touch on $\mathcal{C}$, then the four points P_i, P_i', P_j, P_j' are distinct. We will now investigate the ways that $\mathcal{C}_i$ and $\mathcal{C}_j$ might touch on $\mathcal{C}$.

Suppose that $\mathcal{C}_i$ and $\mathcal{C}_j$ have a common point P on $\mathcal{C}$. Then the line $l = \pi_i \cap \pi_j$ meets $\mathcal{E}_3$ at P and possibly another point P'. There are various possibilities for l and P'.

(i) If $l = \pi_i \cap \pi_j$ is a tangent to $\mathcal{E}_3$, then its polar Q_iQ_j is also tangent to $\mathcal{E}_3$ at P, by Theorem 15.3.10. So the three points P, Q_i, and Q_j of $\mathcal{K}$ are collinear, which is impossible.

(ii) If P' is a point of $\mathcal{C}\backslash\{P\}$, then the three planes π, π_i, and π_j all contain l; so their poles Q, Q_i, and Q_j are collinear on the polar of l, which again contradicts that $\mathcal{K}$ is a cap.

(iii) If P' is not on $\mathcal{C}$, then the conics $\mathcal{C}_i$ and $\mathcal{C}_j$ touch in neither P nor P'. So, consider the cone $Q\mathcal{C}_i$. If the tangent line at P to $\mathcal{C}_i$ is l_i, then the tangent plane at P to $Q\mathcal{C}_i$ is Ql_i. Since P lies on $\mathcal{C}$, it is conjugate to Q and so Ql_i is also the tangent plane to $\mathcal{E}_3$ at P. As P' is not on $\mathcal{C}$, the tangent planes to $Q\mathcal{C}_i$ and $\mathcal{E}_3$ at P' are distinct. So the cone $Q\mathcal{C}_i$ meets the conic $\mathcal{C}_j$ twice at P and once at P', and so at a further point R_j.

(a) Let R_j be distinct from P and P'. Then the generator QR_j of $Q\mathcal{C}_i$ meets $\mathcal{C}_i$ in a point R_i distinct from R_j, P, and P'. If either R_i or R_j lies on $\mathcal{C}$, then R_iR_j touches $\mathcal{E}_3$ at this point since R_iR_j contains Q; so both R_i and R_j coincide with this point. As the construction excludes this, neither R_i nor R_j lies on $\mathcal{C}$. However, since R_i is on $\mathcal{C}_i$ and R_j on $\mathcal{C}_j$, they are both points of $\mathcal{K}_1$ collinear with Q: a contradiction.

(b) Now we have that R_j coincides with P or P'. If $R_j = P'$, then the tangent plane at P' to $\mathcal{E}_3$ contains Q and so P' is on $\mathcal{C}$, contrary to the hypothesis. Therefore $R_j = P$ and so the cone $Q\mathcal{C}_i$ meets $\mathcal{E}_3$ in $\mathcal{C}_i$ and a second conic $\mathcal{C}_i'$ tangent to $\mathcal{C}_j$ at P. Hence, as in (i), the pole Q_i' of the plane of $\mathcal{C}_i'$ is collinear with Q_j and P. So PQ_i' meets $\mathcal{K}$ in at most one point Q_j.

We conclude therefore that a point P on $\mathcal{C}$ can belong to at most two distinct conics $\mathcal{C}_i$. So among the $2r$ points $P_1, P_1', \ldots, P_r, P_r'$ on $\mathcal{C}$, at least r are distinct. So $r \leq q+1$ and $k = k_1 + r \leq (q^2+3q+6)/2$. □

Corollary: *In $PG(3, q)$, $q>2$, there exists a complete cap which is not an ovaloid.*

Proof: For $q>4$, $(q^2+3q+6)/2<q^2+1$: so the theorem gives an example. For $q=3$ and $q=4$, Lemmas 18.2.2 and 18.2.3 give examples. □

Theorem 18.2.5: *In $PG(3, q)$ with $q \equiv -1 \pmod 4$, there exists a complete k-cap $\mathcal{K}$ with $k=(q^2+q+4)/2$.*

Proof: We show that, in the previous theorem, $\mathcal{K}_1$ can be chosen so that $\mathcal{K}=\mathcal{K}_1$.

In Theorem 9.4.4, complete plane $(q+5)/2$-arcs were constructed. Let A_0A_1 and A_0A_2 be tangents to the conic $\mathcal{P}_2$ with A_1 and A_2 on $\mathcal{P}_2$. Then $\mathcal{P}_2 \backslash \{A_1, A_2\} = \mathcal{S} \cup \mathcal{N}$ where $|\mathcal{S}|=|\mathcal{N}|=(q-1)/2$ and each bisecant of $\mathcal{P}_2$ through A_0 contains a point of $\mathcal{S}$ and a point of $\mathcal{N}$. It was shown how to choose $\mathcal{S}$ and $\mathcal{N}$ so that both $\mathcal{S} \cup \{A_0, A_1, A_2\}$ and $\mathcal{N} \cup \{A_0, A_1, A_2\}$ were complete.

Let P_0 be a point of $\mathcal{C}$, and let π_0 and l_0 be respectively the tangent plane to $\mathcal{E}_3$ and tangent line to $\mathcal{C}$ at P_0. Let π' be another plane through l_0 and let $\mathcal{C}'$ be the conic $\pi' \cap \mathcal{E}_3$. Let π'_i, $i \in \mathbf{N}_q$, be the planes other than π_0 through QP_0; then $\pi'_i \cap \mathcal{E}_3$ is a conic $\mathcal{C}'_i$, by Theorem 15.3.9(f). As $\mathcal{C}'_i$ is tangent to neither $\mathcal{C}$ nor $\mathcal{C}'$, let $\mathcal{C}'_i \cap \mathcal{C} = \{P_0, M_i\}$ and $\mathcal{C}'_i \cap \mathcal{C}' = \{P_0, N_i\}$.

Each $\mathcal{C}'_i$ determines two complete plane $(q+5)/2$-arcs $\mathcal{K}_i$ and $\mathcal{K}'_i$ by taking $A_0=Q$, $A_1=P_0$, $A_2=M_i$; let $\mathcal{K}_i$ be the arc containing N_i. Then $\{Q\} \cup (\bigcup_i \mathcal{K}_i)$ is a k_1-cap $\mathcal{K}_1$ with $k_1=(q^2+q+4)/2$, as in the previous theorem: also $\mathcal{K}_1$ contains both $\mathcal{C}$ and $\mathcal{C}'$.

A point on l_0 or QP_0 is on a bisecant of $\mathcal{K}_1$. If P is a point not in π_0, then there is a bisecant through P of the arc $\mathcal{K}_i$ cut out on $\mathcal{K}$ by the plane QP_0P. Now take P in π_0 but not on l_0 or QP_0. Since P is not on l_0, it is not in π. So consider the cone $P\mathcal{C}$; it meets $\mathcal{E}_3$ in $\mathcal{C}$ and another conic $\mathcal{C}_0$ containing P_0. Since P is not on QP_0, the conic $\mathcal{C}_0$ is not tangent to $\mathcal{C}$ at P_0; this also means that $\mathcal{C}_0$ is not tangent to $\mathcal{C}'$ at P_0. However, $\mathcal{C}_0$ and $\mathcal{C}'$ will have a common point P' other than P_0. Then the line PP' meets $\mathcal{C}$ and $\mathcal{C}'$ in distinct points and so is a bisecant of $\mathcal{K}_1$. Hence $\mathcal{K}=\mathcal{K}_1$. □

18.3 Caps in ovaloids for q even

In this section we obtain, for q even, a numerical condition that the only complete cap containing a given k-cap is an ovaloid. The results depend very much on Theorem 10.3.3, which stated that, in $PG(2, q)$ with q even, the only complete arc containing a given k-arc is an oval if $k>q-\sqrt{q}+1$. We first require a lemma which might more properly have been included in § 16.1, but is given here since it depends on the not-so-elementary Theorem 10.3.1. As in § 16.1, let $\sigma_2(Q)$ and $\sigma_1(Q)$ be the respective numbers of bisecants and tangents to a k-cap $\mathcal{K}$ through a point Q off $\mathcal{K}$; also, t is the number of tangents through a point P in $\mathcal{K}$.

Lemma 18.3.1: *When $\mathcal{K}$ is a k-cap and Q is a point off $\mathcal{K}$ such that $\sigma_2(Q) \geq 1$, then $\sigma_1(Q) \leq t$.*

Proof: As $\sigma_2(Q) \geq 1$, there exist P_1 and P_2 in $\mathcal{K}$ such that QP_1P_2 is a bisecant. Let π_i, $i = 1, \ldots, q+1$, be the planes through QP_1P_2. Then $\pi_i \cap \mathcal{K} = \mathcal{K}_i$ is a k_i-arc with $2 \leq k_i \leq q+2$.

Now, let $\sigma_1^{(i)}(Q)$ be the number of tangents through Q to $\mathcal{K}_i$. Then

$$\sum_i \sigma_1^{(i)}(Q) = \sigma_1(Q). \tag{18.4}$$

Also, let t_i be the number of tangents to $\mathcal{K}_i$ in π_i through P_1; then

$$\sum_i t_i = t. \tag{18.5}$$

Therefore, it suffices to show that, for all i,

$$\sigma_1^{(i)}(Q) \leq t_i. \tag{18.6}$$

If $k_i = 2$, then $\sigma_1^{(i)}(Q) = 1$ and $t_i = q$. If $k_i = q+2$, then $\sigma_1^{(i)}(Q) = 0$ and $t_i = 0$. So in both these cases, (18.6) holds.

Finally, let $3 \leq k_i \leq q+1$. Then, by Theorem 10.3.1, the tangents to $\mathcal{K}_i$ in π_i belong to an algebraic envelope Γ of class t_i. As Q lies on the bisecant P_1P_2 of $\mathcal{K}$, it also follows that the pencil through Q in π_i is not a component of Γ. Hence through Q there are at most t_i tangents of $\mathcal{K}$; that is, $\sigma_1^{(i)}(Q) \leq t_i$. So, in all cases (18.6) holds. Hence, by (18.4) and (18.5),

$$\sigma_1(Q) = \sum \sigma_1^{(i)}(Q) \leq \sum t_i = t. \quad \square$$

Theorem 18.3.2: *In $PG(3, q)$ with q even, a complete k-cap $\mathcal{K}$ with $k < q^2+1$ satisfies*

$$k \leq q^2 - \sqrt{q}/2 + 1.$$

Proof: Since, by Theorem 18.2.1, a complete k-cap for $q = 2$ has $k = 5$ or 8, the case of $q = 2$ may be disregarded.

As above, t is the number of tangents through a point of $\mathcal{K}$. Write $k = q+1-d$; then $t = q+1+d$. It must be shown that $d \geq \sqrt{q}/2$.

Take P in $\mathcal{K}$ and let the tangents at P be $l_1, l_2, \ldots, l_t$. Either (i) there exists a plane π_0 containing exactly one l_i, in which case $\mathcal{K} \cap \pi_0$ is a $(q+1)$-arc, or (ii) there exists no such plane.

(i) $\mathcal{K}_0 = \mathcal{K} \cap \pi_0$ is a $(q+1)$-arc. By Lemma 8.1.4, the arc $\mathcal{K}_0$ has a nucleus N. Then, by Lemma 18.3.1, we have $\sigma_1(N) \leq t = q+1+d$. The $q+1$ lines through N in π_0 are tangents to $\mathcal{K}$. So there are at most d tangents through N not in π_0.

As $\mathcal{K}$ is complete, there is at least one bisecant l of $\mathcal{K}$ through N. A plane π' through l meets $\mathcal{K}_0$ in a point P, where NP is tangent to $\mathcal{K}$.

Hence, if $\mathcal{K}' = \pi' \cap \mathcal{K}$ is a k'-arc, then $k' \leq q+1$. Let $k' = q+1-m'$; then $m' \geq 0$. So

$$k = q^2+1-d = 2+\sum_{\pi'} (q-1-m')$$
$$= q^2+1-\sum m';$$

whence

$$\sum m' = d, \qquad m' \leq d.$$

Let us fix P in $\mathcal{K}_0$ and let the planes through PN other than π_0 be π_i, $i \in \mathbf{N}_q$. There are two cases:

(a) there exists among the π_i other than π_0, a plane π_1, say, such that no bisecant of $\mathcal{K}_1 = \mathcal{K} \cap \pi_1$ contains N;

(b) every plane π_i other than π_0 contains some bisecant of $\mathcal{K}_i = \mathcal{K} \cap \pi_i$ through N, where $\mathcal{K}_i$ is a k_i-arc.

(a) The lines through N to points of $\mathcal{K}_1$ other than P are all tangents and hence number at most d. Hence, if $\mathcal{K}_1$ is a k_1-arc, then $k_1 \leq d+1$. The number of bisecants to $\mathcal{K}_1$ in π_1 through P is $k_1 - 1 \leq d$. So the number of tangents to $\mathcal{K}$ through P in π_1 is

$$(q+1)-(k_1-1) \geq q+1-d.$$

There is no tangent to $\mathcal{K}$ in π_0 through P apart from PN. As there are $q+1+d$ tangents to $\mathcal{K}$ altogether, the number of tangents to $\mathcal{K}$ through P in neither π_0 nor π_1 is at most $2d$.

Let l' be one such tangent. Each of the $q+1$ planes π'_i through l' has at most $2d$ tangents at P to $\mathcal{K}$ in neither π_0 nor π_1. Let $\pi'_i \cap \mathcal{K} = \mathcal{K}'_i$, where $\mathcal{K}'_i$ is a k'_i-arc with $t'_i = q+2-k'_i$. If π'_i meets π_0 and π_1 in NP, then, this is also a tangent at P; so $t'_i \leq 2d+1$. If π'_i does not contain NP, it may meet π_1 but not π_0 in a tangent at P; so, again, $t'_i \leq 2d+1$. Therefore, in both cases, $k'_i \geq q-2d+1$.

Suppose that $d < \sqrt{q}/2$ so that $2d+1 < \sqrt{q}+1$ and $k'_i > q-\sqrt{q}+1$. Then, by Theorem 10.3.3, $\mathcal{K}'_i$ is contained in an oval which meets l in a point Q_i other than P such that Q_i is not in $\mathcal{K}$. As there are $q+1$ planes π'_i and only q possible points for the Q_i, two points coincide, say $Q_1 = Q_2 = Q$. The lines joining Q to the points of $\mathcal{K}'_1$ and $\mathcal{K}'_2$ are all tangent to them and hence to $\mathcal{K}$. Thus

$$\begin{aligned}\sigma_1(Q) &\geq 1+(k'_1-1)+(k'_2-1)\\ &\geq 1+2(q-2d)\\ &> q+d+1 \quad (\text{if } q>5d)\\ &= t.\end{aligned}$$

But as $d < \sqrt{q}/2$, so $q > 5d$ when $q > 5\sqrt{q}/2$; that is, $q > 4$. So, if $q > 4$, then $\sigma_1(Q) > t$, contradicting that $\mathcal{K}$ is complete, by Lemma 18.3.1. If

$q=4$, then $\sqrt{q}/2=1$ and $d=0$, contradicting that $k<q^2+1$. Therefore, in case (a), it has been shown that $d \geqslant \sqrt{q}/2$.

(b) In each plane π_i other than π_0, there is a line NP_iP_i' with P_i and P_i' in $\mathcal{K}_i$. Also

$$k_i = q+1-m_i = q+2-t_i \geqslant q+1-d;$$

so

$$m_i \leqslant d, \qquad t_i \leqslant d+1.$$

Suppose that $d<\sqrt{q}/2$ and so $d<\sqrt{q}$; therefore $k_i<q-\sqrt{q}+1$. Then, by Theorem 10.3.3, Corollary 2, $\mathcal{K}_i$ is contained in an oval $\mathcal{O}_i$ with $\mathcal{O}_i \cap PN = \{P, Q_i\}$. Also $Q_i \neq N$, as every line Q_iR for R in $\mathcal{K}_i$ is a tangent to $\mathcal{K}_i$, whereas NP_i is a bisecant. So there are q planes π_i other than π_0, but only $q-1$ possible points for the Q_i. So $Q_1=Q_2=Q$, say. Then the lines joining Q to the points of $\mathcal{K}_1$ and $\mathcal{K}_2$ are tangents to them and to $\mathcal{K}$. So

$$\begin{aligned}\sigma_1(Q) &\geqslant 1+(k_1-1)+(k_2-1)\\ &\geqslant 1+2(q-d)\\ &> q+1+d \quad (\text{if } q>3d)\\ &= t.\end{aligned}$$

Now, since $d<\sqrt{q}/2$, we have that $q>3d$ providing $q>3\sqrt{q}/2$; that is, $q>2$. So it has been shown that $\sigma_1(Q)>t$, which again contradicts Lemma 18.3.1. So, also in case (b), it has been shown that $d \geqslant \sqrt{q}/2$.

(ii) Let the $q+1$ planes through l_1, say, be $\pi_1, \ldots, \pi_{q+1}$. Then π_i contains r_i other lines l_j, where $r_i \geqslant 1$. Since $l_1, l_2, \ldots, l_t$ are all the tangents at P to $\mathcal{K}$, it follows that

$$1+\sum r_i = t = q+d+1;$$

so

$$\sum (r_i-1) = d-1, \qquad r_i \leqslant d.$$

If $\pi_i \cap \mathcal{K} = \mathcal{K}_i$, then $\mathcal{K}_i$ is a k_i-arc with t_i tangents at each point; then $t_i = r_i+1 \leqslant d+1$.

Suppose that $d<\sqrt{q}/2$ and so $d<\sqrt{q}$. Then $t_i \leqslant d+1<\sqrt{q}+1$ and $k_i=q+2-t_i \geqslant q+1-d>q-\sqrt{q}+1$. So, by Theorem 10.3.3, Corollary 2, $\mathcal{K}_i$ is contained in an oval $\mathcal{O}_i$, which meets l_1 in a point Q_i other than P. There are $q+1$ planes π_i but only q distinct points for the Q_i. So $Q_1=Q_2=Q$, say. The lines joining Q to the points of $\mathcal{K}_1$ and $\mathcal{K}_2$ are tangents to $\mathcal{K}$, whence

$$\begin{aligned}\sigma_1(Q) &\geqslant 1+(k_1-1)+(k_2-1)\\ &\geqslant 1+2(q-d)\\ &> q+d+1 \quad (\text{if } q>3d)\\ &= t.\end{aligned}$$

Since $d < \sqrt{q}/2$, we have that $q > 3d$ if $q > 3\sqrt{q}/2$; that is, $q > 2$. So $\sigma_1(Q) > t$, again a contradiction. Therefore, in all cases, $d \geq \sqrt{q}/2$. □

Theorem 18.3.3: *In $PG(3, q)$ with q even and $q > 2$, a k-cap $\mathscr{K}$ satisfying $q^2 - \sqrt{q}/2 + 1 < k < q^2 + 1$ is incomplete and lies in exactly one complete cap, which is an ovaloid.*

Proof: By the previous theorem, any complete cap containing $\mathscr{K}$ is an ovaloid. So, if $\mathscr{O}$ is an ovaloid containing $\mathscr{K}$, then any cap containing $\mathscr{K}$ lies in $\mathscr{O}$ providing $k \geq (q^2 + q + 4)/2$, by Lemma 18.1.8. However,

$$q^2 - \sqrt{q}/2 + 1 \geq (q^2 + q + 4)/2$$

when

$$q^2 \geq q + \sqrt{q} + 2,$$

which is true for $q > 2$. So the only complete cap containing $\mathscr{K}$ is $\mathscr{O}$. □

Corollary: *If $\mathscr{K}$ is a k-cap in $PG(3, q)$, q even and $q > 2$, such that either* (a) *at least one plane section of $\mathscr{K}$ is an oval, or* (b) *at least one point P of $\mathscr{K}$ has no pencil among the tangents to $\mathscr{K}$ at P, then $\mathscr{K}$ is not contained in an ovaloid and $k \leq q^2 - \sqrt{q}/2 + 1$.* □

18.4 Caps in ovaloids for q odd

To consider when a k-cap is necessarily contained in an elliptic quadric, there are several methods, whose difficulty depends on how good a bound on k is required. The best bound is obtained using the deep results of § 10.4; this however does not yield results for all q and will be examined later. Firstly we obtain some results for all odd q by elementary methods.

Theorem 18.4.1: *In $PG(3, q)$ with q odd and $q \geq 7$, a k-cap $\mathscr{K}$ with $k \geq q^2 - q + 7$ lies on a unique elliptic quadric.*

Proof: If $k = q^2 + 1$, then the result is that of Theorem 16.1.7. So let $k = q^2 + 1 - d$ with $1 \leq d \leq q - 6$.

Since $q + 1 - d \geq 7$, through any bisecant of $\mathscr{K}$ there are at least seven planes meeting $\mathscr{K}$ in a conic, by Lemma 18.1.6(ii). Let P_1 and P_2 be points of $\mathscr{K}$, and let π_1 and π_2 be planes meeting $\mathscr{K}$ in the respective conics $\mathscr{C}_1$ and $\mathscr{C}_2$. Take points Q_1 on $\mathscr{C}_1$ and Q_2 on $\mathscr{C}_2$ both different from P_1 and P_2. Let π_3 be a plane through Q_1Q_2 containing neither P_1 nor P_2 and neither of the tangents to $\mathscr{C}_1$ at Q_1 and to $\mathscr{C}_2$ at Q_2 such that $\pi_3 \cap \mathscr{K}$ is a conic $\mathscr{C}_3$. Let $\mathscr{C}_3 \cap \mathscr{C}_1 = \{Q_1, R_1\}$ and $\mathscr{C}_3 \cap \mathscr{C}_2 = \{Q_2, R_2\}$.

Let $\mathscr{F}$ be a quadric containing P_1, P_2, Q_1, Q_2, R_1, R_2, and a further point on each of $\mathscr{C}_1$, $\mathscr{C}_2$, and $\mathscr{C}_3$. Then each $\mathscr{C}_i$ contains five points of $\mathscr{F}$ and so lies entirely on $\mathscr{F}$.

Let R_3 be a point on $\mathscr{C}_3$ other than the five already considered. Take a plane π_4 through P_1R_3 meeting $\mathcal{K}$ in a conic $\mathcal{C}_4$, but containing none of P_2, Q_1, Q_2, R_1, and R_2. Then $\mathcal{C}_4$ contains P_1, R_3, and three other points of $\mathcal{F}$, one from each of $\mathcal{C}_1$, $\mathcal{C}_2$, and $\mathcal{C}_3$. So $\mathcal{C}_4$ lies on $\mathcal{F}$.

Similarly, if we begin with a different point R_4 on $\mathcal{C}_3$, there is a conic $\mathcal{C}_5$ containing P_1 and R_4 but none of the other six points P_2, Q_1, Q_2, R_1, R_2, and R_3, such that $\mathcal{C}_5$ lies on both $\mathcal{K}$ and $\mathcal{F}$.

Now let P be a point of $\mathcal{K}$ on none of the conics $\mathcal{C}_1$, $\mathcal{C}_2$, $\mathcal{C}_3$, $\mathcal{C}_4$, $\mathcal{C}_5$. The conics $\mathcal{C}_1$, $\mathcal{C}_2$, $\mathcal{C}_4$, and $\mathcal{C}_5$ all contain P_1 and so intersect in at most six other points. Therefore there is a plane through PP_1 containing none of these six points and meeting $\mathcal{K}$ in a conic $\mathcal{C}$. Then $\mathcal{C}$ contains P_1 and four other points of $\mathcal{F}$, one on each of the conics $\mathcal{C}_1$, $\mathcal{C}_2$, $\mathcal{C}_4$, and $\mathcal{C}_5$. So $\mathcal{C}$ lies on $\mathcal{F}$ as does P. So $\mathcal{K}$ is contained in $\mathcal{F}$. By the corollary to Lemma 18.1.8, $\mathcal{F}$ is elliptic and the only elliptic quadric containing $\mathcal{K}$. □

Only the values $q=3$ and $q=5$ were omitted from the Theorem 18.4.1. The following theorem completes the result for all odd q.

Theorem 18.4.2: *In $PG(3, q)$ with $q=3$ or $q=5$, every q^2-cap $\mathcal{K}$ is contained in a unique elliptic quadric.*

Proof: For $q=3$, the cap $\mathcal{K}$ has nine points and so lies on a quadric $\mathcal{F}$. By Lemma 18.1.8, Corollary, $\mathcal{F}$ is elliptic and unique.

Now let $q=5$. Then through any point P of $\mathcal{K}$ there are 24 bisecants and $31-24=7$ tangent lines. If no three of these seven tangents are coplanar, then the seven tangents meet a plane not through P in a 7-arc, contradicting Theorem 8.1.3. If a plane π containing three of the tangents at P contains another point of $\mathcal{K}$, then, by Lemma 18.1.6, it contains at least q points of $\mathcal{K}$. But a plane q-arc has two tangents at each point and a $(q+1)$-arc only one, by Lemma 8.1.1. So π contains no point of $\mathcal{K}$ other than P and therefore six of the seven tangents: it is called the *tangent plane* at P.

Let P' be another point of $\mathcal{K}$ with tangent plane π'. Let $\pi_1, \pi_2, \ldots, \pi_6$ be the six planes through PP'; five of these meet $\mathcal{K}$ in a conic and one in a 5-arc. Suppose $\pi_1 \cap \mathcal{K}$ is a conic $\mathcal{C}_1$. Let l_0 be a line through P in π but not π_1. Let π_0 be a plane through l_0 but not P' meeting $\mathcal{K}$ in a conic $\mathcal{C}_0$. Let $\mathcal{C}_0 \cap \mathcal{C}_1 = \{P, Q\}$ and let Q' and Q'' be other points of $\mathcal{C}_0$.

There is a quadric $\mathcal{F}$ through P, P', Q, Q', Q'' with π and π' as tangent planes at P and P' respectively. Then $\mathcal{F}$ contains both $\mathcal{C}_0$ and $\mathcal{C}_1$. If $\mathcal{F}$ contains lines, there is at least one line l of $\mathcal{F}$ in π'. So l meets π_1 in a point of $\mathcal{C}_1$ contradicting that π' is a tangent plane to $\mathcal{K}$. So $\mathcal{F}$ is elliptic.

If π_i is one of the planes through PP' but not l_0 meeting $\mathcal{K}$ in a conic $\mathcal{C}_i$, then $\mathcal{C}_i$ has the tangents $\pi_i \cap \pi$ at P and $\pi_i \cap \pi'$ at P', and contains the point other than P in which π_i meets $\mathcal{C}_0$. So $\mathcal{C}_i$ lies on $\mathcal{F}$.

It has been shown therefore that all points of $\mathcal{K}$ lie on $\mathcal{F}$ except perhaps those in $\pi_5 = P'l_0$ and π_6, where $\pi_6 \cap \mathcal{K}$ is a q-arc. Let R be any point of π_5, say, and consider the points of $\mathcal{F}$ on lines through R. Let $\mathcal{C}_5=\pi_5\cap\mathcal{F}$, let $\mathcal{F}_6$ be the points on $\mathcal{F}$ on the lines joining R to $\pi_6\cap\mathcal{F}$ and let $\mathcal{F}_6'$ be the points of contact on $\mathcal{F}$ of the tangents through R. Then $|\mathcal{C}_5|=q+1$, $|\mathcal{F}_6|\leq 2(q+1)$ and $|\mathcal{F}_6'|=q+1$. So $\mathcal{F}'=\mathcal{F}\backslash(\mathcal{C}_4\cup\mathcal{F}_5\cup\mathcal{F}_5')$ contains at least $(q^2+1)-(q+1)-2(q+1)-(q+1)+2=q^2-4q-1=4$ points; the points of $\mathcal{F}'$ belong to $\mathcal{F}\cap\mathcal{K}$ and lie in pairs on lines through R. So every point of π_5 not on $\mathcal{F}$ is not on $\mathcal{K}$. For R in π_6, the argument is the same. So $\mathcal{K}$ is contained in $\mathcal{F}$. By the corollary to Lemma 18.1.8, any other elliptic quadric containing $\mathcal{K}$ lies on $\mathcal{F}$. □

During the remainder of this section, an improvement on Theorem 18.4.1 will be given for large q. We wish to show that if $\mathcal{K}$ is a k-cap with k sufficiently large, then $\mathcal{K}$ lies in a unique elliptic quadric. The 'sufficiently large' can be presented in two ways:

(i) $k\geq q^2-cq\sqrt{q}$ for $q\geq q_0$, where c is some constant with $0\leq c<\frac{1}{4}$ and q_0 is an unspecified constant dependent on c;

(ii) $k\geq q^2-q\sqrt{q}/4+R(q)$ for $q\geq q_1$, where $R(q)=a_0q+a_1\sqrt{q}+a_2$ and q_1 will be determined during the proof. In fact, q_1 is taken large enough that $q^2-q\sqrt{q}/4+R(q)\leq q^2+1$.

Way (i) brings out the essential, asymptotic nature of the result. However, way (ii) will be preferred as it is an exact result from which (i) follows immediately. Also, improvements in the details of the method may lead to improvements for $R(q)$ and q_1. Any improvement on $q^2-q\sqrt{q}/4$ will require a different method or an improvement in Theorem 10.4.4, which states that, if a k-arc $\mathcal{K}$ in $PG(2,q)$ with q odd satisfies $k>q-\sqrt{q}/4+7/4$, then $\mathcal{K}$ is contained in a unique conic.

Lemma 18.4.3: *In $PG(3,q)$ with q odd, let $\mathcal{C}$ and $\mathcal{C}'$ be conics in the respective planes π and π' such that the line $l=\pi\cap\pi'$ meets both $\mathcal{C}$ and $\mathcal{C}'$ in the points P and P'. Then the quadrics containing $\mathcal{C}$ and $\mathcal{C}'$ form a pencil $\mathcal{F}=\{\mathcal{F}_\rho\mid\rho\in\gamma^+\}$ such that, if n_2, n_3, n_4^+, and n_4^- are respectively the number of plane pairs, cones, hyperbolic quadrics, and elliptic quadrics in $\mathcal{F}$, then*

$$n_2=1,\qquad n_3=0\ \text{ or }\ 2,$$

$$n_4^+=(q-1-n_3)/2\geq(q-3)/2,\qquad n_4^-=(q+1-n_3)/2\geq(q-1)/2.$$

The plane pair in $\mathcal{F}$ is $\pi+\pi'$, and n_3 is 0 or 2 as a point P on l which is external to $\mathcal{C}$ is internal or external to $\mathcal{C}'$.

Proof: Let $\pi=\mathbf{u}_3$ and $\pi'=\mathbf{u}_2$; let $P=\mathbf{U}_0$ and $Q=\mathbf{U}_1$. Also let

$$\mathcal{C}=\mathbf{V}(x_3,\,x_2^2-x_0x_1)=\{\mathbf{P}(t^2,1,t,0)\mid t\in\gamma^+\};$$

then

$$\begin{aligned}\mathcal{C}'&=\mathbf{V}(x_2,\,bx_0x_3+cx_1x_3+dx_3^2-x_0x_1)\\&=\{\mathbf{P}(cs+d,\,s^2-bs,\,0,\,s-b)\mid s\in\gamma^+\}.\end{aligned}$$

Hence, if $\mathscr{F}_\rho = \mathbf{V}(F_\rho)$ is a quadric containing $\mathscr{C}$ and $\mathscr{C}'$, then for some ρ in γ^+,

$$F_\rho = (x_2^2 - x_0x_1 + bx_0x_3 + cx_1x_3 + dx_3^2) + \rho x_2x_3.$$

So, with $F^{(i)} = \partial F_\rho/\partial x_i$,

$$F^{(0)} = -x_1 + bx_3, \qquad F^{(1)} = -x_0 + cx_3, \qquad F^{(2)} = 2x_2 + \rho x_3,$$
$$F^{(3)} = bx_0 + cx_1 + 2dx_3 + \rho x_2.$$

So $\mathscr{F}_\rho$ is a plane pair when $\rho = \infty$ and otherwise has a singular point where $F^{(0)} = F^{(1)} = F^{(2)} = F^{(3)} = 0$, which occurs at $\mathbf{P}(2c, 2b, -\rho, 2)$ when $\rho^2 = 4(bc + d)$.

Hence $n_2 = 1$, and $n_3 = 2$ or 0 according as $bc + d$ is a square or not. Also $\mathbf{P}(1, -r, 0, 0)$ is external or internal to $\mathscr{C}$ according as r is a square or not, and is external or internal to $\mathscr{C}'$ according as $(bc + d)r$ is a square or not.

It remains to find n_4^+ and n_4^-. If Q and Q' denote respective points of $\mathscr{C}$ and $\mathscr{C}'$ other than P and P', there are $(q-1)^2$ lines QQ', each of which lies on exactly one $\mathscr{F}_\rho$. Also, every line on an irreducible $\mathscr{F}_\rho$ through P or P' is some QQ'. Each cone contains $q-1$ lines QQ' and each hyperbolic quadric $2(q-1)$ lines QQ'. So

$$n_3(q-1) + 2n_4^+(q-1) = (q-1)^2,$$
$$n_3 + n_4^+ + n_4^- = q,$$

whence the result. □

Three conics $\mathscr{C}_1$, $\mathscr{C}_2$, and $\mathscr{C}_3$ in the respective planes π_1, π_2, and π_3 form a *tetrahedral system* if

(i) $\mathscr{C}_1 \cap \mathscr{C}_2 \cap \mathscr{C}_3 = \pi_1 \cap \pi_2 \cap \pi_3 = \{P\}$;

(ii) $\mathscr{C}_2 \cap \mathscr{C}_3 = \{P, P_1\}$, $\mathscr{C}_3 \cap \mathscr{C}_1 = \{P, P_2\}$, $\mathscr{C}_1 \cap \mathscr{C}_2 = \{P, P_3\}$ with P, P_1, P_2, P_3 not coplanar.

The tetrahedral system is *flat* if the tangents l_1, l_2, and l_3 to $\mathscr{C}_1$, $\mathscr{C}_2$, and $\mathscr{C}_3$ at P are coplanar.

Lemma 18.4.4: *The three conics of a tetrahedral system are contained in a quadric if and only if the system is flat; the quadric is then unique and irreducible.*

Proof: If, in the above notation, the conics form a tetrahedral system, then by the previous lemma there is a pencil $\mathscr{F}$ of quadrics $\mathscr{F}_\rho$ through $\mathscr{C}_1$ and $\mathscr{C}_2$. If the system is flat, each $\mathscr{F}_\rho$ has the tangent plane $l_1l_2l_3$ at P. So such an $\mathscr{F}_\rho$ is tangent to $\mathscr{C}_3$ at P and contains the points P_1 and P_2 of $\mathscr{C}_3$. Hence a quadric $\mathscr{F}_\rho$ containing one further point of $\mathscr{C}_3$ contains $\mathscr{C}_3$ and is necessarily irreducible. As two quadrics cannot intersect in a curve of degree more than four, the quadric is unique.

If there is a quadric $\mathscr{F}_\rho$ containing the conics $\mathscr{C}_1$, $\mathscr{C}_2$, and $\mathscr{C}_3$ of a tetrahedral system, then $\mathscr{F}_\rho$ has a simple point at $P=\mathscr{C}_1\cap\mathscr{C}_2\cap\mathscr{C}_3$. So the tangent plane π at P to $\mathscr{F}_\rho$ meets π_1, π_2, and π_3 in the tangents l_1, l_2, and l_3 to the conics; so l_1, l_2, and l_3 are coplanar. □

Lemma 18.4.5: *In $PG(3,q)$ with $q>13$, let $\mathscr{K}_i$ be a k_i-arc with $k_i>q-\sqrt{q}/4+7/4$ and so contained in a conic $\mathscr{C}_i$, $i=1, 2, 3$. If $\mathscr{C}_1$, $\mathscr{C}_2$, and $\mathscr{C}_3$ form a non-flat tetrahedral system, then there is a line meeting $\mathscr{K}_1$, $\mathscr{K}_2$, and $\mathscr{K}_3$ in distinct points. So $\mathscr{K}_1\cup\mathscr{K}_2\cup\mathscr{K}_3$ is not a cap.*

Proof: By Lemma 18.4.3, there is a pencil $\mathscr{F}$ of quadrics $\mathscr{F}_\rho$ through $\mathscr{C}_1$ and $\mathscr{C}_2$. Any such $\mathscr{F}_\rho$ has in common with $\mathscr{C}_3$ the points P, P_1, and P_2, where, as above, $\mathscr{C}_i\cap\mathscr{C}_j=\{P, P_k\}$. As the system is not flat, there is exactly one $\mathscr{F}_\rho$ tangent to $\mathscr{C}_3$ at P, one $\mathscr{F}_\rho$ tangent to $\mathscr{C}_3$ at P_1 and one $\mathscr{F}_\rho$ tangent to $\mathscr{C}_3$ at P_2. As there are $n_4^+\geqslant(q-3)/2$ hyperbolic quadrices in $\mathscr{F}$, at least $(q-3)/2-3=(q-9)/2$ hyperbolic quadrics in $\mathscr{F}$ meet $\mathscr{C}_3$ in a point Q other than P, P_1, and P_2. Let $\mathscr{M}$ be the set of such points Q; then $m=|M|\geqslant(q-9)/2$.

If A, B, and C are three sets such that $C\supset A\cup B$, then, since $|A\cap B|=|A|+|B|-|A\cup B|$,

$$|A\cap B|\geqslant|A|+|B|-|C|. \tag{18.7}$$

This result will be frequently used until the end of the section.

Let $\mathscr{K}'_i=\mathscr{K}_i\backslash\{P, P_1, P_2\}$ and put $r_i=|\mathscr{K}'_i|$; then $r_3>q-\sqrt{q}/4-5/4$ and, for $i=1$ and 2, $r_i>q-\sqrt{q}/4-1/4$. Also, let $\mathscr{M}'=\mathscr{M}\cap\mathscr{K}'_3$ and $m'=|\mathscr{M}'|$; finally, put $\mathscr{C}'_3=\mathscr{C}_3\backslash\{P, P_1, P_2\}$ and so $|\mathscr{C}'_3|=q-2$. Then, by (18.7),

$$\begin{aligned} m'&\geqslant m+r_3-(q-2)\\ &>(q-9)/2+(q-\sqrt{q}/4-5/4)-(q-2)\\ &=(2q-\sqrt{q}-15)/4. \end{aligned}$$

If R_3 is a point of $\mathscr{M}'$, then R_3 is the meet of two generators g and g$'$ of an $\mathscr{H}_3$ in $\mathscr{F}$. Also g and g$'$ meet $\mathscr{C}_2$ in distinct points R_2 other than P and P_1, and meet $\mathscr{C}_1$ in distinct points R_1 other than P and P_2. If g or g$'$ contains P_3, it is coplanar with the tangents to $\mathscr{C}_1$ and $\mathscr{C}_2$ at P_3; these tangents are distinct and their plane meets $\mathscr{C}_3$ in at most two points. If R_3 is not of this type, then g and g$'$ both meet each $\mathscr{C}_i$ in some point other than P, P_1, P_2, and P_3. There are at least $m'-2$ such R_3 each giving two different R_2, each of which is obtained from at most two R_3. Hence the number of such R_2 is

$$\begin{aligned} &\geqslant m'-2\\ &>(2q-\sqrt{q}-23)/4, \end{aligned}$$

which is positive for $q \geq 14$. Hence the number of R_2 also in $\mathcal{K}_2'$ is

$$>(2q-\sqrt{q}-23)/4+(4q-\sqrt{q}-1)/4-(q-2)$$
$$=(q-\sqrt{q}-8)/2,$$

which is positive for $q \geq 13$. Each of these R_2 lies on at least one line R_2R_3 meeting $\mathscr{C}_2$ and $\mathscr{C}_3$ in points of $\mathcal{K}_2$ and $\mathcal{K}_3$, and meeting $\mathscr{C}_1$ in a point R_1. Each such R_1 arises from at most two R_2. So the number of R_1 in $\mathcal{K}_1'$ is

$$>(q-\sqrt{q}-8)/4+(q-\sqrt{q}/4-1/4)-(q-2)$$
$$=(q-2\sqrt{q}-1)/4,$$

which is positive for $q \geq 7$.

Any one of the corresponding lines $R_1R_2R_3$ meets $\mathcal{K}_1$, $\mathcal{K}_2$, and $\mathcal{K}_3$ in distinct points. □

Lemma 18.4.6: *Let $\mathcal{K}$ be a k-cap in $PG(3, q)$, q odd and $q \geq 67$, with*

$$k > q^2 - q\sqrt{q}/4 + R(q)$$

and

$$R(q) = (31q + 14\sqrt{q} - 53)/16$$
$$= 2q - [(\sqrt{q}-7)^2 + 4]/16. \tag{18.8}$$

Let l be any line such that $|l \cap \mathcal{K}| \geq 1$ and let the $q+1$ planes through l be π_i, $i \in \mathbf{N}_{q+1}$. If $\mathcal{K}_i = \pi_i \cap \mathcal{K}$ is a k_i-arc, and, if $k_i > q - \sqrt{q}/4 + 7/4$ only for $i = 1, \ldots, n$, then $n > 3(\sqrt{q}+9)/4$.

Proof: If $|l \cap \mathcal{K}| = r$, then $r = 1$ or 2. As $k_i \leq q+1$ for $i = 1, \ldots, n$ and $k_i \leq q - \sqrt{q}/4 + 7/4$ for $i = n+1, \ldots, q+1$, so

$$\begin{aligned} k &= r + \sum (k_i - r) \\ &\leq \sum k_i - q \\ &\leq n(q+1) + (q+1-n)(q-\sqrt{q}/4+7/4) - q \\ &= n(\sqrt{q}-3)/4 + q^2 - q\sqrt{q}/4 + (7q - \sqrt{q} + 7)/4. \end{aligned}$$

If the lemma is false and $n \leq 3(\sqrt{q}+9)/4$, then

$$\begin{aligned} k &\leq q^2 - q\sqrt{q}/4 + (31q + 14\sqrt{q} - 53)/16 \\ &= q^2 - q\sqrt{q}/4 + R(q), \end{aligned}$$

contradicting (18.8). □

Notes: (i) It is necessary to take $q \geq 67$ only so that $q^2 - q\sqrt{q}/4 + R(q) < q^2 + 1$. It only requires $q \geq 11$ for $q - \sqrt{q}/4 + 7/4 < q+1$, and $q \geq 9$ for $3(\sqrt{q}+9)/4 < q+1$.

(ii) The number $3(\sqrt{q}+9)/4$ is the smallest that is large enough for both the next lemma and the final theorem.

(iii) This is the lemma that determines $R(q)$; it is the smallest value consistent with the required lower bound for n.

Lemma 18.4.7: *If $\mathcal{K}$ is a k-cap in $PG(3, q)$, q odd and $q \geq 67$, with $k > q^2 - q\sqrt{q}/4 + R(q)$, then there exist conics $\mathcal{C}_1$, $\mathcal{C}_2$, and $\mathcal{C}_3$ forming a flat tetrahedral system such that, if $\mathcal{K}_i = \mathcal{K} \cap \mathcal{C}_i$ is a k_i-arc, then $k_i > q - \sqrt{q}/4 + 7/4$ for $i = 1, 2, 3$.*

Proof: As $k > 1$, take two distinct points P and P_3 in $\mathcal{K}$. By the previous lemma, there are $n > 3(\sqrt{q}+9)/4$ planes π_i through PP_3 such that $\mathcal{K} \cap \pi_i = \mathcal{K}_i$ is a k_i-arc with $k_i > q - \sqrt{q}/4 + 7/4$.

Take two such planes π_1 and π_2; then $\mathcal{K}_1$ and $\mathcal{K}_2$ contain P and P_3. Also $\mathcal{K}_1$ is contained in a conic $\mathcal{C}_1$ and $\mathcal{K}_2$ in a conic $\mathcal{C}_2$, by Theorem 10.4.4; hence $\mathcal{K} \cap \mathcal{C}_1 = \mathcal{K}_1$ and $\mathcal{K} \cap \mathcal{C}_2 = \mathcal{K}_2$.

Let P_1 be a point of $\mathcal{K}_2$ other than P and P_3. Through PP_1 there are more than $3(\sqrt{q}+9)/4$ planes π'_i such that $\mathcal{K} \cap \pi'_i = \mathcal{K}'_i$ is a k'_i-arc with $k'_i > q - \sqrt{q}/4 + 7/4$. One of these planes, namely π_2, passes through P_3 and at most one is tangent to $\mathcal{C}_1$ at P. Therefore there are more than $3(\sqrt{q}+9)/4 - 2 = (3\sqrt{q}+19)/4$ planes π'_i each meeting $\mathcal{C}_1$ in a distinct point P_2 other than P and P_3. If there are m of these points P_2, then $m > (3\sqrt{q}+19)/4$. So the number of points P_2 in $\mathcal{K}_1$ is, again by (18.7),

$$\begin{aligned} &\geq m + k_1 - (q+1) \\ &> (3\sqrt{q}+19)/4 + (q - \sqrt{q}/4 + 7/4) - (q+1) \\ &= (\sqrt{q}+11)/2. \end{aligned}$$

Therefore there exists P_2 in $\mathcal{K}_1$ and so the plane $\pi_3 = PP_1P_2$ is one of the planes π'_i. Put $\mathcal{K} \cap \pi_3 = \mathcal{K}_3$; then $\mathcal{K}_3$ is contained in a conic $\mathcal{C}_3$. The three conics $\mathcal{C}_1$, $\mathcal{C}_2$, and $\mathcal{C}_3$ constitute a tetrahedral system. If the system were not flat, then $\mathcal{K}_1 \cup \mathcal{K}_2 \cup \mathcal{K}_3$ would not be a cap, by Lemma 18.4.5. However, as $\mathcal{K}_1 \cup \mathcal{K}_2 \cup \mathcal{K}_3$ is contained in $\mathcal{K}$, it is a cap and therefore the system is flat. □

Theorem 18.4.8: *In $PG(3, q)$ with q odd and $q \geq 67$, if $\mathcal{K}$ is a k-cap with $k > q^2 - q\sqrt{q}/4 + 2q$, then the only complete cap containing $\mathcal{K}$ is an elliptic quadric. In fact, it suffices to take $k > q^2 - q\sqrt{q}/4 + R(q)$, where*

$$R(q) = (31q + 14\sqrt{q} - 53)/16 = 2q - [(\sqrt{q}-7)^2 + 4]/16.$$

Proof: By Lemma 18.4.7, there exists a flat tetrahedral system of conics $\mathcal{C}_1$, $\mathcal{C}_2$, $\mathcal{C}_3$ such that $\mathcal{K} \cap \mathcal{C}_i = \mathcal{K}_i$ is a k_i-arc with $k_i > q - \sqrt{q}/4 + 7/4$. By Lemma 18.4.4, the three conics are contained in a unique quadric $\mathcal{F}$. It is now sufficient to show that $\mathcal{K} \subset \mathcal{F}$, for, as $k \geq (q^2+q+4)/2$, so by the corollary to Lemma 18.1.8, $\mathcal{F}$ is elliptic and the only complete cap containing $\mathcal{K}$.

Suppose there is some point Q in $\mathcal{K}\backslash\mathcal{F}$. As before, $\mathscr{C}_1$, $\mathscr{C}_2$, and $\mathscr{C}_3$ have the point P in common and have the residual intersections P_1, P_2, and P_3. So Q is on no $\mathscr{C}_i$ and $Q \neq P, P_1, P_2, P_3$.

Let m_1 be the number of planes π_i' through PQ meeting $\mathscr{C}_1$, apart from P, in a point Q_1 of $\mathcal{K}_1$ with Q_1 distinct from P, P_2, P_3 and such that $\mathcal{K} \cap \pi_i' = \mathcal{K}_i'$ is a k_i'-arc with $k_i' > q - \sqrt{q}/4 + 7/4$.

Let m_2 be the number of these m_1 planes π_i' meeting $\mathscr{C}_2$ in a point Q_2 of $\mathcal{K}_2$ with Q_2 distinct from P, P_1, P_3. Similarly, let m_3 be the number of these m_2 planes meeting $\mathscr{C}_3$ in a point Q_3 of $\mathcal{K}_3$ with Q_3 distinct from P, P_1, P_2.

It will be shown that $m_3 > 0$. Firstly we note that $|\mathcal{K}_1\backslash\{P, P_2, P_3\}| > q - \sqrt{q}/4 - 5/4$. From Lemma 18.4.6, the number of planes π_i' with $k_i' > q - \sqrt{q}/4 + 7/4$ is more than $3(\sqrt{q}+9)/4$. So we obtain in succession the following, always using (18.7):

$$m_1 > (q - \sqrt{q}/4 - 5/4) + 3(\sqrt{q}+9)/4 - (q+1) = (\sqrt{q}+9)/2,$$
$$m_2 > (q - \sqrt{q}/4 - 5/4) + (\sqrt{q}+9)/2 - (q+1) = (\sqrt{q}+9)/4,$$
$$m_3 > (q - \sqrt{q}/4 - 5/4) + (\sqrt{q}+9)/4 - (q+1) = 0.$$

So there exists a plane π_0', say, containing P and Q, and meeting $\mathscr{C}_1$ in Q_1, $\mathscr{C}_2$ in Q_2 and $\mathscr{C}_3$ in Q_3 with each Q_i distinct from P, P_1, P_2, P_3, such that $\mathcal{K}_0' = \mathcal{K} \cap \pi_0'$ is contained in a conic $\mathscr{C}_0'$.

The four conics $\mathscr{C}_0'$, $\mathscr{C}_1$, $\mathscr{C}_2$, $\mathscr{C}_3$ form four tetrahedral systems. As $\mathscr{C}_0' \cup \mathscr{C}_1 \cup \mathscr{C}_2 \subset \mathcal{K}$, the system of $\mathscr{C}_0'$, $\mathscr{C}_1$, $\mathscr{C}_2$ is flat, by Lemma 18.4.5. The conic $\mathscr{C}_0'$ has at Q a tangent coplanar with the tangents at Q to $\mathscr{C}_1$ and $\mathscr{C}_2$; this plane is the tangent plane to $\mathcal{F}$ at P. So $\mathscr{C}_0'$ is tangent to $\mathcal{F}$ at P and meets $\mathcal{F}$ in Q_1, Q_2 and Q_3. Therefore $\mathscr{C}_0'$ lies on $\mathcal{F}$ and, as Q is on $\mathscr{C}_0'$, so Q is on $\mathcal{F}$. This contradiction completes the proof. □

Corollary: *In $PG(3, q)$, q odd and $q \geq 67$, if $\mathcal{K}$ is a complete k-cap which is not an elliptic quadric, then*

$$k \leq q^2 - q\sqrt{q}/4 + (31q + 14\sqrt{q} - 53)/16$$
$$< q^2 - q\sqrt{q}/4 + 2q. \quad \square$$

Note: This theorem gives a better result than Theorem 18.4.1 for $q \geq 139$.

18.5 Notes and references

§ 18.1. Lemma 18.1.8 and its corollary are due to Segre (1959b).

§ 18.2. Lemmas 18.2.2 and 18.2.3(ii) come from Segre (1959a, § 23). Lemma 18.2.3(i) comes from Segre (1959b). Theorems 18.2.4 and 18.2.5 also follow Segre (1959a). A complete k-cap with $k = (q^2+q+4)/2$ has been constructed by Di Comite (1964a) for $q=2^h$ with h odd and $h>3$, by L. M. Abatangelo (1980) for all even q with $q>8$, and by L. M. Abatangelo and Pertichino (1982) for $q=8$; in the last article, a complete 39-cap is also constructed. For further results, see Di Comite (1965), Migliori (1981), L. M. Abatangelo (1982), V. Abatangelo (1984).

§ 18.3. This is entirely based on Segre (1967, Chapter IV).

§ 18.4. Theorems 18.4.1 and 18.4.2 are due to Barlotti (1956a); see, alternatively, Barlotti (1965). The remainder of the section follows Hirschfeld (1983a), which is a slight improvement on Segre (1967, Chapter V). See also Thas (1968b).

Added in proof: Theorem 18.3.2 has been improved by the author and J. A. Thas as follows.

Theorem 18.5.1: *In $PG(3,q)$, with q even, a complete k-cap for which $k<q^2+1$ satisfies $k\leq q^2-\frac{1}{2}q-\frac{1}{2}\sqrt{q}+2$.* □

19

HERMITIAN SURFACES

19.1 Basic properties

Throughout this chapter, $q = s^2$, so that $\sigma : \gamma \to \gamma$ given by $x^\sigma = x^s = \bar{x}$ is an involutory automorphism of γ. If $B = (b_{ij})$, then $\bar{B} = (\bar{b}_{ij})$. Hermitian varieties were defined in § 5.1. The variety $\mathbf{V}(F)$ is a Hermitian surface in $PG(3, q)$ if

$$F = \rho X A \bar{X}^* \quad \text{and} \quad \bar{A}^* = A,$$

where $X = (x_0, x_1, x_2, x_3)$.

Lemma 19.1.1: *The projectively distinct Hermitian surfaces* $\mathbf{V}(F)$ *in* $PG(3, q)$ *are as listed in Table* 19.1. $\square$

Table 19.1

Symbol	F	$\mathbf{V}(F)$
$\Pi_2 \mathcal{U}_{0,q}$	x_0^{s+1}	The plane $\mathbf{u}_0$
$\Pi_1 \mathcal{U}_{1,q}$	$x_0^{s+1} + x_1^{s+1}$	The joins of the $q+1$ points on $\mathbf{U}_2\mathbf{U}_3$ to the $s+1$ points of $\mathcal{U}_{1,q} = \mathbf{V}(x_0^{s+1} + x_1^{s+1}, x_2, x_3)$ on the line $\mathbf{U}_0\mathbf{U}_1$
$\Pi_0 \mathcal{U}_{2,q}$	$x_0^{s+1} + x_1^{s+1} + x_2^{s+1}$	The joins of $\mathbf{U}_3$ to the points of $\mathcal{U}_{2,q} = \mathbf{V}(x_0^{s+1} + x_1^{s+1} + x_2^{s+1}, x_3)$ in the plane $\mathbf{u}_3$
$\mathcal{U}_{3,q}$	$x_0^{s+1} + x_1^{s+1} + x_2^{s+1} + x_3^{s+1}$	The non-singular Hermitian surface.

$\mathcal{U}_{3,q}$ will also be denoted $\Pi_{-1}\mathcal{U}_{3,q}$, $\mathcal{U}_3$, and $\mathcal{U}$. The set of lines on $\mathcal{U}$ will be denoted by $\mathcal{U}^{(1)}$ and when there is special emphasis on the points of $\mathcal{U}$, it will be denoted by $\mathcal{U}^{(0)}$.

Lemma 19.1.2: *A section of a Hermitian surface* $\mathcal{W}$ *by a subspace* Π_r *of* $PG(3, q)$ *is a Hermitian variety in* Π_r.

Proof: If $\mathcal{W} = \mathbf{V}(F)$, then F is, up to a scalar, a Hermitian form and remains so if, for x_i, a linear form in the other x_j is substituted. $\square$

Lemma 19.1.3: *The numbers of points and lines on* $\Pi_0 \mathcal{U}_{2,q}$ *and* $\Pi_1 \mathcal{U}_{1,q}$

are as follows:

	Singular points	Non-singular points	Total points	Singular lines	Non-singular lines	Total lines
$\Pi_0\mathcal{U}_{2,q}$	1	$q(q\sqrt{q}+1)$	$q^2\sqrt{q}+q+1$	0	$q\sqrt{q}+1$	$q\sqrt{q}+1$
$\Pi_1\mathcal{U}_{1,q}$	$q+1$	$q(\sqrt{q}+1)(q+1)$	$(q+1)(q\sqrt{q}+q+1)$	1	$(q+1)(\sqrt{q}+1)$	$q\sqrt{q}+q+\sqrt{q}+2$

□

The rest of the chapter is devoted to properties of $\mathcal{U}=\mathcal{U}_{3,q}$.

Firstly, from § 1.4, we recall that $x^n=d$ has n roots in γ if $n'=(q-1)/n$ is an integer and $d^{n'}=1$. So, in particular, when $q=s^2$ and $n=s-1$, the equation has $s-1$ roots if $d^{s+1}=1$; that is,

$$\bar{x}/x=d \tag{19.1}$$

has $s-1$ non-zero solutions if $d\bar{d}=1$. Hence $\bar{x}=-x$ has a non-zero solution α. Thus

$$\mathcal{U}_{1,q}=\mathbf{V}(x_0\bar{x}_0+x_1\bar{x}_1)$$

can be transformed to

$$\mathbf{V}(x_0\bar{x}_1-x_1\bar{x}_0)$$

showing that $\mathcal{U}_{1,q}$ is a $PG(1,\sqrt{q})$ on $PG(1,q)$, as in Lemma 6.2.1. Similarly,

$$\mathcal{U}=\mathbf{V}(x_0\bar{x}_0+x_1\bar{x}_1+x_2\bar{x}_2+x_3\bar{x}_3)$$

can be transformed to

$$\mathcal{U}'=\mathbf{V}(x_0\bar{x}_1-x_1\bar{x}_0+x_2\bar{x}_3-x_3\bar{x}_2).$$

Associated to $\mathcal{U}$ is the polarity $\mathfrak{U}$ given by

$$\mathbf{P}(a_0,a_1,a_2,a_3)\leftrightarrow\boldsymbol{\pi}(\bar{a}_0,\bar{a}_1,\bar{a}_2,\bar{a}_3).$$

This is equivalent to saying that the points $\mathbf{P}(X)$ and $\mathbf{P}(Y)$ are conjugate under $\mathfrak{U}$ if $X\bar{Y}^*=0$, where

$$X\bar{Y}^*=x_0\bar{y}_0+x_1\bar{y}_1+x_2\bar{y}_2+x_3\bar{y}_3.$$

Lemma 19.1.4: (i) *The tangent plane at a point P of $\mathcal{U}$ is the polar plane of P, whence all tangent planes are distinct.*

(ii) *If P is on $\mathcal{U}$, then $P\mathfrak{U}\cap\mathcal{U}=P\mathcal{U}_{1,q}$, that is, the $\sqrt{q}+1$ lines joining P to a $\mathcal{U}_{1,q}$. If Q is off $\mathcal{U}$, then $Q\mathfrak{U}\cap\mathcal{U}=\mathcal{U}_{2,q}$, a non-singular Hermitian curve.*

(iii) *If P_1 and P_2 are points of $\mathcal{U}$, then P_1P_2 lies on $\mathcal{U}$ if and only if P_1 is conjugate to P_2.*

(iv) *If P is on $\mathcal{U}$ and Q is off $\mathcal{U}$, then $PQ\cap\mathcal{U}=P$ if and only if PQ lies in the tangent plane at P.*

Proof: (i) Let $P=\mathbf{P}(Y)$ be a point of $\mathcal{U}=\mathbf{V}(X\bar{X}^*)$. Then the tangent plane π at P is

$$\boldsymbol{\pi}\left(\frac{\partial}{\partial y_0}(Y\bar{Y}^*), \frac{\partial}{\partial y_1}(Y\bar{Y}^*), \frac{\partial}{\partial y_2}(Y\bar{Y}^*), \frac{\partial}{\partial y_3}(Y\bar{Y}^*)\right)$$
$$=\boldsymbol{\pi}(\bar{y}_0, \bar{y}_1, \bar{y}_2, \bar{y}_3)$$

which is the polar plane of P.

(ii) Consider $\mathcal{U}'=\mathbf{V}(x_0\bar{x}_1-x_1\bar{x}_0+x_2\bar{x}_3-x_3\bar{x}_2)$ and take $P=\mathbf{U}_0$. Then $P\mathfrak{U}'=\mathbf{u}_1$ meets $\mathcal{U}'$ in $\mathbf{V}(x_1, x_2\bar{x}_3-x_3\bar{x}_2)$, which is the set of joins of P to the $\sqrt{q}+1$ points of $\mathcal{U}_{1,q}=\mathbf{V}(x_0, x_1, x_2\bar{x}_3-x_3\bar{x}_2)$.

Now consider $\mathcal{U}=\mathbf{V}(X\bar{X}^*)$ and take $Q=\mathbf{U}_0$. Then $Q\mathfrak{U}=\mathbf{u}_0$ meets $\mathcal{U}$ in the Hermitian curve $\mathcal{U}_{2,q}=\mathbf{V}(x_0, x_1\bar{x}_1+x_2\bar{x}_2+x_3\bar{x}_3)$.

(iii) Let $P_1=\mathbf{P}(X)$ and $P_2=\mathbf{P}(Y)$ be points of $\mathcal{U}$. If P is on P_1P_2, then $P=\mathbf{P}(\lambda X+\mu Y)$. So P is on $\mathcal{U}$ if and only if

$$(\lambda X+\mu Y)\overline{(\lambda X+\mu Y)}^*=0;$$

that is,

$$\lambda\bar{\lambda}X\bar{X}^*+\bar{\lambda}\mu Y\bar{X}^*+\lambda\bar{\mu}X\bar{Y}^*+\mu\bar{\mu}Y\bar{Y}^*=0.$$

Hence

$$\lambda\bar{\mu}X\bar{Y}^*+\bar{\lambda}\mu Y\bar{X}^*=0. \tag{19.2}$$

If P_1P_2 lies on $\mathcal{U}$, then (19.2) holds for all λ and μ, whence $X\bar{Y}^*=0$; so P_1 and P_2 are conjugate. Conversely, if P_1 and P_2 are conjugate, then $X\bar{Y}^*=Y\bar{X}^*=0$ and (19.2) holds.

(iv) If $P=\mathbf{P}(X)$ and $\mathbf{Q}=P(Y)$, then PQ meets $\mathcal{U}$ in $\mathbf{P}(\lambda X+Y)$ if and only if

$$\lambda X\bar{Y}^*+\bar{\lambda}Y\bar{X}^*+Y\bar{Y}^*=0.$$

But $\bar{x}+x=c$ has $\sqrt{q}$ solutions for each c in $GF(\sqrt{q})$. So $PQ\cap\mathcal{U}=P$ if and only if $X\bar{Y}^*=0$. □

Theorem 19.1.5: *$\mathcal{U}=\mathcal{U}_{3,q}$ consists of $(q+1)(q\sqrt{q}+1)$ points lying on $(\sqrt{q}+1)(q\sqrt{q}+1)$ lines with $\sqrt{q}+1$ lines through each point. Each tangent plane meets $\mathcal{U}$ in $\sqrt{q}+1$ lines through the point of contact and each non-tangent plane meets $\mathcal{U}$ in a non-singular Hermitian curve.*

Proof: The second part follows from parts (i) and (ii) of Lemma 19.1.4. So, if l is a line on $\mathcal{U}$, each plane through l is a tangent plane to $\mathcal{U}$ and contains $q\sqrt{q}$ points off l. So $\mathcal{U}$ contains $q\sqrt{q}(q+1)+q+1=(q+1)(q\sqrt{q}+1)$ points altogether. Since there are $\sqrt{q}+1$ lines on $\mathcal{U}$ through a point of $\mathcal{U}$ and $q+1$ points on a line, $\mathcal{U}$ contains $(\sqrt{q}+1)(q\sqrt{q}+1)$ lines. □

Let $PGU(4, q)$ be the group of all projectivities fixing $\mathcal{U}$; that is, it consists of all projectivities $\mathfrak{T}$ with matrix T such that, if $\mathcal{U} = \mathbf{V}(XA\bar{X}^*)$, then $TA\bar{T}^* = \rho A$. In the canonical form $\mathcal{U} = \mathbf{V}(X\bar{X}^*)$ where $A = I$, we have $T\bar{T}^* = \rho I$.

Let $P\gamma U(4, q)$ be the set of all collineations $\mathfrak{T}$ fixing $\mathcal{U}$ which are either projectivities or associated to the involutory automorphism σ; in the latter case $\mathfrak{T} = \sigma\mathfrak{S}$ where $\mathfrak{S}$ is a projectivity with matrix S and so $\mathbf{P}(X)\mathfrak{T} = \mathbf{P}(X)\sigma\mathfrak{S} = \mathbf{P}(\bar{X})\mathfrak{S} = \mathbf{P}(\bar{X}S)$.

Let $P\Gamma U(4, q)$ be the group of all collineations fixing $\mathcal{U}$. So

$$PGU(4, q) < P\gamma U(4, q) < P\Gamma U(4, q)$$

and the respective indices are 2 and $h/2$. If $q = p^2$, then $P\gamma U(4, q) = P\Gamma U(4, q)$.

Lemma 19.1.6: *With q square,*

$$|PGU(4, q)| = q^3(q-1)(q\sqrt{q}+1)(q^2-1).$$

Proof: Although, in Theorem 5.1.3, the canonical form for a Hermitian variety was determined algebraically, it can also be done geometrically. In $PG(3, q)$ let $\mathcal{F}$ be a Hermitian surface such that $\mathbf{U}_0$, $\mathbf{U}_1$, $\mathbf{U}_2$, $\mathbf{U}_3$ are mutually conjugate, external points of $\mathcal{F}$. Then

$$\mathcal{F} = \mathbf{V}(x_0\bar{x}_0 + a_1x_1\bar{x}_1 + a_2x_2\bar{x}_2 + a_3x_3\bar{x}_3)$$

with a_1, a_2, a_3 in $GF(\sqrt{q})$. Hence there are $(\sqrt{q}+1)^3$ projectivities given by

$$x_0' = x_0, \qquad x_1' = t_1x_1, \qquad x_2' = t_2x_2, \qquad x_3' = t_3x_3$$

with $t_i\bar{t}_i = a_i$ $(i = 1, 2, 3)$ such that $\mathcal{F}$ is transformed to $\mathcal{U}$.

So it remains to determine the number of ordered sets of four points P_0, P_1, P_2, P_3 which are mutually conjugate and all external to $\mathcal{U}$. Firstly, P_0 can be chosen in $\theta(3) - |\mathcal{U}|$ ways, then P_1 can be chosen in $\theta(2) - (q\sqrt{q}+1)$ ways, then P_2 can be chosen in $\theta(1) - (\sqrt{q}+1)$ ways, and finally P_3 is determined. So

$$\begin{aligned}|PGU(4, q)| &= [(q^2+1)(q+1) - (q\sqrt{q}+1)(q+1)][(q^2+q+1)\\ &\quad - (q\sqrt{q}+1)][(q+1) - (\sqrt{q}+1)](\sqrt{q}+1)^3\\ &= (q^2 - q\sqrt{q})(q+1)(q^2 - q\sqrt{q}+q)(q-\sqrt{q})(\sqrt{q}+1)^3\\ &= q^3(q-1)(q\sqrt{q}+1)(q^2-1). \quad \square\end{aligned}$$

Corollary: (i) $|P\gamma U(4, q)| = 2q^3(q-1)(q\sqrt{q}+1)(q^2-1)$.

(ii) $|P\Gamma U(4, q)| = hq^3(q-1)(q\sqrt{q}+1)(q^2-1)$.

(iii) *The number of non-singular Hermitian surfaces in $PG(3, q)$ is $q^3(q+1)(q\sqrt{q}-1)(q^2+1)$.* $\square$

To discuss the different types of subspaces with respect to $\mathcal{U}_3$, we use a similar notation to that used for quadrics in § 17.5.

Let $\Pi_r(\mathcal{V})$ denote a subspace Π_r meeting $\mathcal{U}_3$ in a variety projectively equivalent to $\mathcal{V}$. Below we list all types of subspace together with the notation for the set of subspaces of each type.

$\Pi_0(\Pi_0)$ = point on $\mathcal{U}$:	Ψ_1
$\Pi_0(\Pi_{-1})$ = point off $\mathcal{U}$:	Ψ_0
$\Pi_1(\Pi_1)$ = line on $\mathcal{U}$:	$\mathcal{L}_3$
$\Pi_1(\mathcal{U}_1) = (\sqrt{q}+1)$-secant of $\mathcal{U}$:	$\mathcal{L}_2$
$\Pi_1(\Pi_0)$ = tangent line to $\mathcal{U}$:	$\mathcal{L}_1$
$\Pi_2(\mathcal{U}_2)$ = non-tangent plane:	Φ_3
$\Pi_2(\Pi_0\mathcal{U}_1)$ = tangent plane:	Φ_2

Theorem 19.1.7: *The action of $PGU(4, q)$ on the sets Π, $\mathcal{L}$, Φ partitions them into orbits as follows*:
(i) $\Pi = \Psi_0 \cup \Psi_1$; (ii) $\mathcal{L} = \mathcal{L}_1 \cup \mathcal{L}_2 \cup \mathcal{L}_3$; (iii) $\Phi = \Phi_2 \cup \Phi_3$. □

Theorem 19.1.8: *Under the polarity* $\mathfrak{U}$,
(i) $\Psi_0 \leftrightarrow \Phi_3$; (ii) $\Psi_1 \leftrightarrow \Phi_2$;
(iii) $\mathcal{L}_1 \leftrightarrow \mathcal{L}_1$; (iv) $\mathcal{L}_2 \leftrightarrow \mathcal{L}_2$; (v) $\mathcal{L}_3 \leftrightarrow \mathcal{L}_3$.
Only (v) *is the identity mapping.*

Proof: By Lemma 19.1.4, a point P of $\mathcal{U}$ is interchanged with the tangent plane $P\mathfrak{U}$ at P. Through a point Q off $\mathcal{U}$, there are $q\sqrt{q}+1$ tangent planes whose points of contact lie in a plane, which is $Q\mathfrak{U}$. Conversely, a plane π in Φ_3 meets $\mathcal{U}$ in the $q\sqrt{q}+1$ points of a Hermitian curve, the tangent planes at which are concurrent at $\pi\mathfrak{U}$.

Each line of $\mathcal{L}_3$ is self-polar under $\mathcal{U}$. Then $\mathfrak{U}$ acts on the pencil $\mathcal{P}$ of lines through a point P of $\mathcal{U}$ lying in the tangent plane $P\mathfrak{U}$ as follows. The tangent lines are self-conjugate but not self-polar; they fall into $(q-\sqrt{q})/2$ pairs under $\mathfrak{U}$, while the $\sqrt{q}+1$ generators are fixed. So $\mathfrak{U}$ acts on $\mathcal{P}$ as the collineation induced by the involutory automorphism σ acts on $PG(1, q)$.

If l is in $\mathcal{L}_2$, then the tangent planes at the $\sqrt{q}+1$ points of $l \cap \mathcal{U}$ have a line l' in common, which is $l\mathfrak{U}$. More particularly, through each of the $\sqrt{q}+1$ points of $l \cap \mathcal{U}$, there are $q-\sqrt{q}$ tangents. These meet in sets of $q-\sqrt{q}$ at $\sqrt{q}+1$ points of l', which are the points of $l' \cap \mathcal{U}$. □

Theorem 19.1.9: *Table* 19.2 *is an incidence table for $PG(3, s^2)$ with respect to $\mathcal{U}$, where a 2×2 section with $i<j$*

	$\Pi_i(\mathcal{V})$	$\Pi_j(\mathcal{V}')$
$\Pi_i(\mathcal{V})$	a_{ii}	a_{ij}
$\Pi_j(\mathcal{V}')$	a_{ji}	a_{jj}

Table 19.2 *Incidences of Subspaces of $PG(3,s^2)$ with Respect to $\mathcal{U}_3$*

	$\Pi_0(\Pi_0)$	$\Pi_0(\Pi_{-1})$	$\Pi_1(\Pi_1)$	$\Pi_1(\Pi_0)$	$\Pi_1(\mathcal{U}_1)$	$\Pi_2(\mathcal{U}_2)$	$\Pi_2(\Pi_0\mathcal{U}_1)$
$\Pi_0(\Pi_0)$	$(s^3+1)(s^2+1)$	–	s^2+1	1	$s+1$	s^3+1	s^3+s+1
$\Pi_0(\Pi_{-1})$	–	$s^3(s^2+1)(s-1)$	0	s^2	$s(s-1)$	$s^2(s^2-s+1)$	$s^3(s-1)$
$\Pi_1(\Pi_1)$	$s+1$	0	$(s^3+1)(s+1)$	–	–	0	$s+1$
$\Pi_1(\Pi_0)$	$s(s-1)$	s^3+1	–	$s(s^3+1)(s^2+1)(s-1)$	–	s^3+1	$s(s-1)$
$\Pi_1(\mathcal{U}_1)$	s^4	$s^2(s^2-s+1)$	–	–	$s^4(s^2+1)(s^2-s+1)$	$s^2(s^2-s+1)$	s^4
$\Pi_2(\mathcal{U}_2)$	$s^3(s-1)$	$s^2(s^2-s+1)$	0	s^2	$s(s-1)$	$s^3(s^2+1)(s-1)$	–
$\Pi_2(\Pi_0\mathcal{U}_1)$	s^3+s+1	s^3+1	s^2+1	1	$s+1$	–	$(s^3+1)(s^2+1)$

means that $|\{\Pi_i(\mathcal{V})\}| = a_{ii}$, $|\{\Pi_j(\mathcal{V}')\}| = a_{jj}$, *the number of* $\Pi_i(\mathcal{V})$ *in* $\Pi_j(\mathcal{V}')$ *is* a_{ij}, *and the number of* $\Pi_j(\mathcal{V}')$ *through* $\Pi_i(\mathcal{V})$ *is* a_{ji}. □

To conclude this section we consider the sub-Hermitian surface $\mathcal{U}_3'$ in $PG(3, q^2)$, $q = s^2$, where

$$\mathcal{U}_3' = \mathbf{V}_{3,q^2}(x_0^{s+1} + x_1^{s+1} + x_2^{s+1} + x_2^{s+1});$$

that is, we consider what happens to $\mathcal{U}_3$ when $GF(q)$ is extended to $GF(q^2)$. Regard $PG(3, q)$ as embedded in $PG(3, q^2)$.

Theorem 19.1.10: *In* $PG(3, q^2)$, $q = s^2$,

(i) *every line on* $\mathcal{U}_3'$ *is a line of* $\mathcal{U}_3$;
(ii) *every point of* $\mathcal{U}_3' \backslash \mathcal{U}_3$ *lies on a unique line of* $\mathcal{U}_3$;
(iii) *the number of lines on* $\mathcal{U}_3'$ *is* $(s^3+1)(s+1)$;
(iv) *the number of points on* $\mathcal{U}_3'$ *is* $(s^3+1)(s^5+s^4-s^3+1)$.

Proof: For $P = \mathbf{P}(a_0, a_1, a_2, a_3)$ in $PG(3, q^2)$, let $\tilde{P} = \mathbf{P}(a_0^q, a_1^q, a_2^q, a_3^q)$. If $P \in \mathcal{U}_3' \backslash \mathcal{U}_3$, then $\tilde{P} \neq P$ and $\tilde{P} \in \mathcal{U}_3' \backslash \mathcal{U}_3$. So $P\tilde{P}$ is a line over $GF(q)$ either lying on $\mathcal{U}_3'$ or a $(\sqrt{q}+1)$-secant of $\mathcal{U}_3'$. It cannot be the latter as all points of $l \cap \mathcal{U}_3'$ with $l = \Pi_1(\mathcal{U}_1)$ lie over $GF(q)$. So $P\tilde{P}$ lies on $\mathcal{U}_3'$. This implies (i), (ii), and (iii).

Now, each line of $\mathcal{U}_3'$ contains $q^2 - q = s^4 - s^2$ points not on $\mathcal{U}_3$. So

$$|\mathcal{U}_3'| = (s^3+1)(s+1)(s^4-s^2) + (s^3+1)(s^2+1)$$
$$= (s^3+1)(s^5+s^4-s^3+1). \quad \square$$

19.2 Lines on $\mathcal{U}$

To explore properties of the lines on $\mathcal{U}$, it is useful to obtain their line coordinates.

Lemma 19.2.1: *Let* $\mathcal{U} = \mathbf{V}(x_0\bar{x}_0 + x_1\bar{x}_1 + x_2\bar{x}_2 + x_3\bar{x}_3)$.

(i) *If* $l = \mathbf{l}(L)$ *and* $l' = \mathbf{l}(L')$ *is the polar of* l, *then*

$$\rho L' = \hat{\bar{L}}.$$

(ii) *The lines on* $\mathcal{U}$ *are those lines* $l = \mathbf{l}(L)$ *such that*

$$\rho L = \hat{\bar{L}}.$$

(iii) *Every line* l *on* $\mathcal{U}$ *has a vector* L *such that*

$$L = (x, y, z, \bar{z}, \bar{y}, \bar{x}) \quad \text{with} \quad x\bar{x} + y\bar{y} + z\bar{z} = 0$$

Proof: (i) This was already shown in the corollary to Lemma 15.2.8. However, more directly, $\mathfrak{U}$ is given by

$$\mathbf{P}(y_0, y_1, y_2, y_3) \rightarrow \boldsymbol{\pi}(\bar{y}_0, \bar{y}_1, \bar{y}_2, \bar{y}_3).$$

So, the dual line vector of l' is $\bar{L}$, whence the line vector is $\hat{\bar{L}}$, by Lemma 15.2.1. This proves (i).

Since the lines on $\mathscr{U}$ are the self-polar lines, (ii) follows.

If $\rho L = \hat{\bar{L}}$, then $\bar{\rho}\hat{\bar{L}} = L$, whence $\rho\bar{\rho} = 1$. So, the equation (19.1) can be solved with $d = 1/\rho$ to give μ such that $\bar{\mu}/\mu = 1/\rho$. Therefore $\mu L = \bar{\mu}\hat{\bar{L}}$, and the replacement of μL by L gives $L = \hat{\bar{L}}$. So, with $L = (l_{01}, l_{02}, l_{03}, l_{12}, l_{31}, l_{23})$, we have

$$l_{23} = \bar{l}_{01}, \quad l_{31} = \bar{l}_{02}, \quad l_{12} = \bar{l}_{03}. \quad \square$$

Corollary: *The number of lines on $\mathscr{U}$ is $(q\sqrt{q}+1)(\sqrt{q}+1)$.*

Proof: The number of points $\mathbf{P}(x, y, z)$ satisfying $x\bar{x} + y\bar{y} + z\bar{z} = 0$ is $q\sqrt{q}+1$; see § 7.3. The triple $(\rho x, \rho y, \rho z)$ determines the same line as (x, y, z) exactly when $\bar{\rho} = \rho$ with $\rho \neq 0$. So the number of lines is

$$(q\sqrt{q}+1)(q-1)/(\sqrt{q}-1) = (q\sqrt{q}+1)(\sqrt{q}+1). \quad \square$$

Note: Occasionally, a convenient form for the lines on $\mathscr{U}$ is

$$\mathbf{l}(\bar{\mu}x, \bar{\mu}y, \bar{\mu}z, \nu\bar{z}, \nu\bar{y}, \nu\bar{x}) \quad \text{with} \quad \mu\bar{\mu} = \nu\bar{\nu}$$

or

$$\mathbf{l}(\rho x, \rho y, \rho z, \bar{z}, \bar{y}, \bar{x}) \quad \text{with} \quad \rho\bar{\rho} = 1$$

where $x\bar{x} + y\bar{y} + z\bar{z} = 0$. Then we can run through all lines on $\mathscr{U}$ by taking all $q\sqrt{q}+1$ points $\mathbf{P}(x, y, z)$ on the Hermitian curve $\mathscr{U}_2$ and all $\sqrt{q}+1$ elements in γ with $\rho\bar{\rho} = 1$ or alternatively all $\sqrt{q}+1$ points $\mathbf{P}(\mu, \nu)$ on the subgeometry $PG(1, \sqrt{q})$ (equivalent to the Hermitian set $\mathscr{U}_1$).

Now we consider the effect of the map $\mathfrak{G}: \mathscr{L} \to \mathscr{H}_{5,q}$ on $\mathscr{U}^{(1)}$; see § 15.4.

Theorem 19.2.2: $\mathscr{U}^{(1)}\mathfrak{G} = \mathscr{E}_{5,\sqrt{q}}$.

Proof: The theorem means that the lines on $\mathscr{U}$ are transformed by $\mathfrak{G}$ to the points of an elliptic quadric $\mathscr{E}_{5,\sqrt{q}}$ over $GF(\sqrt{q})$ lying on the Grassmannian $\mathscr{H}_{5,q}$.

Firstly we find a projectivity $\mathfrak{T}$ of $PG(5, q)$ such that any point of $\mathscr{U}^{(1)}\mathfrak{G}\mathfrak{T}$ has all its coordinates in $GF(\sqrt{q})$. Let $P(X)\mathfrak{T} = P(X')$ be given by

$$\begin{aligned} x_0' &= \alpha x_0 + \bar{\alpha} x_5, & x_5' &= \bar{\alpha} x_0 + \alpha x_5, \\ x_1' &= \alpha x_1 + \bar{\alpha} x_4, & x_4' &= \bar{\alpha} x_1 + \alpha x_4, \\ x_2' &= \alpha x_2 + \bar{\alpha} x_3, & x_3' &= \bar{\alpha} x_2 + \alpha x_3, \end{aligned}$$

where $\bar{\alpha}^2 \neq \alpha^2$. So, if $l = \mathbf{l}(x, y, z, \bar{z}, \bar{y}, \bar{x})$, then

$$l\mathfrak{G}\mathfrak{T} = \mathbf{P}(\alpha x + \bar{\alpha}\bar{x}, \alpha y + \bar{\alpha}\bar{y}, \alpha z + \alpha\bar{z}, \bar{\alpha} z + \alpha\bar{z}, \bar{\alpha} y + \alpha\bar{y}, \bar{\alpha} x + \alpha\bar{x}).$$

Since $(\alpha t + \bar{\alpha}\bar{t})\sigma = \bar{\alpha}\bar{t} + \alpha t$ and $(\bar{\alpha} t + \alpha\bar{t})\sigma = \alpha\bar{t} + \bar{\alpha} t$, all the coordinates of $l\mathfrak{G}\mathfrak{T}$ are in $GF(\sqrt{q})$. Hence $\mathscr{U}^{(1)}\mathfrak{G}\mathfrak{T} = \mathscr{E}' \subset PG(5, \sqrt{q}) = \Pi'$.

As $\mathcal{H} = \mathcal{H}_{5,q} = \mathbf{V}(x_0x_5 + x_1x_4 + x_2x_3)$, we consider $\mathcal{H}' = \mathcal{H}\mathfrak{T}$. With $x_i' = \alpha x_i + \bar{\alpha}x_j$, $x_j' = \bar{\alpha}x_i + \alpha x_j$, where $i + j = 5$, then

$$x_i = (\alpha x_i' - \bar{\alpha}x_j')/(\alpha^2 - \bar{\alpha}^2)$$
$$x_j = (\alpha x_j' - \bar{\alpha}x_i')/(\alpha^2 - \bar{\alpha}^2).$$

So

$$\begin{aligned}\mathcal{H}' &= \mathbf{V}((\alpha x_0 - \bar{\alpha}x_5)(\alpha x_5 - \bar{\alpha}x_0) + (\alpha x_1 - \bar{\alpha}x_4)(\alpha x_4 - \bar{\alpha}x_1) \\ &\quad + (\alpha x_2 - \bar{\alpha}x_3)(\alpha x_3 - \bar{\alpha}x_2)) \\ &= \mathbf{V}(\alpha\bar{\alpha}\textstyle\sum x_i^2 - (\alpha^2 + \bar{\alpha}^2)(x_0x_5 + x_1x_4 + x_2x_3)) \\ &= \mathbf{V}(F)\end{aligned}$$

where

$$F = x_0^2 + x_1^2 + x_2^2 + x_3^2 + x_4^2 + x_5^2 - e(x_0x_5 + x_1x_4 + x_2x_3)$$

and

$$e = (\alpha^2 + \bar{\alpha}^2)/(\alpha\bar{\alpha}) = \alpha/\bar{\alpha} + \bar{\alpha}/\alpha.$$

Now, $|\mathcal{U}^{(1)}| = (\sqrt{q} + 1)(q\sqrt{q} + 1)$ by Theorem 19.1.5, and $|\mathscr{E}_{5,\sqrt{q}}| = (\sqrt{q} + 1)(q\sqrt{q} + 1)$ by Theorem 5.2.6.

So it remains to show that, although $\mathbf{V}(F)$ is hyperbolic over $GF(q)$, it is elliptic over $GF(\sqrt{q})$. Now, F can be written

$$F = f(x_0, x_5) + f(x_1, x_4) + f(x_2, x_3)$$

where $f(x, y) = x^2 - exy + y^2$ and f is irreducible over $GF(\sqrt{q})$. So, exactly as in Lemma 17.1.1,

$$\mathbf{V}(f(x_1, x_4) + f(x_2, x_3))$$

is a hyperbolic quadric in $PG(3, \sqrt{q})$ and can therefore be reduced, over $GF(\sqrt{q})$, to the canonical form

$$\mathbf{V}(x_1x_4 + x_2x_3).$$

So $\mathbf{V}(F)$ can be reduced, over $GF(\sqrt{q})$, to

$$\mathbf{V}(f(x_0, x_5) + x_1x_4 + x_2x_3),$$

which, except for a rearrangement of the indeterminates, is the canonical form for an elliptic quadric in $PG(5, \sqrt{q})$, by Theorem 5.2.4. So $\mathscr{E}'$ is an $\mathscr{E}_{5,\sqrt{q}}$ as is $\mathscr{E} = \mathcal{U}^{(1)}\mathfrak{G}$. So $\mathcal{U}^{(1)}\mathfrak{G} = \mathscr{E} \subset PG(5, \sqrt{q}) = \Pi$. $\square$

Note: With $\beta = \alpha/\bar{\alpha}$, we have $e = \beta + \bar{\beta}$ where $\beta^2 \neq 1$ and $\beta\bar{\beta} = 1$. So $f = x^2 - exy + y^2$ is irreducible over $GF(\sqrt{q})$ and has roots β, $\bar{\beta}$ over $GF(q)$. So e can take any value in $GF(\sqrt{q})\backslash\{2, -2\}$ such that f is irreducible.

In particular, for $q=4$, $e=1$ and

$$F=x_0^2+x_1^2+x_2^2+x_3^2+x_4^2+x_5^2+x_0x_5+x_1x_4+x_2x_3;$$

for $q=9$, $e=0$ and

$$F=x_0^2+x_1^2+x_2^2+x_3^2+x_4^2+x_5^2.$$

Also, since the only condition on α is that $\bar{\alpha}^2 \neq \alpha^2$, there are $q-(2\sqrt{q}-1)=(\sqrt{q}-1)^2$ values of α when q is odd and $q-\sqrt{q}$ values of α when q is even.

Corollary: $P\gamma U(4, q) \cong PGO_-(6, \sqrt{q})$.

Proof: By Lemma 15.2.8, any projectivity $\mathfrak{T}$ of $PG(3, q)$ determines a projectivity $\mathfrak{T}'$ of $PG(5, q)$ fixing $\mathcal{H}_{5,q}$. So, by the theorem, a projectivity fixing $\mathcal{U}$ determines a projectivity fixing $\mathscr{E}$. Hence there is a homomorphism

$$\phi: PGU(4, q) \to PGO_-(6, \sqrt{q}).$$

However, the automorphism σ of γ induces a collineation of $PG(3, q)$ fixing $\mathcal{U}$. This becomes under $\mathfrak{G}$ the projectivity $x_0'=x_5$, $x_1'=x_4$, $x_2'=x_3$, $x_3'=x_2$, $x_4'=x_1$, $x_5'=x_0$, fixing $\mathscr{E}$. So we may also write

$$\phi: P\gamma U(4, q) \to PGO_-(6, \sqrt{q}).$$

If $\mathfrak{T}\phi$ is the identity, then $\mathfrak{T}$ fixes all the lines on $\mathcal{U}$ and hence all the points. Hence $\mathfrak{T}$ is the identity in $P\gamma U(4, q)$, whence ϕ is injective. By Appendix III, $|P\gamma U(4, q)|=|PGO_-(6, \sqrt{q})|$. So ϕ is an isomorphism. □

From Theorem 19.2.2, $\mathcal{U}^{(1)}\mathfrak{G}=\mathscr{E}$ is an $\mathscr{E}_{5,\sqrt{q}}$ in a $PG(5, \sqrt{q})=\Pi$. So it is natural to ask what corresponds in $PG(3, q)$ to the points of $\Pi \backslash \mathscr{E}$.

Lemma 19.2.3: (i) *If $\mathfrak{U}$ is a Hermitian polarity and $\mathfrak{T}$ a collineation, then the following are equivalent*:

(a) $\mathfrak{U}\mathfrak{T}=\mathfrak{T}\mathfrak{U}$;

(b) $\mathfrak{T}$ fixes $\mathfrak{U}$;

(c) for every line l on $\mathcal{U}$, the line $l'=l\mathfrak{T}$ is on $\mathcal{U}$.

(ii) If the conditions in (i) hold and $\mathfrak{T}$ is also involutory, then $\mathfrak{U}\mathfrak{T}$ is a polarity which is Hermitian or not as $\mathfrak{T}$ is a projectivity or not.

Proof: (i) If l lies on $\mathcal{U}$, then $l\mathfrak{U}=l$. So $l\mathfrak{U}\mathfrak{T}=l\mathfrak{T}=l'$ and $l\mathfrak{T}\mathfrak{U}=l'\mathfrak{U}$. So if (a) holds, then $l'=l'\mathfrak{U}$, whence (c) holds. Conversely, if (c) holds, $l\mathfrak{U}\mathfrak{T}=l\mathfrak{T}\mathfrak{U}$ for all lines l on $\mathfrak{U}$. So $\mathfrak{U}\mathfrak{T}\mathfrak{U}^{-1}\mathfrak{T}^{-1}$ fixes all lines on $\mathcal{U}$, so all points on $\mathcal{U}$ and so all points of $PG(3, q)$. Therefore $\mathfrak{U}\mathfrak{T}=\mathfrak{T}\mathfrak{U}$. If (b) holds, then for any point P on $\mathcal{U}$, the point $P\mathfrak{T}$ is on $\mathcal{U}$, whence (c) holds and conversely.

(ii) With $\mathfrak{I}$ as the identity, $\mathfrak{T}^2=\mathfrak{U}^2=\mathfrak{I}$ and $\mathfrak{T}\mathfrak{U}=\mathfrak{U}\mathfrak{T}$. So $(\mathfrak{U}\mathfrak{T})^2=I$ and $\{\mathfrak{I}, \mathfrak{U}, \mathfrak{T}, \mathfrak{U}\mathfrak{T}\}$ is a group isomorphic to $\mathbf{Z}_2 \times \mathbf{Z}_2$.

If $\mathfrak{T}$ is a projectivity with matrix T, then $\mathbf{P}(X)\mathfrak{U}\mathfrak{T}=\boldsymbol{\pi}(\bar{X})\mathfrak{T}=\boldsymbol{\pi}(\bar{X}T^{*-1})$, whence $\mathfrak{U}\mathfrak{T}$ is a Hermitian polarity. If $\mathfrak{T}=\sigma\mathfrak{S}$, where $\mathfrak{S}$ is a projectivity with matrix S and σ is the involutory automorphism, then $\mathbf{P}(X)\mathfrak{U}\mathfrak{T}=\mathbf{P}(X)\mathfrak{U}\sigma\mathfrak{S}=\boldsymbol{\pi}(\bar{X})\sigma\mathfrak{S}=\boldsymbol{\pi}(X)\mathfrak{S}=\boldsymbol{\pi}(XS^{*-1})$, whence $\mathfrak{U}\mathfrak{T}$ is a non-Hermitian polarity. □

Corollary: (i) *If $\mathfrak{U}$ is a Hermitian polarity and $\mathfrak{T}$ a polarity, then the following are equivalent*:

(a) $\mathfrak{U}\mathfrak{T}=\mathfrak{T}\mathfrak{U}$;

(b) *for every line l on $\mathscr{U}$, the line $l'=l\mathfrak{T}$ is on $\mathscr{U}$.*

(ii) *If the conditions of* (i) *hold, then $\mathfrak{U}\mathfrak{T}$ is an involutory collineation, which is a projectivity or not according as $\mathfrak{T}$ is Hermitian or not.* □

Now, we investigate in particular the null polarities commuting with the canonical Hermitian polarity $\mathfrak{U}$. As in § 15.2, if $\mathscr{A}=\boldsymbol{\lambda}(A)$ is a linear complex, we write $\mathfrak{A}$ for the corresponding null polarity.

Lemma 19.2.4: *If the null polarity $\mathfrak{A}$ corresponding to the linear complex $\mathscr{A}=\boldsymbol{\lambda}(A)$ commutes with $\mathfrak{U}$, then*

(i) $\mu\hat{A}=\bar{A}$ *with* $\mu\bar{\mu}=1$;

(ii) $\mathscr{A}=\boldsymbol{\lambda}(b,c,d,\bar{d},\bar{c},\bar{b})$ *with* $b\bar{b}+c\bar{c}+d\bar{d}\neq 0$.

Proof: By Lemma 15.2.8, Corollary (i)(a), if $l=\mathbf{l}(L)$ and $l'=\mathbf{l}(L')=l\mathfrak{A}$, then

$$\rho L'=\varpi(A)L-(AL^*)\hat{A}.$$

From Lemma 19.2.1, $\mathbf{l}(L)\mathfrak{U}=\mathbf{l}(\hat{\bar{L}})$. So

$$l\mathfrak{A}\mathfrak{U}=\mathbf{l}(\varpi(A)L-(AL^*)\hat{A})\mathfrak{U}=\mathbf{l}(\overline{\varpi(A)}\hat{\bar{L}}-(\bar{A}\bar{L}^*)\bar{A})$$

and

$$l\mathfrak{U}\mathfrak{A}=\mathbf{l}(\hat{\bar{L}})\mathfrak{A}=\mathbf{l}(\varpi(A)\hat{\bar{L}}-(A\hat{\bar{L}}^*)\hat{A}).$$

Hence $l\mathfrak{A}\mathfrak{U}=l\mathfrak{U}\mathfrak{A}$ if and only if $\mu\hat{A}=\bar{A}$, whence $\bar{\mu}\bar{A}=\hat{A}$ and $\mu\bar{\mu}=1$. Then, exactly as in Lemma 19.2.1, putting $B=\nu A$ with $\mu=\nu/\bar{\nu}$ gives $\hat{B}=\bar{B}$. Hence

$$\mathscr{A}=\boldsymbol{\lambda}(b,c,d,\bar{d},\bar{c},\bar{b}).$$

As $\mathscr{A}$ is non-special, $b\bar{b}+c\bar{c}+d\bar{d}\neq 0$. □

Corollary: *The number of null polarities $\mathfrak{A}$ commuting with the Hermitian polarity $\mathfrak{U}$ is $q(q\sqrt{q}+1)$.*

Proof: For any triple (b,c,d) giving $\mathscr{A}$, the triple $(\alpha b,\alpha c,\alpha d)$ gives the same $\mathscr{A}$ when $\bar{\alpha}=\alpha\neq 0$. So the required number is

$$[(q^2+q+1)-(q\sqrt{q}+1)](\sqrt{q}+1)=q(q\sqrt{q}+1).\quad □$$

Let $\mathcal{N}$ be the set of linear complexes corresponding to null polarities commuting with $\mathfrak{U}$. Then $\mathcal{N}\mathfrak{G}$ is the corresponding set of prime sections of $\mathcal{H}_{5,q}$. Under the polarity $\mathfrak{H}$ of $\mathcal{H}_{5,q}$ we obtain the set $\mathcal{N}\mathfrak{G}\mathfrak{H}$ of points of $PG(5, q)$. It was shown in Theorem 19.2.2 that $\mathcal{U}^{(1)}\mathfrak{G} = \mathcal{E}$, an elliptic quadric $\mathcal{E}_{5,\sqrt{q}}$ lying in a subgeometry $\Pi = PG(5, \sqrt{q})$.

Theorem 19.2.5: $\mathcal{N}\mathfrak{G}\mathfrak{H} = \Pi \backslash \mathcal{E}$.

Proof: Any linear complex $\mathcal{A} = \boldsymbol{\lambda}(A)$ determining a null polarity is non-special; that is, $a_{01}a_{23} + a_{02}a_{31} + a_{03}a_{12} \neq 0$. Now, $\mathcal{N} = \{\mathcal{A} = \boldsymbol{\lambda}(A) \mid \hat{A} = \bar{A}\}$. If $B = \hat{A}$ and $\hat{A} = \bar{A}$, then $\hat{B} = \bar{B}$. Then

$$\mathcal{N}\mathfrak{G}\mathfrak{H} = \{\mathbf{P}(B) \in PG(5, q) \mid \hat{B} = \bar{B}, \mathbf{P}(B) \notin \mathcal{H}_{5,q}\}.$$

If $P \in \mathcal{N}\mathfrak{G}\mathfrak{H}$, then $P = \mathbf{P}(b, c, d, \bar{d}, \bar{c}, \bar{b})$ with $b\bar{b} + c\bar{c} + d\bar{d} \neq 0$. So, with the same $\mathfrak{T}$ as in Theorem 19.2.2,

$$P\mathfrak{T} = \mathbf{P}(\alpha b + \bar{\alpha}\bar{b}, \alpha c + \bar{\alpha}\bar{c}, \alpha d + \bar{\alpha}\bar{d}, \bar{\alpha} d + \alpha\bar{d}, \bar{\alpha} c + \alpha\bar{c}, \bar{\alpha} b + \alpha\bar{b}),$$

which has all its coordinates in $GF(\sqrt{q})$. As $P \notin \mathcal{H}$, so $P\mathfrak{T} \notin \mathcal{H}\mathfrak{T}$. So $\mathcal{N}\mathfrak{G}\mathfrak{H}\mathfrak{T}$ lies in Π', the $PG(5, \sqrt{q})$ containing $\mathcal{U}^{(1)}\mathfrak{G}\mathfrak{T} = \mathcal{E}'$. But, by the corollary to Lemma 19.2.4, $|\mathcal{N}| = |\mathcal{N}\mathfrak{G}\mathfrak{H}\mathfrak{T}| = q(q\sqrt{q}+1)$ and $|\Pi' \backslash \mathcal{E}'| = (q^3-1)/(\sqrt{q}-1) - (\sqrt{q}+1)(q\sqrt{q}+1) = q(q\sqrt{q}+1)$. So $\mathcal{N}\mathfrak{G}\mathfrak{H}\mathfrak{T} = \Pi' \backslash \mathcal{E}'$ and hence $\mathcal{N}\mathfrak{G}\mathfrak{H} = \Pi \backslash \mathcal{E}$. □

Corollary: *If $\mathfrak{A}$ is a null polarity commuting with $\mathfrak{U}$, then the number of lines on $\mathcal{U}$ fixed by $\mathfrak{A}$ is $(q+1)(\sqrt{q}+1)$. The other lines on $\mathcal{U}$ form $q(q-1)/2$ polar pairs with respect to $\mathfrak{A}$.*

Proof: The lines of $\mathcal{A}$ in $\mathcal{U}^{(1)}$ correspond under $\mathfrak{G}\mathfrak{T}$ to the points of a prime section over $GF(\sqrt{q})$ of $\mathcal{E}'$ and so $\mathcal{U}^{(1)} \cap \mathcal{A}$ corresponds to $\mathcal{P}_{4,\sqrt{q}}$, which contains $(q+1)(\sqrt{q}+1)$ points. This leaves $(\sqrt{q}+1)(q\sqrt{q}+1) - (q+1)(\sqrt{q}+1) = q(q-1)$ lines on $\mathcal{U}$. □

19.3 Regular systems of lines on $\mathcal{U}$

A *regular system* $\mathcal{L}_m$ *of order* m on $\mathcal{U}$ is a subset of $\mathcal{U}^{(1)}$ such that through each point of $\mathcal{U}$ there are exactly m lines of $\mathcal{L}_m$. So $\mathcal{U}^{(1)}$ is a regular system of order $\sqrt{q}+1$. If $k = |\mathcal{L}_m|$, then $k = m(q\sqrt{q}+1)$.

To investigate the existence of regular systems, we first require properties of quadrics $\mathcal{H}_{3,q}$ which meet $\mathcal{U}$ in at least three lines of a regulus. A hyperbolic quadric whose complementary reguli are $\mathcal{R}$ and $\mathcal{R}'$ is denoted $\mathcal{H}(\mathcal{R}, \mathcal{R}')$.

Lemma 19.3.1: (i) *If $\mathcal{H}_{3,q} = \mathcal{H}(\mathcal{R}, \mathcal{R}')$ has three skew lines on $\mathcal{U}$, then $\mathcal{U} \cap \mathcal{H}_{3,q}$ consists of the $2(\sqrt{q}+1)$ lines of $\mathcal{R}_0 \cup \mathcal{R}_0'$, where $\mathcal{R}_0 \subset \mathcal{R}$, $\mathcal{R}_0' \subset \mathcal{R}'$*

and $|\mathcal{R}_0| = |\mathcal{R}_0'| = \sqrt{q}+1$. *Further,* $\mathcal{R}_0$ *lies on* $\mathcal{R}$ *and* $\mathcal{R}_0'$ *on* $\mathcal{R}'$ *as* $PG(1, \sqrt{q})$ *on* $PG(1, q)$.

(ii) *There are* $n_q = q^2(q\sqrt{q}+1)(q+1)/2$ *such quadrics* $\mathcal{H}_{3,q}$.

Proof: (i) Under $\mathfrak{G}$ the three lines correspond on $\mathcal{H}_{5,q}$ to three points spanning a plane π meeting $\mathcal{H}_{5,q}$ in a conic $\mathcal{C}$ representing the regulus $\mathcal{R}$, say; the polar plane π' of π meets $\mathcal{H}_{5,q}$ in a conic $\mathcal{C}'$ representing $\mathcal{R}'$. As $\mathcal{U}^{(1)}\mathfrak{G} = \mathcal{E} \subset \mathcal{H}_{5,q}$ (Theorem 19.2.2), the planes π and π' meet $\mathcal{E}$ in conics of the $PG(5, \sqrt{q})$ containing $\mathcal{E}$; more precisely, with $\mathfrak{T}$ as in Theorems 19.2.2 and 19.2.5, the planes $\pi\mathfrak{T}$ and $\pi'\mathfrak{T}$ meet $\mathcal{E}' = \mathcal{E}\mathfrak{T}$ in conics over $GF(\sqrt{q})$. Hence $\mathcal{U} \cap \mathcal{H}_{3,q}$ contains $2(\sqrt{q}+1)$ lines as enunciated. Since the degrees of $\mathcal{U}$ and $\mathcal{H}_{3,q}$ are $\sqrt{q}+1$ and 2 respectively, this is their total intersection.

(ii) We may argue in $PG(3, q)$ or $PG(5, \sqrt{q})$ with equal ease.

In $PG(5, \sqrt{q})$, the quadric $\mathcal{E} = \mathcal{E}_{5,\sqrt{q}}$ contains $(\sqrt{q}+1)(q\sqrt{q}+1)$ points on $(q+1)(q\sqrt{q}+1)$ lines with $q+1$ lines through each point. So, for a point P on $\mathcal{E}$, there are $\sqrt{q}(q+1)$ other points on the lines through P and hence

$$(\sqrt{q}+1)(q\sqrt{q}+1) - \sqrt{q}(q+1) - 1 = q^2$$

points of $\mathcal{E}$ on no line through P.

Given any such point Q, a point R such that RP and RQ are both on $\mathcal{E}$ lies in the intersection of $\mathcal{E}$ with the polar solid of PQ. This intersection is $\mathcal{E}_{3,\sqrt{q}}$ containing $q+1$ points. So the number of points R on $\mathcal{E}$ conjugate to neither P nor Q is

$$(\sqrt{q}+1)(q\sqrt{q}+1) - 2\sqrt{q}(q+1) + (q+1) - 2 = \sqrt{q}(\sqrt{q}-1)(q+1).$$

So the number of planes of $PG(5, \sqrt{q})$ meeting $\mathcal{E}$ in a conic is

$$[(\sqrt{q}+1)(q\sqrt{q}+1) \cdot q^2 \cdot \sqrt{q}(\sqrt{q}-1)(q+1)]/[(\sqrt{q}+1) \cdot \sqrt{q} \cdot (\sqrt{q}-1)]$$
$$= q^2(q+1)(q\sqrt{q}+1).$$

Therefore the number of quadrics $\mathcal{H}_{3,q}$ meeting $\mathcal{U}$ in $2(\sqrt{q}+1)$ lines is $q^2(q+1)(q\sqrt{q}+1)/2$. □

A quadric $\mathcal{H}_{3,q}$ with the property of Lemma 19.3.1 is *permutable with* $\mathcal{U}$. This name arises because, for q odd, the quadrics $\mathcal{H}_{3,q}$ permutable with $\mathcal{U}$ induce ordinary polarities $\mathfrak{H}$ commuting with $\mathfrak{U}$.

If $\mathcal{H} = \mathcal{H}_{3,q}$ is permutable with $\mathcal{U}$, then $\mathcal{C} = \mathcal{U} \cap \mathcal{H}$ is a curve of degree $2(\sqrt{q}+1)$ consisting of that number of lines such that $\mathcal{C} = \mathcal{C}^0 \cup \mathcal{C}^+$, where

$$\mathcal{C}^+ = \{l \cap l' \mid l \in \mathcal{R}_0,\ l' \in \mathcal{R}_0'\}$$

is the set of double points and $\mathcal{C}^0$ is the set of simple points: $|\mathcal{C}^+| = (\sqrt{q}+1)^2$, $|\mathcal{C}^0| = 2(\sqrt{q}+1)(q-\sqrt{q}) = 2\sqrt{q}(q-1)$.

Lemma 19.3.2: *If* $\mathcal{H}$ *is permutable with* $\mathcal{U}$, *then no line of* $\mathcal{U}$ *is external to* $\mathcal{H}$.

Proof: With respect to $\mathscr{H}$, the lines of $\mathscr{U}$ are of at least 3 types:

(i) the lines comprising $\mathscr{C}$, of which there are $2(\sqrt{q}+1)$;

(ii) the lines of $\mathscr{U}\backslash\mathscr{H}$ through the points of $\mathscr{C}^+$, of which there are $(\sqrt{q}-1)(\sqrt{q}+1)^2=(q-1)(\sqrt{q}+1)$ (these lines are all tangents to $\mathscr{H}$);

(iii) the lines of $\mathscr{U}\backslash\mathscr{H}$ through the points of $\mathscr{C}^0$, of which there are $2\sqrt{q}(q-1)\sqrt{q}/2=q(q-1)$ (these lines are all bisecants of $\mathscr{H}$). This gives $2(\sqrt{q}+1)+(q-1)(\sqrt{q}+1)+q(q-1)=(\sqrt{q}+1)(q\sqrt{q}+1)$, which is all the lines of $\mathscr{U}$. □

Lemma 19.3.3: *Let $\mathscr{L}_m$ be a regular system of order m on $\mathscr{U}$ and $\mathscr{H}=\mathscr{H}(\mathscr{R},\mathscr{R}')$ a quadric permutable with $\mathscr{U}$. If $|\mathscr{L}_m\cap\mathscr{R}|=n$ and $|\mathscr{L}_m\cap\mathscr{R}'|=n'$, then $n+n'=2m$.*

Proof: $\mathscr{L}_m$ consists of $k=m(q\sqrt{q}+1)$ lines. By the previous lemma, no line of $\mathscr{L}_m$ is external to $\mathscr{H}$, whence each line of $\mathscr{L}_m$ has at least one point on $\mathscr{H}$ and so on $\mathscr{C}=\mathscr{U}\cap\mathscr{H}$. By considerations analogous to those of the previous lemma, we partition the lines of $\mathscr{L}_m$ into the classes $\mathscr{S}_i$, $i\in\mathbf{N}_6$.

$\mathscr{S}_1=\{\mathscr{L}_m\cap(\mathscr{R}\cup\mathscr{R}')\}$: $|\mathscr{S}_1|=n+n'$. Let $\mathscr{V}$ be the set of points on the lines of $\mathscr{L}_m\cap\mathscr{R}$ and $\mathscr{V}'$ the set of points on the lines of $\mathscr{L}_m\cap\mathscr{R}'$. Let $\mathscr{W}=\mathscr{V}\cup\mathscr{V}'$ and $\mathscr{W}'=\mathscr{V}\cap\mathscr{V}'$. Then $\mathscr{W}'$ consists of nn' points, all in $\mathscr{C}^+$. Each line of $\mathscr{L}_m\cap\mathscr{R}$ contains n' points of $\mathscr{W}'$, $\sqrt{q}+1-n'$ other points of $\mathscr{C}^+$, and $q-\sqrt{q}$ points of $\mathscr{C}^0$, while the corresponding numbers for a line of $\mathscr{L}_m\cap\mathscr{R}'$ are n, $\sqrt{q}+1-n$, and $q-\sqrt{q}$.

$\mathscr{S}_2=\{$lines of $\mathscr{L}_m$ not on $\mathscr{H}$ through a point of $\mathscr{W}'\}$. Each point of $\mathscr{W}'$ lies on $m-2$ such lines, each of which is tangent to $\mathscr{H}$. So $|\mathscr{S}_2|=nn'(m-2)$.

$\mathscr{S}_3=\{$lines of $\mathscr{L}_m$ through a point of $(\mathscr{C}^+\cap\mathscr{W})\backslash\mathscr{W}'\}$. There are $n(\sqrt{q}+1-n')+n'(\sqrt{q}+1-n)=(\sqrt{q}+1)(n+n')-2nn'$ such points and $m-1$ lines of $\mathscr{L}_m$ through each, all of which are tangent to $\mathscr{H}$. So $|\mathscr{S}_3|=(m-1)[(\sqrt{q}+1)(n+n')-2nn']$.

$\mathscr{S}_4=\{$lines of $\mathscr{L}_m$ through a point of $\mathscr{C}^+\backslash\mathscr{W}\}$. The number of such points is $(\sqrt{q}+1)^2-n(\sqrt{q}+1-n')-n'(\sqrt{q}+1-n)-nn'=(\sqrt{q}+1-n)\times(\sqrt{q}+1-n')$. There are m lines of $\mathscr{L}_m$ all tangent to $\mathscr{H}$ through each of these points. So $|\mathscr{S}_4|=m(\sqrt{q}+1-n)(\sqrt{q}+1-n')$.

$\mathscr{S}_5=\{$lines of $\mathscr{L}_m$ through a point of $\mathscr{C}^0\cap\mathscr{W}\}$. The number of such points is $(n+n')(q-\sqrt{q})$. Each lies on $m-1$ lines of $\mathscr{L}_m$, all of which are bisecants to $\mathscr{H}$. So $|\mathscr{S}_5|=(m-1)(n+n')(q-\sqrt{q})/2$.

$\mathscr{S}_6=\{$lines of $\mathscr{L}_m$ through a point of $\mathscr{C}^0\backslash\mathscr{W}\}$. The number of such points is $2\sqrt{q}(q-1)-(n+n')(q-\sqrt{q})=(q-\sqrt{q})[2(\sqrt{q}+1)-(n+n')]$. Each lies on m lines of $\mathscr{L}_m$ which are bisecants of $\mathscr{H}$. So $|\mathscr{S}_6|=m(q-\sqrt{q})\times[2(\sqrt{q}+1)-(n+n')]/2$.

$$\text{Therefore}\quad k=m(q\sqrt{q}+1)=\sum|\mathscr{S}_i| = (\sqrt{q}+1)[m(q+1)-(n+n')\sqrt{q}/2].$$

So $(n+n')\sqrt{q}/2=m\sqrt{q}$, whence $n+n'=2m$. □

Lemma 19.3.4: *Let l_1 and l_2 be skew lines of a regular system $\mathcal{L}_m$ on $\mathcal{U}$ and let N_0 and N_2 be the numbers of lines of $\mathcal{L}_m$ respectively skew to and meeting both l_1 and l_2. Then*

$$N_0 = (q - \sqrt{q})[m(\sqrt{q} - 1) + 2],$$
$$N_2 = m(\sqrt{q} + 1) - 2\sqrt{q}.$$

Proof: Let Ω be the set of quadrics $\mathcal{H}$ permutable with $\mathcal{U}$ and containing both l_1 and l_2. Then, from the proof of Lemma 19.3.1, the number of lines of $\mathcal{U}$ skew to both l_1 and l_2 is $\sqrt{q}(\sqrt{q} - 1)(q + 1)$. So $w = |\Omega| = \sqrt{q}(\sqrt{q} - 1)(q + 1)/(\sqrt{q} - 1) = \sqrt{q}(q + 1)$.

For any $\mathcal{H}$ in Ω, let $n(\mathcal{H})$ and $n'(\mathcal{H})$ respectively denote the numbers of lines of $\mathcal{L}_m$ in the regulus on $\mathcal{H}$ containing l_1 and l_2 and in its complement. Then, by Lemma 19.3.3,

$$n(\mathcal{H}) + n'(\mathcal{H}) = 2m. \tag{19.3}$$

Each line l of $\mathcal{L}_m$ skew to l_1 and l_2 determines an $\mathcal{H}$ in Ω. Since any $\mathcal{H}$ in Ω contains $n(\mathcal{H}) - 2$ lines l, the total number N_0 of such lines is

$$N_0 = \sum_{\mathcal{H} \in \Omega} [n(\mathcal{H}) - 2] = \sum n(\mathcal{H}) - 2w. \tag{19.4}$$

Now, if l' is any line of $\mathcal{U}$ meeting l_1 and l_2, the number of lines of $\mathcal{U}$ meeting l' and skew to l_1 and l_2 is $\sqrt{q}(q - 1)$. So the number of $\mathcal{H}$ in Ω containing l' is $\sqrt{q}(q - 1)/(\sqrt{q} - 1) = q + \sqrt{q}$. So, counting the set

$$\{(l', \mathcal{H}) \mid l' \in \mathcal{L}_m, \quad \mathcal{H} \in \Omega, \quad l' \text{ meets } l_1 \text{ and } l_2\},$$

we have

$$N_2(q + \sqrt{q}) = \sum_{\mathcal{H} \in \Omega} n'(\mathcal{H}). \tag{19.5}$$

So, from (19.3), (19.4), and (19.5),

$$N_0 + N_2(q + \sqrt{q}) = 2mw - 2w = 2(m - 1)\sqrt{q}(q + 1). \tag{19.6}$$

The k lines of $\mathcal{L}_m$ are partitioned into four classes with respect to l_1 and l_2; one consists of l_1 and l_2 themselves and the others consist of those lines of $\mathcal{L}_m$ meeting the pair $\{l_1, l_2\}$ in 0, 1, and 2 points respectively. These four classes contain respectively

$$2, \quad N_0, \quad 2[(q + 1)(m - 1) - N_2], \quad N_2$$

elements. So

$$k = m(q\sqrt{q} + 1) = 2 + N_0 + 2[(q + 1)(m - 1) - N_2] + N_2,$$

whence

$$N_0 - N_2 = m(q\sqrt{q} - 2q - 1) + 2q. \tag{19.7}$$

Solving (19.6) and (19.7) gives the result. □

Theorem 19.3.5: *If $\mathcal{L}_m$ is a regular system of order m on $\mathcal{U}$, then $m=\sqrt{q}+1$ or $m=(\sqrt{q}+1)/2$.*

Proof: Let l_0 be any line of $\mathcal{L}_m$ and let us count the number N of pairs (l, l') of lines of $\mathcal{L}_m$ such that l is skew to l_0 and that l' meets both l_0 and l (Fig. 19.1).

Through each point of l_0 there are $m-1$ lines l' of $\mathcal{L}_m$, which gives $(q+1)(m-1)$ lines l'. For any such l', there are $m-1$ lines l through each of the q points of $l' \backslash l_0$. Hence

$$N=q(q+1)(m-1)^2. \tag{19.8}$$

On the other hand, the number of lines l skew to l_0 is $(k-1)-(m-1)\times(q+1)=mq(\sqrt{q}-1)+q$. Then, for any such l, the number of lines l' meeting l_0 and l_2 is, from Lemma 19.3.4, $N_2=m(\sqrt{q}+1)-2\sqrt{q}$. Hence $N=[mq(\sqrt{q}-1)+q][m(\sqrt{q}+1)-2\sqrt{q}]$. So, equating this value of N with that of (19.8), we have

$$q(q+1)(m-1)^2=m^2q(q-1)-mq(2q-3\sqrt{q}-1)-2q\sqrt{q}.$$

Hence

$$2m^2-3m(\sqrt{q}+1)+(\sqrt{q}+1)^2=0;$$

that is,

$$[2m-(\sqrt{q}+1)][m-(\sqrt{q}+1)]=0,$$

whence

$$m=\sqrt{q}+1 \quad \text{or} \quad m=(\sqrt{q}+1)/2. \quad \square$$

Corollary 1: *$\mathcal{U}_{3,q}$ cannot be partitioned by lines.* □

Corollary 2: *For q even, the only regular system on $\mathcal{U}$ is $\mathcal{U}^{(1)}$.* □

A regular system $\mathcal{L}_m$ with $m=(\sqrt{q}+1)/2$ is a *hemisystem*. By Corollary 2, a hemisystem can only exist for q odd and the remainder of this section is devoted to constructing hemisystems on $\mathcal{U}_{3,9}$.

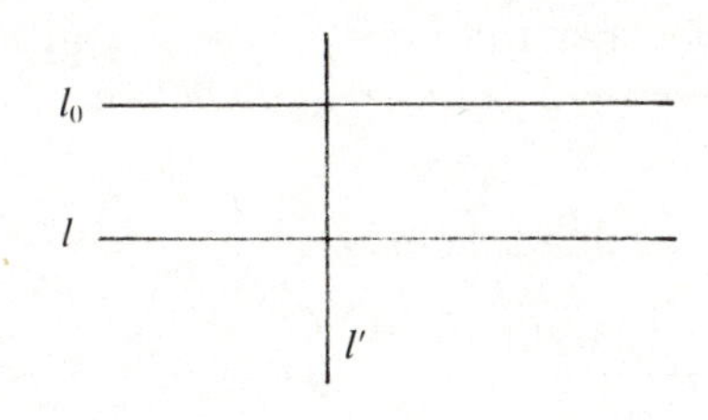

Fig. 19.1

We firstly list some properties of Hermitian varieties over $GF(9)$.

(i) $\mathcal{U}_{1,9}$ is a harmonic and therefore a superharmonic tetrad of points; see § 6.1.

(ii) $\mathcal{U}_{2,9}$ consists of 28 points with a tangent having 4-point contact at each point. The other 63 lines meet $\mathcal{U}_{2,9}$ in a $\mathcal{U}_{1,9}$.

(iii) $\mathcal{U} = \mathcal{U}_{3,9}$ consists of 280 points on 112 lines, 10 points on each line, and 4 lines through each point forming an harmonic set.

(iv) Three skew lines on $\mathcal{U}$ determine a regulus $\mathcal{R}$ with one more line on $\mathcal{U}$ forming a harmonic set: the complementary regulus $\mathcal{R}'$ also has a harmonic tetrad of lines on $\mathcal{U}$. The total number of such tetrads is $2^3 \cdot 3^4 \cdot 5 \cdot 7$ and the number of quadrics $\mathcal{H}(\mathcal{R}, \mathcal{R}')$ permutable with $\mathcal{U}$ is $2^2 \cdot 3^4 \cdot 5 \cdot 7$.

(v) $PGU(4, 9)$ has order $2^9 \cdot 3^6 \cdot 5 \cdot 7$. It is transitive on the set of harmonic tetrads of skew lines each of which has a projective group of order $2^6 \cdot 3^2$ isomorphic to $\mathbf{S}_4 \times \mathbf{S}_4$.

(vi) If a hemisystem $\mathcal{L}_2$ exists on $\mathcal{U}$, then it comprises 56 lines, each of which meets 10 other lines of $\mathcal{L}_2$ and is skew to 45. Any pair of lines of $\mathcal{L}_2$ have two transversals; two skew lines are both skew to 36 lines of $\mathcal{L}_2$.

Lemma 19.3.6: *If l and l' are two skew lines of $\mathcal{U}$, then the ten lines of $\mathcal{U}$ meeting both l and l' define a collineation from l to l'.*

Proof: By Lemma 6.1.1, there are 30 harmonic tetrads on a line. If $\mathcal{R}$ is a regulus containing l and l' and $\mathcal{H}(\mathcal{R}, \mathcal{R}')$ a quadric permutable with $\mathcal{U}$, then the harmonic tetrad of lines $\mathcal{R}' \cap \mathcal{U}^{(1)}$ cuts out a harmonic set of points on each line of $\mathcal{R}$, and in particular on l and l'. From the proof of Lemma 19.3.4, there are $w = 30$ quadrics $\mathcal{H}(\mathcal{R}, \mathcal{R}')$ permutable with $\mathcal{U}$ and such that $\mathcal{R}$ contains l and l'. So there are thirty tetrads of lines $\mathcal{R}' \cup \mathcal{U}^{(1)}$ cutting out all thirty harmonic tetrads on l and l'. So the correspondence from l to l' defined by their ten transversals preserves harmonic sets and is therefore a collineation. □

A hemisystem will be constructed by considering properties of skew quadrilaterals on $\mathcal{U}$.

Lemma 19.3.7: *The lines of $\mathcal{U}$ form $2^2 \cdot 3^6 \cdot 5 \cdot 7$ skew quadrilaterals on which $PGU(4, 9)$ operates transitively. The stabilizer in $PGU(4, 9)$ of a quadrilateral $\mathcal{Q}$ has order 2^7 and contains a normal subgroup of order 2^4 fixing each side of $\mathcal{Q}$.*

Proof: The number of quadrilaterals is

$$\frac{112 \cdot 81}{2} \cdot \frac{10 \cdot 9}{2} \cdot \frac{1}{2} = 2^2 \cdot 3^6 \cdot 5 \cdot 7.$$

So the stabilizer of $\mathcal{Q}$ has order $(2^9 \cdot 3^6 \cdot 5 \cdot 7)/(2^2 \cdot 3^6 \cdot 5 \cdot 7) = 2^7$. The

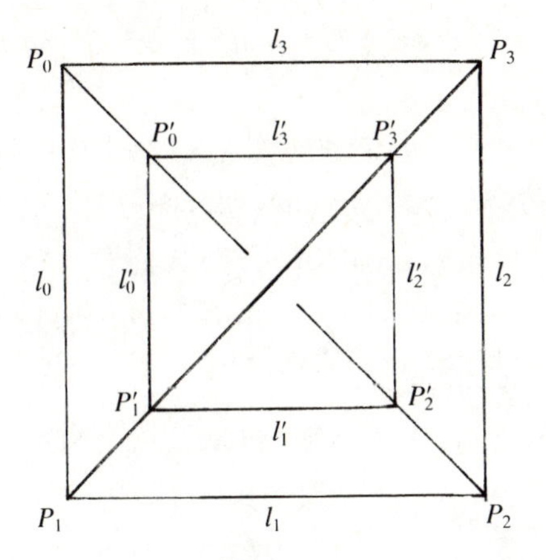

Fig. 19.2

permutation group of $\mathcal{Q}$ is the dihedral group $\mathbf{D}_8$, whence the stabilizer has a subgroup of order 2^4 fixing each side. □

Let P_0, P_1, P_2, P_3 be the vertices in this order of a skew quadrilateral $\mathcal{Q}$ on $\mathcal{U}$ and let $P_0P_2 \cap \mathcal{U} = \{P_0, P_2, P'_0, P'_2\}$, $P_1P_3 \cap \mathcal{U} = \{P_1, P_3, P'_1, P'_3\}$. Write $\mathcal{Q} = P_0P_1P_2P_3$ and $\mathcal{Q}' = P'_0P'_1P'_2P'_3$ (Fig. 19.2).

Lemma 19.3.8: *$\mathcal{Q}'$ is a skew quadrilateral on $\mathcal{U}$, which determines $\mathcal{Q}$ in the same way.*

Proof: Since P_0 and P_2 are both conjugate to P_1 and P_3, the diagonals P_0P_2 and P_1P_3 are polars, whence P'_0 and P'_2 are both conjugate to P'_1 and P'_3. □

The quadrilaterals $\mathcal{Q}$ and $\mathcal{Q}'$ are called *opposites*.

In relation to $\mathcal{Q}$, the set $\mathcal{U}^{(1)}$ is partitioned into five subsets:

$\mathcal{S}_1 = \mathcal{S}_1^{\mathcal{Q}} = \{l_0 = P_0P_1,\ l_1 = P_1P_2,\ l_2 = P_2P_3,\ l_3 = P_3P_0\}$, $\quad |\mathcal{S}_1| = 4$;

$\mathcal{S}_2 = \mathcal{S}_2^{\mathcal{Q}} = \{$lines other than a side through a vertex of $\mathcal{Q}\}$, $\quad |\mathcal{S}_2| = 8$;

$\mathcal{S}_3 = \mathcal{S}_3^{\mathcal{Q}} = \{$lines other than a side meeting a pair of opposite sides of $\mathcal{Q}\}$, $\quad |\mathcal{S}_3| = 16$;

$\mathcal{S}_4 = \mathcal{S}_4^{\mathcal{Q}} = \{$lines meeting exactly one side of $\mathcal{Q}\}$, $\quad |\mathcal{S}_4| = 64$;

$\mathcal{S}_5 = \mathcal{S}_5^{\mathcal{Q}} = \{$lines skew to the sides of $\mathcal{Q}\}$, $\quad |\mathcal{S}_5| = 20$.

Lemma 19.3.9: *The 20 lines of $\mathcal{S}_5$ form five skew quadrilaterals which with $\mathcal{Q}$ comprise three opposite pairs; any two sides of different quadrilaterals are skew.*

Proof: Unlike § 1.7, we write the elements of $GF(9)$ as $0, \pm 1, \pm\nu, \pm\nu^2, \pm\nu^3$, where

$$\nu^2-\nu-1=\nu^3-\nu^2-\nu=\nu^3+\nu^2+1=\nu^3+\nu-1=\nu^4+1=0.$$

Let $\mathcal{Q}=\mathbf{U}_0\mathbf{U}_1\mathbf{U}_2\mathbf{U}_3$. Then we may choose

$$\mathcal{U}=\mathbf{V}(x_0x_2(x_0^2+x_2^2)-x_1x_3(x_1^2+x_3^2)).$$

Then the 16 projectivitics $\mathfrak{T}$ fixing all the vertices of $\mathcal{Q}$ have the set of matrices $\{T=\operatorname{diag}(1, \alpha, e, e\alpha^{-3}) \mid e^2=\alpha^8=1\}$. They form a group isomorphic to $\mathbf{Z}_2\times\mathbf{Z}_8$.

The diagonals of $\mathcal{Q}$ are $\mathbf{V}(x_0, x_2)$ and $\mathbf{V}(x_1, x_3)$. Hence the vertices of $\mathcal{Q}'$ are

$$P_0'=\mathbf{P}(1, 0, \nu^2, 0), \qquad \mathbf{P}_2'=P(1, 0, -\nu^2, 0)$$
$$P_1'=\mathbf{P}(0, 1, 0, \nu^2), \qquad \mathbf{P}_3'=P(0, 1, 0, -\nu^2).$$

The sides of $\mathcal{Q}$ are

$$\mathbf{V}(x_0, x_1), \qquad \mathbf{V}(x_1, x_2), \qquad \mathbf{V}(x_2, x_3), \qquad \mathbf{V}(x_3, x_0),$$

and those of $\mathcal{Q}'$ are

$$\mathbf{V}(x_0\pm\nu^2x_2, x_1\pm\nu^2x_3).$$

Hence the other 16 lines of $\mathcal{S}_5$ are

$$l_{\varepsilon,\alpha}=\mathbf{V}(x_1-\alpha x_0+\varepsilon\alpha^{-3}x_2, x_3-\varepsilon\alpha x_0+\alpha^{-3}x_2)$$
$$=\mathbf{l}(\varepsilon\alpha^{-3}, -1, \alpha^{-3}, -\alpha, \alpha^{-2}, \varepsilon\alpha)$$

with $-\varepsilon^2=\alpha^8=1$. The condition that $l_{\varepsilon,\alpha}$ meets $l_{\rho,\beta}$ is that $\varepsilon\rho(\alpha^4+\beta^4)=(\alpha+\beta)^4$. Hence the sides of one quadrilateral and its opposite have the following parameters (ε, α):

$$(\nu^2, 1), \quad (-\nu^2, \nu^2), \quad (\nu^2, -1), \quad (-\nu^2, -\nu^2);$$
$$(-\nu^2, 1), \quad (\nu^2, \nu^2), \quad (-\nu^2, -1), \quad (\nu^2, -\nu^2).$$

The remaining pair of opposite quadrilaterals is similarly as follows:

$$(\nu^2, \nu), \quad (-\nu^2, \nu^3), \quad (\nu^2, -\nu), \quad (-\nu^2, -\nu^3);$$
$$(-\nu^2, \nu), \quad (\nu^2, \nu^3), \quad (-\nu^2, -\nu), \quad (\nu^2, -\nu^3). \quad \square$$

The three pairs of quadrilaterals are also called an *associated hexad.*

Corollary: *For each quadrilateral of an associated hexad, the set of 24 lines of* $\mathcal{S}_1\cup\mathcal{S}_5$ *is the same.* □

The next lemma relates the sets $\mathcal{S}_2$, $\mathcal{S}_3$, $\mathcal{S}_4$ for different quadrilaterals in a hexad. Firstly, we recall from § 6.4 that a harmonic tetrad on

$PG(1, 9)$ determines a partition of the remaining sextuple of points into three pairs any two of which form a harmonic tetrad.

Lemma 19.3.10: (i) *The lines of $\mathcal{S}_2^{\mathcal{Q}}$ are the lines of $\mathcal{S}_2^{\mathcal{Q}'}$, where $\mathcal{Q}'$ is opposite to $\mathcal{Q}$, but belong to $\mathcal{S}_3^{\mathcal{Q}_i}$ for any of the four other quadrilaterals $\mathcal{Q}_i$ in the hexad. Any such line l meets the four $\mathcal{Q}_i$ in four pairs of points, each of which forms a harmonic tetrad with the vertices of $\mathcal{Q}$ and $\mathcal{Q}'$ on l. The two pairs of opposite quadrilaterals determine complementary harmonic tetrads.*

(ii) *A pair of opposite sides of $\mathcal{Q}$ and a pair of opposite sides of $\mathcal{Q}'$ lie on a quadric $\mathcal{H}$ permutable with $\mathcal{U}$. The other four lines of $\mathcal{H} \cap \mathcal{U}$ are in both $\mathcal{S}_3^{\mathcal{Q}}$ and $\mathcal{S}_3^{\mathcal{Q}'}$. The 16 lines of $\mathcal{S}_3^{\mathcal{Q}}$ are those of $\mathcal{S}_3^{\mathcal{Q}'}$ and are partitioned into four harmonic tetrads, each meeting a pair of opposide sides of $\mathcal{Q}$ and $\mathcal{Q}'$. Also, each of these 16 lines is in $\mathcal{S}_2^{\mathcal{Q}_i}$ for one pair of opposite quadrilaterals $\mathcal{Q}_i$ and in $\mathcal{S}_3^{\mathcal{Q}_i}$ for the other pair.*

(iii) *The 64 lines of $\mathcal{S}_4^{\mathcal{Q}}$ are in $\mathcal{S}_4^{\mathcal{Q}_i}$ for all quadrilaterals $\mathcal{Q}_i$ in the hexad. Also, the three pairs of points on one of these lines l in which it meets the three pairs of opposite quadrilaterals is a partition as above of the sextuple residual to a harmonic tetrad.*

Proof: Since the diagonals $P_0P_2P_0'P_2'$ and $P_1P_3P_1'P_3'$ are polars, the joins of the four points on one to the four points on the other give 16 lines on $\mathcal{U}$. Apart from the eight sides of $\mathcal{Q}$ and $\mathcal{Q}'$, we have eight lines P_iP_j' with $i-j$ odd which form $\mathcal{S}_2^{\mathcal{Q}}$ as well as $\mathcal{S}_2^{\mathcal{Q}'}$.

As above, let $P_0 = \mathbf{U}_0$ and $P_1' = \mathbf{P}(0, 1, 0, \nu^2)$ and so an arbitrary point of P_0P_1' is $\mathbf{P}(\rho, 1, 0, \nu^2)$, which lies on $l_{\varepsilon,\alpha}$ when $\varepsilon = \nu^2$ and $\rho = \alpha^{-1}$. The rest of (i) now follows.

To consider a pair of opposite sides of $\mathcal{Q}$ and a pair of $\mathcal{Q}'$, we may take

$$\mathbf{V}(x_0, x_1), \mathbf{V}(x_2, x_3), \mathbf{V}(x_0 + \nu^2 x_2, x_1 + \nu^2 x_3), \mathbf{V}(x_0 - \nu^2 x_2, x_1 - \nu^2 x_3).$$

These lines all lie on the quadric $\mathcal{H} = \mathbf{V}(x_0x_3 - x_1x_2)$ and so form a harmonic tetrad, by Lemma 19.3.1. The remaining four lines of $\mathcal{H} \cap \mathcal{U}$ are $\mathbf{V}(x_0 - tx_1, x_2 - tx_3)$ with $t^4 = 1$. This gives (ii).

The first part of (iii) follows from (i) and (ii). For the remainder, an arbitrary point on a typical line of $\mathcal{S}_4$ is $Q(\rho) = \mathbf{P}(\rho + \nu^2, \rho - \nu^2, 1, 1)$. Then $Q(\infty)$ is on the side $\mathbf{V}(x_2, x_3)$ of $\mathcal{Q}$ and $Q(0)$ is on the side $\mathbf{V}(x_0 - \nu^2 x_2, x_1 + \nu^2 x_3)$ of $\mathcal{Q}'$. Apart from these, $Q(\rho)$ lies on $l_{\varepsilon,\alpha}$ in the following four cases:

$$\begin{array}{lll} \varepsilon = \nu^2, & \alpha = -1, & \rho = -\nu^2; \\ \varepsilon = -\nu^2, & \alpha = 1, & \rho = \nu^2; \\ \varepsilon = -\nu^2, & \alpha = -\nu^3, & \rho = -1; \\ \varepsilon = -\nu^2, & \alpha = -\nu, & \rho = 1. \quad \square \end{array}$$

Lemma 19.3.11: *The 64 lines of $\mathcal{S}_4$ are the sides of 48 skew quadrilaterals, with each line a side of three quadrilaterals.*

Proof: Take a pair of opposite sides of $\mathscr{Q}$ and of $\mathscr{Q}'$, for example l_0, l_2, l_0', l_2'. As these are mutually skew, the ten transversals of any pair, say l_0 and l_0', define a collineation $\mathfrak{T}(l_0, l_0')$ from l_0 to l_0', by Lemma 19.3.6. Similarly, we have the collineations $\mathfrak{T}(l_0', l_2)$, $\mathfrak{T}(l_2, l_2')$, $\mathfrak{T}(l_2', l_0)$ with the following effects on the vertices of $\mathscr{Q}$ and $\mathscr{Q}'$.

$$\begin{aligned} &\mathfrak{T}(l_0, l_0'): && P_0 \to P_1', && P_1 \to P_0';\\ &\mathfrak{T}(l_0', l_2): && P_1' \to P_2, && P_0' \to P_3;\\ &\mathfrak{T}(l_2, l_2'): && P_2 \to P_3', && P_3 \to P_2';\\ &\mathfrak{T}(l_2', l_0): && P_3' \to P_0, && P_2' \to P_1. \end{aligned}$$

The product of these four collineations is a projectivity $\mathfrak{T}$ from l_0 to itself fixing P_0 and P_1. However, $\mathfrak{T}$ also fixes the four points in which the four transversals of l_0, l_2, l_0', l_2', obtained as in Lemma 19.3.10(ii), meet l_0. So $\mathfrak{T}$ is the identity. Now, let the remaining four points on l_0 be denoted by Q_0, $\tilde{Q}_0$, R_0, $\tilde{R}_0$ so that, as in Lemma 19.3.10(iii), the partition of a sextuple residual to a harmonic tetrad is

$$l_0: (P_0, P_1), (Q_0, \tilde{Q}_0), (R_0, \tilde{R}_0).$$

Similarly, on l_0', l_2, l_2', we have partitions

$$\begin{aligned} &l_0': (P_1', P_0'), (Q_0', \tilde{Q}_0'), (R_0', \tilde{R}_0');\\ &l_2: (P_2, P_3), (Q_2, \tilde{Q}_2), (R_2, \tilde{R}_2);\\ &l_2': (P_3', P_2'), (Q_2', \tilde{Q}_2'), (R_2', \tilde{R}_2'), \end{aligned}$$

obtained by applying the collineations obtained above. This gives the four skew quadrilaterials

$$Q_0Q_0'Q_2Q_2', \qquad \tilde{Q}_0\tilde{Q}_0'\tilde{Q}_2\tilde{Q}_2', \qquad R_0R_0'R_2R_2', \qquad \tilde{R}_0\tilde{R}_0'\tilde{R}_2\tilde{R}_2',$$

whose sides are 16 lines in $\mathscr{S}_4^{\mathscr{Q}}$ and $\mathscr{S}_4^{\mathscr{Q}'}$.

The above process can be carried out in four ways according to the number of choices of a pair of opposite sides of $\mathscr{Q}$ and a pair of $\mathscr{Q}'$. This gives all 64 lines of $\mathscr{S}_4^{\mathscr{Q}}$. But instead of $(\mathscr{Q}, \mathscr{Q}')$ we could use any of the other two pairs of opposite quadrilaterals in the hexad. This gives $4 \cdot 4 \cdot 3 = 48$ skew quadrilaterals formed by the 64 lines. □

Lemma 19.3.12: *A hemisystem $\mathscr{L}_2$ which contains the skew quadrilateral $\mathscr{Q}$ contains the associated hexad. The other 32 lines of $\mathscr{L}_2$ are all in $\mathscr{S}_4$.*

Proof: From property (vi) above of $\mathscr{U}_{3,9}$, any two skew lines of $\mathscr{L}_2$ have two transversals. So the number of skew quadrilaterals is $56 \cdot 45/4 = 630$. Let $\mathscr{Q}$ be one of these. Through each of the 32 points other than the vertices on the sides of $\mathscr{Q}$, there is one line of $\mathscr{L}_2$ other than the side. So

$\mathcal{L}_2$ contains $56-4-32=20$ lines skew to the sides of $\mathcal{Q}$, which are the lines of $\mathcal{S}_5$. □

To construct $\mathcal{L}_2$ we must choose through each of the 32 points on the sides of $\mathcal{Q}$ other than the vertices one of the two lines of $\mathcal{S}_4$.

We consider one of the systems of 16 lines obtained in Lemma 19.3.11 as the sides of four quadrilaterals. Consider one, $Q_0Q_0'Q_2Q_2'$. The two sides through a vertex are the two lines of $\mathcal{S}_4$ through the point, so that exactly one of these sides is in $\mathcal{L}_2$.

Lemma 19.3.13: *If Q_0Q_0' and Q_2Q_2' are in $\mathcal{L}_2$, then so are*

$$\tilde{Q}_0\tilde{Q}_0', \quad \tilde{Q}_2\tilde{Q}_2'; \qquad R_0R_2', \quad R_2R_0'; \qquad \tilde{R}_0\tilde{R}_2', \quad \tilde{R}_2\tilde{R}_0'.$$

Proof: Suppose that $\tilde{Q}_0\tilde{Q}_0'$ is not in $\mathcal{L}_2$. As in Lemma 19.3.10, let $l_0=\mathbf{V}(x_2, x_3)$ and $l_0'=\mathbf{V}(x_0-\nu^2x_2, x_1+\nu^2x_3)$. Then we may take

$$Q_0Q_0'=\mathbf{V}(-\nu^3x_0-x_1+\nu^2x_3, x_2-\nu^2x_3);$$

it lies in both $\mathcal{S}_4^{\mathcal{Q}}$ and $\mathcal{S}_4^{\mathcal{Q}'}$. Also

$$Q_0=\mathbf{P}(1, -\nu^3, 0, 0), Q_0'=\mathbf{P}(-\nu^3, \nu^2, \nu, -1).$$

Since there is the harmonic separation $H(P_0', P_1'; Q_0', \tilde{Q}_0')$, with $P_0'=\mathbf{P}(1, 0, \nu^2, 0)$ and $P_1'=\mathbf{P}(0, 1, 0, \nu^2)$, it follows that $\tilde{Q}_0'=\mathbf{P}(-\nu^3, -\nu^2, \nu, 1)$.

The quadric $\mathcal{H}$ containing P_0P_3, P_1P_2 and Q_0Q_0' is $\mathcal{H}=V(x_1x_3-\nu^2x_0x_2+\nu x_2x_3)$: it also contains $l_0=P_0P_1Q_0$. By construction, $\mathcal{H}$ is permutable with $\mathcal{U}$ and so $\mathcal{H}$ meets $\mathcal{U}$ in eight lines, of which exactly four are in $\mathcal{L}_2$, from Lemma 19.3.3. By hypothesis, these are P_0P_3, P_1P_2, Q_0Q_0', and P_0P_1. If the line of $\mathcal{R}(P_0P_3, P_1P_2, Q_0Q_0')$ on $\mathcal{U}$ not in $\mathcal{L}_2$ is l, then l meets P_0P_1 in a point P such that $H(P_0, P_1; Q_0, P)$; hence $P=\tilde{Q}_0$. Suppose that l is not in $\mathcal{S}_4$ and is therefore in $\mathcal{S}_3$. Then l meets P_2P_3 and so P_2P_3 lies on $\mathcal{H}$ as it has three points on $\mathcal{H}$. As we already have four lines of $\mathcal{L}_2$ on $\mathcal{H}$, this is impossible. So l is a line through $\tilde{Q}_0$ not in $\mathcal{L}_2$ but in $\mathcal{S}_4$. Therefore $l=\tilde{Q}_0\tilde{Q}_0'$. As l lies on $\mathcal{H}$, so does $\tilde{Q}_0'$, which the above coordinates show to be false.

It has been shown that if Q_0Q_0' is in $\mathcal{L}_2$, so is $\tilde{Q}_0\tilde{Q}_0$. Similarly, if Q_2Q_2' is in $\mathcal{L}_2$, so is $\tilde{Q}_2\tilde{Q}_2'$. For the rest, it suffices to prove that none of R_0R_0', R_2R_2', $\tilde{R}_0\tilde{R}_0'$, $\tilde{R}_2\tilde{R}_2'$ is in $\mathcal{L}_2$.

Consider the skew tetrad of lines of $\mathcal{U}$

$$Q_0Q_0', \qquad \tilde{Q}_0\tilde{Q}_0', \qquad R_0R_0', \qquad \tilde{R}_0\tilde{R}_0',$$

which meet $l_0=P_0P_1$ and $l_0'=P_0'P_1'$ in the harmonic tetrads $(Q_0\tilde{Q}_0R_0\tilde{R}_0)$ and $(Q_0'\tilde{Q}_0'R_0'\tilde{R}_0')$. So the tetrad of lines is harmonic. So it has a tetrad of transversals on $\mathcal{U}$. By Lemma 19.3.3, exactly four lines of these two tetrads are in $\mathcal{L}_2$ and these are l_0, l_1, Q_0Q_0', $\tilde{Q}_0\tilde{Q}_0'$. So R_0R_0' and $\tilde{R}_0\tilde{R}_0'$ are not in $\mathcal{L}_2$. Similarly R_2R_2' and $\tilde{R}_2\tilde{R}_2'$ are not in $\mathcal{L}_2$. □

Corollary: *The 64 lines of* $\mathcal{S}_4$ *are partitioned into four sets of 16. Each set of 16 consists of the lines meeting a side of* $\mathcal{Q}$ *and a side of* $\mathcal{Q}'$ *chosen from a pair of sides of* $\mathcal{Q}$ *and a pair of* $\mathcal{Q}'$. *A set of 16 is also partitioned into two octuples such that two lines of the same octuple are skew and two lines of different octuples can meet only on a side of* $\mathcal{Q}$ *or* $\mathcal{Q}'$. *The 64 lines are therefore partitioned into eight octuples such that, if* $\mathcal{L}_2$ *contains* $\mathcal{Q}$ *and a line* l *of* $\mathcal{S}_4$, *then* $\mathcal{L}_2$ *contains the whole octuple in which* l *lies.*

Proof: The sets of 16 are obtained from Lemma 19.3.11 and the octuples from Lemma 19.3.13. Lines from the octuple in the lemma and its residual in the set of 16 can only meet on one of l_0, l_2, l_0', l_2'. □

Let $\mathcal{O}_1$, $\mathcal{O}_1'$ be the octuples obtained from l_0, l_2, l_0', l_2' and let $\mathcal{O}_2$, $\mathcal{O}_2'$ be the octuples obtained from l_1, l_3, l_1', l_3'. So, if $\mathcal{L}_2$ contains $\mathcal{Q}$, it contains one of the four sets

$$\mathcal{O}_1 \cup \mathcal{O}_2, \qquad \mathcal{O}_1' \cup \mathcal{O}_2', \qquad \mathcal{O}_1 \cup \mathcal{O}_2', \qquad \mathcal{O}_1' \cup \mathcal{O}_2,$$

each of which consists of 16 lines of $\mathcal{S}_4$. With calculations similar to those above it can be shown, perhaps with $\mathcal{O}_2$ and $\mathcal{O}_2'$ interchanged, that every line of $\mathcal{O}_1 \cup \mathcal{O}_2$ is skew to every line of $\mathcal{O}_1' \cup \mathcal{O}_2'$ but that each line of $\mathcal{O}_1$ meets six lines of $\mathcal{O}_2'$ and each line of $\mathcal{O}_1'$ meets six lines of $\mathcal{O}_2$.

Lemma 19.3.14: *If* $\mathcal{L}_2$ *contains* $\mathcal{Q}$, *it contains* $\mathcal{O}_1 \cup \mathcal{O}_2$ *or* $\mathcal{O}_1' \cup \mathcal{O}_2'$ *but neither* $\mathcal{O}_1 \cup \mathcal{O}_2'$ *nor* $\mathcal{O}_1' \cup \mathcal{O}_2$.

Proof: If $\mathcal{L}_2$ contains $\mathcal{O}_1 \cup \mathcal{O}_2'$, then a line l of $\mathcal{O}_1$ meets six lines of $\mathcal{O}_2'$ and the six quadrilaterals of the hexad. So there are at least two points on l through which pass three lines of $\mathcal{L}_2$. □

Lemma 19.3.15: *Through each point of* $\mathcal{U}$ *on no side of the hexad of quadrilaterals associated to* $\mathcal{Q}$, *there passes exactly one line of* $\mathcal{O}_1 \cup \mathcal{O}_2$ *and one line of* $\mathcal{O}_1' \cup \mathcal{O}_2'$.

Proof: The 16 lines of $\mathcal{O}_1 \cup \mathcal{O}_2$ contain 160 points. Of these points, there are 16 on each quadrilateral of the hexad: this leaves $160-6\cdot16=64$. Since $\mathcal{U}$ contains 280 points and a skew quadrilateral contains $4+4\cdot8=38$, the number of points of $\mathcal{U}$ external to the six quadrilaterals is $280-6\cdot36=64$. □

We are now almost finished. If, instead of considering $\{l_0, l_2, l_0', l_2'\}$ and $\{l_1, l_3, l_1', l_3'\}$ which gave $\mathcal{O}_1$, $\mathcal{O}_1'$, $\mathcal{O}_2$, $\mathcal{O}_2'$, we consider the other pairs of opposite sides of $\mathcal{Q}$ and $\mathcal{Q}'$, namely $\{l_0, l_2, l_1', l_3'\}$ and $\{l_1, l_3, l_0', l_2'\}$, then the sets $\mathcal{P}_1$, $\mathcal{P}_1'$, $\mathcal{P}_2$, $\mathcal{P}_2'$ are obtained. Let

$$\mathcal{O} = \mathcal{O}_1 \cup \mathcal{O}_2, \qquad \mathcal{O}' = \mathcal{O}_1' \cup \mathcal{O}_2', \qquad \mathcal{P} = \mathcal{P}_1 \cup \mathcal{P}_2, \qquad \mathcal{P}' = \mathcal{P}_1' \cup \mathcal{P}_2'.$$

Theorem 19.3.16: *There exist four hemisystems on* $\mathcal{U}$ *containing a given quadrilateral* $\mathcal{Q}$. *They are obtained by adding to the associated hexad of quadrilaterals one of* $\mathcal{O} \cup \mathcal{P}$, $\mathcal{O} \cup \mathcal{P}'$, $\mathcal{O}' \cup \mathcal{P}$, $\mathcal{O}' \cup \mathcal{P}'$.

Proof: From the previous lemmas each of the four given sets consists of 32 lines meeting each of the quadrilaterals associated to $\mathcal{Q}$ in the 32 points distinct from the vertices. □

Lemma 19.3.17: *If $\mathcal{L}_2$ and $\mathcal{L}_2^0$ are hemisystems on $\mathcal{U}$ containing respectively $\{\mathcal{Q}, l\}$ and $\{\mathcal{Q}^0, l^0\}$, where $l \in \mathcal{S}_4^{\mathcal{Q}}$ and $l^0 \in \mathcal{S}_4^{\mathcal{Q}^0}$, there is a unique projectivity $\mathfrak{T}$ fixing $\mathcal{U}$ such that $\mathcal{L}_2\mathfrak{T} = \mathcal{L}_2^0$, $\mathcal{Q}\mathfrak{T} = \mathcal{Q}^0$, $l\mathfrak{T} = l^0$.*

Proof: By the theorem, there are two hemisystems on $\mathcal{U}$ containing $\mathcal{Q}$ and l. There are 64 lines in $\mathcal{S}_4^{\mathcal{Q}}$ and, by Lemma 19.3.7, the stabilizer of $\mathcal{Q}$ in $PGU(4, 9)$ has order 2^7. So there are two projectivities transforming $\mathcal{Q}$ to $\mathcal{Q}^0$ and l to l^0. □

Theorem 19.3.18:

(i) *There are* 648 *hemisystems on $\mathcal{U}$ divided into* 324 *complementary pairs.*

(ii) *$PGU(4, 9)$ operates transitively on the* 648 *hemisystems.*

(iii) *The group $G(\mathcal{L}_2)$ of projectivities fixing $\mathcal{L}_2$ has order $2^6 \cdot 3^2 \cdot 5 \cdot 7$ and is isomorphic to $PSL(3, 4)$.*

(iv) *The stabilizer of a line of $\mathcal{L}_2$ in $G(\mathcal{L}_2)$ is isomorphic to $PSL(2, 9)$.*

(v) *The group $\Gamma(\mathcal{L}_2)$ of collineations fixing a hemisystem has order $2^7 \cdot 3^2 \cdot 5 \cdot 7$ and is isomorphic to $P\Gamma SL(3, 4)$.*

(vi) *The stabilizer of a line of $\mathcal{L}_2$ in $\Gamma(\mathcal{L}_2)$ is isomorphic to $P\Gamma SL(2, 9)$.*

Proof: (i) By Lemma 19.3.7, there are $2^2 \cdot 3^6 \cdot 5 \cdot 7$ skew quadrilaterals in $\mathcal{U}^{(1)}$ and, from Lemma 19.3.12, $2 \cdot 3^2 \cdot 5 \cdot 7$ skew quadrilaterals in $\mathcal{L}_2$. As each quadrilateral lies in four hemisystems, there are $4 \cdot 2 \cdot 3^4 = 648$ hemisystems.

(ii) This follows from Lemma 19.3.17.

(iii) From Lemma 19.3.17, $|G(\mathcal{L}_2)| = (2 \cdot 3^2 \cdot 5 \cdot 7) \cdot 32 = 2^6 \cdot 3^2 \cdot 5 \cdot 7$. Alternatively $|PGU(4, 9)|/648 = (2^9 \cdot 3^6 \cdot 5 \cdot 7)/(2^3 \cdot 3^4) = 2^6 \cdot 3^2 \cdot 5 \cdot 7$, which is the order of $PSL(3, 4)$. The analogous configuration in $PG(2, 4)$ to a hemisystem is an orbit of 56 ovals under $PSL(3, 4)$; see § 14.3.

(iv) $G(\mathcal{L}_2)$ is transitive on the lines of $\mathcal{L}_2$, whence the stabilizer $G(\mathcal{L}_2)_0$ of a line l has order $2^6 \cdot 3^2 \cdot 5 \cdot 7/56 = 10 \cdot 9 \cdot 8/2 = |PSL(2, 9)|$. Also $G(\mathcal{L}_2)_0$ induces a group of projectivities G_0 on l. Thus we have an epimorphism $\phi: G(\mathcal{L}_2)_0 \to G_0$. However, $\mathcal{L}_2$ consists of l, ten lines meeting l and 45 other lines, one meeting each pair of the ten. So if $\mathfrak{T}\phi$ is the identity in G_0, each of the ten lines meeting l is fixed and so are the remaining 45. Therefore $\mathfrak{T}$ is the identity and ϕ is an isomorphism. As $G_0 < PGL(2, 9)$ and $|G_0| = |PSL(2, 9)|$, so $G(\mathcal{L}_2)_0 \cong PSL(2, 9)$.

(v), (vi) The collineation groups are obtained by including the automorphism σ. □

Corollary: *On an elliptic quadric $\mathscr{E}_{5,3}$ in $PG(5, 3)$ the number of* 56*-caps is* 648, *and each has a projective group isomorphic to $P\Gamma SL(3, 4)$.*

Proof: By Theorem 19.2.2, $\mathcal{U}^{(1)}_{3,9}\mathfrak{G} = \mathcal{E}_{5,3}$. Since a hemisystem $\mathcal{L}_2$ comprises 56 lines, no three of which lie in a pencil, so $\mathcal{L}_2\mathfrak{G}$ is a set of 56 points on $\mathcal{E}_{5,3}$, no three of which are collinear; that is, $\mathcal{L}_2\mathfrak{G}$ is a 56-cap on $\mathcal{E}_{5,3}$ with two points on each generator of $\mathcal{E}_{5,3}$. □

One question that remains to be discussed is how the hyperbolic quadrics permutable with $\mathcal{U}$ are related to a hemisystem $\mathcal{L}_2$. By Lemma 19.3.1, the number of such quadrics is

$$n_9 = 81 \cdot 28 \cdot 10/2 = 2^2 \cdot 3^4 \cdot 5 \cdot 7.$$

By lemma 19.3.3, these quadrics are of three types with respect to $\mathcal{L}_2$, where n and n' are the numbers of lines of $\mathcal{L}_2$ in the reguli $\mathcal{R}$ and $\mathcal{R}'$ respectively of $\mathcal{H}(\mathcal{R}, \mathcal{R}')$:

(i) $n = n' = 2$;
(ii) $n = 3$ and $n' = 1$, or $n = 1$ and $n' = 3$;
(iii) $n = 4$ and $n' = 0$, or $n = 0$ and $n' = 4$.

Let the numbers of these quadrics be respectively $N(2, 2)$, $N(3, 1)$, and $N(4, 0)$.

Lemma 19.3.19:
(i) $N(2, 2) = 2^3 \cdot 3^2 \cdot 5 \cdot 7$;
(ii) $N(3, 1) = 2^6 \cdot 3 \cdot 5 \cdot 7$;
(iii) $N(4, 0) = 2^2 \cdot 3 \cdot 5^2 \cdot 7$.

Proof: (i) As in Lemma 19.3.12, $\mathcal{L}_2$ contains 630 skew quadrilaterals. If $\mathcal{H}$ is a quadric of type (i), then $\mathcal{H}$ meets $\mathcal{L}_2$ in a skew quadrilateral $\mathcal{Q}$ with sides l_0, l_1, l_2, l_3. We consider how many other quadrics meet $\mathcal{L}_2$ in $\mathcal{Q}$. Any such quadric $\mathcal{H}'$ contains two other lines of $\mathcal{U}$ in the same regulus as l_0 and l_2; these four lines meet l_1 and l_3, and form a harmonic tetrad. Given l_0 and l_2, the other eight lines on $\mathcal{U}$ incident with l_1 and l_3 fall into four pairs each of which forms a harmonic tetrad with l_0 and l_2. This is exactly the same as on $PG(1, 9)$, where two points are the double points of a unique hyperbolic involution and each of the other four pairs in the involution form a harmonic tetrad with the double points. Hence $N(2, 2) = 4 \cdot 630 = 2^3 \cdot 3^2 \cdot 5 \cdot 7$.

(ii) To determine $N(3, 1)$ we must find the number of ways of choosing a line and three transversals in $\mathcal{L}_2$. So $N(3, 1) = 56\mathbf{c}(10, 3) = 56 \cdot 10 \cdot 9 \cdot 8/6 = 2^6 \cdot 3 \cdot 5 \cdot 7$.

(iii) $N(4, 0) = n_9 - N(2, 2) - N(3, 1) = 2^2 \cdot 3 \cdot 5^2 \cdot 7 \cdot$ □

Lemma 19.3.20: *The group $G(\mathcal{L}_2)$ is transitive on the quadrics of type* (i) *and* (ii).

Proof: This follows from Lemma 19.3.17. □

We now investigate the quadrics of type (iii) or, equivalently, the harmonic tetrads of skew lines in $\mathscr{L}_2$.

Lemma 19.3.21: *Two opposite sides of a quadrilateral $\mathscr{Q}$ on $\mathscr{U}$ and a side of an associated quadrilateral $\mathscr{Q}_1 \neq \mathscr{Q}'$ have as harmonic conjugate a side of $\mathscr{Q}_1'$.*

Proof: Let $\mathscr{H}$ be the quadric containing the opposite sides $\mathbf{V}(x_0, x_1)$ and $\mathbf{V}(x_2, x_3)$ of $\mathscr{Q}$ as well as the line $l_{\varepsilon,\alpha}$. Then, since the first two lines are on $\mathscr{H}$,

$$\mathscr{H} = \mathbf{V}(a_0x_0x_2 + a_1x_1x_2 + a_2x_0x_3 + a_3x_1x_3);$$

since $l_{\varepsilon,\alpha}$ is on $\mathscr{H}$,

$$a_0 + \alpha a_1 - \alpha^{-3}a_2 = a_2 + \alpha a_3 = \alpha^3 a_1 - a_3 = 0.$$

So $l_{-\varepsilon,\alpha}$ also lies on $\mathscr{H}$. □

Let $\{r_0, r_1, r_2, r_3\}$ be a harmonic tetrad of skew lines and consider the pairs $\{r_0, r_1\}$ and $\{r_2, r_3\}$. By Lemma 19.3.4, there are exactly two transversals s_0 and s_1 in $\mathscr{L}_2$ of r_0 and r_1, and similarly two transversals s_2 and s_3 of r_2 and r_3. If s_0 meets r_2, then s_0 lies on the quadric $\mathscr{H}$ containing the four r_i, which contradicts Lemma 19.3.3 that $\mathscr{L}_2$ meets $\mathscr{H}$ in exactly four lines. Hence s_0 and s_1 are skew to both r_2 and r_3, and s_2 and s_3 are skew to both r_0 and r_1.

The pairs $\{r_0, r_1\}$ and $\{r_2, r_3\}$ therefore define the skew quadrilaterals $r_0s_0r_1s_1$ and $r_2s_2r_3s_3$, which are respectively skew to the lines of the pairs $\{r_2, r_3\}$ and $\{r_0, r_1\}$. Let $\mathscr{Q} = r_0s_0r_1s_1$; then, since $\mathscr{Q}$ is in $\mathscr{L}_2$, the associated hexad is in $\mathscr{L}_2$, by Lemma 19.3.12. With $\mathscr{Q}'$ opposite to $\mathscr{Q}$, we note that if r_2 is a side of $\mathscr{Q}'$, then so is r_3, by Lemma 19.3.10(ii). So there are two cases:

(a) both r_2 and r_3 are sides of $\mathscr{Q}'$;

(b) neither r_2 nor r_3 is a side of $\mathscr{Q}'$.

In case (a), $\mathscr{Q}' = r_2s_2r_3s_3$, and s_0 and s_1 are skew to s_2 and s_3. In case (b), r_2 and r_3 are sides of two opposite quadrilaterals, both associated to $\mathscr{Q}$, by Lemma 19.3.21. So the quadrilaterals $r_0s_0r_1s_1$ and $r_2s_2r_3s_3$ are not associated, whence s_0 and s_1 are not skew to s_2 and s_3.

Let N' and N'' be the numbers of sets of two pairs $\{\{r_0, r_1\}, \{r_2, r_3\}\}$ for cases (a) and (b) respectively.

Lemma 19.3.22: $N' = 2^2 \cdot 3^2 \cdot 5 \cdot 7$, $N'' = 2^4 \cdot 3^2 \cdot 5 \cdot 7$.

Proof: The pair $\{r_0, r_1\}$ can be chosen in $56 \cdot 45/2 = 2^2 \cdot 3^2 \cdot 5 \cdot 7$ ways, and, given such a pair, the quadrilateral $\mathscr{Q} = r_0s_0r_1s_1$ is determined. Then case (a) or (b) arises as r_2 and r_3 are sides of the opposite quadrilateral $\mathscr{Q}'$ or of the other four quadrilaterals in the associated hexad. So $N' = (2^2 \cdot 3^2 \cdot 5 \cdot 7)2/2 = 2^2 \cdot 3^2 \cdot 5 \cdot 7$ and $N'' = (2^2 \cdot 3^2 \cdot 5 \cdot 7)8/2 = 2^4 \cdot 3^2 \cdot 5 \cdot 7$. □

Theorem 19.3.23: *Under $G(\mathscr{L}_2)$ there are two orbits of quadrics $\mathscr{H}$ of type* (iii) *corresponding to cases* (a) *and* (b) *of the harmonic tetrad $\mathscr{H} \cap \mathscr{L}_2$. The numbers of quadrics are respectively $M' = 2^2 \cdot 3 \cdot 5 \cdot 7$ and $M'' = 2^4 \cdot 3 \cdot 5 \cdot 7$.*

Proof: We firstly suppose that a harmonic tetrad of lines gives both (a) and (b) using different partitions into two pairs. Then $N'/N'' = r(3-r)$ with $r = 1$ or 2. Hence $N'/N'' = \frac{1}{2}$ or 2, contradicting that $N'/N'' = \frac{1}{4}$. So any partition of a harmonic tetrad always gives the same case.

Finally $M' = N'/3 = 2^2 \cdot 3 \cdot 5 \cdot 7$ and $M'' = N''/3 = 2^4 \cdot 3 \cdot 5 \cdot 7$. We note that $M' + M'' = 2^2 \cdot 3 \cdot 5^2 \cdot 7 = N(4, 0)$, by Lemma 19.3.19. □

19.4 Sets of type $(1, n, q+1)$

The object of this section is to characterize Hermitian surfaces solely in terms of the numbers of points in which a line can meet them.

In $PG(d, q)$ a subset $\mathscr{K}$ is a $k_{n,d,q}$ if n is a fixed integer satisfying $1 \leq n \leq q$ such that

(i) $|\mathscr{K}| = k$;
(ii) $|l \cap \mathscr{K}| = 1$, n or $q+1$ for each line l;
(iii) $|l \cap \mathscr{K}| = n$ for some line l.

A set $\mathscr{K}$ of *type* $(1, n, q+1)$ is one satisfying (ii). Here we particularly wish to classify all $k_{n,3,q}$, but this requires that we classify all $k_{n,2,q}$ as well. The first few lemmas are applied to $k_{n,d,q}$ with arbitrary d, as the general case is no more difficult than the particular.

The restriction $1 \leq n \leq q$ means that $\mathscr{K}$ cannot be the whole space: a subset $\mathscr{S}$ of $PG(d, q)$ such that $|l \cap \mathscr{S}| = q+1$ for all lines l is just $PG(d, q)$ itself.

The line l is as usual called an *i-secant* of $\mathscr{K}$ if $|l \cap \mathscr{K}| = i$. Also 1-secants are sometimes called *unisecants* and $(q+1)$-secants are *lines of* $\mathscr{K}$.

Lemma 19.4.1: *If $\mathscr{K}$ is a $k_{n,d,q}$, then the section of $\mathscr{K}$ by a subspace Π_r is either a subspace Π_s or a $k'_{n,r,q}$.*

Proof: Let $\mathscr{K}' = \Pi_r \cap \mathscr{K}$. Then if a line l meets $\mathscr{K}'$ in two points, it lies in Π_r and is either an n-secant or $(q+1)$-secant of $\mathscr{K}$ and hence of $\mathscr{K}'$. □

There is one important definition to distinguish points of $\mathscr{K}$, our $k_{n,d,q}$. A point of $\mathscr{K}$ is *singular* if there is no n-secant through it. Then $\mathscr{K}$ is called *singular* or *non-singular* as it has singular points or not.

In $PG(d, q)$, a cone $\Pi_r\mathscr{S}$ is the set of points on the joins of Π_r with points of $\mathscr{S}$, where $\mathscr{S}$ is contained in a subspace Π_s skew to Π_r. The set Π_r is the *vertex* and the set $\mathscr{S}$ is a *base* of the cone.

Lemma 19.4.2: *If $\mathscr{K}$ is a $k_{n,d,q}$, then the singular points of $\mathscr{K}$ form a subspace of $PG(d, q)$.*

Proof: If P_1 and P_2 are singular points of $\mathcal{K}$, then the line P_1P_2 lies in $\mathcal{K}$. If P_3 is any other point of P_1P_2 and l_3 is a line other than P_1P_2 through P_3, then either $l_3 \cap \mathcal{K} = \{P_3\}$ or there is some other point Q of $\mathcal{K}$ on l_3. In the latter case, P_2Q lies in $\mathcal{K}$, whence every line through P_1 in the plane $\pi = P_1P_2Q$ belongs to $\mathcal{K}$; hence π lies entirely in $\mathcal{K}$ and so does l_3. Hence l_3 is either a unisecant or a line of $\mathcal{K}$: so P_3 is singular. Therefore the set of singular points of $\mathcal{K}$ is a subspace. □

The subspace of singular points of $\mathcal{K}$ is the *singular space* of $\mathcal{K}$.

Corollary 1: *If $\mathcal{K}$ is a $k_{n,d,q}$, then the following are equivalent*:

(i) $n = 1$;
(ii) *$\mathcal{K}$ is a prime Π_{d-1}*;
(iii) *all points of $\mathcal{K}$ are singular.* □

Corollary 2:

(i) *A $k_{1,2,q}$ is a line.*
(ii) *A $k_{1,3,q}$ is a plane.* □

Lemma 19.4.3: *If $\mathcal{K}$ is a singular $k_{n,d,q}$, then $\mathcal{K}$ is a prime or a cone $\Pi_r\mathcal{K}'$ where $0 \leq r \leq d-2$ and $\mathcal{K}'$ is a non-singular $k'_{n,d-r-1,q}$.*

Proof: Let P be a non-singular and Q a singular point of $\mathcal{K}$; then PQ lies in $\mathcal{K}$. So, if Π_r is the singular space of $\mathcal{K}$, the space $\Pi_{r+1} = P\Pi_r$ lies in $\mathcal{K}$. Hence, if $r = d-1$, then $\Pi_{r+1} = PG(d, q)$, a contradiction; so there is no point P and $\mathcal{K} = \Pi_{d-1}$. If $r < d-1$, then $\mathcal{K}$ is a cone with vertex Π_r.

Now, let Π'_{d+r-1} be skew to Π_r. Then $\Pi'_{d-r-1} \cap \mathcal{K} = \mathcal{K}'$, which is a $k'_{n,d-r-1,q}$. Each point P' of $\mathcal{K}'$ is non-singular on $\mathcal{K}$ and hence there is an n-secant l of $\mathcal{K}$ through P' skew to Π_r. The projection l' of l from Π_r onto Π'_{d-r-1}, that is, $l' = l\Pi_r \cap \Pi'_{d-r-1}$, is an n-secant through P' of $\mathcal{K}'$. So P' is non-singular on $\mathcal{K}'$. □

Corollary 1: *If $\mathcal{K}$ is a singular $k_{n,2,q}$, then $\mathcal{K}$ consists of a single line or n lines through a point.* □

Corollary 2: *If $\mathcal{K}$ is a singular $k_{n,3,q}$, then $\mathcal{K}$ consists of a plane or n planes through a line or $\mathcal{K} = \Pi_0\mathcal{K}'$, where $\mathcal{K}'$ is a non-singular $k'_{n,2,q}$.* □

Theorem 19.4.4: *If $\mathcal{K}$ is a $k_{n,2,q}$, then $\mathcal{K}$ is one of the following, of which* I–IV *are non-singular.*

I. *A Hermitian arc, that is a $(q\sqrt{q}+1; \sqrt{q}+1)$-arc of type $(1, \sqrt{q}+1)$*:

$$k = q\sqrt{q}+1, \qquad n = \sqrt{q}+1.$$

II. *A subplane $PG(2, \sqrt{q})$, that is a $(q+\sqrt{q}+1, \sqrt{q}+1)$-arc of type $(1, \sqrt{q}+1)$*:

$$k = q+\sqrt{q}+1, \qquad n = \sqrt{q}+1.$$

III. *An* $((n-2)q+n-1; n-1)$*-arc plus an external line*:

$$k=(n-1)q+n, \qquad q\equiv 0 \mod (n-1).$$

IV. *The complement in* $PG(2,q)$ *of a* $((q-n)q+q-n+1;\ q-n+1)$*-arc*:

$$k=n(q+1), \qquad q\equiv 0 \mod (q-n+1).$$

V. n *lines through a point*:

$$k=nq+1.$$

VI. *A single line*:

$$k=q+1, \qquad n=1.$$

Proof: If $\mathcal{K}$ does not contain a line, then it is a $(k;n)$-arc of type $(1,n)$ and so is of type I or II, by Theorem 12.3.6, Corollary 1.

If $\mathcal{K}$ contains a line l, then $\mathcal{K}\backslash l$ is a $(k';n-1)$-arc of type $(0,n-1)$ and so is a maximal arc. Hence $k'=(n-2)q+n-1$ and $q\equiv 0 \mod (n-1)$, by Theorem 12.2.1. Hence we have type III.

If $\mathcal{K}$ contains at least two lines, then either $\mathcal{K}$ contains three lines of a triangle or not. In the former case, each line of the plane meets $\mathcal{K}$ in at least two points and so belongs to $\mathcal{K}$ or is an n-secant. If $\mathcal{K}'=PG(2,q)\backslash\mathcal{K}$, then every line is external to $\mathcal{K}'$ or is a $(q+1-n)$-secant. Hence $\mathcal{K}'$ is maximal and so, by Theorem 12.2.1, we have type IV.

Finally, if $\mathcal{K}$ contains at least two lines l_1 and l_2, but not three independent lines, let P_1 be a point of $l_1\backslash l_2$. Then the q lines through P_1 other than l_1 are all n-secants of $\mathcal{K}$; so $k=q+1+(n-1)q=nq+1$. If $P=l_1\cap l_2$ and Q is any point not in $\mathcal{K}$, then the q lines through Q other than QP are all n-secants of $\mathcal{K}$. If $k'=|PQ\cap\mathcal{K}|$, then $k=k'+nq$. So $k'=1$. Hence PQ is a unisecant of $\mathcal{K}$. So there are no n-secants through P. Hence $\mathcal{K}$ consists of n lines through P. □

This theorem is the essential tool in what follows and the plane sections of types I–VI will be continually referred to. For theorems on the existence of maximal arcs as in III and IV, see § 12.2.

Corollary 1: *If* $\mathcal{K}$ *is a* $k_{n,3,q}$, *then a plane of* $PG(3,q)$ *not contained in* $\mathcal{K}$ *meets it in one of the sets* I–VI, *with the same value for* n *except in case* VI. □

Corollary 2: *If* $\mathcal{K}$ *is a singular* $k_{n,3,q}$ *with singular space* Π_r, *then*

(i) *if* $r=2$, $\mathcal{K}$ *is a plane*;

(ii) *if* $r=1$, $\mathcal{K}$ *is a set of* n *planes through a line*;

(iii) *if* $r=0$, $\mathcal{K}$ *is a cone* $\Pi_0\mathcal{K}'$ *where* $\mathcal{K}'$ *is one of the sets* I, II, III, IV *of the theorem.*

Proof: This follows from the theorem and Lemma 19.4.3. □

Lemma 19.4.5: *If $\mathcal{K}$ is a $k_{2,3,q}$, then $\mathcal{K}$ consists of a plane plus either a point, a line, or a plane.*

Proof: Take a point Q not on $\mathcal{K}$. Then every line through Q meets $\mathcal{K}$ in at least one point. So $k \geq q^2+q+1$ and $\mathcal{K}$ contains a plane π, by Theorem 3.2.1.

Let P_1 and P_2 be points of $\mathcal{K}\backslash\pi$. Then $P_1P_2 \cap \pi$ is a third point of $\mathcal{K}$ on P_1P_2, whence P_1P_2 lies in $\mathcal{K}$. If P_3 is a point of $\mathcal{K}\backslash\pi$ not on P_1P_2, then P_2P_3 lies in $\mathcal{K}$ and so does the join of P_1 to every point of P_2P_3 except perhaps $P = P_2P_3 \cap \pi$. However, by considering the lines of $\pi' = P_1P_2P_3$ through P_2, we see that π' lies entirely in $\mathcal{K}$. Then if $\mathcal{K}$ contains a point Q not on π or π', a similar argument about lines through Q makes $\mathcal{K} = PG(3, q)$. □

Lemma 19.4.6: *If $\mathcal{K}$ is a $k_{q,3,q}$ in $\Pi_3 = PG(3, q)$, then $\mathcal{K}$ is one of the following*:

(i) $(\Pi_3\backslash\Pi_2) \cup \Pi_1$, $\Pi_1 \subset \Pi_2$;
(ii) $(\Pi_3\backslash\Pi_1) \cup \Pi_0$, $\Pi_0 \subset \Pi_1$;
(iii) $\Pi_3\backslash\Pi_0$.

Proof: Let Q be a point of $\Pi_3\backslash\mathcal{K}$. If l_1 and l_2 are 1-secants of $\mathcal{K}$ through Q with $P_1 = l_1 \cap \mathcal{K}$ and $P_2 = l_2 \cap \mathcal{K}$, then either P_1P_2 lies in $\mathcal{K}$ or is a q-secant. If, now, P_3 is a third point on P_1P_2 in $\mathcal{K}$, then every line l in the plane $\pi = l_1l_2$ through P_3 other than P_1P_2 and QP_3 contains 2 points $l \cap l_1$ and $l \cap l_2$ off $\mathcal{K}$; so such an l is a 1-secant of $\mathcal{K}$. Hence all the lines of π through P_1 other than P_1P_2 are 1-secants. So the only points of $\pi \cap \mathcal{K}$ lie on P_1P_2.

If P_1P_2 were a q-secant, then the join of Q to the point Q' of $P_1P_2\backslash\mathcal{K}$ would be a 0-secant. So $\pi \cap \mathcal{K} = P_1P_2$. Thus it has been shown that, if two lines through Q are 1-secants, all the lines of the pencil containing these two are 1-secants and the plane of the pencil meets $\mathcal{K}$ in a line. So the 1-secants through Q form a subspace Π_r meeting $\mathcal{K}$ in a subspace Π_{r-1}, with $r = 0$, 1, or 2.

Since each line through Q not in Π_r is a q-secant, $\mathcal{K}$ consists of the points in Π_{r-1} and those not in Π_r. This give the three possibilities listed. □

Since the previous two lemmas characterize the sets $k_{n,3,q}$ for $n = 2$ and $n = q$, we can now restrict our attention to $3 \leq n \leq q-1$.

Lemma 19.4.7: *In $PG(3, q)$, there is no k-set $\mathcal{K}$ such that every line meets $\mathcal{K}$ in 0 or n points, where n is a fixed integer with $2 \leq n \leq q-1$.*

Proof: Let P be a point of such a set $\mathcal{K}$. Then, as each line through P is an n-secant,

$$k = (n-1)(q^2+q+1)+1.$$

Every plane meeting $\mathcal{K}$ meets it in a maximal arc, that is an $((n-1)q+n; n)$-arc. So in any such plane there are 0-secants. Let l be one and let N be the number of planes through l meeting $\mathcal{K}$. Then

$$k = N[(n-1)q+n]$$

whence

$$N = [(n-1)(q^2+q+1)+1]/[(n-1)q+n].$$

So

$$N-(q-1) = [(n-2)q+2n]/[(n-1)q+n];$$

that is, $0 < N-(q-1) < 1$, a contradiction. □

Corollary: *If $\mathcal{K}$ is a $k_{n,3,q}$ with $2 \leq n \leq q-1$, then $\mathcal{K}$ has a unisecant.*

Proof: If there is such a $\mathcal{K}$ with no unisecants, then $\mathcal{K}' = PG(3, q) \backslash \mathcal{K}$ is a set with i-secants only for $i=0$ and $i=q+1-n$. As $2 \leq q+1-n \leq q-1$, the lemma shows that $\mathcal{K}'$ does not exist, whence $\mathcal{K}$ does not. So $\mathcal{K}$ has a unisecant. □

In the subsequent theorems we make continual use of Theorem 19.4.4, Corollary 1. It is important to note that, if $\mathcal{K}$ is a $k_{n,3,q}$, then a plane section of type I–V has the same n as $\mathcal{K}$. For ease of reference, a plane lying on $\mathcal{K}$ will sometimes be called a *plane (section) of type* VII.

It will also be useful to have a table showing the number of i-secants in π through a point P of $\pi \cap \mathcal{K}$ and the number through a point Q of $\pi \backslash \mathcal{K}$ in the cases that π is of types III, IV and V, and $n = q/2+1$. If π is of type III, it meets $\mathcal{K}$ in a $(q(q-1)/2;\ q/2)$-arc $\mathcal{K}'$ plus a line l. If π is of type IV, it meets $\mathcal{K}$ in the complement of a $(q(q-1)/2;\ q/2)$-arc: $\pi \cap \mathcal{K}$ can also be described as the dual of a plane $(q+2)$-arc. If π is of type V, it meets $\mathcal{K}$ in $q/2+1$ lines of a pencil with vertex P_0. Then, using Theorem 12.2.1, we obtain the numbers listed in Table 19.3.

Table 19.3

		1-secants	n-secants	$(q+1)$-secants
III	$P \in l$	1	$q-1$	1
	$P \in \mathcal{K}'$	0	$q+1$	0
	Q	2	$q-1$	0
IV	P	0	$q-1$	2
	Q	0	$q+1$	0
V	$P = P_0$	$q/2$	0	$q/2+1$
	$P \neq P_0$	0	q	1
	Q	1	q	0

Theorem 19.4.8: *If $\mathscr{K}$ is a $k_{n,3,q}$ with $3 \leq n \leq q-1$ and $\mathscr{K}$ has a section of type* IV *by a plane π, then one of the following occurs.*

(i) *$\mathscr{K}$ is singular and $\mathscr{K} = \Pi_0\mathscr{K}'$ where $\mathscr{K}' = \pi \cap \mathscr{K}$.*

(ii) *$\mathscr{K}$ is non-singular, $q = 2^h$, and $n-1 = 2^{h-1} = q/2$: also* (a) *if $h > 2$, that is $q > 4$, then $\mathscr{K} = \mathscr{K}_0$, $\mathscr{K}_1$, or $\mathscr{K}_2$ where $\mathscr{K}_r$ contains r planes;* (b) *if $h = 2$, that is $q = 4$, and $\mathscr{K}$ contains no sections of type* I *or* II, *then $\mathscr{K} = \mathscr{K}_0$, $\mathscr{K}_1$, or $\mathscr{K}_2$.*

Proof: $\pi \cap \mathscr{K}$ is a $k'_{n,2,q}$, which is the complement in π of a $((q-n)q+q-n+1;\ q-n+1)$-arc. This arc is maximal and so, by Theorem 12.2.1, $q \equiv 0 \bmod (q-n+1)$. Therefore, with $q = p^h$, we have $q-n+1 = p^m$ with $0 < m < h$, and so $n-1 = p^m(p^{h-m}-1)$.

There are now the following possibilities:

(i) $n-1 \neq p^m$ in which case every plane section of $\mathscr{K}$ is of type IV, V, VI, or VII;

(ii) $n-1 = p^m$, whence $p = 2$, $m = h-1$, $n-1 = 2^{h-1} = q/2$: (a) if also $h > 2$ so that $n-1 \neq \sqrt{q}$, then a section of $\mathscr{K}$ is of type III, IV, V, VI, or VII; and (b) if $h = 2$ so that $n-1 = q/2 = \sqrt{q} = 2$, then a section of $\mathscr{K}$ of each type I–VII is admissible.

(i) By the corollary to Lemma 19.4.7, there is a unisecant l_1 to $\mathscr{K}$ with point of contact P. We show that P is singular.

Let l be another line through P and let $\pi_1 = ll_1$. Then $\pi_1 \cap \mathscr{K}$ is not of type IV, since maximal arcs have no unisecants. Also $\pi_1 \cap \mathscr{K}$ is not of type VII. So $\pi_1 \cap \mathscr{K}$ is of type V or VI, that is n lines of a pencil or a single line. In both cases, l is a unisecant or a line of $\mathscr{K}$: so P is singular. As every point of $\pi \cap \mathscr{K}$ is non-singular, P is not in π. Therefore, by Lemma 19.4.3, $\mathscr{K}$ is the required cone.

(ii)(a) As in (i), let l_1 be a unisecant to $\mathscr{K}$ with point of contact P and suppose that P is not in π.

If P is non-singular, there exists an n-secant l_2 through P. Then as the plane $\pi' = l_1l_2$ contains an n-secant and a 1-secant meeting at P, it is of type III. Since $n = q/2+1$, the set $\pi' \cap \mathscr{K}$ is a $(q(q-1)/2;\ q/2)$-arc plus a line l_0 external to the arc. Through each point of l_0, there is one 0-secant and $q-1$ lines $(q/2)$-secant to the arc. So l_0 is a line of $\mathscr{K}$ through P. From Table 19.3, there are two lines of $\mathscr{K}$ in π through $l_0 \cap \pi$. So the joins of l_0 with these two lines are planes of type IV, V, or VII. Of the planes through l_0, there are therefore 0, 1, or 2 of type VII giving respectively $\mathscr{K} = \mathscr{K}_0$, $\mathscr{K}_1$, or $\mathscr{K}_2$.

Suppose now that P is in π. Then a plane π' through l_1 and an n-secant through P in π is of type III. Hence there is a line l_0 of $\mathscr{K}$ in π' through P, and so there is a unisecant through each point P' of l_0. By taking $P' \neq P$, we are in the same position as before.

(ii)(b) When $q = 4$ and sections of types I and II are excluded, then the argument in (ii)(a) applies. □

Let $\mathcal{K}$ be a $k_{n,3,q}$ and let ρ_i, σ_i, and τ_i be the respective number of i-secants through a point of $\mathcal{K}$, through a point off $\mathcal{K}$ and in total. Then

$$\left.\begin{aligned} n\sigma_n+\sigma_1&=k\\ \sigma_n+\sigma_1&=q^2+q+1\end{aligned}\right\} \tag{19.9}$$

$$\left.\begin{aligned} q\rho_{q+1}+(n-1)\rho_n&=k-1\\ \rho_{q+1}+\rho_n+\rho_1&=q^2+q+1\end{aligned}\right\} \tag{19.10}$$

Theorem 19.4.9:
(i) *$\mathcal{K}_0$ does not exist.*
(ii) *If $\mathcal{K}_1$ exists, then $k=q^3/2+q^2+q+1$.*
(iii) *$\mathcal{K}_2$ does not exist.*

Proof: We proceed from Theorem 19.4.8 where π is a plane of type IV for $\mathcal{K}$ and $\pi'=l_0l_1l_2$ is a plane of type III. Here l_0 is a line of $\mathcal{K}$, l_1 a unisecant and l_2 an n-secant. $\mathcal{K}$ is a $k_{n,3,q}$ with $n=q/2+1$. See Fig. 19.3.

We now find the diophantine equations obtained by considering the plane sections of $\mathcal{K}$ through l_0, l_1, and l_2. Through a point of $\pi\cap\mathcal{K}$ there are two lines of $\mathcal{K}$ and $q-1$ lines n-secant to $\mathcal{K}$, and through a point of $\pi\backslash\mathcal{K}$ there are $q+1$ lines n-secant to $\mathcal{K}$. So no section of $\mathcal{K}$ through l_0, l_1, or l_2 if of type VI. As sections of type IV contain no unisecants, there is no plane of type IV through l_1. Also there is no section of type VII through l_1 or l_2. Therefore, let n_3, n_4, n_5, and n_7 be the numbers of plane sections through l_0 of respective types III, IV, V, and VII. Similarly, let n_3' and n_5' be the numbers through l_1, and n_3'', n_4'', and n_5'' be the numbers through l_2. Then, with $n_3\geq 1$, $n_3'\geq 1$, and $n_3''\geq 1$,

$$\left.\begin{aligned} (q+1)+n_3q(q-1)/2+n_4q(q+1)/2+n_5q^2/2+n_7q^2&=k\\ n_3+n_4+n_5+n_7&=q+1\end{aligned}\right\} \tag{19.11}$$

$$\left.\begin{aligned} 1+n_3'q(q+1)/2+n_5'q(q+2)/2&=k\\ n_3'+n_5'&=q+1\end{aligned}\right\} \tag{19.12}$$

$$\left.\begin{aligned} (q+2)/2+n_3''q^2/2+n_4''q(q+2)/2+n_5''q(q+1)/2&=k\\ n_3''+n_4''+n_5''&=q+1.\end{aligned}\right\} \tag{19.13}$$

From (19.9) with $n=q/2+1$, we have

$$k-1=q(q^2+3q+3-\sigma_1)/2. \tag{19.14}$$

From (19.12),

$$k-1=q(q^2+2q+1+n_5')/2 \tag{19.15}$$

whence

$$q(q^2+2q+1)/2\leq k-1\leq q(q^2+3q+1)/2. \tag{19.16}$$

From (19.13),

$$k-1=q(q^2+q+1+2n_4''+n_5'')/2 \qquad (19.17)$$

whence

$$q(q^2+q+1)/2 \leqslant k-1 \leqslant q(q^2+3q+1)/2. \qquad (19.18)$$

Through $P_0 = l_0 \cap \pi$, there are two lines of $\mathcal{K}$ in π. So the planes joining l_0 to these lines are of type IV, V, or VII. Hence $n_4+n_5+n_7 \geqslant 2$ and $n_7 \leqslant 2$. As deduced in the previous theorem, we have three cases as $n_7=0$, 1, or 2 and $\mathcal{K}$ is correspondingly denoted $\mathcal{K}_0$, $\mathcal{K}_1$, or $\mathcal{K}_2$.

From (19.11),

$$k-1=q[q^2+1+2n_4+n_5+(q+1)n_7]/2. \qquad (19.19)$$

(*i*) n_7 *equal to* 0.

Here $2 \leqslant 2n_4+n_5 \leqslant 2q$ and, from (19.19),

$$q(q^2+3)/2 \leqslant k-1 \leqslant q(q^2+2q+1)/2. \qquad (19.20)$$

Hence, from (19.16),

$$k-1=q(q^2+2q+1)/2.$$

Also, from (19.14) and (19.15), $\sigma_1=q+2$, $n_5'=0$, $n_3'=q+1$. As the maximum value is attained in (19.20), so $n_3=1$, $n_4=q$, $n_5=0$.

In a plane of type III, there are exactly two unisecants through a point off $\mathcal{K}_0$. Since $n_3'=q+1$, the $q+2$ unisecants through a point off $\mathcal{K}_0$ are no three coplanar.

From Table 19.3 and again from the fact that $n_3'=q+1$, we deduce that, for P,

$$\rho_1=1, \qquad \rho_n=q^2-1, \qquad \rho_{q+1}=q+1.$$

Since $\sigma_1=q+2$ and $k=(q^2+1)(q+2)/2$, so

$$\begin{aligned}\tau_1&=[(q^2+1)(q+1)-k]\sigma_1/q\\&=(q^2+1)(q+2)/2\\&=k.\end{aligned}$$

Thus, for any point P' of $\mathcal{K}_0$, we have $\rho_1=1$, whence, by (19.10), $\rho_n=q^2-1$ and $\rho_{q+1}=q+1$ for P'. As there is a unisecant through each point of $\mathcal{K}$, every property deduced for P applies to the other points.

Since $n_3=1$, there is a unique plane of type III through each point of $\mathcal{K}$; such a section consists of a line plus a maximal $(q(q-1)/2;\, q/2)$-arc. Let $g_0, g_1, \ldots, g_q$ be the $q+1$ lines of $\mathcal{K}$ through P. As $n_3'=q+1$, the unique planes of type III through $g_0, g_1, \ldots, g_q$ respectively all contain the unique unisecant l_1 at P.

Now consider π again. It meets $\mathcal{K}_0$ in a section of type IV, which is the dual of a plane $(q+2)$-arc. Let $g'_0, g'_1, \ldots, g'_{q+1}$ be the lines of $\pi \cap \mathcal{K}_0$. Through each g'_i there is a unique plane π_i of type III.

Suppose four of these planes $\pi_0, \pi_1, \pi_2, \pi_3$ have a point Q in common. Let $g'_0 \cap g'_i = Q_i$, $i = 1, 2, 3$. Then $\pi_0 \cap \pi_i = QQ_i$, which by the above is the unisecant to $\mathcal{K}$ through Q_i. So the three unisecants QQ_1, QQ_2, QQ_3 lie in the plane π_0: a contradiction. Thus no four of the $q+2$ planes $\pi_0, \pi_1, \ldots, \pi_{q+1}$ have a point in common. Therefore these $q+2$ planes form the dual in $PG(3, q)$ of a $(q+2)$-arc; see § 21.2. But, by Theorem 21.3.8, such a $(q+2)$-arc does not exist for $q \geq 4$. So $\mathcal{K}_0$ does not exist.

(ii) n_7 *equal to* 1.
Here $1 \leq 2n_4 + n_5 \leq 2q-2$ and, from (19.19),

$$q(q^2+q+3)/2 \leq k-1 \leq q(q^2+3q)/2.$$

Hence, from (19.16),

$$q(q^2+2q+1)/2 \leq k-1 \leq q(q^2+3q)/2; \tag{19.21}$$

correspondingly, from (19.14),

$$3 \leq \sigma_1 \leq q+2. \tag{19.22}$$

Let π_0 be the plane through l_0 lying in $\mathcal{K}_1$. Define the *residual* $\tilde{\mathcal{K}}_1$ of $\mathcal{K}_1$ by

$$\tilde{\mathcal{K}}_1 = (PG(3, q) \backslash \mathcal{K}_1) \cup \pi_0.$$

$PG(3, q) \backslash \mathcal{K}_1$ has i-secants for $i = 0$, $q/2$, q. So $\tilde{\mathcal{K}}_1$ has i-secants for $i = 1$, $q/2+1$, $q+1$. Hence $\tilde{\mathcal{K}}_1$ is a $\tilde{k}_{n,3,q}$ containing the plane π_0 with $n = q/2+1$ and

$$\begin{aligned} \tilde{k} &= (q^2+1)(q+1) - k + (q^2+q+1) \\ &= q^3 + 2(q^2+q+1) - k. \end{aligned} \tag{19.23}$$

Further, a plane π_3 of type III meets $\mathcal{K}_1$ in a line of π_0 and a $(q(q-1)/2;\, q/2)$-arc. So π_3 meets $\tilde{\mathcal{K}}_1$ in the complement of the $(q(q-1)/2;\, q/2)$-arc. Thus a plane of type III for $\mathcal{K}_1$ is of type IV for $\tilde{\mathcal{K}}_1$ and, vice versa, a plane of type IV for $\mathcal{K}_1$ is of type III for $\tilde{\mathcal{K}}_1$. In particular, as π and π' are of type IV and III respectively for $\mathcal{K}_1$, they are of type III and IV respectively for $\tilde{\mathcal{K}}_1$. So any property of $\mathcal{K}_1$ gives a corresponding property for $\tilde{\mathcal{K}}_1$ and vice versa.

From (19.21) and (19.23),

$$q(q^2+q+4)/2 \leq \tilde{k}-1 \leq q(q^2+2q+3)/2.$$

Using this result with k for $\tilde{k}$ and (19.21), we have

$$q(q^2+2q+1)/2 \leq k-1 \leq q(q^2+2q+3)/2;$$

correspondingly,

$$q \leq \sigma_1 \leq q+2.$$

Suppose that $\sigma_1 = q+2$. Then $k-1 = q(q^2+2q+1)/2$ and, from (19.15), $n_5' = 0$, $n_3' = q+1$. So, for P,

$$\rho_1 = 1, \qquad \rho_n = q^2 - 1, \qquad \rho_{q+1} = q+1.$$

As for $\mathcal{K}_0$, we have $\sigma_1 = q+2$ and $k = (q^2+1)(q+2)/2$, whence

$$\tau_1 = [(q^2+1)(q+1) - k]\sigma_1/q = k.$$

But every unisecant must have its point of contact in π_0. The point P was defined by the fact that it is the point of contact of a unisecant. So, for any point in π_0, we have $\rho_1 \leq 1$; therefore $\tau_1 \leq q^2+q+1$. Hence

$$k = q(q+1)^2/2 + 1 \leq q^2 + q + 1$$

and so $q \leq 1$: a contradiction. So $\sigma_1 \neq q+2$ and, from the residual $\tilde{\mathcal{K}}_1$, we have $\sigma_1 \neq q$. Therefore $\sigma_1 = q+1$ and $k-1 = q(q^2+2q+2)/2$; that is, $k = q^3/2 + q^2 + q + 1$.

(*iii*) n_7 *equal to* 2.

Consider $P_1 = l_1 \cap \pi$, which is not on $\mathcal{K}_2$. Every line l through P_1 not meeting l_0 contains two points of $\mathcal{K}_2$, one in each of the two planes of $\mathcal{K}_2$ through l_0; so l is a $(q/2+1)$-secant. Thus, of the planes through l_1, only π' is of type III. Hence $n_3' = 1$, $n_5' = q$. From (19.15),

$$k - 1 = q(q^2+3q+1)/2.$$

In each plane through l_2 apart from π', there are two lines of $\mathcal{K}_2$ through P; so such a plane is of type IV. Hence $n_3'' = 1$, $n_4'' = q$, $n_5' = 0$.

Since $n_3' = 1$ and $n_5' = q$, for P, $\rho_{q+1} = 1 + nq$. But, since $n_3'' = 1$ and $n_4'' = q$, for P, $\rho_{q+1} = 1 + 2q$. Hence $n = q = 2$, a contradiction. Thus $\mathcal{K}_2$ does not exist. □

In order to give a characterization of $\mathcal{K}_1$, we first develop further properties using the notation of the previous two theorems. In Fig. 19.3, the plane π is of type IV, the plane π' is of type III, and π_0 is of type VII.

(1) Since $k-1 = q(q^2+2q+2)/2$, so from (19.12) and (19.15) we have $n_5' = 1$, $n_3' = q$. Hence Table 19.3 gives that, for P,

$$\rho_1 = \tfrac{1}{2}q, \qquad \rho_n = q(q-1), \qquad \rho_{q+1} = q + (\tfrac{1}{2}q+1) = \tfrac{3}{2}q + 1. \quad (19.24)$$

(2) For the plane sections of type V through l_0, suppose m have their vertex at P. Then, for P,

$$\rho_1 = n_3 + mq/2, \qquad \rho_n = n_3(q-1) + n_4(q-1) + (n_5 - m)q,$$
$$\rho_{q+1} = q + 1 + n_4 + mq/2.$$

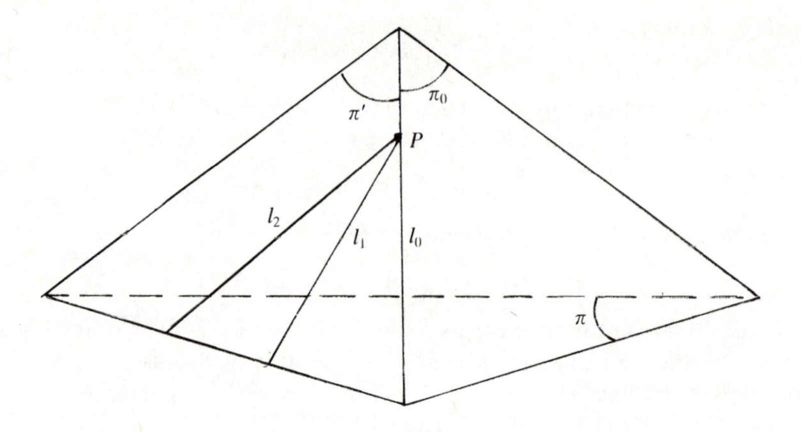

Fig. 19.3

Hence $q/2 = n_3 + mq/2$. As $n_3 \geqslant 1$, so $m = 0$ and

$$n_3 = \tfrac{1}{2}q, \qquad n_4 = \tfrac{1}{2}q, \qquad n_5 = 0, \qquad n_7 = 1.$$

(3) No plane of type V through the n-secant l_2 can contain a unisecant through P. Hence, since $\rho_1 = \tfrac{1}{2}q$, also $n_3'' = \tfrac{1}{2}q$. Since from (19.13) and (19.17) we have that $n_3'' + n_4'' + n_5'' = 2n_4'' + n_5'' = q + 1$, so

$$n_3'' = \tfrac{1}{2}q, \qquad n_4'' = \tfrac{1}{2}q, \qquad n_5'' = 1.$$

(4) From (19.9) and (19.14),

$$\sigma_1 = q + 1, \qquad \sigma_n = q^2.$$

(5) Now,

$$\begin{aligned} \tau_1 &= \tfrac{1}{2}q^3\sigma_1/q = q^2(q+1)/2, \\ \tau_n &= \tfrac{1}{2}q^3\sigma_n/(\tfrac{1}{2}q) = q^4, \\ \tau_{q+1} &= (q^2+1)(q^2+q+1) - \tau_1 - \tau_n \\ &= q^2 + q + 1 + q^2(q+1)/2. \end{aligned}$$

(6) The point P was only distinguished as the point of contact of a unisecant. Thus if there is a unisecant through a point P' of π_0, then $\rho_1 = \tfrac{1}{2}q$ for P'. Hence, if N of the points of π_0 are points of contact for unisecants, then $Nq/2 = \tau_1$, whence

$$N = q^2 + q.$$

So there is exactly one point Q_0 of π_0 through which no unisecants pass. Note that Q_0 plays the same role for $\tilde{\mathcal{K}}_1$ as for $\mathcal{K}_1$.

(7) For Q_0,

$$\rho_1 = 0, \qquad \rho_n = q^2, \qquad \rho_{q+1} = q + 1.$$

Table 19.4

	ρ_1	ρ_n	ρ_{q+1}
$R\in\mathcal{K}_1\backslash\pi_0$	0	q^2	$q+1$
$R=Q_0$	0	q^2	$q+1$
$R\in\pi_0\backslash\{Q_0\}$	$\frac{1}{2}q$	$q(q-1)$	$\frac{3}{2}q+1$

Hence, apart from π_0, all the planes through Q_0 are of type V.

(8) There are q planes of type III through any unisecant of $\mathcal{K}_1$. Hence the $q+1$ unisecants through a point Q' off $\mathcal{K}_1$ have no three coplanar. So the cone of unisecants through Q' meets π_0 in a $(q+1)$-arc $\mathscr{C}$, whose nucleus is Q_0. For if Q_0 were not the nucleus of $\mathscr{C}$, then there would be a line l' through Q_0 external to $\mathscr{C}$ and a plane π_1 of type V through l' meeting the complement of $\mathcal{K}_1$ in a $(q+1)$-arc; there is not sufficient room in π_1 to have $q/2+1$ concurrent lines and a $(q+1)$-arc for which these lines are all external. This means that π_1 meets π_0 in a line through Q_0 unisecant to $\mathscr{C}$.

(9) The values of ρ_i for all points R of $\mathcal{K}_1$ are given in Table 19.4.

(10) There is now sufficient information to list, in Table 19.5, the number of planes of each type through each type of line l.

(11) Let N_{III}, N_{IV}, N_{V}, N_{VII} be the total number of sections of the respective types III, IV, V, VII. Then, from the first two rows of Table 19.5,

$$N_{\text{III}}=N_{\text{IV}}=q^3/2, \qquad N_{\text{V}}=q(q+1), \qquad N_{\text{VII}}=1.$$

(12) The residual $\tilde{\mathcal{K}}_1$ of $\mathcal{K}_1$ has all the same properties as $\tilde{\mathcal{K}}_1$, and Table 19.5 is identical for $\tilde{\mathcal{K}}_1$. Consider the star of lines of $PG(3,q)$ through Q_0. The pencil in π_0 lies in both $\mathcal{K}_1$ and $\tilde{\mathcal{K}}_1$: every other line of the star has half its points other than Q_0 in $\mathcal{K}_1$ and half in $\tilde{\mathcal{K}}_1$.

Lemma 19.4.10: (i) *To each plane α through Q_0 other than π_0 there corresponds a unique point P_α in $\pi_0\backslash\{Q_0\}$.*

Table 19.5

$\lvert l\cap\mathcal{K}_1\rvert$	Conditions on l	III	IV	V	VII
$q+1$	$l\subset\pi_0$, $Q_0\notin l$	$q/2$	$q/2$	0	1
$q+1$	$l\subset\pi_0$, $Q_0\in l$	0	0	q	1
$q+1$	$l\not\subset\pi_0$	0	q	1	0
$q/2+1$	$Q_0\notin l$	$q/2$	$q/2$	1	0
$q/2+1$	$Q_0\in l$	0	0	$q+1$	0
1		q	0	1	0

(ii) *To each point* R *of* $\mathcal{K}_1\backslash\pi_0$ *there corresponds a* $(q+1)$*-arc* $\mathscr{C}_R$ *in* π_0 *with nucleus* Q_0 *such that the lines* PR *for* P *in* $\mathscr{C}_R$ *are the lines of* $\mathcal{K}_1$ *through* R.

(iii) *Each line* l *through* Q_0 *not in* π_0 *is an* n*-secant to which there corresponds a* $(q+1)$*-arc* $\mathscr{C}_l$ *such that* $\mathscr{C}_R=\mathscr{C}_l$ *for all* R *in* $l\cap\mathcal{K}_1\backslash\{Q_0\}$ *and* $\mathscr{C}_l=\{P_\alpha \mid l\subset\alpha\}$.

(iv) $\mathcal{K}_1=\pi_0\cup\bigcup R\mathscr{C}_l$, *where* R *varies in* $l\cap\mathcal{K}_1\backslash\{Q_0\}$; *that is,* $\mathcal{K}_1$ *is the union of the plane* π_0 *and the* $\frac{1}{2}q$ *cones with base* $\mathscr{C}_l$ *and vertices the points of* $\mathcal{K}_1$ *on* l *other than* Q_0. *So* $\mathcal{K}_1$ *is completely determined by the* $(q+1)$*-arc* $\mathscr{C}_l$ *and the* $\frac{1}{2}q$ *points of* $l\cap\mathcal{K}_1/\{Q_0\}$.

Proof: (i) From property (7), each plane α through Q_0 other than π_0 is of type V and meets $\mathcal{K}_1$ in $\frac{1}{2}q+1$ lines concurrent at a point P_α. From Table 19.4, there are $\frac{1}{2}q$ lines of $\mathcal{K}_1\backslash\pi_0$ through P_α, whence P_α is unique.

(ii) From (9), there are $q+1$ lines of $\mathcal{K}_1$ through R. Since R is off $\tilde{\mathcal{K}}_1$, by (8) the $(q+1)$-arc $\mathscr{C}_R$ exists.

(iii) Since all planes through $l=RQ_0$ are of type V, there are $(\frac{1}{2}q+1)(q+1)$ lines of $\mathcal{K}_1$ meeting l. Hence $\mathscr{C}_R=\mathscr{C}_S$ for all R and S in $l\cap\mathcal{K}_1\backslash\{Q_0\}$.

(iv) This follows from (iii). □

To find out what $\mathcal{K}_1$ is precisely, it is necessary to consider the quadric $\mathscr{P}_4=\mathbf{V}(x_0^2+x_1x_2+x_3x_4)$ in $PG(4,q)$. Let Q be any point of $PG(4,q)$ other than the nucleus N of $\mathscr{P}_4$. Let Π_3 be any solid of $PG(4,q)$ not containing Q. The projection $\mathscr{R}_3=\{P'=PQ\cap\Pi_3 \mid P\in\mathscr{P}_4\}$.

Lemma 19.4.11: *For* q *even, the projection* $\mathscr{R}_3$ *of* $\mathscr{P}_4$ *from* Q *onto* Π_3 *is a non-singular* $k_{n,3,q}$ *with* $n=\frac{1}{2}q+1$ *and* $k=\frac{1}{2}q^3+q^2+q+1$.

Proof: Let l be a line in Π_3. The plane Ql meets $\mathscr{P}_4$ in a point, a line, a line pair or a conic $\mathscr{P}_2$, by Theorem 7.2.1. In the last case, either Q is the nucleus of $\mathscr{P}_2$ in which case the lines joining Q to the points of $\mathscr{P}_2$ are $q+1$ distinct tangents, or the lines joining Q to $\mathscr{P}_2$ are $q/2$ bisecants and one tangent. Correspondingly, we have Table 19.6.

Hence $\mathscr{R}_3$ is a $k_{n,3,q}$ with $n=q/2+1$.

The tangents to $\mathscr{P}_4$ through Q meet $\mathscr{P}_4$ in a cone $\Pi_0\mathscr{P}_2$, which contains

Table 19.6

$Ql\cap\mathscr{P}_4$	$\lvert l\cap\mathscr{R}_3\rvert$
point	1
line	$q+1$
line pair	$q+1$
conic	$q+1$ or $q/2+1$

q^2+q+1 points. Hence through Q there are also $\frac{1}{2}q^3$ bisecants and $\frac{1}{2}q^3$ external lines. Each bisecant and each tangent of $\mathscr{P}_4$ give one point of $\mathscr{R}_3$, whence the value for k.

To show that $\mathscr{R}_3$ is non-singular it suffices to show that for any point P of $\mathscr{P}_4$, joined to Q by a Π_1, there exists a Π_2 through Π_1 meeting $\mathscr{P}_4$ in a $\mathscr{P}_2$ for which Q is not the nucleus; if P projects to P', then Π_2 projects to an n-secant of $\mathscr{R}_3$ through P'.

Let Π_1 be a bisecant of $\mathscr{P}_4$. Then every plane through Π_1 meets $\mathscr{P}_4$ in a conic $\mathscr{P}_2$ or a line pair $\Pi_0\mathscr{H}_1$. If there are b_0 of the former and b_1 of the latter,

$$b_0+b_1=q^2+q+1$$
$$(q-1)b_0+(2q-1)b_1+2=(q^2+1)(q+1),$$

whence $b_0=q^2>0$. □

Next we give a description of $\mathscr{R}_3$ close to that of Lemma 19.4.10(iv) for $\mathscr{K}_1$. First an algebraic result is required.

Lemma 19.4.12: (i) *There are $q-1$ additive subgroups of order $\frac{1}{2}q$ in $GF(q)$, q even.*

(ii) *If G is any such subgroup, the full set is $\{G, \delta G, \ldots, \delta^{q-2}G\}$, where δ is a primitive element of $GF(q)$.*

Proof: (i) If $q=2^h$, then $GF(q)$ is an elementary abelian 2-group. Any subgroup of index two is the kernel of a surjective homomorphism $\phi: GF(q)\to \mathbf{Z}_2$. As $GF(q)\cong \mathbf{Z}_2^h$, the number of such homomorphisms is $2^h-1=q-1$.

(ii) Since G is a subgroup, so is λG for any λ in γ_0. Suppose, for some r with $1\leq r\leq q-2$, that $\delta^r G=G$. Then $\delta^r(G\backslash\{0\})=G\backslash\{0\}$. So δ^r acts as a permutation on $G\backslash\{0\}$. The orbits of the group generated by δ^r form a partition of $G\backslash\{0\}$, and any such orbit has the form $\{t, t\delta^r, \ldots, t\delta^{(s-1)r}\}$ with $s=(q-1)/(q-1,r)$. Hence s divides $\frac{1}{2}q-1$, whence $q-1$ divides $(\frac{1}{2}q-1)(q-1,r)$. Since $(q-1,\frac{1}{2}q-1)=1$, so $q-1$ divides $(q-1,r)$, a contradiction. □

In $PG(3,q)$, q even, we consider the pencil $\mathscr{F}$ of quadrics $\mathscr{F}_\lambda$, $\lambda\in\gamma^+$, where $\mathscr{F}_\lambda=\mathbf{V}(F_\lambda)$ and

$$F_\lambda=x_0^2+x_1x_2+\lambda x_3^2.$$

So $\mathscr{F}_\infty$ is the plane $\mathbf{u}_3$ and each $\mathscr{F}_\lambda$ for λ in γ is a cone $\Pi_0\mathscr{P}_2$. Let H be a subgroup of γ of index 2.

Theorem 19.4.13:

$$\mathscr{R}_3=\bigcup_{\lambda\in H\cup\{\infty\}}\mathscr{F}_\lambda.$$

Proof: We use coordinates $\mathbf{P}(y_0, y_1, y_2, y_3, y_4)$ in $PG(4, q)$. The quadric $\mathscr{P}_4 = \mathbf{V}(y_0^2 + y_1y_2 + y_3y_4)$. Let $Q = \mathbf{P}(0, 0, 0, 1, 1)$ and consider the pencil of solids through the plane $\mathbf{V}(y_3, y_4)$. Let

$$\begin{aligned}\mathscr{V}_t &= \mathbf{V}(ty_3 + y_4) \cap \mathscr{P}_4\\ &= \mathbf{V}(y_0^2 + y_1y_2 + ty_3^2, ty_3 + y_4).\end{aligned}$$

In particular, $\mathscr{V}_0 = \mathbf{V}(y_4) \cap \mathscr{P}_4$, $\mathscr{V}_\infty = \mathbf{V}(y_3) \cap \mathscr{P}_4$. For all t, the variety $\mathscr{V}_t$ is a cone $\Pi_0\mathscr{P}_2$. If $A = \mathbf{P}(a_0, a_1, a_2, a_3, a_4)$ lies on $\mathscr{P}_4$, then QA meets $\mathscr{P}_4$ again at $A' = \mathbf{P}(a_0, a_1, a_2, a_4, a_3)$. When A also lies in $\mathscr{V}_t$, then

$$A = \mathbf{P}(a_0, a_1, a_2, a_3, ta_3), \quad A' = \mathbf{P}(a_0, a_1, a_2, ta_3, a_3).$$

So A' lies in $\mathscr{V}_{1/t}$. Also QA is a tangent when $t = 1$ and the tangents through Q meet $\mathscr{P}_4$ in $\mathscr{V}_1$.

Now, project $\mathscr{P}_4$ from Q onto $\mathbf{u}_4$:

$$A = \mathbf{P}(a_0, a_1, a_2, a_3, ta_3), \quad A' = \mathbf{P}(a_0, a_1, a_2, ta_3, a_3).$$

Hence $\mathscr{V}_1 \to \mathbf{V}(y_3, y_4)$ and, for $t \neq 1$,

$$\{\mathscr{V}_t, \mathscr{V}_{1/t}\} \to \mathscr{W}_t = \mathbf{V}\left(y_0^2 + y_1y_2 + \frac{t}{t^2+1}\, y_3^2, y_4\right). \tag{19.25}$$

For, regarding the projection as $\mathbf{P}(Y) \to \mathbf{P}(Y')$, we have

$$y_0' = y_0, \qquad y_1' = y_1, \qquad y_2' = y_2, \qquad y_3' = y_3 + y_4, \qquad y_4' = 0.$$

So, if $\mathbf{P}(Y) \in \mathscr{V}_t$, then $y_4 = ty_3$, whence $y_3 = y_3'/(t+1)$. Notice that, included in (19.25), is $\{\mathscr{V}_0, \mathscr{V}_\infty\} \to \mathscr{W}_0$.

Let $K = \{t/(t^2+1) \mid t \in \gamma^+ \backslash \{1\}\}$. Then $|K| = \frac{1}{2}q$ and K is a subgroup of $GF(q)$ since

$$\frac{t}{t^2+1} + \frac{s}{s^2+1} = \frac{(st+1)/(s+t)}{[(st+1)/(s+t)]^2+1}.$$

We have shown that $\mathscr{R}_3 = \bigcup_{\lambda \in K} \mathscr{W}_\lambda \cup \mathbf{V}(y_3, y_4)$.

In the solid $\mathbf{u}_4 = \mathbf{V}(y_4)$ take coordinates $\mathbf{P}(x_0, x_1, x_2, x_3)$ such that $x_i = y_i$, $i \in \bar{N}_3$. Thus $\mathscr{R}_3 = \bigcup_{\lambda \in K} \mathscr{F}_\lambda \cup \mathbf{V}(x_3)$. Also by Lemma 19.4.12, the subgroup $K = \beta H$ for some β. So the projectivity $\mathbf{P}(X) \to \mathbf{P}(X')$ given by $x_0' = x_0$, $x_1' = x_1$, $x_2' = x_2$, $x_3' = x_3\sqrt{\beta}$ transforms $\mathscr{R}_3$ to the required form. □

Corollary: *With the plane π_0 having parameter ∞, the parameters of the vertices of the $\frac{1}{2}q$ cones $\mathscr{F}_\lambda$ form a subgroup of γ.*

Proof: The vertex of $\mathscr{F}_\lambda$ is $\mathbf{P}(\sqrt{\lambda}, 0, 0, 1)$ and $\sqrt{\lambda} + \sqrt{\mu} = \sqrt{(\lambda + \mu)}$. □

To show that $\mathscr{K}_1$ is in fact $\mathscr{R}_3$, we need to convert the description of $\mathscr{K}_1$ in Lemma 19.4.10(iv) as a union of cones plus a plane to the more precise description of $\mathcal{R}_3$ in which the base of the cones is a conic and the parameters defining the cones form a subgroup of γ.

Lemma 19.4.14: *If l is any line through Q_0 not in the plane π_0, then the following sets are both parametrized by a subgroup of index 2 of γ:*

(a) *$l \cap \mathcal{K}_1 \backslash \{Q_0\}$, with Q_0 parametrized by ∞;*

(b) *the bisecants of the $(q+1)$-arc $\mathscr{C}_l$ through any point P of $\pi_0 \backslash (\mathscr{C}_l \cup \{Q_0\})$, with the tangent from P to $\mathscr{C}_l$ parametrized by ∞.*

Proof: Let α be a plane other than π_0 through Q_0 but not containing l. By Lemma 19.4.10(i) there is a point P_α in $\alpha \cap \pi_0$ such that $\mathcal{K}_1 \cap \alpha$ is the union of $\frac{1}{2}q+1$ lines. Since l is not in α, so P_α is not on $\mathscr{C}_l$ and there are $\frac{1}{2}q$ bisecants of $\mathscr{C}_l$ through P_α; let them be $b_1, b_2, \ldots, b_{\frac{1}{2}q}$. Also let $\mathcal{K}_1 \cap l = \{Q_0, L_1, L_2, \ldots, L_{\frac{1}{2}q}\}$ and let the set of lines in $\mathcal{K}_1 \cap \alpha$ be $\{P_\alpha Q_0, m_1, m_2, \ldots, m_{\frac{1}{2}q}\}$. Then the planes $L_i b_j$ and $b_j m_k$ are both of type IV, whence $L_i b_j \cap \alpha = m_r$ for some r in $\mathbf{N}_{\frac{1}{2}q}$ and $b_j m_k \cap l = L_s$ for some s in $\mathbf{N}_{\frac{1}{2}q}$. See Fig. 19.4.

Let $\pi_0 = \mathbf{u}_3$, let $Q_0 = \mathbf{U}_0$, and let $l = \mathbf{U}_0\mathbf{U}_3$. Then take $L_i = \mathbf{P}(\lambda_i, 0, 0, 1)$ noting that $\lambda_i = \infty$ gives Q_0. Assume also that $\mathbf{U}_3 \in \mathcal{K}_1$ and that $\lambda_1 = 0$. Then take $\alpha = \mathbf{V}(x_1 + x_3)$ and $m_1 = \mathbf{V}(x_0 + x_1, x_1 + x_3)$ so that $P_\alpha = \pi_0 \cap m_1 = \mathbf{U}_2$.

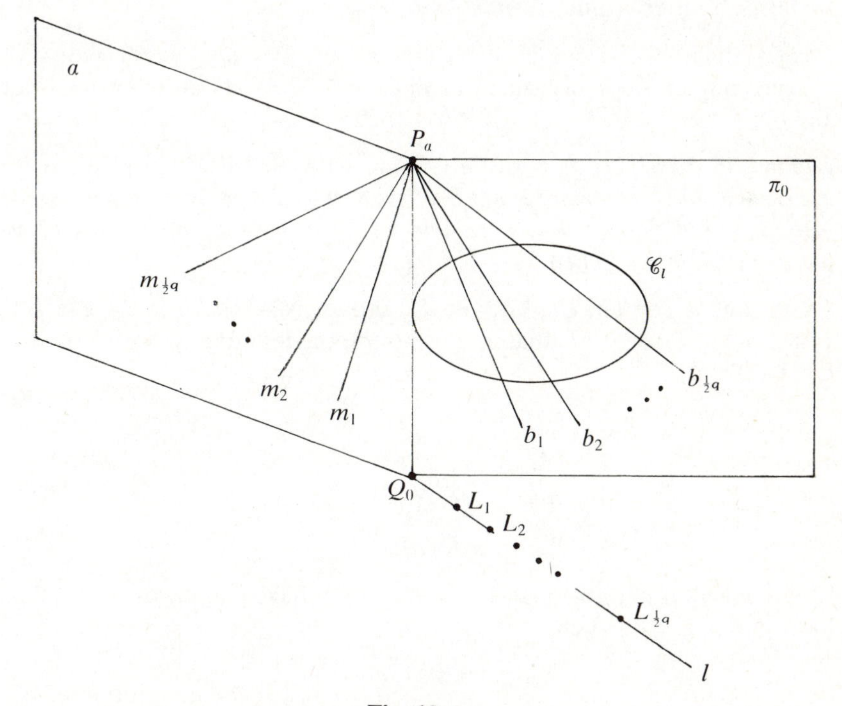

Fig. 19.4

Now, take $b_i = \pi_0 \cap L_i m_1$ so that

$$b_i = \mathbf{V}(x_3, x_0 + x_1 + \lambda_i(x_1 + x_3))$$
$$= \mathbf{V}(x_3, x_0 + (\lambda_i + 1)x_1).$$

In particular, $b_1 = \mathbf{V}(x_3, x_0 + (\lambda_1 + 1)x_1) = \mathbf{V}(x_3, x_0 + x_1)$. Define $m_j = \alpha \cap L_j b_1$, whence

$$m_j = \mathbf{V}(x_1 + x_3, x_0 + x_1 + \lambda_j x_3).$$

Then

$$b_i m_j = \mathbf{V}(x_0 + (\lambda_i + 1)x_1 + (\lambda_i + \lambda_j)x_3),$$

whence

$$b_i m_j \cap l = \mathbf{P}(\lambda_i + \lambda_j, 0, 0, 1).$$

This gives the subgroup for (a).

From above, $b_i = \mathbf{V}(x_3, x_0 + x_1 + \lambda_i x_1)$ and $P_\alpha Q_0 = \mathbf{U}_2\mathbf{U}_0 = \mathbf{V}(x_3, x_1)$. So, with $P_\alpha Q_0$ having parameter ∞, the line b_i has parameter λ_i. By the previous part, $\{\lambda_i \mid i \in \mathbf{N}_{\frac{1}{2}q}\}$ is a subgroup of γ. $\square$

Lemma 19.4.15: *Let $\mathscr{C}_1$ and $\mathscr{C}_2$ be two k-arcs, $k > 2$, in $PG(2, q)$ with the same set $\mathscr{B}$ of bisecants; then $\mathscr{C}_1 = \mathscr{C}_2$.*

Proof: Given $\mathscr{B}$, the points of $\mathscr{C}_1$ and $\mathscr{C}_2$ are identified as the points of concurrency of $k-1$ bisecants; at most $k/2$ bisecants are concurrent at any other point. As $k-1 > k/2$ for $k > 2$, so $\mathscr{C}_1 = \mathscr{C}_2$. $\square$

Lemma 19.4.16: *Let $\mathscr{K}$ be a $(q+1)$-arc with nucleus N in $PG(2, q)$, q even, such that every point P not in $\mathscr{K} \cup \{N\}$ has the property that the set of bisecants of $\mathscr{K}$ through P is parametrized by a subgroup of γ when PN is parametrized by ∞. Then $\mathscr{K}$ is a conic.*

Proof: Let $\mathscr{K}$ contain $\mathbf{U}_0$, $\mathbf{U}_1$, and $\mathbf{U}$, and let $N = \mathbf{U}_2$. Take $P = \mathbf{P}(e, f, 1)$ with $e, f \neq 0, 1$ and $e \neq f$; that is, P is not on the join of any two of $\mathbf{U}_0$, $\mathbf{U}_1$, $\mathbf{U}_2$, $\mathbf{U}$. Then

$$P\mathbf{U}_0 = \boldsymbol{\pi}(0, 1, f),$$
$$P\mathbf{U}_1 = \boldsymbol{\pi}(1, 0, e),$$
$$P\mathbf{U} = \boldsymbol{\pi}(f+1, e+1, e+f),$$
$$P\mathbf{U}_2 = \boldsymbol{\pi}(f, e, 0).$$

In the pencil of lines through P, let $P\mathbf{U}_2 = PN$ have parameter ∞ and $P\mathbf{U}_0$ have parameter 0. Then, since

$$fe^{-1}(1, 0, e) = (0, 1, f) + e^{-1}(f, e, 0),$$
$$f(e+f)^{-1}(f+1, e+1, e+f) = (0, 1, f) + [(f+1)/(e+f)](f, e, 0),$$

so PU_1 and PU have respective parameters

$$1/e \quad \text{and} \quad (f+1)/(e+f).$$

Thus, the line with parameter

$$1/e+(f+1)/(e+f)=f(e+1)/[e(e+f)],$$

that is, the line

$$l(e,f)=\boldsymbol{\pi}(f^2(e+1),\, e^2(f+1),\, ef(e+f))$$

is a bisecant of $\mathcal{K}$. But

$$l(e,f)=\boldsymbol{\pi}\left(1,\frac{e^2(f+1)}{f^2(e+1)},\frac{e(e+f)}{f(e+1)}\right)$$
$$=\boldsymbol{\pi}(1,\, st,\, s+t)$$

with $s=e/f$ and $t=e(f+1)/[f(e+1)]$. Now,

$$\boldsymbol{\pi}(1,\, st,\, s+t)=\mathbf{P}(s^2, 1, s)\mathbf{P}(t^2, 1, t),$$

which is a bisecant of the conic $\mathcal{P}_2=\mathbf{V}(x_0x_1+x_2^2)$. The set

$$\{l(e,f) \mid e, f\neq 0, 1;\, e\neq f\}$$

has $(q-2)(q-3)/2$ members, all of which are bisecants of $\mathcal{P}_2$ and none of which pass through $\mathbf{U}_0$, $\mathbf{U}_1$, or $\mathbf{U}$. Hence the $3q-3$ lines through $\mathbf{U}_0$, $\mathbf{U}_1$, or $\mathbf{U}$ other than $N\mathbf{U}_0$, $N\mathbf{U}_1$, and $N\mathbf{U}$ are all bisecants of $\mathcal{K}$ and bisecants of $\mathcal{P}_2$. So the set of bisecants of $\mathcal{K}$ is the set of bisecants of $\mathcal{P}_2$. Thus Lemma 19.4.15 implies that $\mathcal{K}=\mathcal{P}_2$. □

Theorem 19.4.17: $\mathcal{K}_1=\mathcal{R}_3$.

Proof: From Lemma 19.4.10, the set $\mathcal{K}_1$ is the union of a plane and $\frac{1}{2}q$ cones with base a $(q+1)$-arc. From Lemma 19.4.16, the $(q+1)$-arc is a conic and from Lemma 19.4.14, the cones are parametrized by a subgroup of γ. This coincides with the description of $\mathcal{R}_3$ given in Theorem 19.4.13. □

Theorem 19.4.18: *If $\mathcal{K}$ is a $k_{n,3,q}$ with $3\leq n\leq q-1$ containing a plane π_0, then one of the following occurs:*

(i) *$\mathcal{K}$ is singular and is one of*
 (a) *n planes in a pencil;*
 (b) *a cone $\Pi_0\mathcal{K}'$ where $\mathcal{K}'$ is a section of type* III;
 (c) *a cone $\Pi_0\mathcal{K}''$ where $\mathcal{K}''$ is a section of type* IV;

(ii) *$\mathcal{K}$ is non-singular, q is even, $n=\frac{1}{2}q+1$ and $\mathcal{K}=\mathcal{R}_3$.*

Proof: (i) If $\mathcal{K}$ is singular then, by Theorem 19.4.4, Corollary 2, cases (a), (b), and (c) are the only possibilities.

(ii) Let $\mathcal{K}$ be non-singular. It contains no sections of type I or II, since every plane contains a line of $\mathcal{K}$.

$\mathcal{K}$ contains a line l not in π_0, as otherwise $\mathcal{K}\backslash\pi_0$ has only 0-secants and $(n-1)$-secants in contradiction to Lemma 19.4.7. Let $P=l\cap\pi_0$. Since $\mathcal{K}$ is non-singular, there is an n-secant l' through P. The plane $\pi=ll'$ meets π_0 in a line l_0 of $\mathcal{K}$. So, through P in π, there are two lines of $\mathcal{K}$ and an n-secant. So π is of type IV. Hence, by Theorems 19.4.8, 19.4.9, and 19.4.17, $\mathcal{K}=\mathcal{R}_3$. □

Theorem 19.4.19: *Let $\mathcal{K}$ be a non-singular $k_{n,3,q}$ with $3\leq n\leq q-1$. If $q>4$ and $\mathcal{K}\neq\mathcal{R}_3$, then every plane section of $\mathcal{K}$ is of type* I, II, III, V, *or* VI, *but not* IV *or* VII.

Proof: By Theorems 19.4.8, 19.4.9, and 19.4.17, if $q>4$ and $\mathcal{K}$ is non-singular, then a section of type IV implies a section of type VII and that $\mathcal{K}=\mathcal{R}_3$. So, if $\mathcal{K}\neq\mathcal{R}_3$, there is no section of type IV or VII. □

19.5 The characterization of Hermitian surfaces

We now turn our attention to non-singular sets $k_{n,3,q}$ with $3\leq n\leq q-1$ and not characterized in § 19.4. Only sections of type IV and type VII of a $k_{n,3,q}$ contain triangles. So, if $\mathcal{K}$ is a $k_{n,3,q}$, it will be called *regular* if

(a) $\mathcal{K}$ is non-singular; (b) $3\leq n\leq q-1$;
(c) for q even with $q>4$, $\mathcal{K}\neq\mathcal{R}_3$;
(d) for $q=4$, $\mathcal{K}$ is triangle-free.

Lemma 19.5.1: *If $\mathcal{K}$ is a regular $k_{n,3,q}$, then the number of lines of $\mathcal{K}$ through a point P of $\mathcal{K}$ is* 0, 1, *or n lying in a plane.*

Proof: Suppose that there are three non-coplanar lines l_1, l_2, and l_3 of $\mathcal{K}$ through P. Let π be a plane not containing P and let $P_i=\pi\cap l_i$, $i\in\mathbf{N}_3$.

Now, suppose that π contains a line l of $\mathcal{K}$. Then at least one of the P_i, say P_1, does not lie on l. The lines P_1P_2 and P_1P_3 meet l in the distinct points P_2' and P_3' respectively. The plane PP_1P_2 contains the lines l_1 and l_2, and therefore meets $\mathcal{K}$ in n lines of a pencil, by Theorem 19.4.19. As P_2' lies in this plane, PP_2' is one of these n lines. Similarly PP_3' is on $\mathcal{K}$. But then $PP_2'P_3'$ meets $\mathcal{K}$ in the triangle with sides l, PP_2', and PP_3', contradicting Theorem 19.4.19.

If π contains no line of $\mathcal{K}$, then π meets $\mathcal{K}$ in a Hermitian arc or a subplane $PG(2,\sqrt{q})$: in both cases $n=\sqrt{q}+1$. Since $\mathcal{K}$ is non-singular, there exists an n-secant l' through P. Any plane containing l' and a line of $\mathcal{K}$ through P meets $\mathcal{K}$ in a section of type III or V, and so contains no other line through P. So, if N is the number of lines on $\mathcal{K}$ through P, $N\leq q+1$. However, the plane PP_2P_3 contains two lines of K through P and therefore n lines of $\mathcal{K}$ through P. Similarly, the n planes containing PP_1 and these n lines in PP_2P_3 all meet $\mathcal{K}$ in n lines through P. Hence $N\geq n(n-1)+1=\sqrt{q}(\sqrt{q}+1)+1$. So $q+1\geq N\geq q+\sqrt{q}+1$, a contradiction. □

Lemma 19.5.2: *If $\mathcal{K}$ is a regular $k_{n,3,q}$, then there is at least one plane section of $\mathcal{K}$ which is either a Hermitian arc or a $PG(2, \sqrt{q})$. Hence q is square and $n = \sqrt{q}+1$.*

Proof: By the corollary to Lemma 19.4.7, $\mathcal{K}$ has a 1-secant l; let $\mathcal{K} \cap l = \{P\}$. If each of the planes through l contains a line of $\mathcal{K}$, then there are at least $q+1$ lines of $\mathcal{K}$ through P, contradicting Lemma 19.5.1. So one of the planes through l contains no line of $\mathcal{K}$ and therefore meets $\mathcal{K}$ in a Hermitian arc or a $PG(2, \sqrt{q})$. □

Lemma 19.5.3: *If $\mathcal{K}$ is a regular $k_{n,3,q}$ and l is one of its n-secants, there exists at least one line on $\mathcal{K}$ meeting l.*

Proof: Suppose there is no line on $\mathcal{K}$ meeting l. Then, by Theorem 19.4.19, a plane section of $\mathcal{K}$ through l is either a Hermitian arc or a $PG(2, \sqrt{q})$. If N of these sections are Hermitian arcs, then

$$N[(q\sqrt{q}+1)-(\sqrt{q}+1)]+(q+1-N)[(q+\sqrt{q}+1)-(\sqrt{q}+1)] = k-(\sqrt{q}+1),$$

whence

$$N\sqrt{q}(q-\sqrt{q}-1) = k-(q^2+q+\sqrt{q}+1). \tag{19.26}$$

In each of these planes through l, there is at least one unisecant through a point P of $l \cap \mathcal{K}$; let l' be one. A line of $\mathcal{K}$ in any plane through l' would pass through P. So all plane sections of $\mathcal{K}$ through l' are Hermitian arcs or subplanes. If there are N' of the former,

$$N'[(q\sqrt{q}+1)-1]+(q+1-N')[(q+\sqrt{q}+1)-1] = k-1,$$

whence

$$N'\sqrt{q}(q-\sqrt{q}-1) = k-(q^2+q\sqrt{q}+q+\sqrt{q}+1). \tag{19.27}$$

Elimination of k from (19.26) and (19.27) gives

$$N-N' = q/(q-\sqrt{q}-1). \tag{19.28}$$

So $q/(q-\sqrt{q}-1)$ is an integer, which is false for $q-\sqrt{q}-1>1$, that is, for $q>4$.

It therefore remains to deal with the case $q=4$. Equations (19.26), (19.27), and (19.28) respectively become

$$k = 2N+23, \tag{19.29}$$

$$k = 2N'+31, \tag{19.30}$$

$$N = N'+4. \tag{19.31}$$

Since $N \leqslant 5$, we have either (a) $N'=0$, $N=4$ or (b) $N'=1$, $N=5$.

(a) In this case, $k=31$ and there is exactly one plane through l meeting $\mathcal{K}$ in a subplane, whilst all five planes through l' meet $\mathcal{K}$ in subplanes. In particular, $\pi = ll'$ meets $\mathcal{K}$ in a subplane. Let P' be a point other than P

of $l \cap \mathcal{K}$ and let π' be a plane other than π through l. Then $\pi' \cap \mathcal{K}$ is a Hermitian arc which has a unique 1-secant l'' at P' in π'. Let N'' be the number of planes through l'' whose section of $\mathcal{K}$ is a Hermitian arc. Then, as for (19.30), we have $k = 2N'' + 31$, which, with (19.29), gives $N = N'' + 4$. As $N = 4$, so $N'' = 0$, contradicting the existence of π'.

(b) In this case $k = 33$ and every plane through l meets $\mathcal{K}$ in a Hermitian arc while only one plane through l', namely $\pi = ll'$, meets $\mathcal{K}$ in a Hermitian arc. Let π' be a plane other than π through l'. Then π' meets $\mathcal{K}$ in a subplane and so there is a 3-secant l'' of $\mathcal{K}$ through P in π'.

Consider the sections of $\mathcal{K}$ through l''. Such a section is not of type VI, since l'' is a 3-secant, and not of type V, since then P would lie on a line of $\mathcal{K}$. If the section were of type III, namely a line l_0 plus a 6-arc, then the two planes joining l_0 with the two other points of $l \cap \mathcal{K}$ would meet $\mathcal{K}$ in l_0 plus a 6-arc. If n_3 and n_5 are respectively the number of sections of $\mathcal{K}$ through l_0 of types III and V, then

$$6n_3 + 8n_5 = k - 5 = 28.$$

So

$$3n_3 + 4n_5 = 14, \qquad n_3 \geq 3. \tag{19.32}$$

As (19.32) implies that n_3 is even and as $3 \leq n_3 \leq 5$, so $n_3 = 4$. But then $2n_5 = 1$, a contradiction.

This means that the plane sections of $\mathcal{K}$ through l'' are Hermitian arcs or subplanes. If there are N'' of the former, then as in (19.29), $k = 2N'' + 23$, whence $N'' = 5$, contradicting the existence of π'. □

Lemma 19.5.4: *If $\mathcal{K}$ is a regular $k_{n,3,q}$, then*

(i) *$\mathcal{K}$ contains at least two skew lines;*

(ii) *$q\sqrt{q} + 1 - (q - \sqrt{q}) \leq k/(q+1) \leq q\sqrt{q} + 1$, with equality on the right if and only if every plane through a line of $\mathcal{K}$ meets $\mathcal{K}$ in n concurrent lines.*

Proof: (i) $\mathcal{K}$ has an n-secant l_1. By Lemma 19.5.3, there exists a line l_2 on $\mathcal{K}$ meeting l_1. If P is a point of $l_1 \backslash l_2$, then, in each plane through l_1, there is another n-secant through P. So taking a plane other than $l_1 l_2$, we have an n-secant l_3 skew to l_2 and so there is a line l_4 of $\mathcal{K}$ meeting l_3 which is not l_2.

Suppose now that all lines of $\mathcal{K}$ meet. Then, as $\mathcal{K}$ contains no triangles, the lines are concurrent and so form n lines through P in the plane π, by Lemma 19.5.1. Let π' be a plane meeting π in a unisecant. As π' contains no lines of $\mathcal{K}$, so $\pi' \cap \mathcal{K}$ is a Hermitian arc or a subplane. Let l be an n-secant of $\mathcal{K}$ in π' not through P; so l meets π in a point of $\mathcal{K}$. By Lemma 19.5.3, l meets a line l' of $\mathcal{K}$. As l' lies in π, we have a contradiction.

(ii) Now let l_1 and l_2 be two skew lines of $\mathcal{K}$. Each of the $q+1$ planes through l_1 meets l_2. So, by Theorem 19.4.19, each of these planes meets

$\mathcal{K}$ in a section of type III or V, that is in l_1 plus an $((n-2)q+n-1; n-1)$-arc or in n lines of a pencil. Hence

$$(q+1)[(n-2)q+(n-1)] \leq k-(q+1) \leq (q+1)(n-1)q.$$

Putting $n=\sqrt{q}+1$ gives the inequalities required and shows that $k=(q+1)(q\sqrt{q}+1)$ if and only if every plane through l meets $\mathcal{K}$ in a section of type V. □

Lemma 19.5.5: *If $\mathcal{K}$ is a regular $k_{n,3,q}$, then $\mathcal{K}$ contains at least two incident lines.*

Proof: Suppose that all lines of $\mathcal{K}$ are skew to one another. By the previous lemma, there are at least two, l_1 and l_2. Then every section through l_1 is of type III, that is, l_1 plus *a* $(q\sqrt{q}-q+\sqrt{q}; \sqrt{q})$-arc. Hence

$$k=(q+1)(q\sqrt{q}-q+\sqrt{q}+1). \tag{19.33}$$

Let π be a plane through l_1 and let l be a unisecant of $\mathcal{K}$ in π: its point of contact P_1 is on l_1. The q planes through l other than π contain no lines, as any such line would have to pass through P_1 and so meet l_1. Hence, by Theorem 19.4.19 there are N planes meeting $\mathcal{K}$ in a Hermitian arc and $q-N$ meeting it in a subplane. Hence

$$k=(q+1)+(q\sqrt{q}-q+\sqrt{q})+Nq\sqrt{q}+(q-N)(q+\sqrt{q}).$$

With the value of k from (19.33), we obtain

$$q^2(\sqrt{q}-2)=N\sqrt{q}(q-\sqrt{q}-1)$$

and hence

$$N=q-\sqrt{q}-\sqrt{q}/(q-\sqrt{q}-1). \tag{19.34}$$

Therefore $\sqrt{q}/(q-\sqrt{q}-1)$ is an integer, which is impossible for $q>4$.

Now consider the case $q=4$. Suppose that $\mathcal{K}$ satisfies the following:

(i) the lines of $\mathcal{K}$ are mutually skew;

(ii) there is a point P of $\mathcal{K}$ on none of its lines.

From (19.33) and (19.34), $k=35$ and $N=0$. So every section of $\mathcal{K}$ through l is a subplane, other than the section $\pi \cap \mathcal{K}$.

As P is on neither l_1 nor l_2, there is a unique transversal l' of l_1 and l_2 through P. By (ii), l' is not on $\mathcal{K}$ and is therefore a 3-secant. The planes through l' other than Pl_1 and Pl_2 contain no lines of $\mathcal{K}$, as otherwise such a line would contain either $l' \cap l_1$ or $l' \cap l_2$: so these sections of $\mathcal{K}$ are Hermitian arcs or subplanes. The section of $\mathcal{K}$ by Pl_i is l_i plus a 6-arc. So, if N' is the number of Hermitian arcs, then

$$3+16+6N'+4(3-N')=35,$$

whence $N'=2$. Let π' be one of these planes through l' with $\pi' \cap \mathcal{K}$ of type I. Then there is a unisecant at $l' \cap l_1$ to K in π', which may be taken as l. But it was shown that no planes through l meet $\mathcal{K}$ in a Hermitian arc: a contradiction.

We know therefore that $\mathcal{K}$ cannot satisfy (i) and (ii). Suppose then that (i) is true so that (ii) is not. Then through each point of $\mathcal{K}$ there is a unique line. As $k=35$, so $\mathcal{K}$ consists of seven skew lines. Let l_1, l_2, l_3 be three of them and let l_4 be another line of the regulus $\mathcal{R}(l_1, l_2, l_3)$. Then l_4 meets $\mathcal{K}$ in at least one point Q, through which there is a line l_4' of the complementary regulus. As l_4' meets l_1, l_2, l_3 and contains Q, it contains four points of $\mathcal{K}$ and therefore lies on $\mathcal{K}$, contradicting that the lines of $\mathcal{K}$ are mutually skew. □

Lemma 19.5.6: *If $\mathcal{K}$ is a regular $k_{3,3,4}$, then $k=37$ or $k=45$.*

Proof: Let l_1 and l_2 be two lines of $\mathcal{K}$ meeting at the point P and lying in the plane π. Then π meets $\mathcal{K}$ in a third line l_3 through P. So any line through P not in π is a 1-secant or a 3-secant to $\mathcal{K}$, by Lemma 19.5.1.

Suppose therefore that l is a 1-secant through P not in π. The lines l' and l'' through P in π other than the l_i are also 1-secants to $\mathcal{K}$. Each plane ll_i contains no line of $\mathcal{K}$ other than l_i and so meets $\mathcal{K}$ in the line l_i only or in l_i plus a 6-arc. The planes ll' and ll'' containing no line of $\mathcal{K}$ meet it in a Hermitian arc, which by the corollary to Theorem 11.1.1 is a Hermitian curve $\mathcal{U}_2$, or a $PG(2, 2)$. But only a $PG(2, 2)$ has two 1-secants through a point; so $ll' \cap \mathcal{K}$ and $ll'' \cap \mathcal{K}$ are both subplanes.

Let N be the number of lines l_i for which $ll_i \cap \mathcal{K}$ is l_i plus a 6-arc. Then, considering the planes through l, we have

$$k = 1 + 2 \cdot 6 + 10N + 4(3-N) = 25 + 6N.$$

For $N=0$ and $N=1$, we have that $k=25$ and $k=31$ respectively, contradicting the estimate $k \geqslant 35$ of Lemma 19.5.4.

For $N=3$, $k=43$. Since $ll_2 \cap \mathcal{K}$ is l_2 plus a 6-arc, the five lines through P in ll_2 comprise l_2, l, and three 3-secants. The joins of l_1 with these five lines comprise π, ll_1, and three planes whose section of $\mathcal{K}$ is of type III or V. If M is the total number of points of $\mathcal{K} \backslash l_1$ in the latter three planes, then

$$13 + 6 + M = 43,$$

whence $M=24$. So each of these three plane sections is three concurrent lines as is $\pi \cap \mathcal{K}$. Similarly, the planes through l_2 and l_3 other than ll_2 and ll_3 meet $\mathcal{K}$ in three concurrent lines.

Let l_0 be a 3-secant of $\mathcal{K}$ through P in ll_3. The plane $l_0 l_1$ meets $\mathcal{K}$ in l_1 and two other lines concurrent at $P_1 \neq P$. Let l_1' be one of these lines and let $Q = l_0 \cap l_1'$; then Q is in $\mathcal{K}$ and $Q \neq P$. Similarly, $l_0 l_2$ meets $\mathcal{K}$ in l_2 and two other lines concurrent at $P_2 \neq P$; one of these, l_2' say, must pass through Q. Hence there are two lines, l_1' and l_2', of $\mathcal{K}$ through Q, whence $\pi' = l_1' l_2'$ contains a third line l_3' of $\mathcal{K}$ through Q. On the other hand, the three lines of $\pi' \cap \mathcal{K}$ can be obtained by joining Q to the three points of $P_1P_2 \cap \mathcal{K}$, namely P_1, P_2, and $P_3 = l_3 \cap P_1P_2$. Hence $l'_1 = QP_1$, $l'_2 = QP_2$, $l'_3 = QP_3$. So l'_3 belongs to $\mathcal{K}$ and the plane ll_3, since P_3 is on l_3, Q is on l_0 and l_0 lies in ll_3. But this contradicts the fact that $ll_3 \cap \mathcal{K}$ is l_3 plus a 6-arc.

For $N = 2$, let $ll_1 \cap \mathcal{K}$ consist of l_1 only, so that $ll_2 \cap \mathcal{K}$ is l_2 plus a 6-arc and $ll_3 \cap \mathcal{K}$ is l_3 plus a 6-arc. Then, by a similar counting argument as for $N = 3$, each of the four planes through l_1 other than ll_1 meets $\mathcal{K}$ in two other lines concurrent at a point of l_1. So $\mathcal{K}$ consists of l_1 and eight other lines meeting in pairs at four points of l_1 such that l_1 lies in the pencil defined by each pair; this confirms that $k = 5 + 8 \cdot 4 = 37$.

If each line through P not in π is a 3-secant of $\mathcal{K}$, then $k = 13 + 2 \cdot 16 = 45$. □

Corollary: *If $\mathcal{K}$ is a regular $37_{3,3,4}$, then it has the following properties.*

(i) *There exists a line l_0 in $\mathcal{K}$ and a point P_0 on l_0 such that through each point of $\mathcal{K} \backslash l_0$ there is exactly one line of $\mathcal{K}$.*

(ii) *Four of the five planes through l_0 meet $\mathcal{K}$ in l_0 and two other lines concurrent at a point of l_0, while the fifth plane through l_0 meets $\mathcal{K}$ in l_0 only; hence $\mathcal{K}$ comprises l_0 and eight lines meeting it.*

(iii) *Each plane through P_0 not containing l_0 meets $\mathcal{K}$ in a Hermitian curve $\mathcal{U}_2$; each plane through exactly one line of $\mathcal{K}$ other than l_0 meets $\mathcal{K}$ in that line plus a 6-arc; each plane containing neither a line of $\mathcal{K}$ nor P_0 meets $\mathcal{K}$ in a subplane $PG(2, 2)$.*

Proof: Parts (i) and (ii) were shown in the proof of the lemma.

A plane π through P_0 not containing l_0 contains no lines of $\mathcal{K}$. So $\pi \cap \mathcal{K}$ consists of P_0 and one point on each of the eight lines other than l_0. Hence, by Theorem 19.4.19, $\pi \cap \mathcal{K}$ is a Hermitian arc, which by the corollary to Theorem 11.1.1 is a Hermitian curve.

If $l \neq l_0$ is a line of $\mathcal{K}$ and $\pi \neq ll_0$ is a plane through l, then π meets the six lines of $\mathcal{K}$ not in ll_0 in distinct points. So $\pi \cap \mathcal{K}$ is l plus a 6-arc.

If a plane π contains no line of $\mathcal{K}$ and meets l_0 in $P \neq P_0$, then $\pi \cap \mathcal{K}$ consists of P plus one point on each of the six lines of $\mathcal{K}$ not through P. So $\pi \cap \mathcal{K}$ consists of seven points and is therefore a $PG(2, 2)$. □

It has not yet been shown that a regular $37_{3,3,4}$ actually exists. We recall that a regular $37_{3,3,4}$ is non-singular and triangle-free.

Let $\mathcal{K}^*$ in $PG(3, 4)$ consist of the points on a line l_0 and two lines l_i and l'_i through each of four points P_i, $i \in \mathbf{N}_4$, of l_0 such that $\pi_i = l_0 l_i l'_i$ is a plane (Fig. 19.5). So $\mathcal{K}^*$ comprises $5 + 8 \cdot 4 = 37$ points. Let P_0 be the remaining point on l_0 and π_0 the remaining plane through l_0.

Theorem 19.5.7: (i) *$\mathcal{K}^*$ is a regular $37_{3,3,4}$ if and only if each set of four of the eight lines other than l_0 with one taken from each of the four pairs (l_i, l'_i) has only the transversal l_0.*

(ii) *A regular* $37_{3,3,4}$ *exists and is projectively unique.*

Proof: (i)

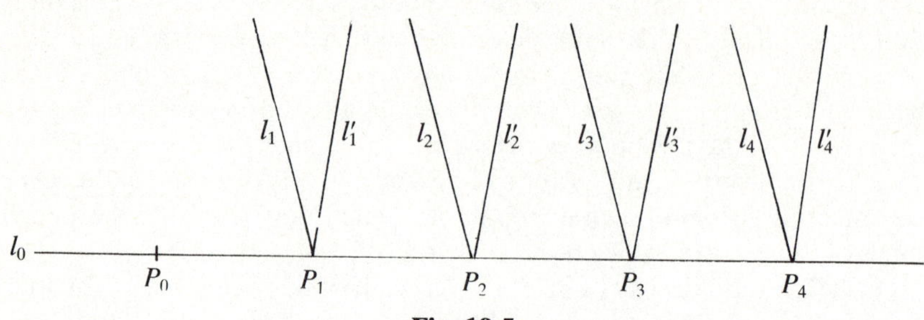

Fig. 19.5

The condition that each set of four skew lines meeting l_0 has only l_0 as transversal is necessary since any other transversal would be a 4-secant of $\mathcal{K}^*$.

To prove the sufficiency, consider the various ways a line l not contained in $\mathcal{K}^*$ can meet it. We must show that, if l contains two points of $\mathcal{K}^*$, then it is a 3-secant, and also that the 1-secants, 3-secants, and 5-secants of $\mathcal{K}^*$ exhaust all 357 lines of $PG(3, 4)$.

If l contains P_0 and meets no line of $\mathcal{K}^*$ other than l_0, then l is a 1-secant of $\mathcal{K}^*$. If l contains P_0 and meets l_1, say, then it lies in π_1, meets l_1' and is a 3-secant of $\mathcal{K}^*$. Similarly, if l contains P_1, either it lies in π_0 or π_1 and is a 1-secant or it lies in π_2, say, meets l_2 and l_2', and is a 3-secant.

Now consider a line l meeting l_1 and l_2; if l met l_0, it is l_0, which has been excluded. No two of the quadrics containing the reguli $\mathcal{R}(l_1, l_2, l_3)$, $\mathcal{R}(l_1, l_2, l_3')$, $\mathcal{R}(l_1, l_2, l_4)$, $\mathcal{R}(l_1, l_2, l_4')$ have any intersection other than l_0, l_1, and l_2, since any further intersection would be a line lying in the complementary reguli and so meeting four of the eight lines other than l_0, contrary to construction. So these four quadrics touch along l_0 and together contain 16 transverals of l_1 and l_2 other than l_0. These 16 transversals are therefore the 16 lines which meet l_1 and l_2 but contain neither P_1 nor P_2. Hence each of these transversals meets exactly one of l_3, l_3', l_4, l_4' and so is a 3-secant of $\mathcal{K}^*$.

The number of 3-secants of $\mathcal{K}^*$ not meeting l_0 is therefore $4 \cdot 2^3 \cdot 4 = 128$ with 12 through each point of $\mathcal{K}^* \backslash l_0$. The number of 3-secants of $\mathcal{K}^*$ in each $\pi_i \neq \pi_0$ is $4 \cdot 4 = 16$, giving 64 altogether with four through each point of $\mathcal{K}^* \backslash l_0$. So through each point of $\mathcal{K}^* \backslash l_0$ there are 16 trisecants and one line of $\mathcal{K}^*$ leaving four unisecants. Through each point $P_i \neq P_0$ there are 12 trisecants and three lines of $\mathcal{K}$ leaving six unisecants. Through P_0 there are 16 trisecants and one line of $\mathcal{K}$, leaving four unisecants. So

$$\tau_1 = 32 \cdot 4 + 4 \cdot 6 + 4 = 156, \qquad \tau_3 = 128 + 64 = 192, \qquad \tau_5 = 9,$$

whence $\tau_1+\tau_3+\tau_5=156+192+9=357$, which is all the lines of $PG(3,4)$.

(ii) To construct $\mathcal{K}^*$, take $l_0=\mathbf{V}(x_2,x_3)=\mathbf{l}(1,0,0,0,0,0)$, where line coordinates are as in § 15.2. Let $P_1=\mathbf{P}(1,0,0,0)$ $P_2=\mathbf{P}(0,1,0,0)$, $P_3=\mathbf{P}(1,1,0,0)$, $P_4=\mathbf{P}(\omega,1,0,0)$, $P_0=\mathbf{P}(\omega^2,1,0,0)$. Also, without loss of generality, let

$$\begin{aligned} l_1&=\mathbf{l}(0,1,0,0,0,0), & l_1'&=\mathbf{l}(1,1,0,0,0,0),\\ l_2&=\mathbf{l}(0,0,0,0,1,0), & l_2'&=\mathbf{l}(b,0,0,0,1,0),\\ l_3&=\mathbf{l}(0,1,1,1,1,0), & l_3'&=\mathbf{l}(c,1,1,1,1,0),\\ l_4&=\mathbf{l}(d,\omega^2,\omega,\omega,1,0), & l_4'&=\mathbf{l}(d',\omega^2,\omega,\omega,1,0). \end{aligned}$$

If neither l_4 nor l_4' lies in any of the eight reguli $\mathcal{R}(t_1,t_2,t_3)$, where t_i is l_i or l_i', then l_4 and l_4' both lie in the tangent plane at P_4 to the quadric containing the regulus. So, if t_4 is l_4 or l_4', then t_1, t_2, t_3, and t_4 have just the transversal l_0, since the five lines are linearly dependent, by Theorem 15.2.7.

Now let us find the line through P_4 of all the eight reguli $\mathcal{R}(t_1,t_2,t_3)$. In each case it is $\mathbf{l}(D,\omega^2,\omega,\omega,1,0)$. However, $t_1=\mathbf{l}(A,1,0,0,0,0)$, $t_2=\mathbf{l}(B,0,0,0,1,0)$, $t_3=\mathbf{l}(C,1,1,1,1,0)$. So $D=A+\omega^2B+\omega C$, which gives the following values:

A	0	0	0	0	1	1	1	1
B	0	0	b	b	0	0	b	b
C	0	c	0	c	0	c	0	c
D	0	ωc	$\omega^2 b$	$\omega^2 b+\omega c$	1	$1+\omega c$	$1+\omega^2 b$	$1+\omega^2 b+\omega c$

To obtain only two values of D, we require $b=\omega$ and $c=\omega^2$: then $D=0$ or 1. Hence $d=\omega$ and $d'=\omega^2$ so that

$$l_4=\mathbf{l}(\omega,\omega^2,\omega,\omega,1,0), \qquad l_4'=\mathbf{l}(\omega^2,\omega^2,\omega,\omega,1,0). \quad \square$$

Theorem 19.5.8: *Let $\mathcal{K}$ be a regular $k_{n,3,q}$ with $q>4$.*

(i) *If $\mathcal{K}$ contains two lines l_1 and l_2 meeting at the point P, then every line through P not in the plane $\pi=l_1l_2$ is an n-secant of $\mathcal{K}$.*

(ii) $k=(q\sqrt{q}+1)(q+1)$.

Proof: (i) Let l be any line through P not in π. Since π meets $\mathcal{K}$ in n lines through P, the line l cannot lie in $\mathcal{K}$, by Lemma 19.5.1. Suppose that l is a unisecant. Then there are $n=\sqrt{q}+1$ planes through l meeting π in a line of $\mathcal{K}$ and $q-\sqrt{q}$ planes meeting π in a unisecant through P. In the former case such a plane π' contains no line of $\mathcal{K}$ other than $\pi\cap\pi'$ as any such line would be an $(n+1)$-th through P; so π' either meets $\mathcal{K}$ in

$\pi \cap \pi'$ only or in this line plus a $(q\sqrt{q}-q+\sqrt{q};\ \sqrt{q})$-arc, by Theorem 19.4.19. In the latter case, a plane π'' also contains no line of $\mathscr{H}$ and as $\pi'' \cap \mathscr{H}$ contains two unisecants, l and $\pi'' \cap \pi$, through P, it is a $PG(2, \sqrt{q})$. Hence

$$\begin{aligned} k &< 1+(\sqrt{q}+1)q+(\sqrt{q}+1)(q\sqrt{q}-q+\sqrt{q})+(q-\sqrt{q})(q+\sqrt{q}) \\ &= 2q^2+q\sqrt{q}+\sqrt{q}+1. \end{aligned}$$

However, by Lemma 19.5 4,

$$k \geqslant (q+1)(q\sqrt{q}-q+\sqrt{q}+1).$$

This gives $q-3\sqrt{q}+1 \leqslant 0$, which is impossible for $q>4$. So l must be an n-secant.

(ii) The lines through P consist of $\sqrt{q}+1$ lines on $\mathscr{H}$, $q-\sqrt{q}$ unisecants and q^2 lines $(\sqrt{q}+1)$-secant to $\mathscr{H}$. So

$$k = 1+(\sqrt{q}+1)q+q^2\sqrt{q} = (q\sqrt{q}+1)(q+1). \quad \square$$

We are now in a position to conclude the characterization of Hermitian surfaces. A $k_{n,3,q}$ will be called *normal* if it is regular and not equal to $\mathscr{H}^*$. So, if $\mathscr{H}$ is a normal $k_{n,3,q}$, then $3 \leqslant n \leqslant q-1$ and

(i) for q odd, $\mathscr{H}$ is non-singular;

(ii) for q even and $q>4$, $\mathscr{H}$ is non-singular and $\mathscr{H} \neq \mathscr{R}_3$;

(iii) for $q=4$, $\mathscr{H}$ is non-singular and triangle-free, and $\mathscr{H} \neq \mathscr{H}^*$.

If $\mathscr{H}$ is a normal $k_{n,3,q}$, then a *tangent plane* to $\mathscr{H}$ at a point P is a plane meeting $\mathscr{H}$ in n lines through P.

Lemma 19.5.9: *If $\mathscr{H}$ is a normal $k_{n,3,q}$, then*

(i) $n = \sqrt{q}+1$;

(ii) $k = (q\sqrt{q}+1)(q+1)$;

(iii) *at each point P of $\mathscr{H}$ there is a distinct tangent plane π;*

(iv) *the lines through P in π consist of n lines of $\mathscr{H}$ and $q-\sqrt{q}$ unisecants;*

(v) *the lines through P not in π are all n-secants;*

(vi) *each non-tangent plane meets $\mathscr{H}$ in a Hermitian arc.*

Proof: Part (i) was proved in Lemma 19.5.2. Part (ii) was proved in Lemma 19.5.6 for $q=4$ and in Theorem 19.5.8 for $q>4$.

If l is a line of $\mathscr{H}$, then each plane through l meets $\mathscr{H}$ in n lines of a pencil with centre on l, from Lemma 19.5.4. By Lemma 19.5.1, each of the $q+1$ planes through l gives a distinct centre on l. So for each point on a line of $\mathscr{H}$, part (iii) holds. But every point of $\mathscr{H}$ lies on one of the plane sections through l and so on some line of $\mathscr{H}$. So (iii) holds, whence (iv) does. Part (v) follows immediately and is also a consequence of Theorem 19.5.8(i).

If π' is a non-tangent plane, then π' contains no lines of $\mathcal{H}$. By Theorem 19.4.19, $\pi' \cap \mathcal{H} \neq \emptyset$. So let P be a point of $\pi' \cap \mathcal{H}$ and let π be the tangent plane at P. Then, by (iv), the line $\pi \cap \pi'$ is a 1-secant of $\mathcal{H}$, whence, by (v), the other q lines of π' through P are all n-secants. So $|\pi' \cap \mathcal{H}| = q(n-1)+1 = q\sqrt{q}+1$ and therefore $\pi' \cap \mathcal{H}$ is a Hermitian arc, by Theorem 19.4.19. □

Lemma 19.5.10: *Let $\mathcal{H}$ be a normal $k_{n,3,q}$ and let l be an n-secant of $\mathcal{H}$.*

(i) *The n tangent planes at the n points of $l \cap \mathcal{H}$ all pass through a line l', which is an n-secant skew to l.*

(ii) *The mapping $\mathfrak{R}$ on the set of n-secants defined by $l\mathfrak{R} = l'$ is bijective and involutory.*

Proof: Of the $q+1$ planes through l, let N be the number of tangent planes so that $q+1-N$ meet $\mathcal{H}$ in Hermitian arcs. Then

$$N[n(q-1)+1]+(q+1-N)[q\sqrt{q}+1-n]=k-n;$$

with $k=(q\sqrt{q}+1)(q+1)$ and $n=\sqrt{q}+1$, this gives $N=n$.

Let π_1 and π_2 be distinct tangent planes through l with respective points of contact P_1 and P_2, which are distinct by the previous lemma. The line $l'=P_1P_2$ is skew to l as otherwise $\pi_1=\pi_2=ll'$. Since l' lies in neither π_1 nor π_2, it is an n-secant of $\mathcal{H}$, by Lemma 19.5.9(v). Let P be a point of $l \cap \mathcal{H}$; then the lines PP_1 and PP_2 lie in π_1 and π_2 respectively and so on $\mathcal{H}$. Hence the tangent plane to $\mathcal{H}$ at P contains PP_1 and PP_2, and therefore l'. This proves (i), and (ii) follows immediately. □

With l and l' as in the lemma, l' is *the polar of l with respect to $\mathcal{H}$.*

Lemma 19.5.11: *Let $\mathcal{H}$ be a normal $k_{n,3,q}$ and let Q be a point not on $\mathcal{H}$.*

(i) *The points of contact of the unisecants through Q form a Hermitian arc lying in a plane π not through Q.*

(ii) *The mapping $\mathfrak{R}$ given by $Q\mathfrak{R} = \pi$ is a bijection from the set of points off $\mathcal{H}$ to the set of non-tangent planes.*

Proof: Let N be the number of 1-secants through Q so that the other $q^2+q+1-N$ lines through Q are all n-secants. Then $N+(q^2+q+1-n)=k$, whence $N=q\sqrt{q}+1$. Let $\mathcal{H}'$ be the set of points of contact of these N unisecants.

If P_1 and P_2 are points of $\mathcal{H}'$, then the plane QP_1P_2 does not meet $\mathcal{H}$ in n lines of a pencil, since there are two unisecants, QP_1 and QP_2, in the plane through a point off $\mathcal{H}$. So, by Lemma 19.5.9(vi), the plane QP_1P_2 meets $\mathcal{H}$ in a Hermitian arc, whence $l=P_1P_2$ is an n-secant of $\mathcal{H}$. Also, since the tangent planes at P_1 and P_2 contain the respective 1-secants QP_1 and QP_2, by Lemma 19.5.9(v), so the polar line l' of l contains Q, by Lemma 19.5.10. If P is any point of $l \cap \mathcal{H}$, the tangent plane at P

contains l' and Q, whence QP is a 1-secant. So P is in $\mathcal{K}'$, which means that all n points of $l \cap \mathcal{K}$ are in $\mathcal{K}'$.

Since $\mathcal{K}'$ consists of $q\sqrt{q}+1$ points, it certainly contains three non-collinear points P_1, P_2, P_3. Let π be the plane $P_1P_2P_3$. As P_1P_2 is an n-secant of $\mathcal{K}$, the joins of P_3 to the n points of $P_1P_2 \cap \mathcal{K}$ are all n-secants of $\mathcal{K}$ whose points on $\mathcal{K}$ lie in $\mathcal{K}'$. So $\mathcal{K}' \cap \pi$ contains at least $n(n-1)+1=q+\sqrt{q}+1$ points.

Let P be any point of $\mathcal{K} \cap \pi$. Since $\mathcal{K}' \cap \pi$ contains at least $q+\sqrt{q}+1$ points, there is at least one line through P containing two points of $\mathcal{K}' \cap \pi$. So P is in $\mathcal{K}'$, whence $\mathcal{K} \cap \pi \subset \mathcal{K}'$. Hence $\mathcal{K} \cap \pi$ is not n lines of a pencil, as otherwise $|\mathcal{K} \cap \pi| > |\mathcal{K}'|$. So $\mathcal{K} \cap \pi = \mathcal{K}'$. Also Q is not in π, as otherwise there would be $q\sqrt{q}+1$ lines through Q in π. This proves (i).

It remains to show that if Q and Q' are points off $\mathcal{K}$, then the respective planes π and π' containing the points of contact through Q and Q' are distinct. If $\pi=\pi'$, then the tangent plane at every point P of $\pi \cap \mathcal{K}$ contains both Q and Q', and therefore QQ'. But at each point of $\pi \cap \mathcal{K}$ there is a distinct tangent plane, which gives $q\sqrt{q}+1$ planes through QQ', a contradiction. Since the number of points off $\mathcal{K}$ is the same as the number of tangent planes, $\mathfrak{K}$ is bijective. □

With Q and π as in the lemma, π is *the polar plane of* Q *with respect to* $\mathcal{K}$.

Theorem 19.5.12: *If* $\mathcal{K}$ *is a normal* $k_{n,3,q}$, *then the mapping* $\mathfrak{K}$ *which associates a point of* $\mathcal{K}$ *to its tangent plane and a point off* $\mathcal{K}$ *to its polar plane is a Hermitian polarity, whence* $\mathcal{K}$ *is a non-singular Hermitian surface.*

Proof: It must be shown that if $\pi = P\mathfrak{K}$ and P' is any point of π, then the plane $\pi' = P'\mathfrak{K}$ contains P; that is, $\mathfrak{K}$ preserves incidence.

Firstly suppose that P is in $\mathcal{K}$ so that π is the tangent plane at P. If P' is in $\mathcal{K}$ and $P' \neq P$, then PP' is a line of $\mathcal{K}$ and π' contains PP'. If P' is not on $\mathcal{K}$, then PP' is a unisecant and so, by Lemma 19.5.11(i), the polar plane π' of P' contains P.

If P is not in $\mathcal{K}$, then π does not contain P and meets $\mathcal{K}$ in a Hermitian arc $\mathcal{K}'$. If P' is in $\mathcal{K}'$, then π' is the tangent plane at P'; so, as PP' is a unisecant, π' contains PP', by Lemma 19.5.9(iv), (v). Now suppose that P' is not in $\mathcal{K}'$ and let l' be an n-secant of $\mathcal{K}'$ through P'. If P_0 is any point of $l' \cap \mathcal{K}'$, then PP_0 is a 1-secant and so P lies in the tangent plane at P_0. So P lies on the polar line l of l', by Lemma 19.5.10, and l is an n-secant. If, now, P_1 and P_2 are points of $l \cap \mathcal{K}$, then the tangent planes at P_1 and P_2 contain l' and therefore P', whence P_1P' and P_2P' are 1-secants; that is, P_1 and P_2 lie in π'. So π' contains l, which in turn contains P.

It has been shown that $\mathfrak{K}$ is a polarity, where, if P is in $\mathcal{K}$, the polar plane $P\mathfrak{K}$ is the tangent plane to $\mathcal{K}$ at P; if P is not in $\mathcal{K}$, the polar plane $P\mathfrak{K}$ is the polar plane with respect to $\mathcal{K}$. The self-conjugate points, that is those points for which P is in $P\mathfrak{K}$, are the points of $\mathcal{K}$. As $k = (q\sqrt{q}+1)(q+1)$, so $\mathfrak{K}$ is a Hermitian polarity and $\mathcal{K}$ is a non-singular Hermitian surface, since the number of self-conjugate points in a non-Hermitian polarity is one of $(q+1)^2$, (q^2+1), $(q+1)(q^2+1)$, q^2+q+1; see § 15.2. □

The results of §§ 19.4–19.5 can be summarized as follows, where $\mathcal{U}_{3,q}$ is a non-singular Hermitian surface.

Theorem 19.5.13: *Let $\mathcal{K}$ be a $k_{n,3,q}$ in $\Pi_3 = PG(3, q)$.*

(i) *If $n = 1$, $\mathcal{K}$ is a plane.*

(ii) *If $n = 2$, then $\mathcal{K}$ is one of* (a) $\Pi_2 \cup \Pi_0$, (b) $\Pi_2 \cup \Pi_1$, (c) $\Pi_2 \cup \Pi_2'$,

(iii) *If $n = q$, then $\mathcal{K}$ is one of* (a) $(\Pi_3 \backslash \Pi_2) \cup \Pi_1$ with $\Pi_1 \subset \Pi_2$, (b) $(\Pi_3 \backslash \Pi_1) \cup \Pi_0$ with $\Pi_0 \subset \Pi_1$, (c) $\Pi_3 \backslash \Pi_0$.

(iv) *If $3 \leqslant n \leqslant q-1$, then one of the following occurs.* (a) *If $\mathcal{K}$ is singular, then $\mathcal{K}$ is n planes through a line or a cone $\Pi_0\mathcal{K}'$ with base $\mathcal{K}'$ of type* I, II, III, *or* IV *as in Theorem* 19.4.4. (*b*) *If $\mathcal{K}$ is non-singular, then* (1) *for q odd, $n = \sqrt{q}+1$ and $\mathcal{K} = \mathcal{U}_{3,q}$;* (2) *for q even and $q > 4$, either $n = \sqrt{q}+1$ and $\mathcal{K} = \mathcal{U}_{3,q}$ or $n = \frac{1}{2}q+1$ and $\mathcal{K} = \mathcal{R}_3$;* (3) *for $q = 4$, $\mathcal{K} = \mathcal{U}_{3,4}$ or $\mathcal{K} = \mathcal{R}_3$ or $\mathcal{K} = \mathcal{K}^*$ or $\mathcal{K}$ contains a triangle but not a plane.* □

The complete resolution of the case $q = 4$ is given in the next section.

19.6 Sets of odd type in $PG(3, 4)$

From § 19.5, it remains to classify the non-singular $k_{3,3,4}$ which contain a section of type IV, no section of type VII and at least one section of type I or II.

The situation is, however, far more interesting. First, we note that a $k_{3,3,4}$ is a set of type $(1, 3, 5)$; that is, every line meets it in 1, 3, or 5 points. Accordingly a set $\mathcal{K}$ in $PG(d, 4)$ is a set of *odd type* if every line meets $\mathcal{K}$ in an odd number of points. This immediately implies a closure operation on such sets.

Let O_d be the set of all sets of odd type in $\Pi_d = PG(d, 4)$. For $\mathcal{K}$, $\mathcal{K}'$ in O_d, define

$$\mathcal{K} \nabla \mathcal{K}' = \Pi_d \backslash (\mathcal{K} \Delta \mathcal{K}');$$

that is, $\mathcal{K} \nabla \mathcal{K}'$ comprises the points in $\mathcal{K} \cap \mathcal{K}'$ and the points in $\Pi_d \backslash (\mathcal{K} \cup \mathcal{K}')$. Let U_d be the set of Hermitian varieties in Π_d, where $\Pi_d = \mathbf{V}(0)$ is included in the set.

Lemma 19.6.1: *For $\mathcal{K}$, $\mathcal{K}'$ in O_d,*

(i) *$\mathcal{K}\nabla\mathcal{K}'$ is a set of odd type;*
(ii) $|\mathcal{K}\nabla\mathcal{K}'| = |\Pi_d| - |\mathcal{K}| - |\mathcal{K}'| + 2|\mathcal{K}\cap\mathcal{K}'|$;
(iii) *O_d is a vector space over $GF(2)$ whose zero is Π_d;*
(iv) *U_d is a subspace of dimension $(d+1)^2$ of O_d.*

Proof: (i) Π_d is a disjoint union of the sets $\mathcal{K}\cap\mathcal{K}'$, $\mathcal{K}\backslash\mathcal{K}'$, $\mathcal{K}'\backslash\mathcal{K}$, $\Pi_d\backslash(\mathcal{K}\cup\mathcal{K}')$. If l is any line in Π_d, then $|l| = 5$ and the numbers $|l\cap\mathcal{K}|$ and $|l\cap\mathcal{K}'|$ are both odd. So we obtain the table

$\|l\cap\mathcal{K}\cap\mathcal{K}'\|$	$\|l\cap(\mathcal{K}\backslash\mathcal{K}')\|$	$\|l\cap(\mathcal{K}'\backslash\mathcal{K})\|$	$\|l\cap(\Pi_d\backslash(\mathcal{K}\cup\mathcal{K}'))\|$
odd	even	even	even
even	odd	odd	odd

For each row, the sum of the first and last columns is odd; that is, $\mathcal{K}\nabla\mathcal{K}'$ is a set of odd type.

(ii)
$$|\mathcal{K}\nabla\mathcal{K}'| = |\Pi_d\backslash(\mathcal{K}\cup\mathcal{K}')| + |\mathcal{K}\cap\mathcal{K}'|,$$
$$|\Pi_d\backslash(\mathcal{K}\cup\mathcal{K}'| = |\Pi_d| - |\mathcal{K}| - |\mathcal{K}'| + |\mathcal{K}\cap\mathcal{K}'|$$

imply the result.

(iii) $\mathcal{K}\nabla\mathcal{K} = \Pi_d$, and all the other axioms for a vector space over $GF(2)$ are easily verified. If $E_d = \{\mathcal{K}^c \mid \mathcal{K}\in O_d\}$, which is the set of sets of *even type* in Π_d, then E_d is closed under symmetric difference; if $\mathcal{M}, \mathcal{M}'\in E_d$, so is $\mathcal{M}\Delta\mathcal{M}'$. Then O_d being a binary vector space under ∇ is equivalent to E_d being a binary vector space under Δ with the empty set as zero.

(iv) The Hermitian varieties form a subspace under the addition of the corresponding Hermitian forms. If $\mathcal{K} = \mathbf{V}(F)$, then $F(X) = 0$ or 1 and $\mathbf{P}(X)$ lies on $\mathcal{K}$ or not accordingly. Hence, if also $\mathcal{K}' = \mathbf{V}(G)$, we have that $(F+G)(X) = 0$ if and only if $F(X) = G(X) = 0$ or $F(X) = G(X) = 1$. Hence U_d is a subspace of O_d.

As, with $a_i \in GF(2)$,

$$F = \sum a_i x_i \bar{x}_i + \sum (b_{ij} x_i \bar{x}_j + \bar{b}_{ij} \bar{x}_i x_j),$$

there are two choices for each a_i and four for each b_{ij}, so

$$|U_d| = 2^{d+1}\cdot 4^{\mathbf{h}(d,2)} = 2^{d+1}\cdot 2^{d(d+1)} = 2^{(d+1)^2},$$

whence the result. $\square$

The closure operation previously appeared in Theorem 19.4.9, where $\tilde{\mathcal{K}}_1$ is in fact $\mathcal{K}_1\nabla\pi_0$.

For $q = 4$, Theorem 19.4.4 can be stated more precisely.

Theorem 19.6.2: *The projectively distinct sets $\mathcal{K}$ of odd type in $PG(2, 4)$ are given in Table 19.7, where $k = |\mathcal{K}|$, $N(\mathcal{K})$ is the total number of $\mathcal{K}$ of*

Table 19.7

Type	Description of $\mathcal{K}$	k	$N(\mathcal{K})$	G	$\|G\|$	τ_1	τ_3	τ_5
I	$\mathcal{U}_{2,4}$	9	280	$PGU(3, 4)$	216	9	12	0
II	$PG(2, 2)$	7	360	$PGL(3, 2)$	168	14	7	0
III	Oval + line	11	1008	$\mathbf{A}_5$	60	5	15	1
IV	Complement of oval	15	168	$\mathbf{A}_6$	360	0	15	6
V	$\Pi_0\mathcal{U}_{1,4}$: 3 lines	13	210	$(\mathbf{Z}_3 \times \mathbf{S}_3)\mathbf{Z}_2^4$	288	2	16	3
VI	Π_1	5	21	$GL(2, 4)\mathbf{Z}_2^4$	2880	20	0	1
VII	$PG(2, 4)$	21	1	$PGL(3, 4)$	60480	0	0	21
			$2048 = 2^{11}$					

that type, G is the projective group of $\mathcal{K}$, and τ_i is the number of i-secants of $\mathcal{K}$. □

The next lemma incorporates various results on sets of odd type from § 19.5. Let $\mathcal{K}_{\mathrm{I}}, \mathcal{K}_{\mathrm{II}}, \ldots, \mathcal{K}_{\mathrm{VII}}$ respectively denote plane odd sets of type I, II, ..., VII.

Lemma 19.6.3: *Let $\mathcal{K}$ be a set of odd type in $PG(3, 4)$ with $k = |\mathcal{K}|$.*

(i) *If $\mathcal{K}$ is singular, then $\mathcal{K}$ is a cone $\Pi_0\mathcal{K}_{\mathrm{I}}, \ldots, \Pi_0\mathcal{K}_{\mathrm{VII}}$.*

(ii) *If $\mathcal{K}$ is non-singular, has a section of type* IV *and has no section of type* I *or* II, *then it has exactly one section of type* VII, *$k = 53$ and $\mathcal{K} = \mathcal{R}_3$.*

(iii) *If $\mathcal{K}$ is non-singular with a section of type* VII, *then $k = 53$ and $\mathcal{K} = \mathcal{R}_3$.*

(iv) *If $\mathcal{K}$ is non-singular and has no section of type* IV, *then $k = 37$ and $\mathcal{K} = \mathcal{K}^*$ or $k = 45$ and $\mathcal{K} = \mathcal{U}_{3,4}$.*

Proof: (i) This follows from Lemma 19.4.3 and Theorem 19.4.4.

(ii) Theorems 19.4.8, 19.4.9, and 19.4.17 imply this result.

(iii) This is part of Theorem 19.4.18.

(iv) This follows from Lemma 19.5.6 and Theorems 19.5.7 and 19.5.12. □

So far we have seven non-singular sets and three non-singular sets of odd type as in List I. To construct the remainder, the closure operation with one of the sets a plane is used.

List I. *Partial List of Sets of Odd Type in $PG(3, 4)$*

Type	1	2	3	4	5	6	7	8	9	10
Names	Π_3	Π_2	$\Pi_1\mathcal{U}_1$	$\Pi_0\mathcal{U}_2$	$\mathcal{P}_{\mathrm{II}}$	$\mathcal{P}_{\mathrm{III}}$	$\mathcal{P}_{\mathrm{IV}}$	$\mathcal{U}_3$	$\mathcal{R}_3$	$\mathcal{K}^*$
	$\Pi_0\mathcal{K}_{\mathrm{VII}}$	$\Pi_0\mathcal{K}_{\mathrm{VI}}$	$\Pi_0\mathcal{K}_{\mathrm{V}}$	$\Pi_0\mathcal{K}_{\mathrm{I}}$	$\Pi_0\mathcal{K}_{\mathrm{II}}$	$\Pi_0\mathcal{K}_{\mathrm{III}}$	$\Pi_0\mathcal{K}_{\mathrm{IV}}$	$\mathcal{U}_{3,4}$		
Hermitian (=H)	H	H	H	H	–	–	–	H	–	–
Singular space	Π_3	Π_2	Π_1	Π_0	Π_0	Π_0	Π_0	–	–	–

To begin more generally, in $PG(d, 4)$ let π be a prime and define the map $\delta_\pi : O_d \to O_d$ by $\mathcal{K}\delta_\pi = \mathcal{K}\nabla\pi$; the map δ_π is called a *disflection* and we say that $\mathcal{K}$ has been disflected in the prime π. To see better the effect of π, it is appropriate to consider the restriction of δ_π to a prime π'; then in π' its image is

$$(\mathcal{K} \cap \pi')\delta_\pi = (\mathcal{K} \cap \pi')\nabla(\pi \cap \pi').$$

In $PG(3, 4)$, a disflection δ_π may be written $\delta_{\mathrm{I}}, \ldots, \delta_{\mathrm{VII}}$ as π is respectively of type I, . . . , VII. Similarly, in $PG(2, 4)$, a disflection δ_l is written $\delta_1, \delta_3, \delta_5$ as $|l \cap \mathcal{K}| = 1, 3, 5$.

Lemma 19.6.4: *The seven plane sets of odd type are disflected according to Table 19.8, where $k = |\mathcal{K}|$ and, for each type of $\mathcal{K}$ and each value of $|l \cap \mathcal{K}|$, the table gives the type of $\mathcal{K}\delta_l$.*

Proof: The type of each odd set in $PG(2, 4)$ is determined by the number of its points. From Lemma 19.6.1(ii),

$$\begin{aligned} |\mathcal{K} \nabla l| &= 21 - 5 - |\mathcal{K}| + 2\,|\mathcal{K} \cap l| \\ &= 16 - |\mathcal{K}| + 2\,|\mathcal{K} \cap l|. \end{aligned}$$

Hence there is the comparison

$\lvert\mathcal{K} \cap l\rvert$	1	3	5
$\lvert\mathcal{K} \nabla l\rvert$	$18 - k$	$22 - k$	$26 - k$. □

Table 19.8 *Disflections of Odd Sets of $PG(2, 4)$*

$l \cap \mathcal{K}$	k:	9	13	5	21	7	11	15
	$\mathcal{K}$:	I	V	VI	VII	II	III	IV
1		I	VI	V	–	III	II	–
3		V	I	–	–	IV	III	II
5		–	V	VII	VI	–	IV	III

Corollary: *Disflection fixes U_2 and $O_2 \backslash U_2$.* □

The information of Lemma 19.6.4 is now used to investigate the possibilities for sections of odd sets in $PG(3, 4)$.

Lemma 19.6.5: *No set in O_3 can have all its plane sections non-Hermitian, that is all of types* II, III, *and* IV.

Proof: All ten $\mathcal{K}$ in List I have some Hermitian section, that is of type I, V, VI, or VII; see Table 19.10. So if $\mathcal{K}$ has only sections of type II, III, and IV, it is not in List I and, from Lemma 19.6.3, it is non-singular with a section by π of type IV and a section by π' of type II. So $\pi \cap \pi'$ is a 3-secant. Through a point P of $\pi \cap \pi' \cap \mathcal{K}$ there is a unisecant l' in π' and a line l of $\mathcal{K}$ in π, whence ll' is a plane of type III.

Through a line l of $\mathcal{K}$, suppose that there are x planes of type III and $5-x$ of type IV. Then, with $k=|\mathcal{K}|$,

$$k = 5+6x+10(5-x) = 55-4x. \tag{19.35}$$

As k is fixed, so is x and hence $1 \leqslant x \leqslant 4$, since there are sections of each type.

Similarly, suppose there are y planes of type III and $5-y$ of type I through a unisecant l'. Thus

$$k = 1+10y+6(5-y) = 31+4y \tag{19.36}$$

with $1 \leqslant y \leqslant 4$. Comparing (19.35) and (19.36) gives $y = 6-x$, whence $2 \leqslant x \leqslant 4$ and $k = 39$, 43, or 47.

For an arbitrary plane π, let $\mathcal{K}' = \mathcal{K} \nabla \pi$, with $k' = |\mathcal{K}'|$ and $t = |\mathcal{K} \cap \pi|$. Then Lemma 19.6.1(ii) becomes

$$k' = 64-k+2t. \tag{19.37}$$

By Table 19.8, $\mathcal{K}'$ also has only sections of type II, III, and IV. From (19.37),

$$k=39 \quad \text{and} \quad t=15 \Rightarrow k'=55;$$
$$k=43 \quad \text{and} \quad t=15 \Rightarrow k'=51;$$
$$k=47 \quad \text{and} \quad t=7 \Rightarrow k'=31.$$

As $\{39, 43, 47\} \cap \{31, 51, 55\} = \varnothing$, we have a contradiction. □

Lemma 19.6.6: *Each set $\mathcal{K}$ in O_3 can be obtained from a set $\mathcal{K}'$ in O_3 containing a plane by at most three disflections.*

Proof: By the previous lemma, each $\mathcal{K}$ in O_3 has a section of type I, V, VI, or VII.

If $\mathcal{K}$ has a section of type VI by a plane π_6, and π is a plane through the line $\pi_6 \cap \mathcal{K}$, then, from Table 19.8, π_6 lies on $\mathcal{K}' = \mathcal{K} \nabla \pi$, whence $\mathcal{K} = \mathcal{K}' \nabla \pi$.

If $\mathcal{K}$ has a section of type V by a plane π_5 and π is a plane through a 1-secant of $\mathcal{K}$ in π_5, then, again from Table 19.8, π_5 is a plane of type VI for $\mathcal{K} \nabla \pi$. By the above, a further disflection in a plane π' gives $\mathcal{K}' = \mathcal{K} \nabla \pi \nabla \pi'$ containing π_5, whence $\mathcal{K} = \mathcal{K}' \nabla \pi' \nabla \pi$.

If $\mathcal{K}$ has a section of type I by a plane π_1 and π is a plane through a 3-secant of $\mathcal{K}$ in π_1, then Table 19.8 tells us that π_1 is a section of type V for $\mathcal{K} \nabla \pi$, whence the previous paragraph implies the existence of planes π' and π'' such that $\mathcal{K}' = \mathcal{K} \nabla \pi \nabla \pi' \nabla \pi''$ contains the plane π_1. □

Theorem 19.6.7: *There are seven singular and seven non-singular projectively distinct sets in O_3.*

Proof: By Lemma 19.6.6, it is sufficient to consider the sets containing a plane. By Lemma 19.6.3 these are all in List I and are Π_3, Π_2, $\Pi_1\mathscr{U}_1$, $\mathscr{P}_{\mathrm{III}}$, $\mathscr{P}_{\mathrm{IV}}$, $\mathscr{R}_3$. Disflections of Hermitian varieties only produce other Hermitian varieties, by Lemma 19.6.1(iv). So it is only necessary to consider $\mathscr{P}_{\mathrm{III}}$, $\mathscr{P}_{\mathrm{IV}}$ and $\mathscr{R}_3$. Disflections produce four new sets

$$\mathscr{S}_{\mathrm{IV}} = \mathscr{P}_{\mathrm{IV}}\delta_{\mathrm{IV}},$$
$$\mathscr{S}_{\mathrm{III}} = \mathscr{P}_{\mathrm{III}}\delta_{\mathrm{III}},$$
$$\mathscr{S}_{\mathrm{II}} = \mathscr{S}_{\mathrm{IV}}\delta_{\mathrm{I}} = \mathscr{P}_{\mathrm{II}}\delta_{\mathrm{II}},$$
$$\mathscr{T} = \mathscr{S}_{\mathrm{IV}}\delta_{\mathrm{II}} = \mathscr{K}^*\delta_{\mathrm{I}},$$

and no more; they are respectively called type 11, 12, 13, 14. Table 19.9 gives the images of all types of disflection of each type of odd set. □

Table 19.9 *Disflections of Sets in O_3*

Type	Name	δ: I	V	VI	VII	II	III	IV
1	Π_3	–	–	–	2	–	–	–
2	Π_2	–	–	3	1	–	–	–
3	$\Pi_1\mathscr{U}_1$	–	4	2	3	–	–	–
4	$\Pi_0\mathscr{U}_2$	8	3	4	–	–	–	–
8	$\mathscr{U}_3$	4	8	–	–	–	–	–
5	$\mathscr{P}_{\mathrm{II}}$	–	7	6	–	13	–	–
6	$\mathscr{P}_{\mathrm{III}}$	–	6	5	7	–	12	–
7	$\mathscr{P}_{\mathrm{IV}}$	–	5	–	6	–	–	11
9	$\mathscr{R}_3$	–	10	–	9	–	11	12
10	$\mathscr{K}^*$	14	9	10	–	12	13	–
11	$\mathscr{S}_{\mathrm{IV}}$	13	–	12	–	14	9	7
12	$\mathscr{S}_{\mathrm{III}}$	12	13	11	–	10	6,14	9
13	$\mathscr{S}_{\mathrm{II}}$	11	12	–	–	5	10	14
14	$\mathscr{T}$	10	14	–	–	11	12	13

Theorem 19.6.8: *The vector space O_3 has dimension 24. For each of the 14 projectively distinct types of set $\mathscr{K}$ in O_3, Table 19.10 gives $k = |\mathscr{K}|$, the total number N of each type, the order $|G|$ of the projective group G of $\mathscr{K}$, the sizes of the orbits of the points of $\mathscr{K}$ under G, the numbers $N_{\mathrm{I}}, \ldots, N_{\mathrm{VII}}$ of each type of plane section of $\mathscr{K}$, and the numbers τ_1, τ_3, and τ_5 of unisecants, trisecants, and lines of $\mathscr{K}$.*

Table 19.10 *Sets of Odd Type in* $PG(3, 4)$

Type	$\mathcal{K}$	k	$85\text{-}k$	N	$\lvert G \rvert$	Orbit lengths	N_{I}	N_{II}	N_{III}	N_{IV}	N_{V}	N_{VI}	N_{VII}	τ_1	τ_3	τ_5
1	Π_3	85	0	1	$85 \cdot 2^{12} \cdot 3^4 \cdot 5 \cdot 7$	85	–	–	–	–	–	–	85	–	–	357
2	Π_2	21	64	85	$2^{12} \cdot 3^4 \cdot 5 \cdot 7$	21	–	–	–	–	–	84	1	336	–	21
3	$\Pi_1\mathcal{U}_1$	53	32	$85 \cdot 2 \cdot 3 \cdot 7$	$2^{11} \cdot 3^3 \cdot 5$	5, 48	–	–	–	–	80	2	3	40	256	61
4	$\Pi_0\mathcal{U}_2$	37	48	$85 \cdot 2^3 \cdot 5 \cdot 7$	$2^9 \cdot 3^4$	1, 36	64	–	–	–	12	9	–	156	192	9
5	$\mathcal{P}_{\mathrm{II}}$	29	56	$85 \cdot 2^3 \cdot 3^2 \cdot 5$	$2^9 \cdot 3^2 \cdot 7$	1, 28	–	64	–	–	7	14	–	238	112	7
6	$\mathcal{P}_{\mathrm{III}}$	45	40	$85 \cdot 2^4 \cdot 3^2 \cdot 7$	$2^8 \cdot 3^2 \cdot 5$	1, 20, 24	–	–	64	–	15	5	1	90	240	27
7	$\mathcal{P}_{\mathrm{IV}}$	61	24	$85 \cdot 2^3 \cdot 3 \cdot 7$	$2^9 \cdot 3^3 \cdot 5$	1, 60	–	–	–	64	15	–	6	6	240	111
8	$\mathcal{U}_3$	45	40	$85 \cdot 2^6 \cdot 7$	$2^6 \cdot 3^4 \cdot 5$	45	40	–	–	–	45	–	–	90	240	27
9	$\mathcal{R}_3$	53	32	$85 \cdot 2^5 \cdot 3^3 \cdot 7$	$2^7 \cdot 3 \cdot 5$	1, 20, 32	–	–	32	32	20	–	1	40	256	61
10	$\mathcal{K}^*$	37	48	$85 \cdot 2^5 \cdot 3^3 \cdot 5 \cdot 7$	$2^7 \cdot 3$	1, 4, 32	16	32	32	–	4	1	–	156	192	9
11	$\mathcal{S}_{\mathrm{IV}}$	33	52	$85 \cdot 2^9 \cdot 3 \cdot 7$	$2^3 \cdot 3^3 \cdot 5$	15, 18	15	45	18	1	–	6	–	195	156	6
12	$\mathcal{S}_{\mathrm{III}}$	41	44	$85 \cdot 2^{10} \cdot 3^2 \cdot 7$	$2^2 \cdot 3^2 \cdot 5$	5, 6, 30	15	15	1+45	3	5	1	–	121	220	16
13	$\mathcal{S}_{\mathrm{II}}$	49	36	$85 \cdot 2^9 \cdot 3^2 \cdot 5$	$2^3 \cdot 3^2 \cdot 7$	7, 42	7	1	42	21	14	–	–	63	252	42
14	$\mathcal{T}$	45	40	$85 \cdot 2^6 \cdot 3^3 \cdot 5 \cdot 7$	$2^6 \cdot 3$	1, 6, 6, 32	8	8	48	8	1+12	–	–	90	240	27
				Total: $16,777,216 = 2^{24}$												

Proof: In each case, there is a straightforward calculation for N from the geometrical properties of $\mathcal{K}$. Since the type of a plane section is uniquely determined by the number of its points, the numbers $N_{\mathrm{I}}, \ldots, N_{\mathrm{VII}}$ are calculated without difficulty. If σ_i is the number of i-secants through a point off $\mathcal{K}$, then from (19.9),

$$3\sigma_3+\sigma_1=k, \qquad \sigma_3+\sigma_1=21,$$

whence

$$\sigma_1=(63-k)/2, \qquad \sigma_3=(k-21)/2.$$

Hence

$$\tau_1=(85-k)\sigma_1/4=(85-k)(63-k)/8,$$
$$\tau_3=(85-k)\sigma_3/2=(85-k)(k-21)/4,$$
$$\tau_5=357-\tau_1-\tau_3.$$

So k determines all τ_i. □

Corollary 1: $85-k$ *is divisible by* 4 *for all sets* $\mathcal{K}$ *in* O_3. □

Corollary 2: *The number* m_k *of projectively distinct* $\mathcal{K}$ *with* $k=|\mathcal{K}|$ *and the number* n_k *of projectively distinct, non-singular* $\mathcal{K}$ *with* $k=|\mathcal{K}|$ *are as follows*:

k	85	61	53	49	45	41	37	33	29	21
m_k	1	1	2	1	3	1	2	1	1	1
n_k	0	0	1	1	2	1	1	1	0	0. □

Corollary 3: *The smallest odd set containing no plane is* $\mathcal{P}_{\mathrm{II}}$. □

Corollary 4: *Table* 19.10 *implies the following characterizations, some previously given, of non-singular odd sets* $\mathcal{K}$.

(i) *If* $\mathcal{K}$ *has only two types of plane section, then* $\mathcal{K}=\mathcal{U}_3$.
(ii) *If* $\mathcal{K}$ *has at most three types of plane section, then* $\mathcal{K}=\mathcal{U}_3$.
(iii) *If* $\mathcal{K}$ *contains a plane, then* $\mathcal{K}=\mathcal{R}_3$.
(iv) *If* $\mathcal{K}$ *has no section of type* I *or* II, *then* $\mathcal{K}=\mathcal{R}_3$.
(v) *If every section of* $\mathcal{K}$ *contains a line, then* $\mathcal{K}=\mathcal{R}_3$.
(vi) *If* $\mathcal{K}$ *has no section of type* IV *and has a section of type* II, III, *or* VI, *then* $\mathcal{K}=\mathcal{K}^*$.
(vii) *If* $\mathcal{K}$ *has no section of type* V, *then* $\mathcal{K}=\mathcal{S}_{\mathrm{IV}}$.
(viii) *If* $\mathcal{K}$ *has a section of type* IV *and no lines other than those in that section, then* $\mathcal{K}=\mathcal{S}_{\mathrm{IV}}$.
(ix) *If* $\mathcal{K}$ *has sections of all types other than* VII, *then* $\mathcal{K}=\mathcal{S}_{\mathrm{III}}$.
(x) *If* $\mathcal{K}$ *has a unique section of type* VI *and at least one section of type* IV, *then* $\mathcal{K}=\mathcal{S}_{\mathrm{III}}$.

(xi) *If $\mathcal{K}$ has a unique section of type* II, *then* $\mathcal{K} = \mathcal{S}_{\mathrm{II}}$.
(xii) *If $\mathcal{K}$ has no section of type* VI *or* VII *and has more than one section of type* II, *then* $\mathcal{K} = \mathcal{T}$.
(xiii) *If $\mathcal{K}$ has more than two types of plane section and there exists $\mathcal{K}'$ of different type with* $|\mathcal{K}'| = |\mathcal{K}|$, *then* $\mathcal{K} = \mathcal{T}$. □

For each of the seven non-singular $\mathcal{K}$, some remarks on the configuration of lines on $\mathcal{K}$ are appended.
Type 8, $\mathcal{K} = \mathcal{U}_3$. Each of the 27 lines l meets ten others, two through each point of l with each such pair coplanar with l; this gives a $(45_3, 27_5)$ configuration. For a more elaborate exploration of this famous configuration, see §§ 20.1–20.3 as well as § 19.1.
Type 9, $\mathcal{K} = \mathcal{R}_3$. Apart from the 21 lines in the plane π_0 on $\mathcal{K}$, the remaining 40 lines pass two through each of 20 points on π_0 and five through each of the other 32 points. So $\mathcal{K} \backslash \pi_0$ and the lines not in π_0 form a $(32_5, 40_4)$ configuration. See also § 19.4.
Type 10, $\mathcal{K} = \mathcal{K}^*$. The configuration of nine lines is described in Theorem 19.5.7.
Type 11, $\mathcal{K} = \mathcal{S}_{\mathrm{IV}}$. The six lines form a dual oval.
Type 12, $\mathcal{K} = \mathcal{S}_{\mathrm{III}}$. One of the lines l_0 meets all 15 others, three through each of its points; these three are coplanar. Three planes through l_0 meet $\mathcal{K}$ in a dual oval, whence each line other than l_0 meets five others including l_0. These 15 lines and the points on them not on l_0 form a $(30_2, 15_4)$ configuration.
Type 13, $\mathcal{K} = \mathcal{S}_{\mathrm{II}}$. Through each of the seven points of the unique section of type II, there are six lines in two coplanar sets of three. Through each of the 42 points of $\mathcal{K}$ not in the plane of type II, there are four of the 42 lines, thus giving a $(42_4, 42_4)$ configuration.
Type 14, $\mathcal{K} = \mathcal{T}$. There is one special section by a plane π of type V meeting $\mathcal{K}$ in three lines l_1, l_2, l_3 such that every other plane through each l_i meets $\mathcal{K}$ in a section of type V, giving the other 24 lines. These 24 lines meet π in six points, two on each l_i and forming the vertices of a quadrilateral: together with $P = l_1 \cap l_2 \cap l_3$, these six points form a $PG(2, 2)$ in π. So, through P, there are three lines of $\mathcal{K}$, through each of the other points of the $PG(2, 2)$ five lines of $\mathcal{K}$, through the remaining points of the l_i just one line of $\mathcal{K}$, and through each point of $\mathcal{K} \backslash \pi$ three lines of $\mathcal{K}$. The last give a $(32_3, 24_4)$ configuration.

The sets of odd type can be coordinatized and this we proceed to do. This will also give an independent check on the dimension of O_d. A set $\mathcal{K}$ in O_d is called *standardized* if each edge $\mathbf{U}_i\mathbf{U}_j$ of the simplex of reference S is a line of $\mathcal{K}$.

Lemma 19.6.9: *Each $\mathcal{K}$ in O_d can be changed by a sequence of disflections into a standardized one.*

Proof: For each $\mathbf{U}_i$ not in $\mathscr{K}$, disflect $\mathscr{K}$ in the prime $\mathbf{u}_i$ to obtain $\mathscr{K}'$; thus $\mathbf{U}_i \in \mathscr{K}'$ and every other $\mathbf{U}_j$ is in both or neither of $\mathscr{K}$ and $\mathscr{K}'$. Continuing this process, we obtain an odd set $\mathscr{K}$ containing all $\mathbf{U}_i$. So each $\mathbf{U}_i\mathbf{U}_j$ is a trisecant or a line of $\mathscr{K}$.

Suppose $\mathbf{U}_i\mathbf{U}_j$ is a 3-secant and that $\mathbf{U}_i\mathbf{U}_j \backslash \mathscr{K} = \{P, P'\}$. Let $\pi = P(\mathbf{u}_i \cap \mathbf{u}_j)$ and $\pi' = P'(\mathbf{u}_i \cap \mathbf{u}_j)$, and define $\mathscr{K}' = \mathscr{K}\delta_\pi\delta_{\pi'}$. Then $\mathbf{u}_i \cap \mathbf{u}_j$ is fixed by $\delta_\pi\delta_{\pi'}$ as are the edges $\mathbf{U}_i\mathbf{U}_k$ and $\mathbf{U}_j\mathbf{U}_k$ for $k \neq i, j$, while $\mathbf{U}_i\mathbf{U}_j$ becomes a line of $\mathscr{K}'$. Thus, after a sequence of such products, an odd set containing all lines $\mathbf{U}_i\mathbf{U}_j$ is obtained. □

Corollary: *Each plane face* $\mathbf{U}_i\mathbf{U}_j\mathbf{U}_k$ *of* S *is a section of type* VII *or of type* IV *of a standardized odd set.* □

Lemma 19.6.10: *A standardized set in* O_2 *has equation*

$$\alpha x_0 x_1 x_2(\alpha x_0 x_1 x_2 + 1) = 0 \tag{19.38}$$

for some α *in* $GF(4)$.

Proof: In $PG(2, 4)$ there are three ovals containing a 3-arc, § 14.3. Now,

$$\begin{aligned}
&\alpha x_0 x_1 x_2(\alpha x_0 x_1 x_2 + 1)\\
&\quad = \alpha^2 x_0^2 x_1^2 x_2^2 + \alpha x_0 x_1 x_2 + \alpha x_0 x_1 x_2 + \alpha x_0 x_1 x_2\\
&\quad = \alpha^2 x_0^2 x_1^2 x_2^2 + \alpha^4 x_0^4 x_1 x_2 + \alpha x_0 x_1^4 x_2 + \alpha x_0 x_1 x_2^4\\
&\quad = \alpha x_0 x_1 x_2(\alpha^3 x_0^3 + x_1^3 + x_2^3 + \alpha x_0 x_1 x_2)\\
&\quad = \alpha x_0 x_1 x_2(\alpha x_0 + x_1 + x_2)(\alpha x_0 + \omega x_1 + \omega^2 x_2)(\alpha x_0 + \omega^2 x_1 + \omega x_2).
\end{aligned}$$

Thus, for $\alpha = 1, \omega, \omega^2$, equation (19.38) represents the three dual ovals containing the lines $\mathbf{u}_0, \mathbf{u}_1, \mathbf{u}_2$. □

Lemma 19.6.11: *The equation*

$$\sum{}'' c_{ijk} x_i x_j x_k = 0 \quad \text{or} \quad 1 \tag{19.39}$$

defines a standardized set in O_d, *where the summation is from* 0 *to* d *with* $i < j < k$.

Proof: The line PQ with $P = \mathbf{P}(A)$ and $Q = \mathbf{P}(B)$ also contains the points $\mathbf{P}(A + B)$, $\mathbf{P}(A + \omega B)$, $\mathbf{P}(A + \omega^2 B)$. On these five points the expression $x_i x_j x_k$ takes on the respective values

$$\begin{gathered}
a_i a_j a_k, \qquad b_i b_j b_k, \qquad (a_i + b_i)(a_j + b_j)(a_k + b_k),\\
(a_i + \omega b_i)(a_j + \omega b_j)(a_k + \omega b_k), \qquad (a_i + \omega^2 b_i)(a_j + \omega^2 b_j)(a_k + \omega^2 b_k);
\end{gathered}$$

the sum of these values is zero. In $GF(4)$, the only ways to obtain zero sums are by combinations of $1 + 1$, $\omega + \omega$, $\omega^2 + \omega^2$, $1 + \omega + \omega^2$. So an even number of the five values are in $GF(4) \backslash GF(2)$.

This property is then valid for $c_{ijk}x_ix_jx_k$, for the sum of any two such expressions and hence for $\sum'' c_{ijk}x_ix_jx_k$. Thus $\sum'' c_{ijk}x_ix_jx_k = 0$ or 1 for either 1 or 3 or 5 points of PQ. So the set is odd and, as each line $\mathbf{U}_i\mathbf{U}_j$ satisfies (19.39), it is standardized. □

Lemma 19.6.12: *Given a set $\mathcal{K}$ in O_d there exists a set $\mathcal{M}$ in O_d such that $\mathcal{M}$ coincides with $\mathcal{K}$ at all planes $\mathbf{U}_i\mathbf{U}_j\mathbf{U}_k$ and*

$$\mathcal{M} = \mathbf{V}(E^2 + E + H),$$

where $E = \sum'' c_{ijk}x_ix_jx_k$ and H is Hermitian.

Proof: By a sequence of disflections in the primes $\mathbf{V}(L_i)$, $i = 1, \ldots, r$, the set $\mathcal{K}$ is transformed to a standardized set $\mathcal{K}'$. If the section $\mathbf{U}_i\mathbf{U}_j\mathbf{U}_k \cap \mathcal{K}'$ has equation $c_{ijk}x_ix_jx_k = 0$ or 1, then $\mathcal{K}'' = \mathbf{V}(E^2 + E)$, where $E = \sum'' c_{ijk}x_ix_jx_k$, coincides with $\mathcal{K}'$ on all planes $\mathbf{U}_i\mathbf{U}_j\mathbf{U}_k$. Hence, with $H = \sum_{i=1}^r L_i^3$, the set $\mathcal{M} = \mathbf{V}(E^2 + E + H)$ is the one required. □

Theorem 19.6.13: *Every set of odd type in $PG(d, 4)$ is uniquely expressible as $\mathbf{V}(E^2 + E + H)$ where $E = \sum'' c_{ijk}x_ix_jx_k$ and H is Hermitian.*

Proof: It must be shown that a standardized set $\mathcal{K}$ is uniquely determined by its sections by the plane faces of the simplex S. Once the uniqueness is settled for O_3, the induction to O_d for $d > 3$ is immediate. For $d = 3$, it is necessary to consider separately the cases that i faces are of type VII and $4 - i$ are of type IV for $i = 0, 1, 2, 3, 4$. The details are omitted; see § 19.7. □

Corollary 1: *The dimension of O_d as a binary vector space is $(d^3 + 3d^2 + 5d + 3)/3$.*

Proof: It was shown in Lemma 19.6.1 that dim $U_d = (d+1)^2$. The number of forms E is $4^{c(d+1,3)} = 2^{d(d^2-1)/3}$. Then $(d+1)^2 + d(d^2-1)/3$ is the required answer. □

This result gives dim $O_2 = 11$ and dim $O_3 = 24$, which coincide with the numbers obtained in Tables 19.7 and 19.10.

Corollary 2: *The rank over $GF(2)$ of the incidence matrix of points and lines in $PG(d, 4)$ is $(4^{d+1} - d^3 - 3d^2 - 5d - 4)/3$.*

Proof: Let M_d be the $r_d \times c_d$ incidence matrix of points and lines in $PG(d, 4)$, where $c_d = \theta(d) = (4^{d+1} - 1)/3$ and $r_d = \theta(d)\theta(d-1)/5 = (4^{d+1} - 1)(4^d - 1)/45$. Thus, if $PG(d, 4) = \{P_i \mid i \in \mathbf{N}_{c_d}\}$ and $PG^{(1)}(d, 4) = \{l_i \mid i \in \mathbf{N}_{r_d}\}$, we take $M_d = (m_{ij})$ where $m_{ij} = 1$ if $P_j \in l_i$ and $m_{ij} = 0$ if $P_j \notin l_i$. Let the i-th row of M_d be m_i. Now, each subset $\mathcal{K}$ of $PG(d, 4)$ can be represented by the vector $K = (a_1, \ldots, a_{c_d})$, where $a_i = 0$ if $P_i \in \mathcal{K}$ and

$a_i = 1$ if $P_i \notin \mathcal{K}$. Therefore

$$\begin{aligned} P_i \in l_j \cap \mathcal{K} &\Leftrightarrow a_i = 0, \qquad m_{ji} = 1;\\ P_i \in l_j \backslash \mathcal{K} &\Leftrightarrow a_i = 1, \qquad m_{ji} = 1;\\ P_i \in \mathcal{K} \backslash l_j &\Leftrightarrow a_i = 0, \qquad m_{ji} = 0;\\ P_i \notin \mathcal{K} \cup l_j &\Leftrightarrow a_i = 1, \qquad m_{ji} = 0. \end{aligned}$$

So $a_i m_{ji} = 1 \Leftrightarrow P_i \in l_j \backslash \mathcal{K}$. As $|l_j| = 5$, so $|l_j \cap \mathcal{K}|$ is odd if and only if $|l_j \backslash \mathcal{K}|$ is even, which occurs if and only if $Km_j^* = 0$. Thus the set $\mathcal{K}$ is of odd type if and only if the scalar product of K with each row of M_d is zero (over $GF(2)$). Hence, if S_d is the vector space over $GF(2)$ generated by the rows of M_d, then rank $M_d = \dim S_d$ and O_d is the orthogonal complement of S_d. Hence

$$\begin{aligned} \text{rank } M_d &= c_d - \dim O_d \\ &= (4^{d+1} - 1)/3 - (d^3 + 3d^2 + 5d + 3)/3. \quad \square \end{aligned}$$

Example: To illustrate Theorem 19.6.13, we give a form $F = E^2 + E + H$ for each type of element in O_2; see Table 19.7.

I. $F = x_0^3 + x_1^3 + x_2^3$
II. $F = x_0^2 x_1^2 x_2^2 + x_0 x_1 x_2 + x_0^3 + x_1^3 + x_2^3 + (x_0 + x_1 + x_2)^3$
III. $F = x_0^2 x_1^2 x_2^2 + x_0 x_1 x_2 + (x_0 + x_1 + x_2)^3$
IV. $F = x_0^2 x_1^2 x_2^2 + x_0 x_1 x_2$
V. $F = x_0^3 + x_1^3$
VI. $F = x_0^3$
VII. $F = 0$.

Exercise: Using Table 19.9, find a form F for each of the fourteen types of sets in O_3.

19.7 Notes and references

§ 19.1. The most detailed description of Hermitian varieties is given by Segre (1965a). See also Segre (1967), Bose and Chakravarti (1966).

§ 19.2. Theorem 19.2.2 comes from Bruen and Hirschfeld (1978), although it is implicit in treatments of the classical isomorphism of Corollary 1 given, for example, by Dieudonné (1971, p. 109).

Another consequence of this theorem is worth putting in a more general setting. A *generalized quadrangle in $PG(n, q)$ is a pair $(\mathcal{P}, \mathcal{B})$* where (i) $\mathcal{B} \subset PG^{(1)}(n, q)$, (ii) $\mathcal{P}$ consists of all the points on the lines of $\mathcal{B}$, (iii) through every point in $\mathcal{P}$ there are r lines of $\mathcal{B}$, (iv) for (P, l) in $(\mathcal{P}, \mathcal{B})$ with P not on l there exists a unique l' in $\mathcal{B}$ meeting l and containing P. Buekenhout and Lefèvre (1974) and Olanda (1972*, 1973*, 1977*) showed that the only generalized quadrangles in $PG(n, q)$ are

Table 19.11 *Generalized Quadrangles in PG(n, q)*

Symbol	Description	v	b	r
$W(q)$	$\mathscr{P} = PG^{(0)}(3, q)$ $\mathscr{B}$ = linear complex	$(q+1)(q^2+1)$	$(q+1)(q^2+1)$	$q+1$
$H(3, q)$	$\mathscr{P} = \mathscr{H}_{3,q}$ $\mathscr{B} = \mathscr{H}^{(1)}_{3,q}$	$(q+1)^2$	$2(q+1)$	2
$U(3, q)$	$\mathscr{P} = \mathscr{U}_{3,q}$ $\mathscr{B} = \mathscr{U}^{(1)}_{3,q}$	$(q+1)(q\sqrt{q}+1)$	$(\sqrt{q}+1)(q\sqrt{q}+1)$	$\sqrt{q}+1$
$P(4, q)$	$\mathscr{P} = \mathscr{P}_{4,q}$ $\mathscr{B} = \mathscr{P}^{(1)}_{4,q}$	$(q+1)(q^2+1)$	$(q+1)(q^2+1)$	$q+1$
$U(4, q)$	$\mathscr{P} = \mathscr{U}_{4,q}$ $\mathscr{B} = \mathscr{U}^{(1)}_{4,q}$	$(q+1)(q^2\sqrt{q}+1)$	$(q\sqrt{q}+1)(q^2\sqrt{q}+1)$	$q\sqrt{q}+1$
$E(5, q)$	$\mathscr{P} = \mathscr{E}_{5,q}$ $\mathscr{B} = \mathscr{E}^{(1)}_{5,q}$	$(q+1)(q^3+1)$	$(q^2+1)(q^3+1)$	q^2+1

those of Table 19.11. Here $v = |\mathscr{P}|$, $b = |\mathscr{B}|$ and, if $\mathscr{V}$ is an algebraic variety, then $\mathscr{V}^{(1)}$ is the set of lines lying on it.

Theorem 19.7.1: (i) *$W(q)$ is isomorphic to the dual of $P(4, q)$;*
(ii) *$U(3, q)$ is isomorphic to the dual of $E(5, \sqrt{q})$.*

Proof: (i) If we apply the Grassmann map $\mathfrak{G}: PG^{(1)}(3, q) \to \mathscr{H}_{5,q}$, then as in § 15.4 the image of a linear complex $\mathscr{A}$ is $\mathscr{P}_4$ in a prime Π_4, which meets each plane on $\mathscr{H}_5$ in a line. So to each point of $PG(3, q)$ there corresponds a line on $\mathscr{P}_4$, whence the result.

(ii) Theorem 19.2.2 showed that $\mathscr{U}^{(1)}_{3,q}\mathfrak{G} = \mathscr{E}_{5,\sqrt{q}}$.

Since the lines on $\mathscr{U}_3$ through a point P lie in a pencil (Theorem 19.1.5), their images under $\mathfrak{G}$ lie on a line. □

See also Payne and Thas (1984).

§ 19.3. Theorem 19.3.5 on the existence of regular systems essentially follows Segre (1965a). There the result is established only for q odd. Here the result is proved for all q by varying the definition of permutable. A briefer proof was given by Thas (1981a). Segre (1965a) is also followed to prove the existence of the hemisystem $\mathscr{L}_2$ on $\mathscr{U}_{3,9}$.

The existence of a 56-cap in $PG(5, 3)$ was first given by Hill (1973), who also established that it is the largest k-cap in $PG(5, 3)$ and projectively unique, (1978a); see also Bruen and Hirschfeld (1978). An equivalent configuration is a set of 56 ovals in $PG(2, 4)$ any two of which have an even number of points in common; see § 14.3. See also Jónsson (1973), Hall, Lane and Wales (1970), Gewirtz (1969a, 1969b), Montague (1970), Wales (1969), Goethals and Seidel (1970), Hubaut (1975), Denniston (1980), Edge (1965). This gives another viewpoint of the hemisystem $\mathscr{L}_2$, explored by Cameron, Delsarte and Goethals (1979). In

$PG(2, 4)$, take two of the three sets of 56 ovals, which are orbits under $PSL(3, 4)$, and call this set of 112 ovals $\mathscr{P}$. Let $\mathscr{B}$ be the set of Hermitian curves; then $|\mathscr{B}| = 280$ as in Table 19.7. Now, in § 14.3, it is shown that the 12 points of the plane residual to any Hermitian curve $\mathscr{U}_2$ are partitioned into four 3-arcs any two of which form an oval; these are the only ovals disjoint from $\mathscr{U}_2$. Then, in the structure $\mathscr{S} = (\mathscr{P}, \mathscr{B})$, define an oval and a Hermitian curve to be 'incident' if they are disjoint. It is shown that $\mathscr{S}$ is a generalized quadrangle isomorphic to $E(5, 3)$ and so the dual of $U(3, 9)$.

§§ 19.4 – 19.5. The first characterization of a Hermitian variety was given by Barlotti (1966b), who showed that, if $\mathscr{K}$ is a set of type $(1, 3, 5)$ in $PG(d, 4)$ such that $|\mathscr{K}| = |\mathscr{U}_{d,4}|$ and that the unisecants and lines of $\mathscr{K}$ through each point of $\mathscr{K}$ lie in a prime, then $\mathscr{K} = \mathscr{U}_{d,4}$. In these sections we follow the programme of Tallini Scafati (1967a) who classified singular $k_{n,3,q}$, and non-singular $k_{n,3,q}$ with $n = 2$ and q. She also showed that, if $\mathscr{K}$ is a non-singular $k_{n,3,q}$ with $3 \leqslant n \leqslant q-1$ and $n \neq \frac{1}{2}q + 1$, then $\mathscr{K} = \mathscr{U}_{3,q}$. Theorems 19.4.8 and 19.4.9 as well as the exegesis of $\mathscr{K}_1$ and $\mathscr{R}_3$ follow Hirschfeld and Thas (1980a). Proposition 10 of Tallini Scafati (1967a) omits this.

The identification of $\mathscr{K}_1$ and $\mathscr{R}_3$ followed in Lemmas 19.4.10, 19.4.14, and 19.4.16, and in Theorem 19.4.17 is due to Glynn (1983a). Lemma 19.5.6 and its corollary follow Russo (1971).

For sets of type $(1, n, q+1)$ in $PG(d, q)$ for $d > 3$, see Tallini Scafati (1967a) and Hirschfeld and Thas (1980b).

For other characterizations of Hermitian varieties via generalized quadrangles, see Thas (1977*, 1978b, 1981a, 1981c, 1982), Payne and Thas (1984). For characterizations via polar spaces, see Tits (1974*), Buekenhout and Shult (1974), Mazzocca (1974b), Buekenhout and Lefèvre (1976*), Lefèvre-Percsy (1977a*, 1977b*, 1981a, 1981c, 1981d, 1981g). For a generalization to sets of type $(0, 1, n, q+1)$, see Tallini (1976b*), Lefèvre-Percsy (1981b, 1981f, 1982b, 1983). It is hoped to explore some of these topics in Volume 3.

§ 19.6. Theorem 19.6.8 is due to Hirschfeld and Hubaut (1980), although, from Lemma 19.6.4, the treatment is based on Sherman (1983). The vector space O_d, or equivalently E_d, is known in Coding Theory as the *projective geometry code*, and the dimension of O_d has been much studied. See Goethals and Delsarte (1968*), Smith (1969), Hamada (1968, 1973), Delsarte and Goethals (1970), Delsarte, Goethals, and MacWilliams (1970*), Peterson and Weldon (1972*, Chapter 10), Goethals (1973), Liebler (1981), Beth (1981). Corollary 1 of Theorem 19.6.8, that says that codewords in E_3 have weight divisible by four, follows from a theorem of McEliece (1972*) and does not seem at all obvious by geometrical methods; see Hirschfeld and Hubaut (1980) for an explanation of the result and Delsarte (1971*) for a more general setting of McEliece's theorem. The result is, however, an immediate consequence of Lemma 19.6.6 and (19.37) once it is known that a set $\mathcal{K}$ in O_3 containing a plane has this property, that is $|\mathcal{K}| \equiv 1 \pmod 4$. Corollary 3 fits in with the theorem of Beutelspacher (1980), a special case of which says that the smallest set $\mathcal{K}$ in $PG(3, q)$, q square, such that every line meets $\mathcal{K}$ and no plane is contained in $\mathcal{K}$ is a cone $\Pi_0\mathcal{K}'$, where $\mathcal{K}'$ is a $PG(2, \sqrt{q})$ in a plane skew to Π_0. Corollary 4 (viii) of Theorem 19.6.8 was proved directly by Hirschfeld and Hubaut (1980).

Other papers relevant to this chapter are Bruen and Thas (1976), Buekenhout (1976a), De Meur (1975*), Dienst (1974a*, 1974b*, 1979a, 1979b), Dienst and Mäurer (1974), Farmer (1981), Segre (1965b), Smit Ghinelli (1969, 1973*, 1976), Tallini (1976), G. E. Wall (1963), Wan and Yang (1964, 1965), de Resmini (1984b).

20

CUBIC SURFACES WITH 27 LINES

20.1 Existence of a double-six

A *double-six* in $PG(3, K)$ is a set of 12 lines

$$\begin{matrix} a_1 & a_2 & a_3 & a_4 & a_5 & a_6 \\ b_1 & b_2 & b_3 & b_4 & b_5 & b_6 \end{matrix}$$

such that each line meets only the five lines not in the same row or column.

Lemma 20.1.1: *A double-six lies on a unique cubic surface* $\mathscr{F}$, *which contains* 15 *further lines.*

Proof: A cubic surface is determined by $\mathbf{h}(4,3)-1=19$ conditions. If four points of a line l lie on a cubic surface, then the whole line lies on it. To put a_1, b_2, b_3, b_4, b_5, b_6 on a cubic surface requires $4+5\cdot 3=19$ conditions. So there is a surface $\mathscr{F}$ containing these lines. Then, as each of a_2, a_3, a_4, a_5, a_6 meets four lines of $\mathscr{F}$, these lines lie on $\mathscr{F}$; finally b_1 meets these lines and therefore lies on $\mathscr{F}$.

If there was another cubic surface $\mathscr{F}'$ containing these lines, then $\mathscr{F}$ and $\mathscr{F}'$ would intersect in a curve of degree at least 12, which is impossible unless $\mathscr{F}$ and $\mathscr{F}'$ have a common component of lower order. It is not possible however to fit a double-six into a plane and a quadric or into three planes.

$\mathscr{F}$ also contains the 15 lines c_{ij}, $i, j=1,\ldots,6$, $i\neq j$, where $c_{ij}=a_ib_j\cap a_jb_i$. □

The lemma shows that cubic surfaces and double-sixes are closely connected. To find out when double-sixes exist, we firstly consider when a skew hexagon is self-polar.

Lemma 20.1.2: *In* $PG(3, q)$, *let* H *be the skew hexagon* $l_1m_2l_3m_1l_2m_3$. *Then there exists a non-Hermitian polarity* $\mathfrak{T}$ *for which* H *is self-polar; that is,* $l_i\mathfrak{T}=m_i$ *for* $i=1,2,3$.

(i) *For* q *even,* $\mathfrak{T}$ *is unique and is null or pseudo as the two sets of alternative vertices are collinear or not.*

(ii) *For* q *odd, if the sets of alternate vertices are not both collinear,* $\mathfrak{T}$ *is unique and an ordinary polarity; otherwise, there exists both an ordinary and a null polarity* $\mathfrak{T}$.

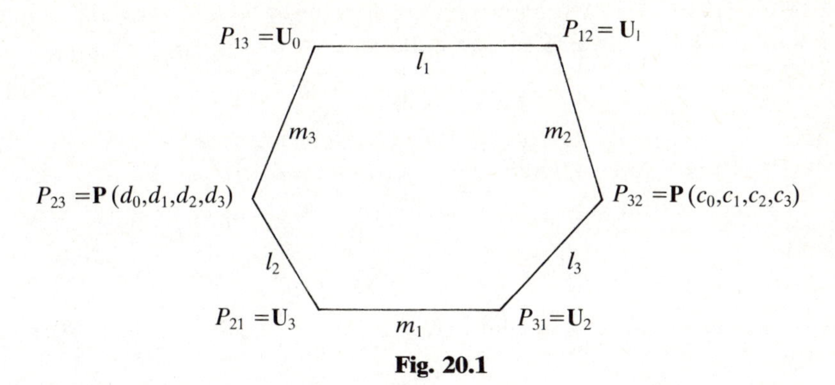

Fig. 20.1

Proof: Let $P_{ij} = l_i \cap m_j$. Then the vertices of H may be given coordinates as in Fig. 20.1.

For H to be self-polar, each vertex must be conjugate to the three non-adjacent vertices. So there are $6 \cdot 3/2 = 9$ conditions, which must be shown to be independent. Let $\mathfrak{T}$ be given by

$$F(X, Y) = \sum{}'' t_{ij}(x_i y_j + x_j y_i).$$

Then, as $\mathbf{U}_0$ and $\mathbf{U}_1$ are both conjugate to $\mathbf{U}_2$ and to $\mathbf{U}_3$, so

$$t_{02} = t_{03} = t_{12} = t_{13} = 0.$$

As P_{32} is conjugate to $\mathbf{U}_0$ and $\mathbf{U}_3$, and as P_{23} is conjugate to $\mathbf{U}_1$ and $\mathbf{U}_2$, we have

$$c_0 t_{00} + c_1 t_{01} = c_2 t_{23} + c_3 t_{33} = 0,$$

$$d_0 t_{01} + d_1 t_{11} = d_2 t_{22} + d_3 t_{23} = 0.$$

Hence

$$(t_{00}, t_{11}, t_{01}) = \lambda(c_1 d_1, c_0 d_0, -c_0 d_1),$$

$$(t_{22}, t_{33}, t_{23}) = \mu(c_3 d_3, c_2 d_2, -c_3 d_2).$$

Finally, as P_{23} and P_{32} are conjugate,

$$c_0 d_0 t_{00} + c_1 d_1 t_{11} + c_2 d_2 t_{22} + c_3 d_3 t_{33} + (c_0 d_1 + c_1 d_0) t_{01} + (c_2 d_3 + c_3 d_2) t_{23} = 0,$$

whence

$$\lambda c_0 d_1 (c_1 d_0 - c_0 d_1) + \mu c_3 d_2 (c_2 d_3 - c_3 d_2) = 0.$$

The fact that H is a skew hexagon means that l_1, l_2, and l_3 are mutually skew as are m_1, m_2, and m_3. These six conditions mean that

$$c_0 c_3 d_1 d_2 (c_0 d_1 - c_1 d_0)(c_2 d_3 - c_3 d_2) \neq 0.$$

Hence $\lambda/\mu \neq 0$ or ∞, and $\mathfrak{T}$ is unique.

For q even, $\mathfrak{T}$ is a pseudo polarity if some t_{ii} is non-zero and a null polarity if $t_{ii}=0$ for all i. In the latter case,

$$c_1=c_2=d_0=d_3=0, \tag{20.1}$$

and the two sets of alternate vertices are collinear; that is, $P_{12}P_{23}P_{31}$ and $P_{21}P_{32}P_{13}$ are lines. If $\mathbf{U}$ is chosen on $P_{23}P_{32}$, then

$$P_{23}=\mathbf{P}(0,1,1,0), \qquad P_{32}=\mathbf{P}(1,0,0,1),$$

and

$$F(X,Y)=(x_0y_1+x_1y_0)+(x_2y_3+x_3y_2).$$

Conversely, if the sets of alternate vertices are collinear, then (20.1) is satisfied and $\mathfrak{T}$ is null.

For q odd, $\mathfrak{T}$ is an ordinary polarity. If the alternate vertices are collinear and $\mathbf{U}$ is selected on $P_{23}P_{32}$, then

$$F(X,Y)=(x_0y_1+x_1y_0)-(x_2y_3+x_3y_2).$$

If we consider a null polarity $\mathfrak{T}$ given by

$$F(X,Y)=\sum{}'' t_{ij}(x_iy_j-x_jy_i),$$

then, for H to be self-polar, each $P_{ij}=l_i \cap m_j$ lies in the plane l_jm_i. Hence (20.1) is satisfied and the alternate vertices are collinear. With $\mathbf{U}$ on $P_{23}P_{32}$,

$$F(X,Y)=(x_0y_1-x_1y_0)+(x_2y_3-x_3y_2). \qquad \square$$

Notes: (i) An ordinary or pseudo polarity is determined by nine conjugacy conditions and a null polarity by five conjugacy conditions.

(ii) For q square, by exactly the same procedure, there is a unique Hermitian polarity for which H is self-polar.

Lemma 20.1.3: *In $PG(3,q)$, four skew lines a_1, a_2, a_3, a_4 not in a regulus have exactly one transversal if and only $|W|=0$, where $W=(\varpi_{ij})$ and $\varpi_{ij}=\varpi(a_i,a_j)$.*

Proof: Let $b=\mathbf{l}(B)$ and $a_i=\mathbf{l}(A_i)$, $i\in\mathbf{N}_4$. Then b is the unique transversal of a_1, a_2, a_3, a_4 if and only if, under the representation $\mathfrak{G}:\mathscr{L}\to\mathscr{H}_5$ of § 15.4, the point $b\mathfrak{G}$ is linearly dependent on the four independent points $a_1\mathfrak{G}$, $a_2\mathfrak{G}$, $a_3\mathfrak{G}$, $a_4\mathfrak{G}$; that is, if and only if the matrix

$$T=(B^*, A_1^*, A_2^*, A_3^*, A_4^*)$$

has rank 4. In other words, this occurs when b is linearly dependent on a_1, a_2, a_3, a_4.

Now let b be a transversal of a_1, a_2, a_3, a_4 and let $c=\mathbf{l}(C)$ be a line such that $\varpi(b,c)=1$; write $\varpi(a_i,c)=\varpi_i$. Let

$$S=(\hat{C}^*, \hat{A}_1^*, \hat{A}_2^*, \hat{A}_3^*, \hat{A}_4^*).$$

Then

$$T^*S=\begin{bmatrix} b_{01} & \dots & b_{23} \\ a_{01}^{(1)} & \dots & a_{23}^{(1)} \\ \vdots & & \vdots \\ a_{01}^{(4)} & \dots & a_{23}^{(4)} \end{bmatrix}\begin{bmatrix} c_{23} & a_{23}^{(1)} & \dots & a_{23}^{(4)} \\ \vdots & \vdots & & \vdots \\ c_{01} & a_{01}^{(1)} & \dots & a_{01}^{(4)} \end{bmatrix}=\begin{bmatrix} 1 & 0 & 0 & 0 & 0 \\ \varpi_1 & & & & \\ \varpi_2 & & & & \\ \varpi_3 & & W & & \\ \varpi_3 & & & & \end{bmatrix}.$$

If T has rank 4, then T^*S has rank at most 4, whence $|W|=0$.

To prove the sufficiency of the condition, we examine the following sufficiently general case. Take

$$\begin{aligned} B&=(1,0,0,0,0,0),\\ A_1&=(0,1,0,0,0,0),\\ A_2&=(0,0,0,0,1,0),\\ A_3&=(0,1,1,-1,1,0),\\ A_4&=(l_{01},l_{02},l_{03},l_{12},l_{31},0). \end{aligned}$$

Then T has rank 4 when $l_{12}+l_{03}=0$. On the other hand, $\varpi_{12}=\varpi_{13}=\varpi_{23}=1$, $\varpi_{14}=l_{31}$, $\varpi_{24}=l_{02}$ and $\varpi_{34}=l_{02}-l_{03}+l_{12}+l_{31}$. Hence, after a little manipulation, $|W|=(l_{12}+l_{03})^2$. This also proves the necessity again. □

Lemma 20.1.4: *In $PG(3,q)$, if the lines a_1, a_2, a_3, a_4, a_5 have a transversal b, then $w=0$ with*

$$w=\sum{}'' \varpi_{ij}W_{ij},$$

where $\varpi_{ij}=\varpi(a_i,a_j)$, $W=(\varpi_{ij})$ and W_{ij} is the cofactor of ϖ_{ij}. If q is odd, an equivalent condition is $|W|=0$.

Proof: Let $a_i=\mathbf{l}(a_{01}^{(i)}, a_{02}^{(i)}, a_{03}^{(i)}, a_{12}^{(i)}, a_{31}^{(i)}, a_{23}^{(i)})$ and suppose that $b=\mathbf{l}(B)=\mathbf{l}(b_{01}, b_{02}, b_{03}, b_{12}, b_{31}, b_{23})$ is a transversal of the a_i. Also let $c=\mathbf{l}(c_{01}, c_{02}, c_{03}, c_{12}, c_{31}, c_{23})$ be a line skew to b. Take $\varpi(b,c)=1$ and $\varpi(c,a_i)=\varpi_i$.

The six lines b, a_1, a_2, a_3, a_4, a_5 lie in the special linear complex with axis b and hence are linearly dependent. Therefore the matrix of their coordinate vectors is singular. So

$$\begin{vmatrix} b_{01} & \dots & b_{23} \\ a_{01}^{(1)} & \dots & a_{23}^{(1)} \\ \vdots & & \vdots \\ a_{01}^{(5)} & \dots & a_{23}^{(5)} \end{vmatrix}\begin{vmatrix} c_{23} & a_{23}^{(1)} & \dots & a_{23}^{(5)} \\ \vdots & & & \vdots \\ c_{01} & a_{01}^{(1)} & \dots & a_{01}^{(5)} \end{vmatrix}=0,$$

whence

$$\begin{vmatrix} 1 & 0 \; 0 \; 0 \; 0 \; 0 \\ \varpi_1 & \\ \varpi_2 & \\ \varpi_3 & W \\ \varpi_4 & \\ \varpi_5 & \end{vmatrix} = 0.$$

So $|W| = 0$. Expansion of $|W|$ by row i gives $\sum_j \varpi_{ij} W_{ij} = 0$. Summation of these equations over j gives $2w = 0$, whence for q odd, $w = 0$.

For q even, W is skew-symmetric and therefore singular. So the previous tactic will not do. Instead, from the five equations

$$\varpi(a_i, b) = 0,$$

we can solve for the b_{ij} and substitute in

$$b_{01}b_{23} + b_{02}b_{31} + b_{03}b_{12} = 0. \tag{20.2}$$

This gives

$$\begin{vmatrix} a_{02}^{(1)} & \cdots & a_{23}^{(1)} \\ \vdots & & \vdots \\ a_{02}^{(5)} & \cdots & a_{23}^{(5)} \end{vmatrix} \begin{vmatrix} a_{01}^{(1)} & \cdots & a_{31}^{(1)} \\ \vdots & & \vdots \\ a_{01}^{(5)} & \cdots & a_{31}^{(5)} \end{vmatrix} + \ldots + \ldots = 0.$$

Hence

$$\begin{vmatrix} a_{01}^{(1)} & \cdots & a_{23}^{(1)} \\ \vdots & & \vdots \\ a_{01}^{(5)} & \cdots & a_{23}^{(5)} \\ 1 & 0 \cdots & 0 \end{vmatrix} \begin{vmatrix} a_{23}^{(1)} & \cdots & a_{23}^{(5)} & 1 \\ & \cdots & & 0 \\ \vdots & & & \vdots \\ a_{01}^{(1)} & \cdots & a_{01}^{(5)} & 0 \end{vmatrix} + \ldots + \ldots = 0. \tag{20.3}$$

Hence

$$\begin{vmatrix} & & a_{01}^{(1)} \\ & W & \vdots \\ & & a_{01}^{(5)} \\ a_{23}^{(1)} & \cdots \; a_{23}^{(5)} & 1 \end{vmatrix} + \begin{vmatrix} & & a_{02}^{(1)} \\ & W & \vdots \\ & & a_{02}^{(5)} \\ a_{31}^{(1)} & \cdots \; a_{31}^{(5)} & 1 \end{vmatrix} + \begin{vmatrix} & & a_{03}^{(1)} \\ & W & \vdots \\ & & a_{03}^{(5)} \\ a_{12}^{(1)} & \cdots \; a_{12}^{(5)} & 1 \end{vmatrix} = 0.$$

Expansion by the last row and then by the last column in each determinant gives

$$3|W| + \sum{}'' \varpi_{ij} W_{ij} = 0,$$

whence

$$\sum{}'' \varpi_{ij} W_{ij} = 0.$$

The latter proof also holds for q odd, but more care is required with the signs. □

Corollary: *In $PG(3, q)$, five lines a_1, a_2, a_3, a_4, a_5 have a transversal over $\gamma = GF(q)$ or a quadratic extension γ' if and only if $w = 0$.*

Proof: If the a_i do not have a transversal over γ or γ' they lie in a general linear complex $\boldsymbol{\lambda}(B)$, where B does not satisfy (20.2). Then, by the same procedure as in the theorem, $w \neq 0$. □

The reason that it is necessary to include γ' in the corollary is that the a_i may be linearly dependent. If, for example, the a_i are all skew and linearly dependent but such that no four are linearly dependent, then they have two transversals over γ or γ'.

In particular, over $GF(2)$, if the a_i are skew, then $\varpi_{ij} = W_{ij} = 1$ for $i \neq j$. So $w = 0$ and, by the corollary, the lines have a transversal over $GF(4)$. In fact, the situation is familiar, as the five lines form an elliptic linear congruence, by Lemma 17.1.3, Corollary 1, and so have no transversals over $GF(2)$ but two transversals over $GF(4)$.

Lemma 20.1.5: (i) *If, in $PG(3, q)$, five skew lines a_1, a_2, a_3, a_4, a_5 have a transversal b, then each set of four a_i has a unique, distinct, second transversal if and only if each set of five of the six lines is linearly independent.*

(ii) *The configuration in* (i) *exists if and only if a plane 6-arc not on a conic exists. This occurs for $q \neq 2$, 3, or 5.*

Proof: (i) From § 15.4, we can make the following statements from which the result follows:

(a) five skew lines with a transversal have a second transversal if and only if the five lines are linearly dependent;

(b) four skew lines with a transversal have no further transversal if and only if the five lines are linearly dependent;

(c) four skew lines have more than two transversals if and only if they are linearly dependent.

(ii) Map the lines $a_1, \ldots, a_5$, and b to points $A_1, \ldots, A_5$, and B, respectively, of the quadric $\mathscr{H}_5$; see § 15.4. The A_i all lie in the tangent prime section at B, which is a cone $B\mathscr{H}_3$ with vertex B and base $\mathscr{H}_3$ in a solid Π_3. Project $A_1, \ldots, A_5$ from B to points $A_1', \ldots, A_5'$ of $\mathscr{H}_3$.

Since the a_i are skew, the join of two A_i is not on $\mathscr{H}_5$, and so no two A_i' lie on the same generator of $\mathscr{H}_3$.

If four A_i' were coplanar, then B and the corresponding four A_i would lie in a solid, whence either b is dependent on the four a_i or the four a_i are dependent. Both possibilities are contrary to the hypothesis.

If $A_2', \ldots, A_5'$ are projected from A_1' to points $A_2'', \ldots, A_5''$ of a plane π in Π_3 not containing A_1', where the generators of $\mathscr{H}_3$ through A_1' meet π

in C and D, then $\{C, D, A_2'', A_3'', A_4'', A_5''\}$ is a 6-arc not on a conic; see § 16.3, I.

Conversely, given the 6-arc, we can map its points to the points $A_1', \ldots, A_5'$ and choose $A_1 = A_1'$, $A_2 = A_2'$, $A_3 = A_3'$, $A_4 = A_4'$. Then A_5 can be chosen on BA_5' so that $A_1, \ldots, A_5$ are linearly independent.

Therefore the configuration exists in $PG(3, q)$ if and only if a 6-arc not on a conic exists in $PG(2, q)$. This occurs providing $q \neq 2$, 3, or 5, by Lemma 7.2.3, Corollary 2. □

Theorem 20.1.6: (i) *In* $PG(3, q)$, *given five skew lines* a_1, a_2, a_3, a_4, a_5 *with a transversal* b_6 *such that each five of the six lines are linearly independent, then the second transversals* b_1, b_2, b_3, b_4, b_5 *of sets of four of the* a_i *have themselves a transversal* a_6.

(ii) *There exists a unique non-Hermitian polarity* $\mathfrak{T}$ *such that* $b_i = a_i\mathfrak{T}$, $i \in \mathbf{N}_6$.

(iii) *If the polarity is null, then* $q = 4^k$.

(iv) *If* $q = 4$, *then the polarity is null.*

Proof: Let H be the skew hexagon $a_1b_3a_2b_1a_3b_2$ and $\mathfrak{T}$ a non-Hermitian polarity for which H is self-polar. By Lemma 20.1.2, $\mathfrak{T}$ is unique for q even, but there are one or two polarities for q odd.

Let $b_4' = a_4\mathfrak{T}$, $b_5' = a_5\mathfrak{T}$ and $a_6' = b_6\mathfrak{T}$. The lines a_1, a_2, and a_3 are mutually conjugate. Since each of b_1, b_2, and b_3 meets both a_4 and a_5, so each of a_1, a_2, and a_3 meets both b_4' and b_5'; hence each of a_1, a_2, and a_3 is conjugate to a_4 and a_5.

If it is shown that a_4 and a_5 are conjugate so that b_4' meets a_5 and b_5' meets a_4, then $b_4' = b_4$, $b_5' = b_5$, and a_6' is the required line a_6. If $\mathfrak{T}$ is also shown to be unique, then parts (i) and (ii) will be established.

The method is to apply the condition that pairs of the lines a_i are conjugate to the condition $w = 0$ or to the condition (20.3), both in Lemma 20.1.4.

Firstly, if $\mathfrak{T}$ is a null polarity or a pseudo polarity, let the self-polar or self-conjugate lines respectively be the linear complex $\mathscr{A} = \boldsymbol{\lambda}(l_{01} + l_{23})$, by Lemma 15.2.8, Corollary, parts (ii) and (iii). Then l and m are conjugate lines if $\varpi(l, m) = \mathscr{A}(l)\mathscr{A}(m)$. Write $\mathscr{A}(a_i) = A_i$ for i in $\mathbf{N}_5$; then $\varpi_{ij} = A_iA_j$ except perhaps for $(i, j) = (4, 5)$. Let $\varpi = \varpi_{45}/(A_4A_5)$. So a_4 and a_5 are conjugate if and only if $\varpi = 1$.

For q odd, $|W| = 0$. Dividing row i and column i by A_i for each i gives

$$\begin{vmatrix} 0 & 1 & 1 & 1 & 1 \\ 1 & 0 & 1 & 1 & 1 \\ 1 & 1 & 0 & 1 & 1 \\ 1 & 1 & 1 & 0 & \varpi \\ 1 & 1 & 1 & \varpi & 0 \end{vmatrix} = 0.$$

Hence $2\varpi(\varpi-3)=0$. As a_4 and a_5 are skew, $\varpi\neq 0$ and hence $\varpi=3$. So a_4 and a_5 are not conjugate. So a double-six with a null polarity does not exist for q odd.

For q even, $w=\sum''\varpi_{ij}W_{ij}=0$. The W_{ij} can be calculated from the displayed determinant and, on suppressing the A_i, we find that

$$W_{45}=1, \qquad W_{12}=W_{13}=W_{23}=\varpi^2,$$
$$W_{14}=W_{24}=W_{34}=W_{15}=W_{25}=W_{35}=\varpi.$$

Hence, substitution in $w=0$ gives

$$3\varpi^2+6\varpi+\varpi=0,$$

whence

$$\varpi(\varpi+1)=0.$$

As a_4 and a_5 are skew, so $\varpi\neq 0$. Hence $\varpi=1$, and a_4 and a_5 are conjugate. Hence, for q even, the double-six exists with a null or pseudo polarity.

For q odd, let the ordinary polarity $\mathfrak{T}$ be given by $F(X, Y)=\sum t_i x_i y_i$. Then two lines l and m are conjugate if $\varpi'(l, m)=0$, where

$$\varpi'(l, m)=\sum'' t_i t_j l_{ij} m_{ij},$$

Lemma 15.2.8, Corollary, part (iv). Write

$$\varpi'(a_i, a_j)=\alpha_{ij}, \quad \text{for} \quad i\neq j,$$

and

$$\varpi'(a_i, a_i)=\alpha_i.$$

Then $\alpha_{ij}=0$ for $\{i, j\}\neq\{4, 5\}$. For the condition that the a_i have a transversal, we apply (20.3) in Lemma 20.1.4. This reads

$$\begin{vmatrix} a_{01}^{(1)} & \dots & a_{23}^{(1)} \\ \vdots & & \vdots \\ a_{01}^{(5)} & \dots & a_{23}^{(5)} \\ 1 \quad 0 & \dots & 0 \end{vmatrix}\begin{vmatrix} a_{23}^{(1)} & \dots & a_{23}^{(5)} & 1 \\ & & & 0 \\ \vdots & & \vdots & \vdots \\ & & & \\ a_{01}^{(1)} & \dots & a_{01}^{(5)} & 0 \end{vmatrix}+\dots+\dots=0.$$

Therefore

$$\begin{vmatrix} a_{01}^{(1)} & \dots & a_{23}^{(1)} \\ \vdots & & \vdots \\ a_{01}^{(5)} & \dots & a_{23}^{(5)} \\ 1 \quad 0 & \dots & 0 \end{vmatrix}\begin{vmatrix} t_0 t_1 a_{01}^{(1)} & \dots & t_0 t_1 a_{01}^{(5)} & 0 \\ \vdots & & \vdots & \vdots \\ & \dots & & 0 \\ t_2 t_3 a_{23}^{(1)} & \dots & t_2 t_3 a_{23}^{(5)} & t_2 t_3 \end{vmatrix}+\dots+\dots=0.$$

Hence

$$\begin{vmatrix} \alpha_1 & 0 & 0 & 0 & 0 & t_2t_3a_{23}^{(1)} \\ 0 & \alpha_2 & 0 & 0 & 0 & t_2t_3a_{23}^{(2)} \\ 0 & 0 & \alpha_3 & 0 & 0 & t_2t_3a_{23}^{(3)} \\ 0 & 0 & 0 & \alpha_4 & \alpha_{45} & t_2t_3a_{23}^{(4)} \\ 0 & 0 & 0 & \alpha_{45} & \alpha_5 & t_2t_3a_{23}^{(5)} \\ t_0t_1a_{01}^{(1)} & t_0t_1a_{01}^{(2)} & t_0t_1a_{01}^{(3)} & t_0t_1a_{01}^{(4)} & t_0t_1a_{01}^{(5)} & 0 \end{vmatrix} + \ldots + \ldots = 0.$$

The factor $t_0t_1t_2t_3$ may be cancelled from each determinant. The displayed determinant then equals

$$-(\alpha_4\alpha_5-\alpha_{45}^2)(\alpha_2\alpha_3a_{01}^{(1)}a_{23}^{(1)}+\alpha_1\alpha_3a_{01}^{(2)}a_{23}^{(2)}+\alpha_1\alpha_2a_{01}^{(3)}a_{23}^{(3)})$$
$$-\alpha_1\alpha_2\alpha_3(\alpha_5a_{01}^{(4)}a_{23}^{(4)}+\alpha_4a_{01}^{(5)}a_{23}^{(5)})+\alpha_1\alpha_2\alpha_3\alpha_{45}(a_{01}^{(4)}a_{23}^{(5)}+a_{01}^{(5)}a_{23}^{(4)}).$$

When the other two determinants are added to this and the sum equated to zero, then

$$\alpha_1\alpha_2\alpha_3\alpha_{45}\varpi_{45}=0.$$

But none of a_1, a_2, a_3 is self-conjugate, nor do a_4 and a_5 meet. So $\alpha_{45}=0$, which means that a_4 and a_5 are conjugate.

It has therefore been established that the double-six configuration exists, and has an ordinary polarity for q odd and a null or pseudo polarity for q even.

We now show that, given q even, the null polarity can only occur for $q=4^k$.

The skew hexagon $H=a_1b_2a_3b_1a_2b_3$ with alternate vertices collinear has, as in Lemma 20.1.2, the canonical form in Fig. 20.2. The vector of each vertex and side of H has been included in the diagram. Since b_6

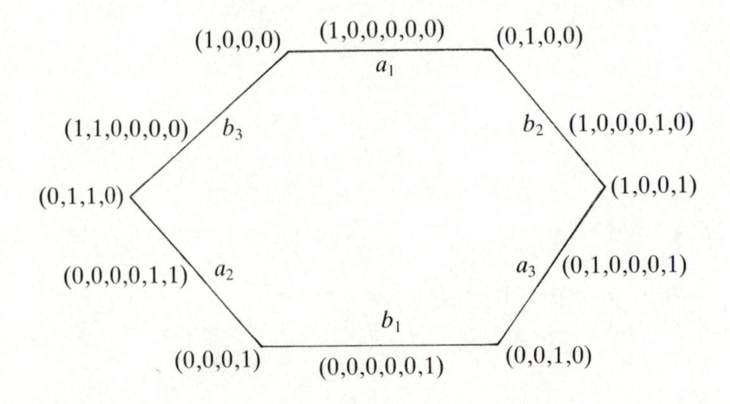

Fig. 20.2

meets a_1, a_2, a_3, so

$$b_6 = \mathbf{l}(t, t, t^2, 1, t, 0), \qquad \text{for some } t \text{ in } \gamma^+.$$

Since a_4 and a_5 meet b_1, b_2, b_3, they have the form

$$\mathbf{l}(0, s, s^2, 1, s, s).$$

Since they meet b_6, so s satisfies $s^2 + st + t^2 = 0$, which only has solutions for $q = 4^k$, § 1.8.

Finally, when $q = 4$, let H be the skew hexagon $a_1b_2a_3b_1a_2b_3$. Each regulus in $PG(3, 4)$ contains five lines. The lines b_4, b_5, b_6 all lie in the regulus $\mathcal{R}'$ complementary to $\mathcal{R}(a_1, a_2, a_3)$ and they meet a_1, a_2, a_3 in no vertex of H. Hence the remaining two lines of $\mathcal{R}'$ contain the six vertices of H. Hence sets of alternate vertices of H are collinear. □

Theorem 20.1.7: *A self-polar double-six and a cubic surface with 27 lines exist in $PG(3, K)$ for every field K except $K = GF(q)$ with $q = 2$, 3, or 5.*

Proof: The proof of the previous theorem did not depend on the finiteness of the field. A field of zero characteristic is treated exactly as a field of odd characteristic.

Lemma 20.1.5(ii) shows that the initial conditions of the previous theorem are satisfied for every field except the three named ones. This gives the existence of the double-six; Lemma 20.1.1 then gives the existence of the cubic surface. □

20.2 Structure of the surface

If $\mathcal{F}$ is any non-singular surface, then $\mathcal{F}$ has a tangent plane at each of its points. From Lemma 2.6.3, we know that, if the point P lies on the line l of $\mathcal{F}$, then l is contained in the tangent plane at P. For any P on $\mathcal{F}$, let $\pi_P(\mathcal{F})$ denote the tangent plane at P.

Lemma 20.2.1: *Let $\mathcal{F}$ be a non-singular cubic surface and let P be a point of $\mathcal{F}$.*

(i) *If P is on no line of $\mathcal{F}$, then $\pi_P(\mathcal{F}) \cap \mathcal{F}$ is an irreducible cubic with a double point at P.*

(ii) *If P is on exactly one line l_1 of F, then $\pi_P(\mathcal{F}) \cap \mathcal{F}$ consists of l_1 and a conic through P.*

(iii) *If P is on exactly two lines l_1 and l_2 of $\mathcal{F}$, then $\pi_P(\mathcal{F}) \cap \mathcal{F}$ consists of l_1, l_2, and a third line forming a triangle.*

(iv) *If P is on exactly three lines l_1, l_2, and l_3 of $\mathcal{F}$, then $\pi_P(\mathcal{F}) \cap \mathcal{F}$ consists of these three lines.* □

Corollary: *If π is a non-tangent plane meeting $\mathcal{F}$ in an irreducible cubic $\mathscr{C}$, then $\mathscr{C}$ is non-singular.* □

In cases (iii) and (iv) of the lemma, $\pi_P(\mathcal{F})$ is called a *tritangent plane*, since, in case (iii), it is the tangent plane at each vertex of the triangle and, in case (iv), it is the tangent plane at the coincident points $l_1 \cap l_2$, $l_2 \cap l_3$, $l_3 \cap l_1$.

In case (iv), P is an *Eckardt point* or *E-point* of $\mathcal{F}$. As we shall see, the number of E-points is determined for small values of q. Let $e_r = e_r(\mathcal{F})$ denote the number of points of $\mathcal{F}$ on exactly r lines of F. So e_3 is the number of E-points. Let N be the total number of points on $\mathcal{F}$.

Lemma 20.2.2: *Let $\mathcal{F}$ be a non-singular cubic surface in $PG(3, q)$ containing a line l_0. Let n and n' be the numbers of planes through l_0 whose residual intersection with $\mathcal{F}$ is respectively a line pair and a single point. Then*

(i) $n + n' \leqslant 5$;
(ii) $N = (q+1)^2 + (n - n')q$.

Proof: Let $\mathcal{F} = \mathbf{V}(F)$ contain $\mathbf{U}_2\mathbf{U}_3$; then

$$F = x_0 F_2(x_0, x_1, x_2, x_3) + x_1 G_2(x_1, x_2, x_3).$$

The plane $\mathbf{V}(\lambda x_1 - x_0)$ meets $\mathcal{F}$ in $\mathbf{U}_2\mathbf{U}_3$ and the quadric

$$\mathcal{F}_\lambda = \mathbf{V}(\lambda F_2(\lambda x_1, x_1, x_2, x_3) + G_2(x_1, x_2, x_3), \lambda x_1 - x_0).$$

If the condition of Theorem 7.2.4 for this quadric to be reducible is applied, then an equation of degree 5 in λ is obtained. This proves (i).

A point $\mathbf{P}(0, 0, a_2, a_3)$ lies in $\mathcal{F}_\lambda$ when

$$\lambda F_2(0, 0, a_2, a_3) + G_2(0, a_2, a_3) = 0. \tag{20.4}$$

So each point of $\mathbf{U}_2\mathbf{U}_3$ lies in exactly one quadric $\mathcal{F}_\lambda$. This may also be expressed by saying that the $\mathcal{F}_\lambda$ define an involution on $\mathbf{U}_2\mathbf{U}_3$; see § 6.3.

Since $\mathcal{F}$ is non-singular, $\mathcal{F}_\lambda$ cannot consist of just a line. So, by Theorem 7.2.1, there are n sections $\mathcal{F}_\lambda$ consisting of a pair of lines, n' consisting of a single point (that is, a line pair over $\gamma' = GF(q^2)$) and $q + 1 - n - n'$ consisting of a conic. So

$$\begin{aligned} N &= n(2q+1) + n' + (q+1-n-n')(q+1) \\ &= (q+1)^2 + (n-n')q. \quad \square \end{aligned}$$

Theorem 20.2.3: *Let $\mathcal{F}$ be a non-singular cubic surface in $PG(3, q)$ containing at least one line. Then the following are equivalent:*

(i) *The $q+1$ residual intersections with $\mathcal{F}$ of the planes through any line of $\mathcal{F}$ contain exactly five line pairs;*
(ii) *$\mathcal{F}$ has 27 lines;*
(iii) *$\mathcal{F}$ has $q^2 + 7q + 1$ points.*

Proof: Substituting $n' = 0$ and $n = 5$ in Lemma 20.2.2 shows that (i) and (iii) are equivalent.

Suppose (i) holds and let l_0 be a line of $\mathcal{F}$. Let $\{l_1, l_2\}$ be one of the five line pairs in planes through l_0. Then through each of l_1 and l_2 there are five planes whose residual intersection with $\mathcal{F}$ is a line pair. So there are eight lines of $\mathcal{F}$ meeting each of l_0, l_1, and l_2 that do not lie in the plane $\pi = l_0l_1l_2$. This gives $3+3\cdot 8=27$ lines. If l is any line of $\mathcal{F}$ other than l_0, l_1, and l_2, it meets π at a point on one of these three lines. So l lies in a plane through one of these three lines and hence has already been counted among the 27.

Conversely, if (ii) holds and there is some pair l_0 and l_1 of intersecting lines on $\mathcal{F}$, then the plane l_0l_1 contains a third line l_2 of $\mathcal{F}$ and the remaining 24 lines all meet one of these three. Hence, to obtain the 27 lines on $\mathcal{F}$, (i) holds for each of l_0, l_1, and l_2. As no line of $\mathcal{F}$ can now be skew to all the others, (i) holds for all 27 lines of $\mathcal{F}$.

If (ii) holds and all 27 lines are mutually skew, then no four can have a transversal l, as l would contain four points of $\mathcal{F}$ and so lie on it. Hence over the quadratic extension γ' of γ, four skew lines on $\mathcal{F}$ do have a transversal l'. Therefore we have a triangle on $\mathcal{F}$ over γ' consisting of say l_0 over γ, and l' and l'' over γ'. So the other 26 lines of $\mathcal{F}$ over γ meet one of l_0, l', and l'', and hence Lemma 20.2.2(i) is contradicted. □

Corollary: *A non-singular cubic surface $\mathcal{F}$ in $PG(3, q)$ contains at most 27 lines. Over some extension of γ, the surface $\mathcal{F}$ contains exactly 27 lines.* □

For the remainder of this section, $\mathcal{F}$ will always denote a non-singular cubic surface with 27 lines. Then $N' = e_3+e_2+e_1$ is the total number of points on the lines of $\mathcal{F}$.

Lemma 20.2.4: (i) $N' = 27(q-4)+e_3$; (ii) $e_3+e_0 = (q-10)^2+9$.

Proof: Let l_i, $i \in \mathbf{N}_{27}$, be the lines of $\mathcal{F}$ and let $e_r^{(i)}$ be the number of points of l_i on exactly r lines of $\mathcal{F}$. So

$$3e_3 = \sum e_3^{(i)}, \qquad 2e_2 = \sum e_2^{(i)}, \qquad e_1 = \sum e_1^{(i)}.$$

Also, since each line meets ten others,

$$2e_3^{(i)} + e_2^{(i)} = 10, \qquad e_3^{(i)} + e_2^{(i)} + e_1^{(i)} = q+1.$$

So $e_2^{(i)}/2 + e_1^{(i)} = q-4$ and $e_2+e_1 = 27(q-4)$. Hence

$$N' = 27(q-4)+e_3.$$

From Theorem 20.2.3, $e_0+N' = q^2+7q+1$, whence

$$\begin{aligned} e_3+e_0 &= q^2+7q+1-27(q-4) \\ &= q^2-20q+109 = (q-10)^2+9. \quad \square \end{aligned}$$

Corollary: *For $q \leqslant 16$, upper and lower bounds for e_3 are given by the Table* 20.1.

Table 20.1

q	4	5	7	8	9	11	13	16
Upper bound for e_3	45	34	18	13	10	10	18	45
Lower bound for e_3	45	36	18	9	0	0	0	0

Proof: The upper bound for e_3 is given by part (ii) of the lemma. Since each line on $\mathscr{F}$ meets ten others, we have, for $q \leqslant 9$, that $e_3 \geqslant 27[10-(q+1)]/3 = 9(9-q)$; this gives the lower bound. □

Note: This corollary shows, as does Theorem 20.1.7, that there is no cubic surface with 27 lines for $q=5$.

Before continuing the investigation of cubic surfaces via their E-points, we need to list the classical structures on $\mathscr{F}$ which can be formulated in terms of the 27 lines.

A *trihedral pair* is a set of six tritangent planes divided into two trihedra, such that the three planes of each trihedron contain the same set of nine distinct lines of $\mathscr{F}$. In this case, we shall allow an abuse of language and permit three planes of a trihedron to be collinear; if this occurs, the common line is the *axis* of the trihedron.

A *vertex* of $\mathscr{F}$ is the intersection of two lines of $\mathscr{F}$. So an E-point is three coincident vertices.

Theorem 20.2.5: *Let $\mathscr{F}$ contain the 27 lines a_i, b_i, c_{ij}, where $i, j \in \mathbf{N}_6$.*

(i) *Each line meets ten others, namely*

$$a_i \text{ meets } b_j,\ c_{ij} \text{ with } j \neq i;$$

$$b_i \text{ meets } a_j,\ c_{ij} \text{ with } j \neq i;$$

$$c_{ij} \text{ meets } a_i,\ a_j,\ b_i,\ b_j,\ c_{rs} \text{ with } r, s \neq i, j.$$

(ii) *There are 45 tritangent planes, namely*

$$30 \quad a_i b_j c_{ij},$$

$$15 \quad c_{ij} c_{kl} c_{mn} \text{ with } \{i, j, k, l, m, n\} = \mathbf{N}_6.$$

(iii) *The 27 lines form 36 double-sixes, namely*

$$\mathscr{D} \quad \begin{matrix} a_1 & a_2 & a_3 & a_4 & a_5 & a_6 \\ b_1 & b_2 & b_3 & b_4 & b_5 & b_6 \end{matrix} \quad 1$$

$$\mathscr{D}_{12} \quad \begin{matrix} a_1 & b_1 & c_{23} & c_{24} & c_{25} & c_{26} \\ a_2 & b_2 & c_{13} & c_{14} & c_{15} & c_{16} \end{matrix} \quad 15$$

$$\mathscr{D}_{123} \quad \begin{matrix} a_1 & a_2 & a_3 & c_{56} & c_{46} & c_{45} \\ c_{23} & c_{13} & c_{12} & b_4 & b_5 & b_6 \end{matrix} \quad 20$$

(iv) *The 45 planes form 120 trihedral pairs whose six planes are given by the rows and columns of the following arrays:*

$$\begin{array}{ccc} & T_{123} & 20 \\ c_{23} & a_3 & b_2 \\ b_3 & c_{13} & a_1 \\ a_2 & b_1 & c_{12} \end{array} \qquad \begin{array}{ccc} & T_{12,34} & 90 \\ a_1 & b_4 & c_{14} \\ b_3 & a_2 & c_{23} \\ c_{13} & c_{24} & c_{56} \end{array} \qquad \begin{array}{ccc} & T_{123,456} & 10 \\ c_{14} & c_{25} & c_{36} \\ c_{26} & c_{34} & c_{15} \\ c_{35} & c_{16} & c_{24}. \end{array}$$

(v) *The 120 trihedral pairs form 40 triads, each of which gives a trichotomy of the 27 lines, namely*

$$T_{123}, T_{456}, T_{123,456}, \quad 10; \qquad T_{12,34}, T_{12,56}, T_{34,56}, \quad 30.$$

(vi) *Among the lines, vertices, tritangent planes, double-sixes, trihedral pairs, and triads (of trihedral pairs), there are the following tactical configurations:*

(a) *lines, vertices:* $(27_{10}, 135_2)$,
(b) *lines, tritangent planes:* $(27_5, 45_3)$,
(c) *lines, double-sixes:* $(27_{16}, 36_{12})$,
(d) *lines, trihedral pairs:* $(27_{40}, 120_9)$,
(e) *lines, triads:* $(27_{40}, 40_{27})$,
(f) *vertices, tritangent planes:* $(135_1, 45_3)$,
(g) *vertices, double-sixes:* $(135_8, 36_{30})$,
(h) *vertices, trihedral pairs:* $(135_{16}, 120_{18})$,
(i) *vertices, triads:* $(135_{16}, 40_{54})$,
(j) *tritangent planes, double-sixes:* $(45_{24}, 36_{30})$,
(k) *tritangent planes, trihedral pairs:* $(45_{16}, 120_6)$,
(l) *tritangent planes, triads:* $(45_{16}, 40_{18})$,
(m) *double-sixes, trihedral pairs:* $(36_{20}, 120_6)$,
(n) *double-sixes, triads:* $(36_{10}, 40_9)$,
(o) *trihedral pairs, triads:* $(120_1, 40_3)$.

In (m), *a double-six and a trihedral pair are incident if the nine lines defined by the pair contain six lines of the double-six, occurring necessarily as three opposite pairs. Then, in* (n), *a double-six and a triad are incident if the double-six is incident (in the sense of* (m)) *with a trihedral pair of the triad.* □

Corollary 1: *If any line, a_1 say, and the ten lines meeting it are removed, then the tactical configuration formed by the remaining lines and the depleted double-sixes which had contained a_1 is a $(16_6, 16_6)$, known as Kummer's configuration. In the depleted figure, the lines and vertices form a $(16_5, 40_2)$.* □

Corollary 2: *Disregarding possible E-points, the group of permutations of the 27 lines of $\mathcal{F}$ preserving incidences has order* 51 840.

Proof: Since a double-six determines the remaining 15 lines, we consider the double-sixes. We may permute the six lines of one half of the double-six $\mathcal{D}$ in any way giving a group isomorphic to $\mathbf{S}_6$. Then the two halves may be interchanged, which means that the group fixing a double-six has order $2 \cdot 6! = 1440$. However, $\mathcal{D}$ may be transformed to any other double-six, whence the full group has order $36 \cdot 2 \cdot 6! = 51\,840$. □

Further properties of the group are given in § 20.3.

Lemma 20.2.6: *Let the planes of the trihedra of a pair belonging to $\mathcal{F}$ be given by $\mathbf{V}(F_1)$, $\mathbf{V}(F_2)$, $\mathbf{V}(F_3)$ and $\mathbf{V}(G_1)$, $\mathbf{V}(G_2)$, $\mathbf{V}(G_3)$. Then $\mathcal{F} = \mathbf{V}(F_1F_2F_3 + \lambda G_1G_2G_3)$ for some λ in γ_0.*

Proof: The nine lines $\mathbf{V}(F_i, G_j)$ form the total intersection of the cubic surfaces $\mathbf{V}(F_1F_2F_3)$ and $\mathbf{V}(G_1G_2G_3)$. Now, each of these nine lines meets four others; so the number of points of intersection of pairs of these lines is $9 \cdot 4/2 = 18$. A cubic surface $\mathcal{F}'$ contains the nine lines if and only if these eighteen points lie on $\mathcal{F}'$; also, none of these points lie on $\mathcal{F}$ as a consequence of the others doing so. As $\mathcal{F}'$ is determined by $\mathbf{h}(4, 3) - 1 =$ 19 conditions, $\mathcal{F}'$ lies in the pencil of cubic surfaces containing these nine lines. Hence $\mathcal{F} = \mathbf{V}(F_1F_2F_3 + \lambda G_1G_2G_3)$. □

Lemma 20.2.7: *Let l be a line on $\mathcal{F}$.*

(i) *The residual plane sections $\mathcal{F}_\lambda$ define an involution on l.*

(ii) *Since $\mathcal{F}_\lambda$ is either a line pair or a conic, there are five possible configurations for $(l, \mathcal{F}_\lambda)$ as in Figure 20.3.*

(iii) *For q odd, l contains 0, 1, or 2 E-points and $\mathcal{F}$ at most 18 E-points.*

(iv) *For q even, l contains 0, 1, or 5 E-points and $\mathcal{F}$ at most 45 E-points.*

Proof: As was mentioned in the proof of Lemma 20.2.2, equation (20.4)

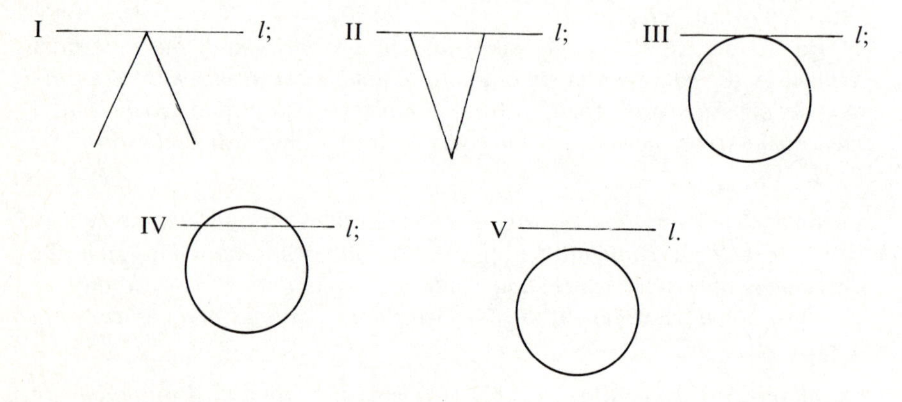

Fig. 20.3

defines the involution. The double points of the involution are given by configurations I and III.

From § 6.3, there can be at most two double points of an involution for q odd and hence 0, 1, or 2 E-points. If each line contains two E-points, then $\mathscr{F}$ has $27 \cdot 2/3 = 18$. For q even, if an involution contains more than one double point, then it is the identity. Lemma 20.2.2(i) shows that l contains at most five E-points. If every line contains five E-points, then $\mathscr{F}$ contains $27 \cdot 5/3 = 45$. □

Corollary: *The number of each type* I–V *of configurations* $(l, \mathscr{F}_\lambda)$ *for the different involutions on* l *are given by Table* 20.2. □

Table 20.2

q	type of involution	I	II	III	IV	V
odd	elliptic	0	5	0	$(q-9)/2$	$(q+1)/2$
odd	hyperbolic	2	3	0	$(q-7)/2$	$(q-1)/2$
odd	hyperbolic	1	4	1	$(q-9)/2$	$(q-1)/2$
odd	hyperbolic	0	5	2	$(q-11)/2$	$(q-1)/2$
even	identity	5	0	$q-4$	0	0
even	parabolic	1	4	0	$(q-8)/2$	$q/2$
even	parabolic	0	5	1	$(q-10)/2$	$q/2$

Lemma 20.2.8: *If two lines of a double-six on* $\mathscr{F}$ *meet at* P, *then* P *is an* E-*point on* $\mathscr{F}$ *if and only if* P *is self-conjugate with respect to the polarity of the double-six.*

Proof: Let $P = a_1 \cap b_2$ and let D be the double-six. The polar plane of P is $\pi = a_2 b_1 c_{12}$. If P is self-conjugate, P is in π and so on one of a_2, b_1, c_{12}. As a_1 and b_2 are both skew to a_2 and b_1, so P lies on c_{12}. Conversely, if $P = a_1 \cap b_2 \cap c_{12}$, then P lies on c_{12} and so in π. □

This lemma can be used to give another proof of Lemma 20.2.5(iii) and (iv).

Corollary: *A line of* $\mathscr{F}$ *contains two* E-*points if and only if it is self-polar with respect to the polarities of the two double-sixes containing the other four lines through the* E-*points.* □

Let us denote the putative E-points as follows:

$$E_{ij} = a_i \cap b_j \cap c_{ij}, \qquad E_{ij,kl,mn} = c_{ij} \cap c_{kl} \cap c_{mn}.$$

Lemma 20.2.9: *Any two* E-*points not on the same line of* $\mathscr{F}$ *are collinear with a third* E-*point.*

Proof: Let $E_{12} = a_1 \cap b_2 \cap c_{12}$ and $E_{23} = a_2 \cap b_3 \cap c_{23}$ be, with no loss of generality, the E-points. Consider the trihedral pair

$$\begin{array}{llll} T_{123} & c_{23} & a_3 & b_2 \\ & b_3 & c_{13} & a_1 \\ & a_2 & b_1 & c_{12}. \end{array}$$

Both E_{12} and E_{23} lie in all three planes $c_{23}a_3b_2$, $b_3c_{13}a_1$, and $a_2b_1c_{12}$. So these three planes have the line $E_{12}E_{23}$ in common. Now $P=a_3\cap b_1$ lies in both $c_{23}a_3b_2$ and $a_2b_1c_{12}$. So P lies on $E_{12}E_{23}$ and in the plane $b_3c_{13}a_1$. Hence P lies on $a_3c_{13}b_1\cap b_3c_{13}a_1=c_{13}$. So $P=E_{13}$ and E_{12}, E_{23}, and E_{31} are collinear. □

Corollary: *The possible collinearities among the E-points are*

$$\begin{array}{llll} E_{12} & E_{23} & E_{31} & 40 \\ E_{12} & E_{34} & E_{14,23,56} & 180 \\ E_{14,25,36} & E_{15,26,34} & E_{16,24,35} & 20. \ \square \end{array}$$

Lemma 20.2.10: *If $\mathcal{F}$ contains exactly r E-points, they form the following exemplified configurations C_r.*

C_0:
C_1: E_{12}.
C_2: E_{12}, E_{13}.
C_3: E_{12}, E_{23}, E_{31}.
C_4: E_{12}, E_{23}, E_{31}, E_{21}.
C_5: *Five on a line such as* E_{12}, E_{13}, E_{14}, E_{15}, E_{16} *on* a_1.
C_6: *Six vertices of a plane quadrilateral (Fig. 20.4).*
C_6': *Six vertices of a skew hexagon with sets of alternate vertices collinear (Fig. 20.5).*
C_9: $AG(2,3)=\mathcal{U}_{2,4}=$ *the nine inflexions of a plane cubic curve lying by threes on twelve lines as in the rows, columns, and diagonals*

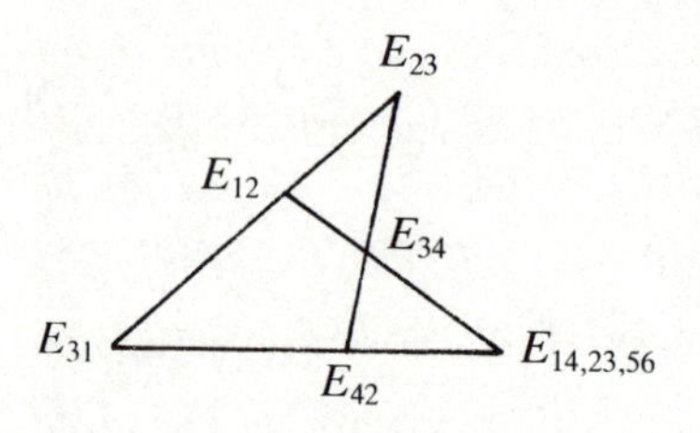

Fig. 20.4

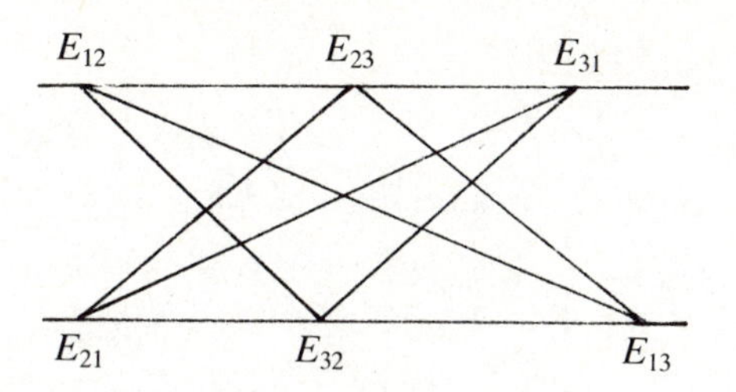

Fig. 20.5

(determinantal products) of the array

$$\begin{matrix} E_{12} & E_{45} & E_{15,24,36} \\ E_{23} & E_{56} & E_{14,26,35} \\ E_{31} & E_{64} & E_{16,25,34}. \end{matrix}$$

Each of the 27 lines has one E-point.

C_{10}: *The ten vertices of a pentahedron* $\mathcal{P}$ *lying two on each of the 15 lines residual to a double-six and whose collinear triples lie on the ten edges of* $\mathcal{P}$; *that is, a non-planar Desargues configuration* (*Fig.* 20.6). *The faces of* $\mathcal{P}$ *are* $c_{12}c_{34}c_{56}$, $c_{13}c_{25}c_{46}$, $c_{14}c_{26}c_{35}$, $c_{15}c_{24}c_{36}$, $c_{16}c_{23}c_{45}$.

C_{13}: *Five on each of three coplanar lines concurrent at an E-point.*

C_{18}: *Two on each of the 27 lines lying by threes on the six axes of a*

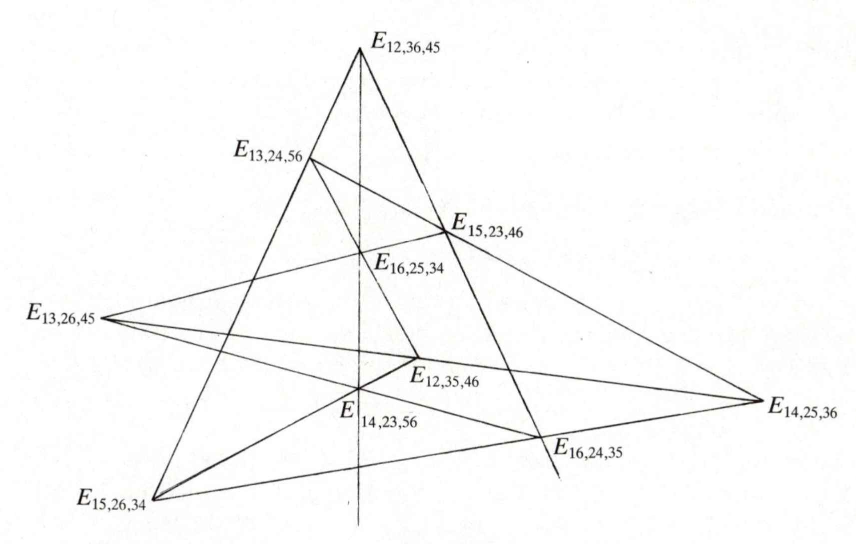

Fig. 20.6

triad of trihedral pairs

$$T_{123}, \quad T_{456}, \quad T_{123,456}$$

C_{45}: *All possible E-points, five on each of the 27 lines.*

Proof: This follows from Lemma 20.2.7, parts (iii) and (iv), and Lemma 20.2.9. □

Corollary: *If a plane π meets $\mathscr{F}$ in an irreducible cubic curve $\mathscr{C}$, then*
(i) *$\mathscr{C}$ contains one of C_0, C_1, C_3, C_9;*
(ii) *the E-points on $\mathscr{C}$ are inflexions.*

Proof: By the corollary to Lemma 20.2.1, $\mathscr{C}$ is non-singular. The cubic $\mathscr{C}$ can contain at most one E-point on any line of $\mathscr{F}$, as otherwise it would contain a line of $\mathscr{F}$. From the lemma, this only allows C_0, C_1, C_3, C_9. If P is an E-point on $\mathscr{C}$ and $\pi_P(\mathscr{F})$ the tangent plane at P, then $\pi_P(\mathscr{F}) \cap \mathscr{F}$ is the three lines of $\mathscr{F}$ through P. Hence $\pi_P(\mathscr{F}) \cap \pi$ is an inflexional tangent to $\mathscr{C}$ at P. The possibilities of 0, 1, 3, and 9 inflexions for a non-singular cubic agree with those of Theorem 11.4.5. □

Lemma 20.2.11: (i) *If $\mathscr{F}$ contains C_{18}, then, in a suitable coordinate system,*

$$\mathscr{F} = \mathbf{V}(x_0^3 + x_1^3 + x_2^3 + x_3^3).$$

(ii) *C_{18} exists in $PG(3, q)$ if and only if $q \equiv 1 \pmod 3$; that is, when $x^2 + x + 1$ has two roots in $GF(q)$.*
(iii) *If $\mathscr{F}$ contains C_6', it contains C_{18}.*

Proof: Suppose $\mathscr{F}$ contains C_6' as in Lemma 20.2.10. Then both trihedra of T_{123} have an axis. The two axes are necessarily skew. Hence we may take the axes as $\mathbf{U}_0\mathbf{U}_1$ and $\mathbf{U}_2\mathbf{U}_3$, and we may take

$$\mathscr{F} = \mathbf{V}(x_0x_1(x_0 + x_1) - cx_2x_3(x_2 + x_3)).$$

The plane $\mathbf{V}(x_0 - \lambda x_2)$ meets $\mathscr{F}$ in $\mathbf{V}(x_0, x_2)$ and

$$\mathscr{F}_\lambda = \mathbf{V}(x_0 - \lambda x_2, \lambda x_1^2 + \lambda^2 x_1 x_2 - cx_2x_3 - cx_3^2).$$

From Theorem 7.2.4, the condition that $\mathscr{F}_\lambda$ is a line pair is $\lambda(\lambda^3 - c) = 0$. So there are five solutions for λ ($\lambda = \infty$ is a solution) if and only if $c = d^3$ and $q \equiv 1 \pmod 3$; see § 1.5. Hence $\mathscr{F}$ can be transformed to

$$\mathscr{F} = \mathbf{V}(x_0^3 + x_1^3 + x_2^3 + x_3^3) \tag{20.5}$$

where the two trihedra are $\mathbf{V}(x_0 + x_1)$, $\mathbf{V}(x_0 + \omega x_1)$, $\mathbf{V}(x_0 + \omega^2 x_1)$ and $\mathbf{V}(x_2 + x_3)$, $\mathbf{V}(x_2 + \omega x_3)$, $\mathbf{V}(x_2 + \omega^2 x_3)$: the axes have been left unchanged. So the nine lines a_1, a_2, a_3, b_1, b_2, b_3, c_{12}, c_{13}, c_{23} are the meets of pairs of planes from the two trihedra. But, from the symmetrical form of the surface, T_{456} may be taken with axes $\mathbf{U}_0\mathbf{U}_2$ and $\mathbf{U}_1\mathbf{U}_3$, and $T_{123,456}$ with axes $\mathbf{U}_0\mathbf{U}_3$ and $\mathbf{U}_1\mathbf{U}_2$. So $\mathscr{F}$ contains C_{18}. This proves (i), (ii), and (iii). □

When $\mathscr{F}$ has the form of (20.5), it is called an *equianharmonic* surface.

Corollary: *The group $G(\mathscr{F})$ of projectivities of an equianharmonic surface $\mathscr{F}$ containing exactly* 18 *E-points is isomorphic to a split extension of* $\mathbf{Z}_3 \times \mathbf{Z}_3 \times \mathbf{Z}_3$ by $\mathbf{S}_4$ *and has order* 648.

Proof: The axes of the three trihedral pairs on which the E-points lie may be taken as edges of the tetrahedron $\mathscr{T}$ of reference, which is fixed. Then $\mathscr{F}$ has the form (20.5). The group $G_{\mathscr{T}}$ fixing each face of $\mathscr{T}$ is given by

$$\{\text{diag}(1, a_1, a_2, a_3) \mid a_1^3 = a_2^3 = a_3^3 = 1\}$$

and so $G_{\mathscr{T}} \cong \mathbf{Z}_3 \times \mathbf{Z}_3 \times \mathbf{Z}_3$, whence $|G_{\mathscr{T}}| = 27$. As all permutations of the faces of $\mathscr{T}$ can be achieved by projectivities, $G(\mathscr{F}) \cong G_{\mathscr{T}}\mathbf{S}_4$ and $|G(\mathscr{F})| = 27 \cdot 4! = 648$. □

Lemma 20.2.12: (i) *If $\mathscr{F}$ contains C_{10}, then, in a suitable coordinate system,*

$$\mathscr{F} = \mathbf{V}(x_0^3 + x_1^3 + x_2^3 + x_3^3 + x_4^3) \quad \text{for} \quad p \neq 3$$

$$\mathscr{F} = \mathbf{V}\left(\sum{}'' x_i x_j x_k\right) \quad \text{for} \quad p = 3. \tag{20.6}$$

where in both cases $x_0 + x_1 + x_2 + x_3 + x_4 = 0$.

(ii) *C_{10} exists if and only if $x^2 - x - 1$ has two roots in γ.*

Proof: Let the pentahedron defining C_{10} be that of the coordinate system; that is, its faces are $\mathbf{u}_0$, $\mathbf{u}_1$, $\mathbf{u}_2$, $\mathbf{u}_3$, and $\mathbf{u}$. Then, from Lemma 20.2.8, the lines c_{ij} are joins of a vertex to one of the 3 vertices on the opposite edge. This gives 15 lines $\mathbf{V}(x_i, x_j + x_k)$, $i, j, k \in \bar{\mathbf{N}}_4$.

For $\mathscr{F}$ to contain these 15 lines, it necessarily has the form required. Using the technique of the previous theorem, we find that there are five tritangent planes through any of these 15 lines exactly when $x^2 - x - 1$ has two roots in γ. □

When $\mathscr{F}$ has the form of (20.6), it is called a *diagonal* surface.

Corollary: *The group of projectivities of a diagonal surface with exactly ten E-points is isomorphic to* $\mathbf{S}_5$ *of order* 120. □

Theorem 20.2.13: (i) *The possible configurations of E-points on $\mathscr{F}$ are as follows*

(a) *for q odd,* C_0, C_1, C_2, C_3, C_4, C_6, C_9, C_{10}, C_{18};

(b) *for q even,* C_0, C_1, C_3, C_5, C_9, C_{13}, C_{45}.

(ii) *If $\mathscr{F}$ contains C_9, then $x^2 + x + 1$ has at least one root in γ; that is,* $q \not\equiv -1 \pmod 3$.

(iii) *$\mathcal{F}$ containing C_{10} exists if and only if x^2-x-1 has two roots in γ; that is,* $q\equiv\pm1 \pmod 5$.

(iv) *$\mathcal{F}$ containing C_{18} exists if and only if x^2+x+1 has two roots in γ; that is,* $q\equiv1 \pmod 3$.

(v) *$\mathcal{F}$ containing C_{45} exists if and only if $q=4^m$.*

Proof: Part (i) follows from Lemmas 20.2.7 and 20.2.10. Theorem 11.1.1 means that C_9 can exist only when $(q+1, 3)=1$, that is, $q\equiv0$ or $1 \pmod 3$: this gives (ii). Parts (iii) and (iv) merely repeat Lemmas 20.2.12 and 20.2.11 respectively.

Part (v) can be deduced in several ways from the above. If $\mathcal{F}$ contains C_{45}, it contains C_{18} and γ has characteristic 2, whence, by (iv), γ contains $GF(4)$. Conversely, if $q=4^m$, the equianharmonic surface contains two E-points on each of the 27 lines and therefore five. □

20.3 Surfaces over small fields

By Theorem 20.1.7, there is a cubic surface with 27 lines in $PG(3, q)$ unless $q=2$, 3, or 5. If, in $PG(3, q)$, there is a cubic surface $\mathcal{F}$ with exactly n points off the lines that is projectively unique, then $\mathcal{F}$ will be denoted by $\mathcal{F}_q^n$ or simply by $\mathcal{F}_q$ if n is also unique. The group of projectivities $G(\mathcal{F})$ fixing $\mathcal{F}$ will be denoted accordingly by G_q^n or G_q and the group $\Gamma(\mathcal{F})$ of collineations by Γ_q^n or Γ_q.

I. $q=4$

From the corollary to Lemma 20.2.4, $\mathcal{F}$ has 45 E-points, whence, by Lemmas 20.2.11 and 20.2.12, it is both equianharmonic and diagonal, as well as projectively unique. So $\mathcal{F}$ can be called $\mathcal{F}_4$. Since it is equianharmonic, $\mathcal{F}_4$ is Hermitian, by Lemma 19.1.1. From Theorem 20.1.6, the polarities of all 36 double-sixes on $\mathcal{F}_4$ are null and, from the corollaries to Lemmas 19.2.3 and 19.2.4, these are all the null polarities commuting with the Hermitian polarity of $\mathcal{F}_4$. From Theorem 19.5.13(iv)(b)(3), if $\mathcal{K}$ is a $45_{3,3,4}$ such that through each point P of $\mathcal{K}$ there are exactly three coplanar lines of $\mathcal{K}$, then $\mathcal{K}$ is an $\mathcal{F}_4$.

Of the 85 planes of $PG(3, 4)$, 45 are tritangent planes of $\mathcal{F}_4$ and 40 meet $\mathcal{F}_4$ in a Hermitian curve $\mathcal{U}_{2,4}$ consisting of nine points. Apart from the 27 lines of $\mathcal{F}_4$, there are 90 unisecants lying in pairs in the tritangent planes and 240 lines 3-secant to $\mathcal{F}_4$. A summary of the incidences is given by Theorem 19.1.9 with $s=2$.

Next consider $\mathcal{F}_4$ as an equianharmonic surface. If one triad of trihedral pairs and so one C_{18} is fixed, then the group of projectivities is of order 648, by Lemma 20.2.11, Corollary. The six axes of the triad of trihedral pairs are the edges of a tetrahedron self-polar with respect to the Hermitian polarity of $\mathcal{F}_4$. As there are 40 triads of trihedral pairs, so

$|G_4| = 40 \cdot 648 = 25{,}920$ and $|\Gamma_4| = 51{,}840$: this is as expected since $G_4 \cong PGU(4, 4)$ and $\Gamma_4 \cong P\Gamma U(4, 4)$. Also $\mathscr{F}_4$ has 40 self-polar tetrahedra, whence the tetrahedra and the 40 points of $PG(3, 4)\backslash\mathscr{F}_4$ form a $(40_4, 40_4)$ tactical configuration. The edges of the 40 tetrahedra are all the 240 trisecants of $\mathscr{F}_4$.

Now consider $\mathscr{F}_4$ as a diagonal surface. In the form (20.6), the faces $\mathbf{u}_0$, $\mathbf{u}_1$, $\mathbf{u}_2$, $\mathbf{u}_3$, and $\mathbf{u}$ of the pentahedron $\mathscr{P}$ are all tritangent planes of $\mathscr{F}_4$. Through each of the ten edges of the pentahedron there is a further tritangent plane of the form $\mathbf{V}(x_i + x_j)$ with $i, j \in \bar{\mathbf{N}}_3$. These 15 tritangent planes may be taken as those of the form $c_{ij}c_{kl}c_{mn}$ and are all the planes of a $PG(3, 2)$. Take any plane, say $\mathbf{u}_0$, of $\mathscr{P}$; through each of the four edges of $\mathscr{P}$ in $\mathbf{u}_0$, there is one of the ten planes other than the faces of $\mathscr{P}$, namely the $\mathbf{V}(x_0 + x_i)$, $i \in \mathbf{N}_4$. With $\mathbf{u}_0$, these four planes form another pentahedron $\mathscr{P}_0$ for $\mathscr{F}_4$, since

$$x_0^3 + (x_0 + x_1)^3 + (x_0 + x_2)^3 + (x_0 + x_3)^3 + (x_0 + x_4)^3 = x_0^3 + x_1^3 + x_2^3 + x_3^3 + x_4^3,$$

and

$$x_0 + (x_0 + x_1) + (x_0 + x_2) + (x_0 + x_3) + (x_0 + x_4) = x_0 + x_1 + x_2 + x_3 + x_4.$$

Another four pentahedra $\mathscr{P}_1$, $\mathscr{P}_2$, $\mathscr{P}_3$, $\mathscr{P}_4$ may be similarly obtained. Thus the 15 planes determine the set $\mathscr{N}_6 = \{\mathscr{P}, \mathscr{P}_0, \mathscr{P}_1, \mathscr{P}_2, \mathscr{P}_3, \mathscr{P}_4\}$ of six mutually interwoven pentahedra. Since five planes may be chosen arbitrarily in Π_3, there is a group $\mathbf{S}_5$ of projectivities permuting the faces of any given one of the pentahedra and, at the same time, permuting the other five pentahedra among themselves. So there is a group $G(\mathscr{N}_6)$ of projectivities permuting the six pentahedra such that $G(\mathscr{N}_6)$ contains six conjugate subgroups isomorphic to $\mathbf{S}_5$, and $G(\mathscr{N}_6) \cong \mathbf{S}_6$.

There is however a dual construction. The five points of the $PG(3, 2)$ other than the ten vertices of $\mathscr{P}$ form a skew pentastigm $\mathscr{S}$; these five points may also be considered as the points of $\mathscr{F}_4$ other than the 30 on the lines of $\mathscr{D}$ and the ten vertices of $\mathscr{P}$. The vertices of $\mathscr{S}$ are in fact $P_0 = \mathbf{P}(0, 1, 1, 1)$, $P_1 = \mathbf{P}(1, 0, 1, 1)$, $P_2 = \mathbf{P}(1, 1, 0, 1)$, $\mathbf{P}_3 = P(1, 1, 1, 0)$, and $\mathbf{U}$. Given any vertex of $\mathscr{S}$, say P_0, there is another E-point (on no line of $\mathscr{D}$) on the join of P_0 with each vertex of $\mathscr{S}$. These points, namely $\mathbf{P}(0, 0, 1, 1)$, $\mathbf{P}(0, 1, 0, 1)$, $\mathbf{P}(0, 1, 1, 0)$, and $\mathbf{U}_0$, together with P_0 form another pentastigm $\mathscr{S}_0$. So we obtain a set $\mathscr{N}_6'$ of six pentastigms $\mathscr{S}$, $\mathscr{S}_0$, $\mathscr{S}_1$, $\mathscr{S}_2$, $\mathscr{S}_3$, $\mathscr{S}_4$.

Since the 15 faces of the pentahedra are those 15 tritangent planes containing no line of $\mathscr{D}$ (or, dually, since the 15 vertices of the pentastigms are those 15 E-points on no line of $\mathscr{D}$), $G(\mathscr{N}_6)$ fixes $\mathscr{D}$. So the group $G(\mathscr{D})$ of projectivities fixing $\mathscr{D}$ is $G(\mathscr{N}_6)$ and hence isomorphic to $\mathbf{S}_6$. There are 12 subgroups of $G(\mathscr{D})$ isomorphic to $\mathbf{S}_5$, namely six fixing a pentahedron and six fixing a line a_i. This gives 12 subgroups $\mathbf{S}_5$ of $\mathbf{S}_6$ in two conjugate sets of six, in contrast to any other integer n when $\mathbf{S}_n$ has a single conjugate set of n subgroups $\mathbf{S}_{n-1}$. The polarity of $\mathcal{D}$ interchanges $\mathcal{N}_6$ and $\mathcal{N}'_6$ as well as a_i and b_i. The automorphism of $PG(3, 4)$ fixes each pentastigm and each pentahedron while interachanging a_i and b_i, $i \in \mathbf{N}_6$.

The faces of the pentahedra may be taken as listed in Table 20.3.

A typical pair of opposite lines of $\mathcal{D}$ is $\mathbf{V}(\omega x_0 + x_1 + x_2,\ \omega x_1 + x_2 + x_3)$, $\mathbf{V}(\omega^2 x_0 + x_1 + x_2,\ \omega^2 x_1 + x_2 + x_3)$.

Since $G(\mathcal{D})$ has order 6!, so G_4 has order $36 \cdot 6! = 25\,920$ and Γ_4 has order 51 840, as above.

As the polarity of $\mathcal{D}$ is null, the self-polar lines form a linear complex $\mathcal{A}$, which for the diagonal surface (20.6) has the form

$$\mathcal{A} = \boldsymbol{\lambda}(l_{01} + l_{02} + l_{03} + l_{12} + l_{31} + l_{23});$$

see § 15.2. Each element of $G(\mathcal{N}_6)$ induces a projectivity of $PG(3, 2)$, the set of 15 points of $\mathcal{F}_4$ on no line of $\mathcal{D}$. As $\mathcal{D}$ is fixed by such a projectivity, so is $\mathcal{A}$.

The following theorem gives a summary.

Theorem 20.3.1: (i) *In $PG(3, 4)$ there is a projectively unique cubic surface $\mathcal{F}_4$, which can be written in 40 ways as an equianharmonic surface and in 216 ways as a diagonal surface.*

(ii) *$\mathcal{F}_4$ is Hermitian, whence*

$$G_4 = PGU(4, 4) = PSU(4, 4) \quad \textit{and} \quad \Gamma_4 = P\Gamma U(4, 4) = P\gamma U(4, 4).$$

(iii) $|G_4| = 25\,920$; $|\Gamma_4| = 51\,840$.

(iv) $\mathbf{S}_6$ *has 12 subgroups isomorphic to* $\mathbf{S}_5$ *in two conjugate sets of six.*

(v) $\mathbf{S}_6 \cong PGSp(4, 2) = Sp(4, 2) \cong$ *the group of projectivities fixing a double-six on $\mathcal{F}_4$.* □

There is a novel way of embedding $PG(3, 3)$ in $PG(3, 4) \backslash \mathcal{F}_4$. Let us call the 40 points of $PG(3, 4) \backslash \mathcal{F}_4$ *pseudopoints*. Let the 90 unisecants to $\mathcal{F}_4$ be *pseudolines of the first kind*. Let the 40 self-polar tetrahedra regarded as sets of their vertices be *pseudolines of the second kind*. Any of the 40 non-tritangent planes π meets $\mathcal{F}_4$ in a curve $\mathcal{U}_2$ comprising nine

Table 20.3 *Faces of the Pentahedra*

$\mathcal{P}$:	$c_{12}c_{34}c_{56}$,	$c_{13}c_{25}c_{46}$,	$c_{14}c_{26}c_{35}$,	$c_{15}c_{24}c_{36}$,	$c_{16}c_{23}c_{45}$
$\mathcal{P}_0$:	$c_{12}c_{34}c_{56}$,	$c_{13}c_{26}c_{45}$,	$c_{14}c_{25}c_{36}$,	$c_{15}c_{23}c_{46}$,	$c_{16}c_{24}c_{35}$
$\mathcal{P}_1$:	$c_{12}c_{36}c_{45}$,	$c_{13}c_{25}c_{46}$,	$c_{14}c_{23}c_{56}$,	$c_{15}c_{26}c_{34}$,	$c_{16}c_{24}c_{35}$
$\mathcal{P}_2$:	$c_{12}c_{36}c_{45}$,	$c_{13}c_{24}c_{56}$,	$c_{14}c_{26}c_{35}$,	$c_{15}c_{23}c_{46}$,	$c_{16}c_{25}c_{34}$
$\mathcal{P}_3$:	$c_{12}c_{35}c_{46}$,	$c_{13}c_{26}c_{45}$,	$c_{14}c_{23}c_{56}$,	$c_{15}c_{24}c_{36}$,	$c_{16}c_{25}c_{34}$
$\mathcal{P}_4$:	$c_{12}c_{35}c_{46}$,	$c_{13}c_{24}c_{56}$,	$c_{14}c_{25}c_{36}$,	$c_{15}c_{26}c_{34}$,	$c_{16}c_{23}c_{45}$

points; so $\pi\backslash\mathscr{F}_4$ comprises 12 points. With respect to the Hermitian polarity of $\mathscr{F}_4$, the plane π has a pole P which is neither on $\mathscr{F}_4$ nor in π. Then π defines a *pseudoplane* whose points are P and the points of $\pi\backslash\mathscr{F}_4$. So each pseudoline contains four pseudopoints and each pseudoplane contains 13 pseudopoints. Also there are 40 pseudopoints, 130 pseudolines, and 40 pseudoplanes.

Theorem 20.3.2: *The pseudopoints, pseudolines, and pseudoplanes are the points, lines, and planes of a* $PG(3, 3)$, *in which the pseudolines of the second kind form a linear complex.*

Proof: The numbers of pseudospaces are correct for the subspaces of $PG(3, 3)$; see § 15.1. The incidence axioms of § 2.1 for a projective space can now be verified; see § 20.8.

Since a pseudoplane comprises a point P off $\mathscr{F}_4$ and the points of $\pi\backslash\mathscr{F}_4$, where π is polar to P, the self-polar tetrahedra with P as a vertex have opposite face a triangle in π which is self-polar for $\mathscr{U}_2=\pi\cap\mathscr{F}_4$. For example, if $\mathscr{F}_4=\mathbf{V}(x_0^3+x_1^3+x_2^3+x_3^3)$ and $P=\mathbf{U}_3$, then we must determine the self-polar triangles of $\mathscr{U}_2=\mathbf{V}(x_0^3+x_1^3+x_2^3, x_3)$. There are exactly four, namely $\mathscr{T}_1$, $\mathscr{T}_2$, $\mathscr{T}_3$, $\mathscr{T}_4$ as in § 14.3. This shows that the pseudolines of the second kind are a set $\mathscr{A}$ such that in each pseudoplane and through each pseudopoint there are four elements of $\mathscr{A}$. By Theorem 15.2.13, $\mathscr{A}$ is a linear complex in $PG(3, 3)$. □

Corollary: (i) $P\Gamma U(4, 4)\cong PGSp(4, 3)$
(ii) $PGU(4, 4)=PSU(4, 4)\cong PSp(4, 3)$.

Proof: (i) Each collineation fixing $\mathscr{F}_4$ fixes the sets of pseudopoints, pseudolines of the first kind, pseudolines of the second kind, and pseudoplanes. This gives an injective homomorphism between the groups; as their orders are equal (see Appendix III) they are isomorphic.

(ii) These are the simple subgroups of index 2 in the previous groups. □

II. $q=7$

By the corollary to Lemma 20.2.4, $\mathscr{F}$ has 18 E-points and so, by Lemma 20.2.11, it is equianharmonic and projectively unique. Thus $\mathscr{F}$ can be called $\mathscr{F}_7$. Also $G_7=\Gamma_7\cong(\mathbf{Z}_3\times\mathbf{Z}_3\times\mathbf{Z}_3)\mathbf{S}_4$ and $|G_7|=648$. We may note that, in $GF(7)$, x^2+x+1 has roots 2 and -3.

In the notation of Lemma 20.2.4,

$$e_3^{(i)}=2, \qquad e_2^{(i)}=6, \qquad e_1^{(i)}=0,$$

for $i\in\mathbf{N}_{27}$, whence

$$e_3=18, \qquad e_2=81, \qquad e_1=0, \qquad e_0=0, \qquad N=N'=99.$$

Apart from the 45 tritangent planes, there are 81 planes meeting $\mathscr{F}_7$ in a line and a conic, and 274 planes meeting $\mathscr{F}_7$ in an irreducible cubic. If π is one of the last type, then $\pi \cap \mathscr{F}_7 = \mathscr{C}$ contains at most one E-point of each line on $\mathscr{F}_7$. As $e_1 = 0$, so $\mathscr{C}$ contains an odd number of E-points. By the corollary to Lemma 20.2.10, only the configurations C_1, C_3, and C_9 are possible on $\mathscr{C}$, whence $|\mathscr{C}| = 13$, 12, and 9 respectively. In fact, by Theorem 11.9.1, a cubic curve in $PG(2, 7)$ can have at most 13 points. An equianharmonic cubic with at least nine points and at least one inflexion has 13, 12, or 9 points and correspondingly 1, 3, or 9 inflexions. In all three cases, the cubic is projectively unique; see Table 11.21.

Theorem 20.3.3: *In $PG(3, 7)$ there is a projectively unique cubic surface $\mathscr{F}_7$, which is equianharmonic; hence $|G_7| = |\Gamma_7| = 648$.* □

III. $q=8$

From the corollary to Lemma 20.2.4, we have $9 \leq e_3 \leq 13$. However, by Theorem 20.2.13(i)(b), the only possibilities are $e_3 = 9$ and $e_3 = 13$. Part (ii) of the same theorem rules out $e_3 = 9$. So $\mathscr{F}$ has 13 E-points, which lie on three coplanar lines of $\mathscr{F}$, say a_1, b_2, c_{12}, by Lemma 20.2.10. Let $\pi_0 = a_1 b_2 c_{12}$.

In § 6.4, it was shown that four points on $PG(1, 8)$ have an associated point, which is also the associated point of the remaining four points on the line. Unsurprisingly, the four E-points other than E_{12} on each of a_1, b_2, c_{12} have E_{12} as their associated point. For, in the notation of § 7.5,

$$(E_{12}, E_{13}, E_{14}, E_{15}, E_{16}) \overset{E_{21}}{\barwedge} (E_{12}, E_{32}, E_{42}, E_{52}, E_{62}) \overset{E_{12,34,56}}{\barwedge} (E_{12}, E_{14}, E_{13}, E_{16}, E_{15}),$$

where the collinearities of E-points are as in the corollary to Lemma 20.2.9. So E_{12} is the fixed point in the involution with pairs (E_{13}, E_{14}) and (E_{15}, E_{16}). There are 16 lines of three E-points. Through each E-point other than E_{12} there are four lines in π_0 containing no other E-point and hence 48 lines of this type. Together with the pencil of lines through E_{12}, we have $16+48+9=73$ lines, which is all there are in the plane.

Apart from $\pi_0 = a_1 b_2 c_{12}$ and the other 12 tangent planes at E-points, there are 32 tritangent planes. From Theorem 20.2.5, π_0 lies in 16 of the 240 trihedra, whose other faces are pairs of the 32 tritangent planes. These 16 trihedra have their faces collinear on the lines of three E-points. The other 16 trihedra of the corresponding trihedral pairs have faces from the set of 12 tangent planes at the E-points other than E_{12}.

One of the trihedral pairs containing π_0 is

$$\begin{array}{llll} T_{13,24} & a_1 & b_2 & c_{12} \\ & b_4 & a_3 & c_{34} \\ & c_{14} & c_{23} & c_{56} \end{array}$$

and a canonical form for $\mathscr{F}$ is derived from this.

Let $\pi_0 = \mathbf{u}_0$ and let the other faces of the trihedron containing π_0 be $\mathbf{u}_1$ and $\mathbf{V}(x_0 + x_1)$. Two faces of the other trihedron may be taken as $\mathbf{u}_2$ and $\mathbf{u}_3$. So two of the lines of $\mathscr{F}$ in π_0 are $\mathbf{V}(x_0, x_2)$ and $\mathbf{V}(x_0, x_3)$. Hence $E_{12} = \mathbf{U}_1$ and the third line of $\mathscr{F}$ in π_0 is $\mathbf{V}(x_0, a_2x_2 + a_3x_3)$ with $a_2a_3 \neq 0$. Thus the remaining face of the trihedron may be taken as $\mathbf{V}(x_0 + a_2x_2 + a_3x_3)$. Hence

$$\mathscr{F} = \mathbf{V}(cx_0x_1(x_0 + x_1) + x_2x_3(x_0 + a_2x_2 + a_3x_3)),$$

which by a projectivity can be reduced to

$$\mathscr{F} = \mathbf{V}(cx_0x_1(x_0 + x_1) + x_2x_3(x_0 + x_2 + x_3)).$$

The plane $\mathbf{V}(x_0 - \lambda x_2)$ meets $\mathscr{F}$ in $\mathbf{V}(x_0, x_2)$ and

$$\mathbf{V}(x_0 - \lambda x_2,\ c\lambda x_1^2 + x_3^2 + c\lambda^2 x_1x_2 + (\lambda + 1)x_2x_3),$$

which, by Theorem 7.2.4, is a line pair if

$$c\lambda(c\lambda^3 + \lambda^2 + 1) = 0.$$

So $c\lambda^3 + \lambda^2 + 1$ has three distinct roots in $GF(8)$, whence it divides $\lambda^7 + 1$. Therefore $c = 1$. So

$$\begin{aligned} \mathscr{F} &= \mathbf{V}(x_0x_1(x_0 + x_1) + x_2x_3(x_0 + x_2 + x_3)) \\ &= \mathbf{V}(x_0^2x_1 + x_0(x_1^2 + x_2x_3) + x_2x_3(x_2 + x_3)), \end{aligned} \tag{20.7}$$

and $\mathscr{F}$ may be called $\mathscr{F}_8$.

The equation for $\mathscr{F}_8$ enables us to determine the order of G_8. Any projectivity fixing $\mathscr{F}_8$ fixes π_0 and so transforms $T_{13,24}$ into itself or one of the other 15 with π_0 as a face. If $T_{13,24}$ is left fixed, then both its trihedra are as well. For the trihedron containing π_0, the other two faces can be interchanged, while, in the other trihedron, all permutations of the faces can be achieved by projectivities. Hence the group of projectivities of $\mathscr{F}$ fixing $T_{13,24}$ is isomorphic to the direct product $\mathbf{S}_3 \times \mathbf{Z}_2$ of order 12. As there are 16 trihedral pairs containing π_0, so $|G_8| = 16 \cdot 12 = 192$ and $|\Gamma_8| = 3 \cdot 192 = 576$.

Apart from the three lines containing five E-points, the remaining 24 each contain one. Hence, from Lemma 20.2.4,

$$e_3 = 13, \qquad e_2 = 96, \qquad e_1 = 12, \qquad e_0 = 0, \qquad N = N' = 121.$$

The 12 points on only one line of $\mathscr{F}_8$ are all in π_0. A plane π meeting $\mathscr{F}_8$ in an irreducible cubic $\mathscr{C}$ meets π_0 in a line through E_{12} or in a line containing three E-points or in a line not through E_{12} containing one E-point. In the three cases $\mathscr{C}$ contains respectively 13, 12, and 14 points and has 1, 3, and 1 inflexion.

Theorem 20.3.4: *In $PG(3, 8)$ there is a projectively unique cubic surface $\mathscr{F}_8$, which has 13 E-points. Also* $|G_8| = 192$, $|\Gamma_8| = 576$. □

IV. $q = 9$

By Lemma 20.2.4, $e_3 + e_0 = 10$ and from the corollary, $0 \leqslant e_3 \leqslant 10$. First we need to restrict the range for e_3.

Lemma 20.3.5: *For a cubic surface $\mathscr{F}$ in $PG(3, 9)$, only C_6, C_9, and C_{10} are possible.*

Proof: By Theorem 20.2.3, $\mathscr{F}$ contains $N = 145$ points. By Theorem AIV.2.4, the total number of points on an irreducible cubic curve in $PG(2, 9)$ is at most 16.

Suppose that $\mathscr{F}$ has two points P and Q on none of its lines. Suppose further that $PQ \cap \mathscr{F} = \{P, Q\}$. Then every plane section of $\mathscr{F}$ through PQ contains no lines of $\mathscr{F}$ and so is an irreducible cubic. Hence, we have $N \leqslant 10(16-2)+2 = 142$, contradicting that $N = 145$.

Now let $PQ \cap \mathscr{F} = \{P, Q, R\}$. If a plane section of $\mathscr{F}$ through PQ comprises a line l and a conic, then l passes through R. So, if m sections consist of a line and a conic, then $m \leqslant 3$, and, when $m = 3$, then R is an E-point. Also, $N \leqslant 3 + m(20-3) + (10-m)(16-3) = 133 + 4m$. As $N = 145$, so $m = 3$. Hence for every pair of points of $\mathscr{F}$ not on a line of $\mathscr{F}$, there is an E-point on their join.

If $e_0 \geqslant 6$, this implies at least five E-points, which is too many. So $e_0 \leqslant 5$. For $e_0 = 2$, 3, 5 we obtain respectively that $e_3 = 8$, 7, 5, all impossible by Theorem 20.2.13. So $e_0 = 0$, 1, or 4 and $e_3 = 10$, 9, or 6. □

Lemma 20.3.6: *If, in $PG(3, 9)$, $\mathscr{F}$ contains C_6, then $\mathscr{F}$ contains C_{10} and is diagonal.*

Proof: Let the four points of $\mathscr{F}$ on none of its lines be $\mathbf{U}_0$, $\mathbf{U}_1$, $\mathbf{U}_2$, $\mathbf{U}_3$ and let the plane π of C_6 be $\mathbf{u}$ (Fig. 20.7). With $\pi = a_1 b_2 c_{12}$, we have

$$c_{12} = \mathbf{V}(x_0 + x_1, x_2 + x_3), \qquad a_1 = \mathbf{V}(x_0 + x_2, x_1 + x_3),$$
$$b_2 = \mathbf{V}(x_0 + x_3, x_1 + x_2).$$

As π meets $\mathscr{F}$ in these three lines and as $\mathbf{U}_0$, $\mathbf{U}_1$, $\mathbf{U}_2$, $\mathbf{U}_3$ lie on $\mathscr{F}$, so $\mathscr{F} = \mathbf{V}(F)$ where

$$F = (x_0 + x_1 + x_2 + x_3) \sum{}'' a_{ij} x_i x_j + (x_2 + x_3)(x_1 + x_3)(x_1 + x_2).$$

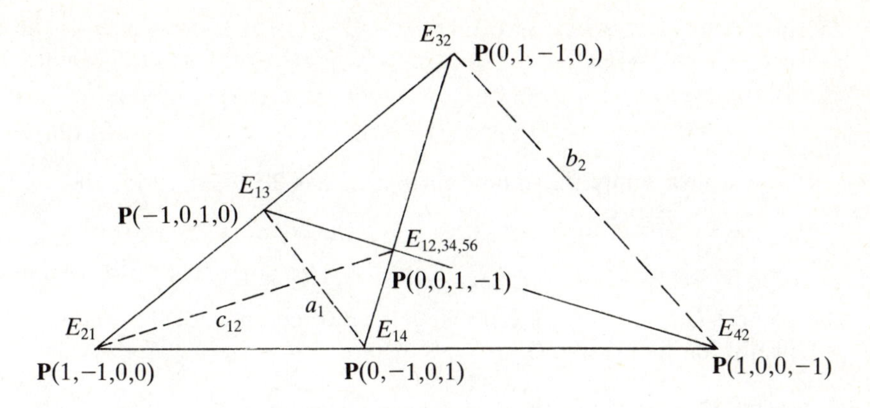

Fig. 20.7

Consider one of the E-points E_{21}. The tangent plane to $\mathscr{F}$ at E_{21} is $a_2b_1c_{12}$ and so is $\mathbf{V}((x_0+x_1)-t(x_2+x_3))$. As each of a_2, b_1, c_{12} lies in a plane through $\mathbf{U}_0\mathbf{U}_1$, so $a_2b_1c_{12}$ meets $\mathscr{F}$ in the cubic

$$\mathbf{V}((x_2+x_3)(t'x_2+x_3)(x_2+t''x_3)).$$

Substituting $x_0=tx_2+tx_3-x_1$ in $\mathscr{F}$ gives $(x_2+x_3)F'$, where

$$F'=(t+1)[(tx_2+tx_3-x_1)(a_{01}x_1+a_{02}x_2+a_{03}x_3)+a_{12}x_1x_2 \\ +a_{13}x_1x_3+a_{23}x_2x_3]+(x_1^2+x_1x_2+x_1x_3+x_2x_3). \quad (20.8)$$

Now, F' can have no terms in x_1^2, x_1x_2, x_1x_3, whence

$$-(t+1)a_{01}+1=0,$$
$$(t+1)(ta_{01}-a_{02}+a_{12})+1=0,$$
$$(t+1)(ta_{01}-a_{03}+a_{13})+1=0.$$

Therefore, since $t\neq-1$,

$$1-a_{02}+a_{12}=0, \qquad 1-a_{03}+a_{13}=0.$$

In exactly the same way, consideration of the tangent plane at E_{13} gives

$$1-a_{01}+a_{12}=0, \qquad 1-a_{03}+a_{23}=0;$$

and from E_{42},

$$1-a_{02}+a_{23}=0, \qquad 1-a_{01}+a_{13}=0.$$

Hence

$$a_{01}=a_{02}=a_{03}=(t+1)^{-1}=c, \qquad a_{12}=a_{13}=a_{23}=c-1.$$

Now, from (20.8),

$$F'=tx_2^2+(t+1)x_2x_3+tx_3^2.$$

The discriminant Δ is $\Delta(F')=(t+1)^2-t^2=1-t=-(1+c)/c$. As F' is the product of two distinct linear terms, Δ is a non-zero square, whence $\Delta^4=1$. So

$$c^3+c+1=0. \tag{20.9}$$

We must now apply the condition that $\mathscr{F}$ has 27 lines. From above,

$$F=(x_0+x_1+x_2+x_3)[cx_0(x_1+x_2+x_3)+(c-1)(x_1x_2+x_1x_3+x_2x_3)] \\ +(x_2+x_3)(x_1+x_3)(x_1+x_2).$$

The plane $\mathbf{V}(x_0+x_1-\lambda(x_2+x_3))$ meets $\mathscr{F}$ in three lines for five values of λ. Substitution in F gives $(x_2+x_3)F''$, where

$$\begin{aligned} F'' &= (\lambda+1)[c(-x_1+\lambda x_2+\lambda x_3)(x_1+x_2+x_3) \\ &\quad +(c-1)(x_1x_2+x_1x_3+x_2x_3)]+(x_1+x_3)(x_1+x_2) \\ &=(1-c-c\lambda)x_1^2+c\lambda(\lambda+1)(x_2^2+x_3^2) \\ &\quad +\lambda(c\lambda+c-1)(x_1x_2+x_1x_3)-(c\lambda^2+\lambda-c)x_2x_3. \end{aligned}$$

Applying the condition of Theorem 7.2.4 that F'' is reducible gives

$$(\lambda+1)(c\lambda+c-1)[(c-1)\lambda+c][c\lambda^2+(c-1)\lambda+c]=0.$$

So $G=c\lambda^2+(c-1)\lambda+c$ has two roots in γ. Also $\Delta(G)=(c-1)^2-c^2=c+1$. As G has two roots in γ, so $\Delta(G)$ is a square and $\Delta(G)^4=1$; whence $c(c^3+c^2+1)=0$. As $c\neq 0$, so $c^3+c^2+1=0$, which, with (20.9), gives $c=1$. Hence

$$\begin{aligned} F &= x_0(x_1+x_2+x_3)(x_0+x_1+x_2+x_3)+(x_2+x_3)(x_1+x_3)(x_1+x_2) \\ &= -\textstyle\sum'' x_ix_jx_k \quad \text{with} \quad x_0+x_1+x_2+x_3+x_4=0. \end{aligned}$$

So $\mathscr{F}$ is the diagonal surface, by Lemma 20.2.12, which contains C_{10} since x^2-x-1 has two roots in $GF(9)$. □

Lemma 20.3.7: *There is a projectively unique surface $\mathscr{F}_9^1$ in $PG(3,9)$ containing C_9.*

Proof: If $\mathscr{F}$ contains C_9, then the nine E-points are inflexions of a plane cubic curve $\mathscr{C}$, by the corollary to Lemma 20.2.10. Then $\mathscr{C}$ necessarily contains a cusp, by Theorem 11.3.10 and Lemma 11.4.1, which is the unique point of $\mathscr{F}$ on none of its lines. The nine E-points are collinear in threes on 12 lines, which are concurrent in threes at 4 points forming a harmonic tetrad on the cuspidal tangent.

It remains to show that $\mathscr{F}$ exists. From Lemma 11.3.3, we may take

$$\mathscr{C}=\mathbf{V}(x_0,\, x_1^3-x_2^2x_3)=\{P(t)=\mathbf{P}(0,t,1,t^3)\mid t\in\gamma^+\};$$

$P(\infty)$ is the cusp. Three other points $P(t_1)$, $P(t_2)$, $P(t_3)$ of $\mathscr{C}$ are collinear if and only if $t_1+t_2+t_3=0$. Then $\mathcal{F}=\mathbf{V}(F)$ where

$$F=x_0\sum' a_{ij}x_ix_j+x_1^3-x_2^2x_3.$$

Let the nine E-points be those of Lemma 20.2.10. Then, of the 40 triads of trihedral pairs, four are special, namely

$$\begin{array}{lll} T_{123} & T_{456} & T_{123,456}, \\ T_{14,25} & T_{25,36} & T_{36,14}, \\ T_{15,26} & T_{26,34} & T_{34,15}, \\ T_{16,24} & T_{24,35} & T_{35,16}. \end{array}$$

One trihedron in each pair has an axis and the three axes belonging to a triad are a concurrent set of three lines through the nine E-points. To show that $\mathcal{F}$ has 27 lines, it suffices to find the equation of $\mathcal{F}$ in terms of each member of a triad and this will be done for one of the four special triads.

Let $P=\mathbf{P}(0, y_1, y_2, y_3)$; then the tangent plane $\pi_P(\mathcal{F})=\mathbf{V}(\sum' a_{ij}y_iy_jx_0+y_2y_3x_2-y_2^2x_3)$. Take

$$GF(9)\backslash\{0\}=\{\pm\nu^i \mid i\in\bar{\mathbf{N}}_3,\ \nu^2-\nu-1=\nu^4+1=0\}.$$

One set of collinear triads of E-points with their lines is

$$\begin{array}{lll} P(0), & P(1), & P(-1):\mathbf{V}(x_0, x_1-x_3); \\ P(\nu^3), & P(-\nu^2), & P(-\nu):\mathbf{V}(x_0, x_1-\nu^2x_2-x_3); \\ P(-\nu^3), & P(\nu^2), & P(\nu)\ :\mathbf{V}(x_0, x_1+\nu^2x_2-x_3). \end{array}$$

Select the tangent planes to $\mathcal{F}$ at $P(0)$, $P(1)$, $P(-1)$ as respectively

$$\mathbf{V}(x_0+x_3),\qquad \mathbf{V}(x_2-x_3),\qquad \mathbf{V}(x_2+x_3).$$

Then, without loss of generality, if T_{123} is the trihedral pair with axis $\mathbf{V}(x_0, x_1-x_3)$,

$$F=(b_1x_0+x_1-x_3)(b_2x_0+x_1-x_3)(b_3x_0+x_1-x_3)$$
$$-(x_0+x_3)(x_2-x_3)(x_2+x_3).$$

The tangent plane at $P(t)=\mathbf{P}(0, t, 1, t^3)$ now is

$$\mathbf{V}([b(t^3-t)^2+t^6-1]x+t^3x_2-x_3),$$

where $b=b_1+b_2+b_3$. Then, if T_{456} has the axis $\mathbf{V}(x_0, x_1-\nu^2x_2-x_3)$,

$$F=(c_1x_0+x_1-\nu^2x_2-x_3)(c_2x_0+x_1-\nu^2x_2-x_3)(c_3x_0+x_1-\nu^2x_2-x_3)$$
$$-\lambda[(\nu-b)x_0+\nu x_2-x_3][(1-b)x_0+\nu^2x_2-x_3][(\nu^3-b)x_0-\nu^3x_2-x_3].$$

Comparing the coefficients of x_3^3 gives $\lambda=1$. Then from the coefficients of

$x_0x_1^2$ and $x_0x_1x_2$,

$$c_1+c_2+c_3=b_1+b_2+b_3=0;$$

from the coefficients of $x_0^2x_1$ and $x_0^2x_2$,

$$c_1c_2+c_1c_3+c_2c_3=b_1b_2+b_1b_3+b_2b_3=1;$$

from the coefficient of x_0^3,

$$c_1c_2c_3+1=b_1b_2b_3.$$

Let $b_1b_2b_3=k$. Then b_1, b_2, b_3 are the roots of x^3+x-k and c_1, c_2, c_3 the roots of $x^3+x-k+1$. However, $x^9-x=(x^3+x)(x^3+x-1)(x^3+x+1)$, whence $k=0$, 1, or -1; that is $k^3=k$. Hence

$$F=kx_0^3+x_0^2(x_1-x_3)-x_0(x_2^2-x_3^2)+x_1^3-x_2^2x_3.$$

The substitution of kx_0+x_1 for x_1, which changes none of the E-points, gives

$$F=x_0^2(x_1-x_3)-x_0(x_2^2-x_3^2)+x_1^3-x_2^2x_3. \tag{20.10}$$

Finally, in terms of the third trihedral pair $T_{123,456}$ with axis $\mathbf{V}(x_0, x_1+\nu^2x_2-x_3)$,

$$F=(x_0+x_1+\nu^2x_2-x_3)(-\nu x_0+x_1+\nu^2x_2-x_3)(-\nu^3x_0+x_1+\nu^2x_2-x_3)$$
$$-(\nu x_0-\nu x_2-x_3)(x_0-\nu^2x_2-x_3)(\nu^3x_0+\nu^3x_2-x_3). \quad \square$$

Corollary: $|G_9^1|=216$, $|\Gamma_9^1|=432$.

Proof: Any projectivity fixing $\mathcal{F}_9^1$ fixes the curve $\mathscr{C}$ containing the E-points and the set of four special triads of trihedral pairs. If T_{123} is left fixed, both its trihedra are. From (20.10),

$$F=(x_1-x_3)(\nu^2x_0+x_1-x_3)(-\nu^2x_0+x_1-x_3)-(x_0+x_3)(x_2+x_3)(x_2-x_3),$$

G_9^1 acts on the collinear faces of the one trihedron as $\mathbf{Z}_3$ and on the non-collinear faces of the other trihedron as $\mathbf{S}_3$; this gives a group isomorphic to $\mathbf{Z}_3\times\mathbf{S}_3$. Then T_{123} can be transformed into the other two trihedral pairs in its triad and this triad can be transformed into the other three special triads. So $|G_9^1|=4\cdot3\cdot6\cdot3=216$ and $|\Gamma_9^1|=2\cdot216=432$. $\square$

There remains the case that $\varrho_3=10$. As x^2-x-1 has two roots over $GF(9)$, a diagonal surface $\mathcal{F}_9^0$ exists.

Theorem 20.3.8: *There are two projectively distinct cubic surfaces $\mathcal{F}_9^0$ and $\mathcal{F}_9^1$ in $PG(3, 9)$.* $\square$

Theorem 20.3.9: *All the projectively distinct cubic surfaces $\mathcal{F}$ with 27 lines in $PG(3, q)$, $q\leq9$, are given in Table 20.4.* $\square$

Table 20.4 *Cubic Surfaces for $q \leqslant 9$*

q	$\mathcal{F}$	e_3	e_2	e_1	e_0	N'	N	$\|G\|$	$\|\Gamma\|$	Equation
4	$\mathcal{F}_4$	45	0	0	0	45	45	25 920	51 840	(20.5), (20.6)
7	$\mathcal{F}_7$	18	81	0	0	99	99	648	648	(20.5)
8	$\mathcal{F}_8$	13	96	12	0	121	121	192	576	(20.7)
9	$\mathcal{F}_9^0$	10	105	30	0	145	145	120	240	(20.6)
9	$\mathcal{F}_9^1$	9	108	27	1	144	145	216	432	(20.10)

Theorem 20.3.10: *All the projectively distinct cubic surfaces $\mathcal{F}$ with 27 lines in $PG(3, q)$ such that all points of $\mathcal{F}$ lie on its lines are given in Table 20.5.*

Proof: By Lemma 20.2.4, if $e_0 = 0$, then $e_3 = (q-10)^2 + 9$. By Lemma 20.2.7, $e_3 \leqslant 18$ for q odd and $e_3 \leqslant 45$ for q even. So, for $e_0 = 0$, we have $q \leqslant 13$ for q odd and $q \leqslant 16$ for q even. The cases of $q \leqslant 9$ are covered by the previous theorem. For $q = 11$, we require $e_3 = 10$. As $x^2 - x - 1$ has the two roots 4 and -3 in $GF(11)$, a diagonal surface $\mathcal{F}$ exists and is the only one containing ten E-points, by Lemma 20.2.12. For $q = 13$, we require $e_3 = 18$. As $x^2 + x + 1$ has the two roots 3 and -4 in $GF(13)$, an equianharmonic surface $\mathcal{F}$ exists and is the only one containing 18 E-points, by Lemma 20.2.11. For $q = 16$, we require $e_3 = 45$. This exists over $GF(16)$ since it exists over $GF(4)$, by Theorem 20.3.1. □

20.4 Mappings onto the plane

As over the complex field, a cubic surface $\mathcal{F}$ with 27 lines in $PG(3, q)$ can be mapped onto the plane in the following way. Let $\mathcal{F} = \mathbf{V}(F)$ be given,

Table 20.5 *Cubic Surfaces with all Points on their 27 Lines*

q	$\mathcal{F}$	e_3	e_2	e_1	e_0	N	$\|G\|$	$\|\Gamma\|$	Equation
4	$\mathcal{F}_4$	45	0	0	0	45	25 920	51 840	(20.5), (20.6)
7	$\mathcal{F}_7$	18	81	0	0	99	648	648	(20.5)
8	$\mathcal{F}_8$	13	96	12	0	121	192	576	(20.7)
9	$\mathcal{F}_9^0$	10	105	30	0	145	120	240	(20.6)
11	$\mathcal{F}_{11}^0$	10	105	64	0	179	120	120	(20.6)
13	$\mathcal{F}_{13}^0$	18	81	162	0	261	648	648	(20.5)
16	$\mathcal{F}_{16}^0$	45	0	324	0	369	25 920	103 680	(20.5), (20.6)

as in Lemma 20.2.6, by

$$F=F_1F_2F_3+G_1G_2G_3=\begin{vmatrix} 0 & F_1 & G_3 \\ G_1 & 0 & F_2 \\ F_3 & G_2 & 0 \end{vmatrix}.$$

So a point $\mathbf{P}(X)$ lies on $\mathscr{F}$ if the determinant is zero; that is, there exists $Y=(y_0, y_1, y_2)$ such that

$$\begin{aligned} y_1G_1(X)+y_2F_3(X)&=0,\\ y_0F_1(X)+y_2G_2(X)&=0,\\ y_0G_3(X)+y_1F_2(X)&=0. \end{aligned}$$

Since the F_i and G_i are linear, solving for the x_i gives

$$x_0:x_1:x_2:x_3=f_0(Y):f_1(Y):f_2(Y):f_3(Y), \tag{20.11}$$

where each f_i is cubic. So we have a mapping

$$\mathfrak{T}:\mathscr{F}\to PG(2,q)$$

given by $\mathbf{P}(X)\mathfrak{T}=\mathbf{P}(Y)$. From (20.11), $\mathfrak{T}$ maps plane sections of $\mathscr{F}$ to cubic curves through the set $\mathscr{B}$ of base points, where

$$\mathscr{B}=\mathbf{V}(f_0, f_1, f_2, f_3).$$

In fact, $\mathscr{B}=\{P_i \mid i\in\mathbf{N}_6\}$ is a 6-arc not on a conic; this agrees with the different construction of Lemma 20.1.5. Then there exists one half of a double-six on $\mathscr{F}$, say $a_1a_2a_3a_4a_5a_6$, such that if $\mathscr{A}$ is the set of points on the lines a_i, then the restriction of $\mathfrak{T}$ is a bijection

$$\mathfrak{T}:\mathscr{F}\backslash\mathscr{A}\to PG(2,q)\backslash\mathscr{B}.$$

Let $\mathscr{C}_i$ be the conic through the 5 points of $\mathscr{B}$ other than P_i. Then

$$\begin{aligned} a_i\mathfrak{T}&=P_i,\\ b_i\mathfrak{T}&=\mathscr{C}_i,\\ c_{ij}\mathfrak{T}&=P_iP_j. \end{aligned}$$

The existence of the E-point E_{12} means that the tangent at P_1 to $\mathscr{C}_2$ is P_1P_2; the E-point $E_{12,34,56}$ means that P_1P_2, P_3P_4, and P_5P_6 are concurrent.

The projective classification of cubic surfaces $\mathscr{F}$ can be done by classifying 6-arcs $\mathscr{B}$ in the plane, although projectively distinct 6-arcs do not necessarily represent projectively distinct surfaces.

There is another classical method for mapping the lines of $\mathscr{F}$ onto a plane. If P is a point on the line l of $\mathscr{F}$, the polar quadric at P contains l. If P is a point on no line of $\mathscr{F}$ and q is odd, then the polar quadric $\mathscr{Q}$ at P

is non-singular and meets $\mathscr{F}$ in a sextic curve $\mathscr{C}^6$ with a double point at P. The projection of $\mathscr{C}^6$ from P onto a plane π is a non-singular quartic curve $\mathscr{C}^4$ with 28 double tangents, 27 of which are projections l_i' of the lines l_i of $\mathscr{F}$: the other is $l' = \pi \cap \pi_P(\mathscr{F})$. The points of contact of l_i' with $\mathscr{C}^4$ are the projections of the double points of the involution on l_i, by Lemma 20.2.7; the points of contact of l' with $\mathscr{C}^4$ are the intersections of the tangents at P to $\mathscr{C}^6$, which are the generators of $\mathscr{Q}$ through P, with π.

With respect to $\mathscr{C}^4$, there is nothing special about l' to distinguish it from the other bitangents l_i'. The group of permutations of the 28 bitangents to $\mathscr{C}^4$ preserving their geometrical properties has order $28 \cdot 51\,840 = 1\,451\,520 = 2^9 \cdot 3^4 \cdot 5 \cdot 7$; see § 20.7.

Of the surfaces $\mathscr{F}$ with $q \leqslant 9$, only $\mathscr{F}_9^1$ has a point $P = \mathbf{U}_3$ on none of its lines; see § 20.3, IV. In this case, $\pi_P(\mathscr{F}) = \mathbf{u}_0$, which is the plane of the E-points. On each line l_i of $\mathscr{F}_9^1$, the double points of the involution are the E-point on l_i and the point on no other line. In this case, $\mathscr{C}^4$ consists of 28 points, comprising the projections of the 27 points on a single l_i and the meet with π of the cuspidal tangent to the curve $\mathscr{C}$ of the E-points. So $\mathscr{C}^4$ is, in fact, the Hermitian curve $\mathscr{U}_{2,9}$, to which all tangents make 4-point contact. From this construction, we would expect that the group of projectivities fixing these 28 bitangents to have order $28\,|G_9^1| = 28 \cdot 216 = 6048$, which is in fact $|PGU(3, 9)|$.

20.5 The representation of $\mathscr{F}_4$ in $PG(5, 2)$

Since $\mathscr{F}_4$ is Hermitian, the image of the set of 27 lines of $\mathscr{F}_4$ under the Grassmann mapping $\mathfrak{G}: \mathscr{L} \to \mathscr{H}_{5,4}$ is $\mathscr{E}_{5,2}$, by Theorem 19.2.2. The $36 = 63 - 27$ points of $PG(5, 2) \backslash \mathscr{E}_{5,2}$ are the poles with respect to $\mathscr{H}_{5,4}$ of the primes whose sections represent the linear complexes of $PG(3, 4)$ defined by the null polarities of the 36 double-sixes of $\mathscr{F}_4$.

In this section we will elaborate this correspondence between $\mathscr{F}_4$ and $\mathscr{E}_{5,2}$, and describe in detail the geometry of $\mathscr{E}_{5,2}$. This gives a good picture of the general cubic surface over the complex numbers. It should also be compared with § 17.5, in which the geometry of $\mathscr{H}_{5,2}$ was described.

As in § 3.1 and § 5.2, $\phi(r; n) = \phi(r; n, q)$ is the number of r-spaces in $PG(n, q)$, and $\psi_+(n)$, $\psi_-(n)$, and $\psi(n)$ are the respective numbers of points on $\mathscr{H}_n$, $\mathscr{E}_n$, and $\mathscr{P}_n$. For $q = 2$,

$\phi(0; 5) = \phi(4; 5) = 63, \quad \phi(1, 5) = \phi(3, 5) = 651, \quad \phi(2; 5) = 1395,$

$\phi(0; 4) = \phi(3, 4) = 31, \quad \phi(1; 4) = \phi(2; 4) = 165,$

$\phi(0; 3) = \phi(2; 3) = 15, \quad \phi(1; 3) = 35,$

$\phi(0; 2) = \phi(1; 2) = 7, \quad \phi(0, 1) = 3,$

$\psi_+(5) = 35, \quad \psi_-(5) = 27, \quad \psi(4) = 15,$

$\psi_+(3) = 9, \quad \psi_-(3) = 5, \quad \psi(2) = 3.$

In the $PG(5,2)$ containing $\mathscr{E}=\mathscr{E}_{5,2}$, let A_i, B_i, C_{ij} be the points representing the lines a_i, b_i, c_{ij} and let D, D_{ij}, D_{ijk} be the points representing the polarities of the double-sixes $\mathscr{D}$, $\mathscr{D}_{ij}$, $\mathscr{D}_{ijk}$ as explained above. In this section, we shall abuse notation and confuse points of $PG(5,2)$ with their vectors.

From the E-points on $\mathscr{F}_4$, we have the following relations, which define the 45 lines on $\mathscr{E}$:

$$A_i+B_j+C_{ij}=0, \quad 30; \qquad C_{ij}+C_{kl}+C_{mn}=0, \quad 15. \tag{20.12}$$

From the polarities of the double-sixes, as in Lemma 15.2.8, Corollary (i)(a), we have the following $6\cdot 36=216$ relations, which define the 2-secants of $\mathscr{E}$:

$$\begin{array}{llll} A_i+B_i+D=0, & 6; & A_i+A_j+D_{ij}=0, & 15; \\ B_i+B_j+D_{ij}=0, & 15; & C_{ij}+C_{jk}+D_{ij}=0, & 60; \\ A_i+C_{jk}+D_{ijk}=0, & 60; & B_l+C_{mn}+D_{ijk}=0, & 60. \end{array} \tag{20.13}$$

From (20.12) and (20.13), we deduce a further 270 relations defining the 1-secants of $\mathscr{E}$:

$$\begin{array}{llll} A_i+D_{jk}+D_{lmn}=0, & 60; & B_i+D_{jk}+D_{ijk}=0, & 60; \\ C_{ij}+D+D_{ij}=0, & 15; & C_{ij}+D_{kl}+D_{mn}=0, & 45; \\ C_{ij}+D_{ikl}+D_{imn}=0, & 90. & & \end{array} \tag{20.14}$$

Also from (20.12) and (20.13), we deduce a final 120 relations defining the 0-secants of $\mathscr{E}$:

$$\begin{array}{llll} D+D_{ijk}+D_{lmn}=0, & 10; & D_{ij}+D_{ikl}+D_{jkl}=0, & 20; \\ D_{ij}+D_{ik}+D_{jk}=0, & 90. & & \end{array} \tag{20.15}$$

This gives all 651 lines of $PG(5,2)$.

The intersections among the 27 lines of $\mathscr{F}_4$ determine all the conjugacy relations among the 63 points of $PG(5,2)$ with respect to $\mathscr{H}=\mathscr{H}_{5,4}$ and hence with respect to $\mathscr{E}$. If $\mathbf{P}(X)$ and $\mathbf{P}(Y)$ are conjugate with respect to $\mathscr{H}$, then $X\hat{Y}^*=0$. Here we shall write $X\hat{Y}^*=X\circ Y$.

From Theorem 20.2.5(i),

$$A_i\circ B_i=0, \qquad A_i\circ C_{ij}=0, \qquad B_i\circ C_{ij}=0, \qquad C_{ij}\circ C_{kl}=0. \tag{20.16}$$

Hence, from (20.13) and (20.16),

$$\begin{array}{llll} D\circ C_{ij}=0, & D_{ij}\circ A_k=0, & D_{ij}\circ B_k=0, & D_{ij}\circ C_{ij}=0, \\ D_{ij}\circ C_{kl}=0, & D_{ijk}\circ A_l=0, & D_{ijk}\circ B_i=0, & D_{ijk}\circ C_{il}=0. \end{array} \tag{20.17}$$

Finally, from (20.13) and (20.17),

$$\begin{aligned} D\circ D_{ij}=0, \quad & D_{ij}\circ D_{kl}=0, \quad D_{ij}\circ D_{ijk}=0, \\ D_{ij}\circ D_{klm}=0, \quad & D_{ijk}\circ D_{ilm}=0. \end{aligned} \tag{20.18}$$

We recall that, since the polarity of $\mathscr{E}$ is null, each point is self-conjugate. Now we list the different types of subspace of $PG(5, 2)$ with respect to $\mathscr{E}$. As in § 17.5, the notation $\Pi_r(\mathscr{V})$ means a Π_r whose intersection with $\mathscr{E}$ is $\mathscr{V}$.

Lemma 20.5.1: *With respect to $\mathscr{E}=\mathscr{E}_{5,2}$, there are two types of point and prime, and four types of line, plane and solid. Listed in polar pairs they are as follows:*

(i) $\Pi_0(\Pi_0)$, $\Pi_4(\Pi_0\mathscr{E}_3)$;
(ii) $\Pi_0(\Pi_{-1})$, $\Pi_4(\mathscr{P}_4)$;
(iii) $\Pi_1(\Pi_1)$, $\Pi_3(\Pi_1\mathscr{E}_1)$;
(iv) $\Pi_1(\mathscr{H}_1)$ $\Pi_3(\mathscr{E}_3)$;
(v) $\Pi_1(\Pi_0\mathscr{P}_0)$, $\Pi_3(\Pi_0\mathscr{P}_2)$;
(vi) $\Pi_1(\mathscr{E}_1)$, $\Pi_3(\mathscr{H}_3)$;
(vii) $\Pi_2(\mathscr{P}_2)$, $\Pi_2'(\mathscr{P}_2)$;
(viii) $\Pi_2(\Pi_0\mathscr{H}_1)$, $\Pi_2(\Pi_0\mathscr{E}_1)$;
(ix) $\Pi_2(\Pi_1\mathscr{P}_0)$, $\Pi_2(\Pi_1\mathscr{P}_0)$. □

The generators of $\mathscr{E}$ are lines. So, in (iii), the polar solid of a line on $\mathscr{E}$ meets $\mathscr{E}$ in that line, and in (ix), the polar plane π' of a plane π meeting $\mathscr{E}$ in a line l only is π itself. In (vii), the conics are distinct but have the same nucleus.

Examples of all the types of spaces are given below.

Theorem 20.5.2: (i) *A point P on $\mathscr{E}$ and a point Q off $\mathscr{E}$ are conjugate if and only if the line of $\mathscr{F}_4$ corresponding to P does not lie in the double-six corresponding to Q. So each point on $\mathscr{E}$ is conjugate to itself and ten other points on $\mathscr{E}$, and to* 20 *points off $\mathscr{E}$; each point off $\mathscr{E}$ is conjugate to itself and* 15 *other points off $\mathscr{E}$, and to* 15 *points on $\mathscr{E}$.*

(ii) *For any double-six, say $\mathscr{D}$, on $\mathscr{F}_4$, there are* 20 *others, namely the $\mathscr{D}_{ijk}$, having six lines in common with $\mathscr{D}$, and there are* 15 *others, namely the $\mathscr{D}_{ij}$, having four lines in common with it. The linear complexes associated to the* 15 $\mathscr{D}_{ij}$ *are all apolar to the linear complex associated to $\mathscr{D}$; also, only five of these* 16 *linear complexes are linearly independent. Similarly the* 15 *apolar complexes associated to $\mathscr{D}_{12}$ arise from*

$\mathscr{D}_1, \mathscr{D}_{34}, \mathscr{D}_{35}, \mathscr{D}_{36}, \mathscr{D}_{45}, \mathscr{D}_{46}, \mathscr{D}_{56}, \mathscr{D}_{123},$
$\mathscr{D}_{124}, \mathscr{D}_{125}, \mathscr{D}_{126}, \mathscr{D}_{345}, \mathscr{D}_{346}, \mathscr{D}_{356}, \mathscr{D}_{456},$

and those associated to $\mathscr{D}_{123}$ arise from

$\mathscr{D}_{12}, \mathscr{D}_{13}, \mathscr{D}_{23}, \mathscr{D}_{45}, \mathscr{D}_{46}, \mathscr{D}_{56}, \mathscr{D}_{145}, \mathscr{D}_{146},$
$\mathscr{D}_{156}, \mathscr{D}_{245}, \mathscr{D}_{246}, \mathscr{D}_{256}, \mathscr{D}_{345}, \mathscr{D}_{346}, \mathscr{D}_{356}.$

(iii) *The two halves of a double-six, say $\mathscr{D}$, on $\mathscr{F}_4$ are represented by two* *simplexes inscribed and circumscribed both to one another and to $\mathscr{E}$, and in perspective from D.*

(iv) *The* 120 *lines external to $\mathscr{E}$ form* 40 *skew triads such that the three polar solids of a triad meet $\mathscr{E}$ in three hyperbolic quadrics $\mathscr{H}_3$, which partition the* 27 *points of $\mathscr{E}$ into three sets of nine.*

Proof: (i) This is a direct consequence of the equations (20.17). Alternatively, since Q is off $\mathscr{E}$, its polar prime meets $\mathscr{E}$ in a $\mathscr{P}_4$ of which Q is the nucleus. Thus the joins of Q to the points of $\mathscr{P}_4$ are tangents to $\mathscr{P}_4$ and to $\mathscr{E}$. With $Q=D$, the 12 points A_i and B_i, representing lines of the double-six $\mathscr{D}$, are joined in pairs to D, by (20.13). Hence these are the 12 points of $\mathscr{E}$ not conjugate to D. The rest now follows.

We remark that, for P on $\mathscr{E}$, the tangent prime at P meets $\mathscr{E}$ in a cone $P\mathscr{E}_3$ consisting of 11 points.

(ii) This is merely a rephrasing of (i) with emphasis on (20.18) rather than (20.17).

(iii) Since any five of the lines a_i are linearly independent (Lemma 20.1.5), the points A_i are vertices of a simplex $\mathscr{A}$ and, similarly, the points B_i are vertices of a simplex $\mathscr{B}$. From (20.13), $\mathscr{A}$ and $\mathscr{B}$ are in perspective from D. Since $A_1 \circ B_i = 0$ for $i \neq 1$, from (20.16), and $A_1 \circ A_1 = 0$, so A_1 lies in the face of $\mathscr{B}$ opposite B_1 and this face is the tangent prime to $\mathscr{E}$ at A_1. So $\mathscr{A}$ is inscribed to $\mathscr{B}$ and $\mathscr{A}$ circumscribes $\mathscr{E}$. The rest follows by symmetry.

(iv) This property represents the 40 ways in which the triads of trihedral pairs give trichotomies of the 27 lines.

To the triad T_{123}, T_{456}, $T_{123,456}$ are associated nine double-sixes, by Theorem 20.2.5(iv), (v), and (vi)(n), which become the three lines

$$D+D_{123}+D_{456}=0, \qquad D_{12}+D_{13}+D_{23}=0, \qquad D_{45}+D_{46}+D_{56}=0. \tag{20.19}$$

The polar solid with respect to $\mathscr{E}$ of the first line meets $\mathscr{E}$ in an $\mathscr{H}_3$ consisting of nine points, which are

$$\begin{matrix} C_{14} & C_{25} & C_{36} \\ C_{26} & C_{34} & C_{15} \\ C_{35} & C_{16} & C_{24}. \end{matrix}$$

These represent the nine lines c_{ij} with the same indices which are in none of the double-sixes $\mathscr{D}$, $\mathscr{D}_{123}$, $\mathscr{D}_{456}$. The nine c_{ij} are just those occurring in $T_{123,456}$. Similarly associated to $D_{12}+D_{13}+D_{23}=0$ is T_{456} and to $D_{45}+D_{46}+D_{56}=0$ is T_{123}.

These are ten triads of lines like (20.19) and 30 like the following with the associated trihedral pairs:

$$D_{56}+D_{345}+D_{346}=0 \quad T_{12,34}$$
$$D_{12}+D_{156}+D_{256}=0 \quad T_{34,56}$$
$$D_{34}+D_{123}+D_{124}=0 \quad T_{56,12}. \quad \square$$

Before the compilation of the incidence table for $\mathscr{E}$, it may be useful to give examples of the different subspaces. Until now, $\mathscr{E}$ has been the set of 27 points

$$\{\mathbf{P}(x, y, z, z^2, y^2, x^2) \mid x^3+y^3+z^3=0\}$$

on $\mathscr{H}=\mathbf{V}_{5,4}(x_0x_5+x_1x_4+x_2x_3)$. However, for the examples below, we take, as in § 5.2,

$$\mathscr{E}=\mathbf{V}_{5,2}(x_0^2+x_0x_1+x_1^2+x_2x_3+x_4x_5).$$

The different types of subspaces, in polar pairs with respect to $\mathscr{E}$, are as follows:

$\Pi_0(\Pi_0)=\mathbf{U}_5$, $\quad\Pi_4(\Pi_0\mathscr{E}_3)=\mathbf{u}_4$;
$\Pi_0(\Pi_{-1})=\mathbf{U}_0$, $\quad\Pi_4(\mathscr{P}_4)=\mathbf{u}_1$;
$\Pi_1(\Pi_1)=\mathbf{U}_3\mathbf{U}_5$, $\quad\Pi_3(\Pi_1\mathscr{E}_1)=\mathbf{V}(x_2, x_4)$;
$\Pi_1(\mathscr{H}_1)=\mathbf{U}_4\mathbf{U}_5$, $\quad\Pi_3(\mathscr{E}_3)=\mathbf{V}(x_4, x_5)$;
$\Pi_1(\Pi_0\mathscr{P}_0)=\mathbf{V}(x_0, x_1, x_2, x_4+x_5)$, $\quad\Pi_3(\Pi_0\mathscr{P}_2)=\mathbf{V}(x_3, x_4+x_5)$;
$\Pi_1(\mathscr{E}_1)=\mathbf{U}_0\mathbf{U}_1$, $\quad\Pi_3(\mathscr{H}_3)=\mathbf{V}(x_0, x_1)$;
$\Pi_2(\mathscr{P}_2)=\mathbf{V}(x_0, x_1, x_2+x_3)$ $\quad\Pi_2'(\mathscr{P}_2)=\mathbf{V}(x_2+x_3, x_4, x_5)$;
$\Pi_2(\Pi_0\mathscr{H}_1)=\mathbf{V}(x_0, x_1, x_2)$, $\quad\Pi_2(\Pi_0\mathscr{E}_1)=\mathbf{V}(x_2, x_4, x_5)$;
$\Pi_2(\Pi_1\mathscr{P}_0)=\mathbf{V}(x_0, x_2, x_4)$, $\quad\Pi_2(\Pi_1\mathscr{P}_0)=\mathbf{V}(x_0, x_2, x_4)$.

The incidences for $\mathscr{E}$ are given in Table 20.6. A part of the table with $r<s$

	$\Pi_r(\mathscr{V})$	$\Pi_s(\mathscr{V}')$
$\Pi_r(\mathscr{V})$	a_{rr}	a_{rs}
$\Pi_s(\mathscr{V}')$	a_{sr}	a_{ss}

has the parameters of a tactical configuration. There are a_{rr} subspaces of type $\Pi_r(\mathscr{V})$ altogether and a_{ss} of type $\Pi_s(\mathscr{V}')$; also there are a_{rs} spaces of type $\Pi_r(\mathscr{V})$ in a space of type $\Pi_s(\mathscr{V}')$ and a_{sr} spaces of type $\Pi_s(\mathscr{V}')$ through a space of type $\Pi_r(\mathscr{V})$. Hence $a_{rr}a_{sr}=a_{rs}a_{ss}$. Some numbers are presented as a sum of two numbers when singular points of certain sections occur.

Table 20.6 *Incidences of Subspaces*

	$\Pi_0(\Pi_0)$	$\Pi_0(\Pi_{-1})$	$\Pi_1(\Pi_1)$	$\Pi_1(\mathscr{H}_1)$	$\Pi_1(\Pi_0\mathscr{P}_0)$	$\Pi_1(\mathscr{E}_1)$	$\Pi_2(\Pi_0\mathscr{H}_1)$	$\Pi_2(\mathscr{P}_2)$
$\Pi_0(\Pi_0)$	27	–	3	2	1	0	1+4	3
$\Pi_0(\Pi_{-1})$	–	36	0	1	2	3	2	4
$\Pi_1(\Pi_1)$	5	0	45	–	–	–	2	0
$\Pi_1(\mathscr{H}_1)$	16	6	–	216	–	–	4	3
$\Pi_1(\Pi_0\mathscr{P}_0)$	10	15	–	–	270	–	1	3
$\Pi_1(\mathscr{E}_1)$	0	10	–	–	–	120	0	1
$\Pi_2(\Pi_0\mathscr{H}_1)$	10+40	15	12	5	1	0	270	–
$\Pi_2(\mathscr{P}_2)$	80	80	0	10	8	6	–	720
$\Pi_2(\Pi_1\mathscr{P}_0)$	15	15	3	0	3	0	–	–
$\Pi_2(\Pi_0\mathscr{E}_1)$	10	45	0	0	3	9	–	–
$\Pi_3(\mathscr{H}_3)$	40	20	16	10	4	2	4	1
$\Pi_3(\Pi_0\mathscr{P}_2)$	10+60	60	18	15	4+12	9	3	3
$\Pi_3(\mathscr{E}_3)$	40	60	0	10	12	18	0	3
$\Pi_3(\Pi_1\mathscr{E}_1)$	5	15	1	0	3	6	0	0
$\Pi_4(\mathscr{P}_4)$	20	16	12	10	8	6	6	4
$\Pi_4(\Pi_0\mathscr{E}_3)$	1+10	15	3	5	1+6	9	1	3

The polarity gives the following symmetry to the table. Any element not in the central two rows or columns can be reflected in one diagonal and then the other to give the same element. The central two rows and columns have respectively vertical and horizontal symmetry (apart from the diagonal elements).

Apart from describing the geometry of $\mathscr{F}_4$, the relations (20.12)–(20.18) also give a good picture of the geometry of a non-singular cubic surface $\mathscr{F}$ over the complex numbers.

Over the complex numbers, an involution on a line is necessarily hyperbolic. The two double points of the involution on a line of $\mathscr{F}$ are *parabolic points*. The 36 quadrics with respect to which the double-sixes of $\mathscr{F}$ are self-polar are necessarily hyperbolic and are called *Schur quadrics*.

Let $A_i = 0$, $B_i = 0$, and $C_{ij} = 0$ be the tangential equations of the pairs of parabolic points on a_i, b_i, and c_{ij} respectively. Let $D = 0$, $D_{ij} = 0$, and $D_{ijk} = 0$ be the tangential equations of the Schur quadrics associated to D, D_{ij}, and D_{ijk} respectively. Then, with some of the plus signs changed to minus signs, the identities (20.12), (20.13), and (20.15) hold, but the identities (20.14) do not. This means that only six of the $27 + 36 = 63$ tangential quadrics are independent, whence we obtain a figure of 63 points in $PG(5, \mathbf{C})$. Over the complex numbers, the equations (20.18) represent the apolarity of the corresponding quadrics.

of PG(5, 2) with Respect to $\mathscr{E}_{5,2}$.

$\Pi_2(\Pi_1\mathscr{P}_0)$	$\Pi_2(\Pi_0\mathscr{E}_1)$	$\Pi_3(\mathscr{H}_3)$	$\Pi_3(\Pi_0\mathscr{P}_2)$	$\Pi_3(\mathscr{E}_3)$	$\Pi_3(\Pi_1\mathscr{E}_1)$	$\Pi_4(\mathscr{P}_4)$	$\Pi_4(\Pi_0\mathscr{E}_3)$
3	1	9	1+6	5	3	15	1+10
4	6	6	8	10	12	16	20
1	0	6	3	0	1	15	5
0	0	18	12	10	0	60	40
6	3	9	4+12	15	18	60	10+60
0	4	2	4	10	16	20	40
–	–	9	3	0	0	45	10
–	–	6	8	10	0	80	80
135	–	0	3	0	3	15	15
–	270	0	1	5	12	15	10+40
0	0	120	–	–	–	10	0
6	1	–	270	–	–	15	10
0	4	–	–	216	–	6	16
1	2	–	–	–	45	0	5
4	2	3	2	1	0	36	–
3	1+4	0	1	2	3	–	27

20.6 The classification of complex singular cubic surfaces by subsets of *PG*(5, 2)

The figure of 27 lines and 36 double-sixes of a cubic surface over any field can also be represented by the 63 points of *PG*(5, 2) if the equations (20.12)–(20.15) are interpreted in the following way. Any of the equations (20.12) is to mean that the three lines of $\mathscr{F}$ are coplanar; any of (20.13) that the two lines of $\mathscr{F}$ are opposite in the double-six; any of (20.14) that the four lines common to the two double-sixes have as their unique transversal among the 27 lines the line of $\mathscr{F}$ occurring in the equation; any of (20.15) that every two of the three double-sixes have six lines in common.

The group G of the 27 lines is isomorphic to the Galois group of a class of equations of degree 27 over the complex numbers. The 27 roots of such an equation can be written

$$\alpha_i = t_i + t/3, \qquad \beta_i = t_i - 2t/3, \qquad \gamma_{ij} = -t_i - t_j + t/3,$$

where $i, j \in \mathbf{N}_6$ and $t = \sum t_i$. Hence, comparable to (20.12),

$$\alpha_i + \beta_i + \gamma_{ij} = 0, \qquad \gamma_{ij} + \gamma_{kl} + \gamma_{mn} = 0. \tag{20.20}$$

The lines a_i, b_i, c_{ij} correspond respectively to α_i, β_i, γ_{ij}.

Theorem 20.6.1: *The different types of cubic surfaces with isolated singularities over the complex numbers correspond to the different subsets of* $PG(5,2)\backslash\mathscr{E}_{5,2}$ *that are linearly closed.*

Proof: Singular surfaces arise from coincidences of lines of a non-singular surface, that is, from coincidences among the roots α_i, β_i, γ_{ij}.

If two skew lines among the 27 coincide, the corresponding roots are equal and hence the equation has six double roots, whence every opposite pair of lines of a double-six coincide. So, in $PG(5,2)$, the two simplexes representing the two halves of a double-six coincide. □

Before giving the list of singular surfaces over the complex numbers, we describe the possible types of singular points on a cubic surface $\mathscr{F}$. Let $\mathbf{U}_3$ be the node on $\mathscr{F}=\mathbf{V}(F)$ so that

$$F=x_3f_2(x_0,x_1,x_2)+f_3(x_0,x_1,x_2).$$

(i) If f_2 is irreducible, the tangent quadric at $\mathbf{U}_3$ is $\mathbf{V}(f_2)$, a cone. Then $\mathbf{U}_3$ is a *conic node*, whose symbol is C^2, signifying that the existence of the node lowers by two the *class* of $\mathscr{F}$, which is the number of tangent planes of $\mathscr{F}$ through an arbitrary point of the space.

Table 20.7 *Complex Singular Cubic Surfaces*

$\mathscr{S}$	$\|\mathscr{S}\|$	$\dim\mathscr{S}$	roots	n	nodes	class	$\|G(\mathscr{S})\|$	$\|G(\mathscr{F})\|$
$\varnothing$	0	-1	27_1	27	–	12	51840	51840
Π_0	1	0	$6_2, 15_1$	21	C^2	10	1440	720
$\Pi_1\backslash\Pi_0$	2	1	$1_4, 8_2, 7_1$	16	$2C^2$	8	192	48
Π_1	3	1	$6_3, 9_1$	15	B^3	9	432	72
$\Pi_0\cup\Pi_0'\cup\Pi_0''$	3	2	$3_4, 6_2, 3_1$	12	$3C^2$	6	96	12
$\Pi_2\backslash\Pi_1$	4	2	$6_4, 3_1$	9	$4C^2$	4	384	24
$\Pi_2\backslash\mathscr{P}_2$	4	2	$1_6, 4_3, 3_2, 3_1$	11	B^3+C^2	7	72	6
$\Pi_2\backslash\Pi_0$	6	2	$1_6, 4_4, 5_1$	10	B^4	8	192	8
$\Pi_1\cup\Pi_0\cup\Pi_0'$	5	3	$2_6, 1_4, 2_3, 2_2, 1_1$	8	B^3+2C^2	5	48	2
$\Pi_3\backslash\mathscr{H}_3$	6	3	$1_9, 6_3$	7	$2B^3$	6	432	12
$(\Pi_2\backslash\Pi_0)\cup\Pi_0'$	7	3	$1_8, 1_6, 2_4, 2_2, 1_1$	7	B^4+C^2	6	96	2
$\Pi_3\backslash\Pi_0\mathscr{P}_2$	8	3	$2_8, 1_6, 1_4, 1_1$	5	B^4+2C^2	4	192	2
$\Pi_3\backslash\mathscr{E}_3$	10	3	$1_{10}, 3_5, 2_1$	6	B^5	7	240	2
$\Pi_3\backslash\Pi_1$	12	3	$3_8, 3_1$	6	U^6	6	1152	6
$\Pi_1\cup\Pi_1'\cup\Pi_0$	7	4	$1_9, 2_6, 2_3$	5	$2B^3+C^2$	4	144	2
$(\Pi_3\backslash\mathscr{E}_3)\cup\Pi_0$	11	4	$2_{10}, 1_5, 1_2$	4	B^5+C^2	5	240	1
$(\Pi_4\backslash\mathscr{P}_4)\backslash\Pi_0$	15	4	$1_{15}, 2_6$	3	B^6	6	1440	2
$\Pi_4\backslash\mathscr{P}_4$	16	4	$1_{15}, 1_{12}$	2	B^6+C^2	4	1440	1
$\Pi_4\backslash\Pi_0\mathscr{E}_3$	20	4	$1_{16}, 1_{10}, 1_1$	3	U^7	5	1920	1
$\Pi_1\cup\Pi_1'\cup\Pi_1''$	9	5	3_9	3	$3B^3$	3	1296	6
$\Pi_5\backslash\mathscr{E}_5$	36	5	1_{27}	1	U^8	4	51840	1

(ii) If $\mathbf{V}(f_2)$ is a pair of planes, then $\mathbf{U}_3$ is a *binode*. Take $f_2 = x_1x_2$ so that the tangent quadric at $\mathbf{U}_3$ is $\mathbf{u}_1 + \mathbf{u}_2$; write $l = \mathbf{u}_1 \cap \mathbf{u}_2$. Also, let $f_3 = \sum d_i x_i^3 + \sum_{i \neq j} d_{ij} x_i^2 x_j + d x_0 x_1 x_2$.

(a) If l does not lie on $\mathscr{F}$, then $\mathbf{U}_3$ is B^3: $d_0 \neq 0$.

(b) If l lies on $\mathscr{F}$, and $\mathbf{u}_1$ and $\mathbf{u}_2$ both meet $\mathscr{F}$ in l and two other lines, then $\mathbf{U}_3$ is B^4: $d_0 = 0$.

(c) If l lies on $\mathscr{F}$, and $\mathbf{u}_1$, say, meets $\mathscr{F}$ in l twice and one other line, then $\mathbf{U}_3$ is B^5: $d_0 = d_{02} = 0$.

(d) If l lies on $\mathscr{F}$, and $\mathbf{u}_1$, say, meets $\mathscr{F}$ in l counted three times, then $\mathbf{U}_3$ is B^6: $d_0 = d_{02} = d_{20} = 0$.

(iii) If $\mathbf{V}(f_2)$ is a repeated plane, then $\mathbf{U}_3$ is a *unode*. Take $f_2 = x_2^2$ so that the tangent quadric at $\mathbf{U}_3$ is $\mathbf{u}_2$. As $\mathbf{u}_2$ meets $\mathscr{F}$ in three distinct lines, two distinct lines, or in only one line, so $\mathbf{U}_3$ is accordingly U^6, U^7, or U^8.

In Table 20.7 we list the different subsets $\mathscr{S}$, the number of points $|\mathscr{S}|$ in $\mathscr{S}$, the dimension of the smallest subspace containing $\mathscr{S}$, the coincidences of the roots of the equation of degree 27 (where m_r means that each of m distinct roots is repeated r times), the number n of lines on $\mathscr{F}$, the nodes on $\mathscr{F}$, the class of $\mathscr{F}$, the order of the group $G(\mathscr{S})$ of projectivities fixing $\mathscr{S}$ (with $\mathscr{E}$ always fixed), and the order of the group $G(\mathscr{F})$ of permutations of the n lines of $\mathscr{F}$.

20.7 The representation in $PG(5, 2)$ of the 28 bitangents of a plane quartic curve

It was explained in § 20.4 how the 27 lines of a cubic surface $\mathscr{F}$ over a field not of even characteristic can be projected from a point P of $\mathscr{F}$ on none of its lines to the 28 bitangents of a plane non-singular quartic curve $\mathscr{C}^4$.

Let us take $\mathscr{C}^4$ in $PG(2, \mathbf{C})$. The bitangents can be denoted T_{ij}, $i, j \in \mathbf{N}_8$, $i < j$, so that their main properties are as follows:

(i) Given any pair of bitangents, five other pairs are uniquely determined so that any two of the six pairs have their eight points of contact with $\mathscr{C}^4$ on a conic. Such a set of six pairs is a *Steiner set*. The six points of contact of three bitangents from distinct pairs do not lie on a conic. There are 63 Steiner sets:

$$28 \text{ like } \{(T_{13}, T_{23}), (T_{14}, T_{24}), (T_{15}, T_{25}), (T_{16}, T_{26}), (T_{17}, T_{27}), (T_{18}, T_{28})\},$$
$$35 \text{ like } \{(T_{12}, T_{34}), (T_{13}, T_{24}), (T_{14}, T_{23}), (T_{56}, T_{78}), (T_{57}, T_{68}), (T_{58}, T_{67})\}.$$

(ii) There exist sets of 7 bitangents, called *Aronhold sets*, such that no three of the seven have their six points of contact on a conic. There are 288 Aronhold sets:

$$8 \text{ like } \{T_{12}, T_{13}, T_{14}, T_{15}, T_{16}, T_{17}, T_{18}\},$$
$$280 \text{ like } \{T_{12}, T_{13}, T_{14}, T_{15}, T_{67}, T_{68}, T_{78}\}.$$

(iii) There are 36 ways in which the bitangents can be arranged, each appearing twice, as the elements of an 8×8 symmetric array excluding the main diagonal such that each row and column is an Aronhold set. This is a *Hesse arrangement* and is, in fact, the basis for the notation T_{ij}. Thus the 288 Aronhold sets fall naturally by eights into 36 Hesse arrangements.

(iv) Any five bitangents of an Aronhold set determine a sixth bitangent such that their twelve points of contact lie on a cubic curve; the six lines also touch a conic. There are 1008 of these hexads of bitangents:

$$168 \text{ like } \{T_{12}, T_{13}, T_{14}, T_{15}, T_{16}, T_{78}\},$$

$$560 \text{ like } \{T_{12}, T_{13}, T_{14}, T_{56}, T_{57}, T_{58}\},$$

$$280 \text{ like } \{T_{12}, T_{13}, T_{23}, T_{45}, T_{46}, T_{56}\}.$$

(v) Given seven general lines in the plane, there exists a unique quartic curve with these as bitangents. The remaining 21 bitangents can be constructed linearly from the initial seven.

A null polarity in $PG(5,2)$ has been previously investigated with respect to $\mathcal{H}_{5,2}$ in § 17.5 and with respect to $\mathcal{E}_{5,2}$ in § 20.5.

We will briefly consider a null polarity $\mathfrak{A}$ in $PG(5,q)$ for arbitrary q. We recall that a subspace Π_r is self-conjugate if Π_r meets its polar $\Pi_r\mathfrak{A}$; see § 2.1 and § 5.3.

Lemma 20.7.1: *Let $\mathfrak{A}$ be a null polarity in $PG(5,q)$.*

(i) *If a line l is self-conjugate, then $l\subset l\mathfrak{A}$.*

(ii) *If a plane π and its polar plane $\pi\mathfrak{A}$ have a line in common, then $\pi=\pi\mathfrak{A}$.*

(iii) *Every plane is self-conjugate.*

Proof: (i) Suppose the point P is in $l\cap l\mathfrak{A}$, then any other point Q on l is conjugate to P. So P is in $Q\mathfrak{A}$ and Q is in $P\mathfrak{A}$. As $\mathfrak{A}$ is null, P is in $P\mathfrak{A}$ and Q is in $Q\mathfrak{A}$. So $l=PQ\subset P\mathfrak{A}\cap Q\mathfrak{A}=l\mathfrak{A}$.

(ii) Let $l\subset\pi\cap\pi\mathfrak{A}$. Take P_1 in $\pi\backslash l$, and let P_2 and P_3 be points of l. Then P_1, P_2, and P_3 are all mutually conjugate. So $\pi=\pi\mathfrak{A}$.

(iii) Let P_1 be a point of the plane π, which is not self-polar. Then $P_1\mathfrak{A}$ meets π in a line or π itself. Let l be a line in $P_1\mathfrak{A}\cap\pi$; then $l\subset l\mathfrak{A}$. Take P_2 in $\pi\backslash l$. Since π is not self-polar, $P_2\mathfrak{A}\cap\pi$ is a line l'. Let $P_3=l\cap l'$. Then P_3 is conjugate to P_2 and all points of l. So $P_3=\pi\cap\pi\mathfrak{A}$. □

Corollary: *With respect to $\mathfrak{A}$, the numbers of the different types of lines and planes are as follows, with the values for $q=2$ also given:*

(i) *self-polar lines*: $(q^3+1)(q^2+1)(q^2+q+1)=315$;

(ii) *other lines*: $q^4(q^4+q^2+1)=336$;

(iii) *self-polar planes*: $(q^3+1)(q^2+1)(q+1)=135$;

(iv) *other planes*: $q^2(q^3+1)(q^2+1)(q^2+q+1)=1260$.

Proof: From Theorem 3.1.1 and the lemma, the numbers are as follows:

(i) $\theta(5)\chi(0, 1; 4, q)/\theta(1) = \theta(5)\theta(3)/\theta(1) = (q^3+1)(q^2+1)(q^2+q+1)$;

(ii) $[\theta(5)\theta(4) - \theta(5)\theta(3)]/\theta(1) = q^4(q^4+q^2+1)$;

(iii) $[\theta(5)\theta(3)/\theta(1)]\theta(1)/\theta(2) = (q^3+1)(q^2+1)(q+1)$;

(iv) $\theta(5)\theta(4)\theta(3)/[\theta(2)\theta(1)] - \theta(5)\theta(3)/\theta(2)$

$$= q^2(q^3+1)(q^2+1)(q^2+q+1). \quad \square$$

Let us now consider the null polarity $\mathfrak{A}$ in $PG(5, 2)$, where $\mathbf{P}(X)$ and $\mathbf{P}(Y)$ are conjugate if $\hat{X}Y^* = X \circ Y = 0$; that is, as in Theorem 5.3.2,

$$x_0y_5 + x_5y_0 + x_1y_4 + x_4y_1 + x_2y_3 + x_3y_2 = 0.$$

The quadrics $\mathscr{F} = \mathbf{V}(F)$ having $\mathfrak{A}$ as their polarity are given by

$$F = \sum c_i x_i^2 + x_0x_5 + x_1x_4 + x_2x_3.$$

Let such a quadric be denoted by $\mathscr{F}_C = \mathbf{V}(F_C)$, where $C = (c_0, c_1, c_2, c_3, c_4, c_5)$. Let $\mathscr{W} = \{\mathscr{F}_C\}$.

Lemma 20.7.2: (i) *There are* 64 *quadrics with the same null polarity. The* 63 *pencils formed by any one with all the others have the* 63 *primes as their other members.*

(ii) $\mathscr{F}_C$ *is an* $\mathscr{H}_{5,2}$ *if* $c_0c_5 + c_1c_4 + c_2c_3 = 0$ *and an* $\mathscr{E}_{5,2}$ *if* $c_0c_5 + c_1c_4 + c_2c_3 = 1$. *Hence* 36 *of the quadrics are hyperbolic and* 28 *elliptic.*

(iii) *Two quadrics in* $\mathscr{W}$ *are of the same or different type as they meet in a tangent or non-tangent prime section of either.*

Proof: (i) As each c_i can be 0 or 1, there are $2^6 = 64$ quadrics $\mathscr{F}_C$ in $\mathscr{W}$. As

$$F_C = \sum c_i x_i^2 + x_0x_5 + x_1x_4 + x_2x_3 = \sum c_i x_i^2 + F_0$$
$$= \sum c_i^2 x_i^2 + F_0 = \left(\sum c_i x_i\right)^2 + F_0 = CX^* + F_0,$$

so $F_C + F_{C'} = (C + C')X^*$.

Hence the pencil of quadrics containing $\mathscr{F}_C$ and $\mathscr{F}_{C'}$ contains $\boldsymbol{\pi}(C + C')$. As C' runs through all elements of $AG(6, 2)\backslash\{C\}$, so $\boldsymbol{\pi}(C + C')$ runs through all primes of $PG(5, 2)$.

(ii) $\mathscr{F}_0$ is $\mathscr{H}_{5,2}$ as in § 17.5 and so contains 35 points. Now take $C \neq 0$. Then the points of $\mathscr{F}_C$ are, from (i), on both the prime $\boldsymbol{\pi}(C)$ and $\mathscr{F}_0$ or on neither. If $c_0c_5 + c_1c_4 + c_2c_3 = 0$, then $\boldsymbol{\pi}(C)$ is a tangent prime to $\mathscr{F}_0$, whence $\boldsymbol{\pi}(C)$ meets $\mathscr{F}_0$ in $\Pi_0\mathscr{H}_{3,2}$, which comprises 19 points. This leaves 12 points of $\boldsymbol{\pi}(C)$ off $\mathscr{F}_0$. So the number of points on neither $\mathscr{F}_0$ nor $\boldsymbol{\pi}(C)$ is $28 - 12 = 16$. Hence $\mathscr{F}_C$ contains $19 + 16 = 35$ points and is hyperbolic.

If $c_0c_5 + c_1c_4 + c_2c_3 = 1$, then $\boldsymbol{\pi}(C)$ is not tangent to $\mathscr{F}_0$ and meets it in $\mathscr{P}_{4,2}$ comprising 15 points. So there are 16 points of $\boldsymbol{\pi}(C)$ off $\mathscr{F}_0$, whence the points on neither $\boldsymbol{\pi}(C)$ nor $\mathscr{F}_0$ number $28-16=12$. Therefore $\mathscr{F}_C$ contains $15+12=27$ points and is elliptic.

(iii) The third member of the pencil containing $\mathscr{F}_C$ and $\mathscr{F}_{C'}$ is $\boldsymbol{\pi}(C+C')$, from (i). This prime is the polar of $\mathbf{P}(\hat{C}+\hat{C}')$ with respect to $\mathfrak{A}$. This point lies on $\mathscr{F}_C$, say, if

$$C(\hat{C}+\hat{C}')^*+F_0(\hat{C}+\hat{C}')=0;$$

that is,

$$c_0c_5+c_1c_4+c_2c_3=c_0'c_5'+c_1'c_4'+c_2'c_3'. \quad \square$$

Corollary: *$PGSp(6,2)$ has 28 conjugate subgroups isomorphic to $PGO_-(6,2)$ and 36 conjugate subgroups isomorphic to $PGO_+(6,2)$. Hence*

$$|PGSp(6,2)|=28\cdot 51\,840=36\cdot 8!=1\,451\,520.$$

Proof: From Lemma 19.1.6 and Theorem 19.2.2, Corollary 1, $|PGO_-(6,2)|=51\,840$; from Theorem 17.5.5, $|PGO_+(6,2)|=8!$. $\square$

It is the 28 elliptic quadrics in $\mathscr{W}$ that correspond to the 28 bitangents and we can now describe the properties in $PG(5,2)$ that correspond to (i)–(v) of pages 223–4 in $PG(2,\mathbf{C})$.

Lemma 20.7.3: *The 63 Steiner sets correspond to the 63 points of $PG(5,2)$ as follows. Any pair of the 28 elliptic quadrics in $\mathscr{W}$ determine five other pairs such that all 12 quadrics have a common point. Any two pairs of such a set of six have one line, self-polar under $\mathfrak{A}$, in common. Three quadrics from distinct pairs of a set of six have only their determining point in common.*

Proof: Let $\mathscr{F}_C$ and $\mathscr{F}_{C'}$ be elliptic quadrics in $\mathscr{W}$. From Lemma 20.7.1(iii), $\mathscr{F}_C$ and $\mathscr{F}_{C'}$ meet where the tangent prime $\boldsymbol{\pi}(C+C')$ meets either; that is, $\mathscr{F}_C$ and $\mathscr{F}_{C'}$ meet in $\Pi_0\mathscr{E}_3$, by § 20.5, where $\Pi_0=\mathbf{P}(\hat{C}+\hat{C}')$. So $\mathscr{F}_C$ and $\mathscr{F}_{C'}$ meet in five lines through Π_0. Neither $\mathbf{P}(\hat{C})$ nor $\mathbf{P}(\hat{C}')$ is on $\mathscr{F}_0$, by Lemma 20.7.1(ii). However Π_0 is determined by six pairs $\{\mathbf{P}(\hat{C}),\mathbf{P}(\hat{C}')\}$ since, from Table 17.3, there are six external lines through a point off $\mathscr{F}_0$ and six tangent lines through a point on $\mathscr{F}_0$. Thus Π_0 arises from six pairs of elliptic quadrics in $\mathscr{W}$, each of which determines the others.

There are 15 lines self-polar under $\mathfrak{A}$ through any point of any of the elliptic quadrics; these are the generators and tangents of $\mathscr{E}$ through the point, by § 20.5. Thus the 15 lines through Π_0 fall into six sets of five, each set of five being the intersection of one of the pairs of elliptic quadrics: each line is common to two pairs of quadrics. As there are 315 self-conjugate lines altogether, by Lemma 20.7.1, Corollary, each line is contained in $28\cdot 45/315=4$ of the 28 elliptic quadrics. So three quadrics from distinct pairs have only Π_0 in common. $\square$

Lemma 20.7.3 corresponds to property (i) of the bitangents. To show how properties (ii) and (iii) of the bitangents occur, let us focus on one of the 36 hyperbolic quadrics in $\mathcal{W}$, say $\mathcal{F}_0$. For any elliptic quadric $\mathcal{F}_C$, the third member of the pencil through $\mathcal{F}_0$ and $\mathcal{F}_C$ is $\boldsymbol{\pi}(C)$, whose pole with respect to $\mathcal{F}_0$ is $\mathbf{P}(\hat{C})$, which is off $\mathcal{F}_0$. So let us consider the 28 points $\mathbf{P}(\hat{C})$, instead of the 28 elliptic quadrics $\mathcal{F}_C$, in relation to $\mathcal{F}_0$. This was exactly the subject of § 17.5. By Lemma 17.5.4, the 28 points form eight heptads any two of which have a point in common, such that the points of a heptad are mutually non-conjugate with respect to $\mathcal{F}_0$ and such that no three points of a heptad are collinear. So, if the eight heptads H_i, $i \in \mathbf{N}_8$, meet in pairs in the points Q_{ij}, then this corresponds to a Hesse arrangement. A heptad corresponds to an Aronhold set. As there are 36 hyperbolic quadrics in $\mathcal{W}$, there are 36 Hesse arrangements and $8 \cdot 36 = 288$ Aronhold sets, as required.

Lemma 20.7.4: *Any five points of a heptad span a non-tangent prime π of $\mathcal{F}_0$. The nucleus P of $\pi \cap \mathcal{F}_0$ is the only point of $\pi \backslash \mathcal{F}_0$ not on the join of any pair of the five points.*

Proof: In the notation of § 17.5, take the points Q_{12}, Q_{13}, Q_{14}, Q_{15}, Q_{16} of the heptad H_1. Then, from the lines given in deduction (vii) following Lemma 17.5.4, the other ten points on the joins of pairs of the five points are Q_{ij}, $i, j = 2, \ldots, 6$. The 15 points of $\mathcal{F}_0$ on joins of pairs of these 15 points Q_{ij} are P_{ijkl}, $i, j, k, l \in \mathbf{N}_6$. Then Q_{78} is the remaining point of the prime π spanning the initial 5 points. So π is not tangent to F_0 and Q_{78} is the nucleus of $\pi \cap \mathcal{F}_0$. □

This lemma corresponds to property (iv) of the bitangents. The hexad $\{Q_{12}, Q_{13}, Q_{14}, Q_{15}, Q_{16}, Q_{78}\}$ is not symmetrical, since only one of the six subsets of five belong to a heptad. As there are 28 non-tangent primes to $\mathcal{F}_0$, the number of corresponding sets of six bitangents is $6 \cdot 28 \cdot 36/6 = 1008$, as required.

Lemma 20.7.5: *Any seven points in $PG(5, 2)$, no six of which lie in a prime, together with the 21 points on joins of pairs of them, are the 28 points off an $\mathcal{H}_{5,2}$.*

Proof: Take the seven points as $\mathbf{U}_0$, $\mathbf{U}_1$, $\mathbf{U}_2$, $\mathbf{U}_3$, $\mathbf{U}_4$, $\mathbf{U}_5$, $\mathbf{U}$. Then, if P_{ij} lies on the join of $\mathbf{U}_i$ and $\mathbf{U}_j$, and P_i lies on the join of $\mathbf{U}_i$ and $\mathbf{U}$, we have, for example,

$$P_{01} = \mathbf{P}(1, 1, 0, 0, 0, 0), \qquad P_0 = \mathbf{P}(0, 1, 1, 1, 1, 1).$$

These 28 points are all the points of $PG(5, 2)$ not on $\mathcal{F} = \mathbf{V}(\sum' x_i x_j)$. For, none of the 28 points lie on $\mathcal{F}$ and the 35 points with three or four 1's in their coordinates do lie on $\mathcal{F}$. □

This lemma corresponds to property (v) of the bitangents.

Theorem 20.7.6: *The group of permutations of the* 28 *bitangents of a quartic curve is isomorphic to* $PGSp(6, 2)$. □

20.8 Notes and references

§§ 20.1–3. These sections are based on Hirschfeld (1964, 1967a). For classical proofs of the double-six theorem and other properties of cubic surfaces, see Baker (1921, Volumes 3, 4). A computational rather than geometric approach has been chosen to be sure of the circumstances in fields of low and of even order.

For deeper results over finite fields, see Manin (1974), Swinnerton-Dyer (1967, 1971, 1981).

A non-singular cubic surface $\mathscr{F}$ over $GF(q)$ with q odd has 27, 15, 9, 7, 5, 3, 2, 1, or 0 lines (Segre 1949, 1951); see Rosati (1957, 1958) for more details. Dickson (1915g) shows that, if $q = 2$, then $\mathscr{F}$ can have 15, 9, 5, 3, 2, 1, or 0 lines; see also Campbell (1933b). Dickson (1915f) gives a projective classification of all non-degenerate cubic surfaces in $PG(3, 2)$. Rosati (1956) gives a formula for the number of points on various types of cubic surface.

Dickson (1915g) shows the existence of $\mathscr{F}_4$ but not its uniqueness. See also Frame (1938), Coxeter (1940, 1959, 1974), Edge (1959, 1965), Biscarini (1976). For the group of the 27 lines, see also Dickson (1905), who lists all the subgroups, Frame (1951), Edge (1955d, 1959). For a study of the collineation groups that can occur over the complex numbers and the connection with Eckardt points, see Segre (1942, Chapters 13, 14). Theorem 20.3.2, apart from the identification of the pseudolines of the second kind as a linear complex, is due to Bose (1971), who gives a full account. For the geometry of $PG(3, 3)$ implicit in the corollary, see Coble (1908). Theorem 20.3.10 follows Hirschfeld (1981, 1982); see also Manin (1974, Chapter 4).

§ 20.5. See also Edge (1959). Similar relations to (20.15) hold for the Schur quadrics of the 36 double-sixes of a cubic surface over the complex field; see Room (1932).

§ 20.6. This section follows Hubaut (1968). Tallini (1959) characterizes some singular cubic surfaces over $GF(q)$.

§ 20.7. This section is based on Coble (1912).

Other papers relevant to this chapter are Carlitz (1957), Small (1982), de Finis and de Resmini (1983).

21

TWISTED CUBICS AND k-ARCS

21.1 Elementary properties of the twisted cubic

Contrary to the consideration of other varieties in Part IV, we consider the generalization of the twisted cubic in an arbitrary number of dimensions before specializing to the case of 3 dimensions.

A *rational curve* $\mathscr{C}_n$ of order n in $PG(r, q)$ is the set of points

$$\{P(t) = \mathbf{P}(g_0(t_0, t_1), \ldots, g_r(t_0, t_1)) \mid t_0, t_1 \in \gamma\}$$

where each g_i is a binary form of degree n and a highest common factor of $g_0, g_1, \ldots, g_r$ is 1. The curve $\mathscr{C}_n$ may also be written

$$\{P(t) = \mathbf{P}(f_0(t), f_1(t), \ldots, f_r(t)) \mid t \in \gamma^+\}$$

where $f_i(t) = g_i(1, t)$. As the g_i have no non-trivial common factor, so at least one f_i has degree n. Also $\mathscr{C}_n$ is *normal* if it is not the projection of a rational curve $\mathscr{C}'_n$ in $PG(r+1, q)$, where $\mathscr{C}'_n$ is not contained in any subspace Π_r.

Theorem 21.1.1: *Let $\mathscr{C}_n$ be a rational normal curve in $PG(r, q)$ not contained in a subspace Π_s with $s < r$. Then*

(i) $q \geq r$;

(ii) $n = r$;

(iii) *$\mathscr{C}_r$ is projectively equivalent to*

$$\{P(t) = \mathbf{P}(t^r, t^{r-1}, \ldots, t, 1) \mid t \in \gamma^+\};$$

(iv) *$\mathscr{C}_r$ consists of $q+1$ points no $r+1$ of which lie in a Π_{r-1};*

(v) *if $q \geq r+2$, then there is a unique $\mathscr{C}_r$ through any $r+3$ points of $PG(r, q)$ no $r+1$ of which lie in a Π_{r-1}.*

Proof: (i) If $q < r$, then all the points of $\mathscr{C}_n$ are contained in a Π_{r-1}, contrary to the hypothesis.

(ii) If $n < r$, the $r+1$ polynomials f_i are linearly dependent and so $\mathscr{C}_n$ lies in a Π_s with $s < r$. If $n > r$, then there exists a polynomial f_{r+1} of degree n independent of $f_0, f_1, \ldots, f_r$. So $\mathscr{C}_n$ is the projection of

$$\mathscr{C}'_n = \{\mathbf{P}(f_0(t), f_1(t), \ldots, f_r(t), f_{r+1}(t)) \mid t \in \gamma^+\}$$

from $\mathbf{U}_{r+1}$ onto $\mathbf{u}_{r+1}$ in $PG(r+1, q)$. Since the $r+2$ polynomials $f_0, f_1, \ldots, f_{r+1}$ are independent, $\mathscr{C}'_n$ does not lie in a Π_r. Hence $\mathscr{C}_n$ is not normal. Thus $n = r$.

(iii) Since $f_0, \ldots, f_r$ are linearly independent, they form a basis for the set of polynomials of degree at most r. Another such basis is $\{t^r, t^{r-1}, \ldots, t, 1\}$. So there is a projectivity giving the required form.

(iv) Since $f_0, \ldots, f_r$ are independent, they have no common factor. A prime $\boldsymbol{\pi}(A)$ meets $\mathscr{C}_r$ in those points $P(t)$ with $\sum a_i f_i(t) = 0$. So $\boldsymbol{\pi}(A)$ meets $\mathscr{C}_r$ in at most r points, corresponding to the solutions of this equation.

(v) For $r = 2$, the result says that for $q \geq 4$ there is a unique conic through five points of $PG(2, q)$; this was shown in § 7.2.

Let $r > 2$ and let $\mathscr{K} = \{\mathbf{U}_0, \mathbf{U}_1, \ldots, \mathbf{U}_r, \mathbf{U}, P(A)\}$ be an $(r+3)$-arc in $PG(r, q)$; that is, no $r+1$ points of $\mathscr{K}$ lie in a Π_{r-1}. Hence, with $A = (a_0, a_1, \ldots, a_r)$, we have $a_i \neq 0$ all i, and $a_i \neq a_j$ for $i \neq j$. Suppose there is a $\mathscr{C}_r$ containing $\mathscr{K}$ for which the points of $\mathscr{K}$ in the given order have respective parameters.

$$\lambda_0, \lambda_1, \ldots, \lambda_r, \lambda, \mu,$$

one of which may be ∞. Since $\mathbf{U}_0, \ldots, \mathbf{U}_r$ are on $\mathscr{C}_r$, so the curve has the form

$$\{\mathbf{P}(c_0/(t - \lambda_0), \ldots, c_r/(t - \lambda_r)) \mid t \in \gamma^+\}.$$

As $t = \lambda$ gives $\mathbf{U}$, so $c_i = c(\lambda - \lambda_i)$; we may take $c = 1$. As $t = \mu$ gives $\mathbf{P}(A)$, so $a_i = d(\lambda - \lambda_i)/(\mu - \lambda_i)$; again let $d = 1$. Then

$$\lambda_i = (\lambda - \mu a_i)/(1 - a_i).$$

So

$$x_i = c_i/(t - \lambda_i) = a_i(\mu - \lambda)/[(1 - a_i)t - \lambda + \mu a_i].$$

With $t = (\lambda - \mu)s + \mu$, we have

$$x_i = a_i/[(a_i - 1)s + 1].$$

Thus

$$\mathscr{C}_r = \left\{\mathbf{P}\left(\frac{a_0}{(a_0 - 1)t + 1}, \ldots, \frac{a_r}{(a_r - 1)t + 1}\right) \,\middle|\, t \in \gamma^+\right\},$$

In particular $\mathbf{U}_i$ is given by $t = 1/(1 - a_i)$, $\mathbf{U}$ by $t = 1$ and $\mathbf{P}(A)$ by $t = 0$. □

A rational normal curve of order 2 is a conic. A rational normal curve of order 3 is a *twisted cubic*. So any twisted cubic $\mathscr{C}_3$ is in $PG(3, q)$ and may be written

$$\mathscr{C}_3 = \{P(t) = \mathbf{P}(f_0(t), f_1(t), f_2(t), f_3(t)) \mid t \in \gamma^+\}. \tag{21.1}$$

From the theorem, $\mathscr{C}_3$ is not contained in a plane for $q \geq 3$. When $q = 2$, the curve $\mathscr{C}_3$ is a conic. However, unless specifically excluded, this case will be included in the treatment below.

The dual of a twisted cubic is a *cubic envelope* Γ. In general,

$$\Gamma_3 = \{\pi(t) = \boldsymbol{\pi}(g_0(t), g_1(t), g_2(t), g_3(t)) \mid t \in \gamma^+\}, \tag{21.2}$$

where $g_0, \ldots, g_3$ are linearly independent polynomials of degree at most 3.

Let γ' be a quadratic extension of γ. Then, with $P(t)$ as in (21.1) and $\pi(t)$ as in (21.2),

$$\mathscr{C}'_3 = \{P(t) \mid t \in \gamma'^+\},$$
$$\Gamma'_3 = \{\pi(t) \mid t \in \gamma'^+\}.$$

In keeping with the terminology of § 11.4, the points $P(t)$ of $\mathscr{C}'_3$ are called *(2-)complex* points of $\mathscr{C}_3$ and the planes $\pi(t)$ of Γ'_3 the *(2-)complex* planes of Γ_3. For emphasis, the points of $\mathscr{C}_3$ and the planes of Γ_3 are sometimes called *real* points of $\mathscr{C}_3$ and *real* planes of Γ_3 respectively. The points $P(t)$ and $P(t')$ are *(2-)complex conjugate* points if t and t' are conjugate over γ; under the same conditions $\pi(t)$ and $\pi(t')$ are *(2-)complex conjugate* planes of Γ_3.

From Theorem 21.1.1, we may take a twisted cubic in the canonical form

$$\mathscr{C} = \{P(t) = \mathbf{P}(t^3, t^2, t, 1) \mid t \in \gamma^+\}: \tag{21.3}$$

$t = \infty$ gives the point $\mathbf{U}_0$. By the theorem, $\mathscr{C}$ comprises $q+1$ points no four of which are coplanar and hence no three of which are collinear.

The plane joining the points $P(t_1)$, $P(t_2)$, $P(t_3)$ is

$$\boldsymbol{\pi}(1, -(t_1+t_2+t_3), t_1t_2+t_1t_3+t_2t_3, -t_1t_2t_3).$$

So, at each point $P(t)$ of $\mathscr{C}$, there is an *osculating plane*

$$\pi(t) = \boldsymbol{\pi}(1, -3t, 3t^2, -t^3) \tag{21.4}$$

meeting $\mathscr{C}$ only at $P(t)$. The osculating planes form the *osculating developable* Γ to $\mathscr{C}$.

A *chord* of $\mathscr{C}$ is a line of $PG(3, q)$ joining either a pair of real points of $\mathscr{C}$, possibly coincident, or a pair of complex conjugate points. Let $l(t_1, t_2) = P(t_1)P(t_2)$; then

$$\begin{aligned} l(t_1, t_2) &= \mathbf{l}(t_1^2t_2^2, t_1t_2(t_1+t_2), t_1^2+t_1t_2+t_2^2, t_1t_2, -(t_1+t_2), 1) \\ &= \mathbf{l}(\alpha_2^2, \alpha_1\alpha_2, \alpha_1^2-\alpha_2, \alpha_2, -\alpha_1, 1), \end{aligned} \tag{21.5}$$

where $\alpha_1 = t_1+t_2$, $\alpha_2 = t_1t_2$. A line is a chord of $\mathscr{C}$ if and only if it has the form (21.5).

There are three types of chord according as $x^2-\alpha_1x+\alpha_2$ has 2, 1, or 0 roots in γ. If $x^2-\alpha_1x+\alpha_2$ has two roots in γ, that is $P(t_1)$ and $P(t_2)$ are real points of $\mathscr{C}$, then $l(t_1, t_2)$ is a *real chord* of $\mathscr{C}$. If $x^2-\alpha_1x+\alpha_2$ has only one root in γ, that is $P(t_1)$ and $P(t_2)$ are coincident, then $l(t_1, t_2)$ is a *tangent.* If $x^2-\alpha_1 x+\alpha_2$ has no roots in γ, that is $P(t_1)$ and $P(t_2)$ are complex conjugate points of $\mathscr{C}$, then $l(t_1, t_2)$ is an *imaginary chord* of $\mathscr{C}$. So a real chord is a bisecant to $\mathscr{C}$, a tangent is a unisecant, and an imaginary chord is an external line (0-secant).

From (21.5), the tangent to $\mathscr{C}$ at $P(t)$ is

$$l(t)=l(t, t)=\mathbf{l}(t^4, 2t^3, 3t^2, t^2, -2t, 1). \tag{21.6}$$

For $p\neq 3$, dual to the chords of $\mathscr{C}$ are the axes of Γ. An *axis* of Γ is a line of $PG(3, q)$ which is the meet of a pair of real planes of Γ, possibly coincident, or of a pair of complex conjugate planes of Γ. Let $l'(v_1, v_2)=\pi(v_1)\cap\pi(v_2)$. Then

$$\left.\begin{aligned} l'(v_1, v_2)&=\mathbf{l}(v_1^2v_2^2, v_1v_2(v_1+v_2), 3v_1v_2, (v_1^2+v_1v_2+v_2^2)/3, -(v_1+v_2), 1)\\ &=\mathbf{l}(\beta_2^2, \beta_1\beta_2, 3\beta_2, (\beta_1^2-\beta_2)/3, -\beta_1, 1),\end{aligned}\right\} \tag{21.7}$$

where $\beta_1=v_1+v_2$, $\beta_2=v_1v_2$. We call $l'(v_1, v_2)$ a *real* axis, a *generator* or an *imaginary* axis of Γ as $x^2-\beta_1 x+\beta_2$ has 2, 1, or 0 roots in γ. The generator in $\pi(t)$ is

$$l'(t, t)=\mathbf{l}(t^4, 2t^3, 3t^2, t^2, -2t, 1)=l(t),$$

the tangent to $\mathscr{C}$ at $P(t)$.

Theorem 21.1.2: *Let $\mathscr{C}$ be the twisted cubic* (21.3) *and Γ its osculating developable in $PG(3, q)$, $q=p^h$.*

(i) *For $p=3$, Γ is a pencil of planes; for $p\neq 3$, Γ is a cubic developable.*

(ii) *The tangents to $\mathscr{C}$ lie in the linear complex $\mathscr{A}=\boldsymbol{\lambda}(l_{03}-3l_{12})$, which is special if and only if $p=3$.*

(iii) *For $p=2$, the tangents form a regulus lying on $\mathbf{V}(x_0x_3+x_1x_2)$; for $p\neq 2$, no four of the tangents lie in a regulus.*

(iv) *For $p\neq 3$, the null polarity $\mathfrak{A}$ defined by $\mathscr{A}$ is given by*

$$\mathbf{P}(a_0, a_1, a_2, a_3)\leftrightarrow\boldsymbol{\pi}(a_3, -3a_2, 3a_1, -a_0).$$

It interchanges $\mathscr{C}$ and Γ, and their corresponding chords and axes.

Proof: (i) This follows from (21.4).

(ii) This follows from (21.6).

(iii) Again from (21.6), when $p=2$,

$$l(t)=\mathbf{l}(t^4, 0, t^2, t^2, 0, 1).$$

So the tangents lie in the regulus whose lines satisfy $l_{02}=l_{31}=l_{03}+l_{12}=0$, whence the first part. If, for $p\neq 2$, the four distinct tangents $l(t_1)$, $l(t_2)$, $l(t_3)$, and $l(t_4)$ were in a regulus, their coordinate vectors would form a matrix of rank 3. So, from (21.6), the matrix with rows

$$(2t_i^3, t_i^2, -2t_i, 1)$$

would have determinant zero. This can only occur if two of the t_i are equal.

(iv) $\mathbf{P}(X)\mathbf{P}(Y)$ is in $\mathscr{A}$, or equivalently $\mathbf{P}(X)$ and $\mathbf{P}(Y)$ are conjugate with respect to $\mathfrak{A}$, if

$$(x_0y_3 - x_3y_0) - 3(x_1y_2 - x_2y_1) = 0;$$

that is,

$$x_0y_3 - 3x_1y_2 + 3x_2y_1 - x_3y_0 = 0. \quad \square$$

For $q = 2, 3, 4$, special care must be taken, as a twisted cubic does not contain the six points necessary to define the cubic as in Theorem 21.1.1(v). In other words, for $q > 4$, any projectivity of the cubic is given by a permutation of the parameters of the points which does not change the degree of the curve.

Let G_q be the group of projectivities in $PG(3, q)$ fixing $\mathscr{C}$.

Lemma 21.1.3: (i) *For* $q \geq 5$, $G_q \cong PGL(2, q)$.

(ii) $G_4 \cong \mathbf{S}_5 \cong P\Gamma L(2, 4)$; $|G_4| = 2\,|PGL(2, 4)| = 120$.

(iii) $G_3 \cong \mathbf{S}_4\mathbf{Z}_2^3$; $|G_3| = 8\,|PGL(2, 3)| = 192$.

(iv) $G_2 \cong \mathbf{S}_3\mathbf{Z}_2^3$; $|G_2| = 8\,|PGL(2, 2)| = 48$.

Proof: (i) By Theorem 21.1.1(v), the cubic through six points is determined as

$$\{\mathbf{P}(f_0(t), f_1(t), f_2(t), f_3(t)) \mid t \in \gamma^+\}$$

where the f_i are determined up to a permutation of the parameters of the points. Since $q \geq 5$, the only permutations of γ^+ permissible are those given by

$$t \mapsto \frac{at+b}{ct+d}, \qquad ad - bc \neq 0.$$

If there were higher degree terms, we would no longer have a cubic. Conversely, with $\mathscr{C}$ as in (21.3), $t \mapsto (at+b)/(ct+d)$ induces

$$(t^3, t^2, t, 1) \mapsto (t^3, t^2, t, 1)\begin{pmatrix} a^3 & a^2c & ac^2 & c^3 \\ 3a^2b & a^2d + 2abc & bc^2 + 2acd & 3c^2d \\ 3ab^2 & b^2c + 2abd & ad^2 + 2bcd & 3cd^2 \\ b^3 & b^2d & bd^2 & d^3 \end{pmatrix}.$$

So $G_q \cong PGL(2, q)$.

(ii) Since $\mathscr{C}$ consists of five points no four coplanar, any permutation of the points is achieved by a unique projectivity; see § 2.1. Hence $G_4 \cong \mathbf{S}_5 \cong P\Gamma L(2,4)$, see § 6.4.

(iii), (iv) These follow similarly. □

Corollary 1: *G_q acts triply transitively on $\mathscr{C}$.* □

Corollary 2: *If $\mathfrak{T}_P : \mathscr{C} \to \pi$ is the projection of $\mathscr{C}$ from P on $\mathscr{C}$ to a plane π, so that $P\mathfrak{T}_P = l \cap \pi$ where l is the tangent to $\mathscr{C}$ at P, then $\mathscr{C}\mathfrak{T}_P$ is a conic.*

Proof: By Corollary 1, P may be taken as any point of $\mathscr{C}$. So let $P = P(\infty) = \mathbf{U}_0$ and let $\pi = \mathbf{u}_0$. Then $l = l(\infty) = \mathbf{l}(1,0,0,0,0,0) = \mathbf{U}_0\mathbf{U}_1$. So $l \cap \mathbf{u}_0 = \mathbf{U}_1$. So

$$\mathscr{C}\mathfrak{T}_P = \{\mathbf{P}(0, t^2, t, 1) \mid t \in \gamma^+\} = \mathbf{V}(x_0, x_2^2 - x_1x_3). \quad \square$$

Corollary 3: *The number of twisted cubics in $PG(3, q)$ is*

$$q^5(q^4-1)(q^3-1) \quad \text{for} \quad q \geqslant 5,$$
$$q^5(q^4-1)(q^3-1)/2 \quad \text{for} \quad q = 4,$$
$$q^5(q^4-1)(q^3-1)/8 \quad \text{for} \quad q = 2, 3.$$

Proof: In each case, the number is $p(4,q)/|G_q|$. □

Corollary 4: *Under G_q, there are five orbits $\mathcal{N}_i$ of planes with $|\mathcal{N}_i| = n_i$ as follows:*

$\mathcal{N}_1 = \Gamma$, $n_1 = q+1$;
$\mathcal{N}_2 = \{$*planes containing exactly two points of* $\mathscr{C}\}$, $n_2 = q(q+1)$;
$\mathcal{N}_3 = \{$*planes through three distinct points of* $\mathscr{C}\}$, $n_3 = q(q^2-1)/6$;
$\mathcal{N}_4 = \{$*planes not in* Γ *through exactly one point of* $\mathscr{C}\}$, $n_4 = q(q^2-1)/2$;
$\mathcal{N}_5 = \{$*planes containing no point of* $\mathscr{C}\}$, $n_5 = q(q^2-1)/3$.

Proof: Let γ' be a quadratic extension and γ'' a cubic extension of γ. Then the five orbits correspond to the possibilities for a cubic equation over γ as in § 1.10. They are (i) three coincident roots in γ; (ii) two coincident roots and one other, all in γ; (iii) three distinct roots in γ; (iv) one root in γ and two in $\gamma' \backslash \gamma$; (v) three roots in $\gamma'' \backslash \gamma$. By Corollary 1, the sets $\mathcal{N}_1$, $\mathcal{N}_2$, and $\mathcal{N}_3$ are orbits. For (iv), let us fix ∞ and try to transform t and $\bar{t}$ to s and $\bar{s}$, where $s, t \in \gamma' \backslash \gamma$ and $\bar{t} = t^q$ with $\bar{\bar{t}} = t$. If

$$s = at + b, \qquad \bar{s} = a\bar{t} + b,$$

then $a = (s-\bar{s})/(t-\bar{t})$, $b = (s\bar{t} - \bar{s}t)/(\bar{t}-t)$. So $\bar{a} = a$ and $\bar{b} = b$, whence $a, b \in \gamma$ and $\mathcal{N}_4$ is an orbit. For (v), the process is similar. Let $s, t \in \gamma'' \backslash \gamma$ with $\bar{t} = t^q$, $\bar{\bar{t}} = t^{q^2}$ so that $\bar{\bar{\bar{t}}} = t$. If

$$cst + ds - at - b = 0,$$
$$c\bar{s}\bar{t} + d\bar{s} - a\bar{t} - b = 0,$$
$$c\bar{\bar{s}}\bar{\bar{t}} + d\bar{\bar{s}} - a\bar{\bar{t}} - b = 0,$$

then these equations can be solved for $a : b : c : d$ with all ratios in γ. □

Corollary 5: *Under G_q, there are five orbits $\mathcal{M}_i$ of points with $|\mathcal{M}_i| = m_i$. For $q \not\equiv 0 \pmod 3$, with $\mathfrak{A}$ the null polarity interchanging $\mathscr{C}$ and Γ, we have $\mathcal{M}_i\mathfrak{A} = \mathcal{N}_i$ and $m_i = n_i$.*

(i) $q \not\equiv 0 \pmod 3$.

$\mathcal{M}_1 = \mathscr{C}$, $m_1 = q+1$;
$\mathcal{M}_2 = \{$*points off* $\mathscr{C}$ *on a tangent*$\}$, $m_2 = q(q+1)$;
$\mathcal{M}_3 = \{$*points on three osculating planes*$\}$, $m_3 = q(q^2-1)/6$;
$\mathcal{M}_4 = \{$*points off* $\mathscr{C}$ *on exactly one osculating plane*$\}$, $m_4 = q(q^2-1)/2$;
$\mathcal{M}_5 = \{$*points on no osculating plane*$\}$, $m_5 = q(q^2-1)/3$.

(ii) $q \equiv 0 \pmod 3$.

$\mathcal{M}_1 = \mathscr{C}$, $m_1 = q+1$;
$\mathcal{M}_2 = \{$*points on all osculating planes*$\}$, $m_2 = q+1$;
$\mathcal{M}_3 = \{$*points off* $\mathscr{C}$ *on a tangent and one osculating plane*$\}$, $m_3 = q^2-1$;
$\mathcal{M}_4 = \{$*points off* $\mathscr{C}$ *on a real chord*$\}$, $m_4 = q(q^2-1)/2$;
$\mathcal{M}_5 = \{$*points on an imaginary chord*$\}$, $m_5 = q(q^2-1)/2$. □

For differences in the cases that $q \equiv 1 \pmod 3$ and $q \equiv -1 \pmod 3$, see Lemma 21.1.11, Corollary.

Lemma 21.1.3 shows that, for $q = 2, 3, 4$, the notation of tangent to the cubic $\mathscr{C}$ is not completely satisfactory, since it depends on the polynomials by which $\mathscr{C}$ is defined. For example, when $q = 4$, the tangents to $\mathscr{C} = \{\mathbf{P}(t^3, t^2, t, 1)\}$ form the regulus with equations $l_{02} = l_{31} = l_{03} + l_{12} = 0$ by (21.6). However,

$$\mathscr{C} = \{\mathbf{U}_0, \mathbf{U}_3, \mathbf{U}, \mathbf{P}(1, \omega^2, \omega, 1), \mathbf{P}(1, \omega, \omega^2, 1)\}$$

and can equally well be presented as

$$\bar{\mathscr{C}} = \{\mathbf{P}(t^3, t, t^2, 1) \mid t \in \gamma^+\}.$$

In this form the procedure of finding the bisecants as in (21.5) and thence the tangents gives the complementary regulus $\mathscr{R}'$. This occurs for all even $q > 2$, since the substitution $t = s^{q/2}$ in (21.3) gives

$$\bar{\mathscr{C}} = \{\mathbf{P}(ss^{q/2}, s, s^{q/2}, 1) \mid s \in \gamma^+\}.$$

However, only for $q = 4$, is $\bar{\mathscr{C}}$ a cubic. Since $|G_4| = 2\,|PGL(2, 4)|$, only the two reguli $\mathscr{R}$ and $\mathscr{R}'$ can occur as tangents. But, in what follows, we shall still refer to the tangents of $\mathscr{C}$ as those $q+1$ lines given by (21.6).

The lines of $PG(3, q)$ can be partitioned into classes, each of which is a union of orbits under G_q, in a natural way. These classes are named $\mathcal{O}_i$ or $\mathcal{O}'_i$ for some i, where $\mathcal{O}'_i = \mathcal{O}_i\mathfrak{A}$ with $\mathfrak{A}$ the null polarity of $\mathscr{C}$ for $p \neq 3$.

Lemma 21.1.4: *$PG^{(1)}(3, q)$ may be partitioned as follows.*

(i) $p \neq 3$, $q > 4$.

$\mathcal{O}_1=\{$*real chords of* $\mathscr{C}\}$, $|\mathcal{O}_1|=q(q+1)/2$;
$\mathcal{O}'_1=\{$*real axes of* $\Gamma\}$, $|\mathcal{O}'_1|=q(q+1)/2$;
$\mathcal{O}_2=\mathcal{O}'_2=\{$*tangents of* $\mathscr{C}\}$, $|\mathcal{O}_2|=q+1$;
$\mathcal{O}_3=\{$*imaginary chords of* $\mathscr{C}\}$, $|\mathcal{O}_3|=q(q-1)/2$;
$\mathcal{O}'_3=\{$*imaginary axes of* $\Gamma\}$, $|\mathcal{O}'_3|=q(q-1)/2$;
$\mathcal{O}_4=\mathcal{O}'_4=\{$*non-tangent unisecants in osculating planes*$\}$, $|\mathcal{O}_4|=q(q+1)$;
$\mathcal{O}_5=\{$*unisecants not in osculating planes*$\}$, $|\mathcal{O}_5|=q(q^2-1)$;
$\mathcal{O}'_5=\{$*external lines in osculating planes*$\}$, $|\mathcal{O}'_5|=q(q^2-1)$;
$\mathcal{O}_6=\mathcal{O}'_6=\{$*external lines, other than chords, not in osculating planes*$\}$, $|\mathcal{O}_6|=q(q-1)(q^2-1)$.

For q even, the lines in the regulus complementary to that of the tangents form an orbit contained in $\mathcal{O}_4$.

(ii) $p=3$, $q>3$.

$\mathcal{O}_1=\{$*real chords of* $\mathscr{C}\}$, $|\mathcal{O}_1|=q(q+1)/2$;
$\mathcal{O}_2=\{$*tangents of* $\mathscr{C}\}$, $|\mathcal{O}_2|=q+1$;
$\mathcal{O}_3=\{$*imaginary chords of* $\mathscr{C}\}$, $|\mathcal{O}_3|=q(q-1)/2$;
$\mathcal{O}_4=\{$*non-tangent unisecants in osculating planes*$\}$, $|\mathcal{O}_4|=q(q+1)$;
$\mathcal{O}_5=\{$*unisecants not in osculating planes*$\}$, $|\mathcal{O}_5|=q(q^2-1)$;
$\mathcal{O}_6=\{$*external lines, other than chords, not in osculating planes*$\}$, $|\mathcal{O}_6|=q(q-1)(q^2-1)$;
$\mathcal{O}_7=\{$*axis of* $\Gamma\}$, $|\mathcal{O}_7|=1$;
$\mathcal{O}_8=\{$*external lines meeting the axis of* $\Gamma\}$, $|\mathcal{O}_8|=(q+1)(q^2-1)$. □

To make clear the nature of the tangents to $\mathscr{C}$, they can be characterized as follows.

Lemma 21.1.5: *For $q>2$, the unisecants of $\mathscr{C}$ such that every plane through such a unisecant meets $\mathscr{C}$ in at most one point other than the point of contact are*

(i) *for q odd, the tangents;*

(ii) *for q even, the tangents and the unisecants in the complementary regulus.*

Proof: Consider the point $\mathbf{U}_0$ on $\mathscr{C}$. Let $\mathbf{P}(a_0, a_1, a_2, a_3)$ be on the line l through $\mathbf{U}_0$. Let $\pi=\boldsymbol{\pi}(c_0, c_1, c_2, c_3)$ be a plane through l. Then

$$c_0=c_1a_1+c_2a_2+c_3a_3=0. \tag{21.8}$$

Now, π contains the points $P(t)$ of $\mathscr{C}$ for which

$$c_1t^2+c_2t+c_3=0.$$

We require that there should be at most one solution t in γ for all c_1, c_2, c_3 satisfying (21.8). Either $c_1=0$ and hence $a_2=a_3=0$, or, if q is even, $c_2=0$ and hence $a_1=a_3=0$. So, for q odd, $\mathbf{U}_0\mathbf{U}_1$ is the only unisecant through $\mathbf{U}_0$ with the required property, and it is the tangent $l(\infty)$ at $\mathbf{U}_0 = P(\infty)$ to $\mathscr{C}$. For q even, $\mathbf{U}_0\mathbf{U}_1$ and $\mathbf{U}_0\mathbf{U}_2$ are the only suitable unisecants through $\mathbf{U}_0$; they are respectively the tangent to $\mathscr{C}$ at $\mathbf{U}_0$ and the line through $\mathbf{U}_0$ of the regulus complementary to that of the tangents. □

Twisted cubics are intimately connected with ruled quadrics. As will be seen, a twisted cubic can also be defined by a pair of quadrics.

Lemma 21.1.6: *In $PG(3, q)$, the quadrics $\mathbf{V}(F)$ containing $\mathscr{C} = \{P(t) = \mathbf{P}(t^3, t^2, t, 1) \mid t \in \gamma^+\}$ are as follows:*

(i) *if $q \geq 7$, $F = F_{\lambda\mu\nu} = \lambda(x_0x_3 - x_1x_2) + \mu(x_0x_2 - x_1^2) + \nu(x_1x_3 - x_2^2)$;*
(ii) *if $q = 5$, $F = F_{\lambda\mu\nu} + \rho(x_0x_1 - x_2x_3)$;*
(iii) *if $q = 4$, $F = F_{\lambda\mu\nu} + \xi(x_2x_3 - x_1^2) + \eta(x_0x_1 - x_2^2)$;*
(iv) *if $q = 3$, $F = a_{11}x_1^2 + a_{22}x_2^2 + \sum'' a_{ij}x_ix_j$ with*

$$a_{01} + a_{03} + a_{12} + a_{23} = a_{02} + a_{11} + a_{13} + a_{22} = 0;$$

(v) *if $q = 2$, $F = a_{11}x_1^2 + a_{22}x_2^2 + \sum'' a_{ij}x_ix_j$ with*

$$a_{11} + a_{22} + \sum'' a_{ij} = 0. \quad \square$$

Corollary 1: *If $q \geq 7$, then the net $\mathscr{N}$ of $q^2 + q + 1$ quadrics containing $\mathscr{C}$ comprises q^2 hyperbolic quadrics and $q + 1$ cones.*

Proof: If $\lambda \neq 0$, then

$$\lambda F_{\lambda\mu\nu} = (\lambda x_0 + \nu x_1)(\mu x_2 + \lambda x_3) - (\lambda x_1 + \nu x_2)(\mu x_1 + \lambda x_2).$$

If $\lambda = 0$, then

$$F_{0\mu\nu} = -x_1(\mu x_1 - \nu x_3) + x_2(\mu x_0 - \nu x_2).$$

So, in either case, the quadric is singular if and only if $\lambda^2 = \mu\nu$. □

Corollary 2: *If $q \geq 7$, then*

(i) *through $\mathscr{C}$ and one of its chords, there is a pencil of ruled quadrics;*

(ii) *through $\mathscr{C}$ and two of its chords, there is a unique ruled quadric, which is hyperbolic or a cone as the chords are skew or not;*

(iii) *through $\mathscr{C}$ and one of its non-tangential unisecants, there is a unique hyperbolic quadric.*

Proof: (i) A quadric must contain seven points of $\mathscr{C}$ to contain it entirely. If l is a chord of $\mathscr{C}$ and if $\mathscr{C}$ lies on the quadric $\mathscr{Q}$, then $\mathscr{Q}$ must contain a further point of l for l to lie on $\mathscr{Q}$. So $\mathscr{Q}$ satisfies eight conditions to contain $\mathscr{C}$ and l. As $\mathscr{Q}$ is determined by nine conditions, there is a pencil of quadrics containing $\mathscr{C}$ and l.

Parts (ii) and (iii) follow by a similar argument.

Alternatively, since G_q is triply transitive on $\mathscr{C}$, by Lemma 21.1.3,

Corollary 1, so the result may be proved by choosing particular chords and unisecants. □

Corollary 3: *For any q, two ruled quadrics through $\mathscr{C}$ meet residually in a chord of $\mathscr{C}$.* □

Corollary 4: *For $q \geq 7$,*

(i) *the number N of twisted cubics on a hyperbolic quadric $\mathscr{H}_3$ is* $2q^3(q^2-1)$;

(ii) *the number N_0 of twisted cubics on $\mathscr{H}_3$ through a point P is* $2q^3(q-1)$.

Proof: (i)

$$\begin{aligned} N &= (\text{number of twisted cubics in } PG(3,q)) \\ &\quad \times (\text{number of hyperbolic quadrics containing a cubic}) \\ &\quad \div (\text{number of hyperbolic quadrics in } PG(3,q)) \\ &= q^5(q^4-1)(q^3-1)q^2/[q^4(q^3-1)(q^2+1)/2] \\ &= 2q^3(q^2-1), \end{aligned}$$

where the numbers are given in Lemma 21.1.2, Corollary 1, Lemma 21.1.6, and Theorem 15.3.16.

(ii)

$$\begin{aligned} N_0 &= N \times (\text{number of points on a twisted cubic}) \\ &\quad \div (\text{number of points on } \mathscr{H}_3) \\ &= 2q^3(q^2-1)(q+1)/(q+1)^2 \\ &= 2q^3(q-1). \quad \square \end{aligned}$$

Lemma 21.1.7: (i) *If the twisted cubic $\mathscr{C}$ lies on the quadric $\mathscr{H} = \mathscr{H}(\mathscr{R}, \mathscr{R}')$, then all the lines of one regulus, say $\mathscr{R}$, are chords of $\mathscr{C}$ and all the lines of the other regulus $\mathscr{R}'$ are non-tangential unisecants of $\mathscr{C}$.*

(ii) *If $\mathscr{C}$ lies on the cone $P\mathscr{P}_2$, then P lies on $\mathscr{C}$ and the generators of $P\mathscr{P}_2$ comprise q real chords and the tangent to $\mathscr{C}$ at P.*

Proof: (i) Consider any tangent plane π of $\mathscr{H}$; then $\pi = ll'$ with $l \in \mathscr{R}$, $l' \in \mathscr{R}'$. Suppose $l \cap l'$ is not a point of $\mathscr{C}$. Since π meets $\mathscr{C}$ in three complex points, one line, say l, meets $\mathscr{C}$ in two complex points P_1 and P_2 and the other line l' meets $\mathscr{C}$ in one complex point P. The points P_1 and P_2 are either both real or coincident or 2-complex conjugate; so P is real. Hence each generator of $\mathscr{H}$ is a chord or a unisecant of $\mathscr{C}$. Then, as all lines of $\mathscr{R}$ meet l', they are chords, and, as all lines of $\mathscr{R}'$ meet l, they are unisecants.

(ii) Suppose $P \notin \mathscr{C}$. Since every plane through a pair of generators of the cone meets $\mathscr{C}$ in three complex points, one generator is a chord and one a unisecant, as in (i). Let l_1, l_2, l_3 be generators of $P\mathscr{P}_2$. Then, as one of l_1 and l_2 is a chord, let it be l_1. So l_2 and l_3 are unisecants (non-tangential). Hence $l_2 l_3$ meets $\mathscr{C}$ in only two points: a contradiction. So P is on $\mathscr{C}$ and all generators of $P\mathscr{P}_2$ are chords of $\mathscr{C}$. □

For q even and $q>4$, this lemma distinguishes the tangents and the unisecants in the complementary regulus.

Lemma 21.1.8: *Let the cubic $\mathscr{C}$ lie on the quadric $\mathscr{H}(\mathscr{R})$ and let $\mathscr{R}$ comprise n_2 real chords, n_1 tangents and n_0 imaginary chords of $\mathscr{C}$.*

(i) *For q odd, either* (a) $n_1=0$, $n_0=n_2=(q+1)/2$ *or* (b) $n_1=2$, $n_0=n_2=(q-1)/2$, *and, when $q\geq 7$, there are $q(q-1)/2$ quadrics for which* (a) *holds and $q(q+1)/2$ for which* (b) *holds.*

(ii) *For q even, either* (a) $n_1=1$, $n_0=n_2=q/2$ *or* (b) $n_1=q+1$, $n_0=n_2=0$, *and, when $q\geq 7$, there are q^2-1 quadrics for which* (a) *holds and one for which* (b) *holds.*

Proof: Counting the lines of $\mathscr{R}$ and the points of $\mathscr{C}$ gives respectively

$$n_0+n_1+n_2=q+1, \qquad n_1+2n_2=q+1.$$

If q is odd, then n_1 is even. As $\mathscr{R}$ cannot contain four tangents, by Theorem 21.1.2(iii), so $n_1=0$ or 2; thus we have (a) or (b). Since $\mathscr{H}$ is determined by containing $\mathscr{C}$ and two of its tangents, by Lemma 21.1.6, Corollary 2(ii), there are $q(q+1)/2$ quadrics of type (a), leaving $q(q-1)/2$ of type (b) to make a total of q^2, by Lemma 21.1.6, Corollary 1.

If q is even, then n_1 is odd. As the tangents to $\mathscr{C}$ form a regulus, if $\mathscr{R}$ contains three tangents, it contains them all. Hence $n_1=1$ or $q+1$. This gives (a) or (b) and also the required number of quadrics. □

Corollary: *For q even, the number N_0' of cubics on a quadric $\mathscr{H}_3$ such that each generator of $\mathscr{H}_3$ contains just one point of the cubic is*

$$2q(q^2-1) \quad \text{for} \quad q\geq 8,$$
$$q(q^2-1) \quad \text{for} \quad q=4.$$

Proof:

$N_0'=$ (number of cubics in $PG(3,q)$)

$\times$(number of quadrics through a cubic with each generator containing a point of the cubic)

$\div$(number of quadrics in $PG(3,q)$).

So, for $q\geq 8$,

$$N_0'=q^5(q^4-1)(q^3-1)\times 1/[q^4(q^3-1)(q^2+1)/2]$$
$$=2q(q^2-1).$$

For $q=4$, there is a factor of 2 in the total number of cubics in $PG(3, q)$. The numbers are obtained from Lemma 21.1.3, Corollary 3, Lemma 21.1.8(ii), and Theorem 15.3.17. □

Theorem 21.1.9: *If $\mathscr{C}$ is a twisted cubic in $PG(3, q)$, then*
(i) *no two chords of $\mathscr{C}$ meet off $\mathscr{C}$;*
(ii) *every point off $\mathscr{C}$ lies on exactly one chord of $\mathscr{C}$.*

Proof: (i) If two chords met off $\mathscr{C}$, their plane π would meet $\mathscr{C}$ in four complex points (not necessarily distinct), contrary to the fact that the parameters of the points of $\pi \cap \mathscr{C}$ are roots of a cubic.

Alternatively, two chords l and l' meet if and only if $\varpi(l, l')=0$, by Lemma 15.2.2. If, in terms of (21.5), $l=l(s, t)$ and $l'=l(s', t')$, then

$$\varpi(l, l')=(s-s')(s-t')(t-s')(t-t').$$

So $\varpi(l, l')=0$ only when l and l' meet on $\mathscr{C}$.

(ii) A real chord, a tangent and an imaginary chord respectively contain $q-1$, q, and $q+1$ points off $\mathscr{C}$. So the total number of points on the chords of $\mathscr{C}$ is

$$\begin{aligned}(q+1)+(q-1)q(q+1)/2+q(q+1)+(q+1)q(q-1)/2 \\ =(q+1)(q^2+1)=|PG(3, q)|,\end{aligned}$$

as required.

Alternatively, given some point P off $\mathscr{C}$, there is a pencil of quadrics containing P and $\mathscr{C}$. If $\mathscr{H}$ is a hyperbolic quadric with this property, then one of the generators of $\mathscr{H}$ through P is a chord of $\mathscr{C}$, by Lemma 21.1.7(i). □

Corollary: *If Γ is the osculating developable to $\mathscr{C}$ in $PG(3, q)$, $p \neq 3$, then*
(i) *no two axes of Γ meet unless they lie in the same plane of Γ;*
(ii) *every plane not in Γ contains exactly one axis of Γ.* □

Lemma 21.1.10: *The points on the tangents to $\mathscr{C}=\{P(t)=\mathbf{P}(t^3, t^2, t, 1) \mid t \in \gamma^+\}$ form the variety*

$$\mathbf{V}((x_0x_3-x_1x_2)^2-4(x_0x_2-x_1^2)(x_1x_3-x_2^2)).$$

Proof: If $\mathbf{P}(X)$ lies on $\mathbf{l}(L)$, then, by Lemma 15.2.3,

$$x_0l_{12}-x_1l_{02}+x_2l_{01}=0,$$
$$x_1l_{23}+x_2l_{31}+x_3l_{12}=0.$$

So, from (21.6), the conditions for $\mathbf{P}(X)$ to lie on $l(t)$ are

$$x_0t^2-2x_1t^3+x_2t^4=0, \qquad x_1-2x_2t+x_3t^2=0.$$

Hence

$$1:2t:t^2=x_1x_3-x_2^2:x_0x_3-x_1x_2:x_0x_2-x_1^2.$$

Eliminating t gives

$$(x_0x_3-x_1x_2)^2=4(x_0x_2-x_1^2)(x_1x_3-x_2^2). \quad \square$$

This lemma applies to all q, although, for q even, the quartic surface merely becomes the quadric $\mathbf{V}(x_0x_3+x_1x_2)$ containing the regulus of tangents of $\mathscr{C}$.

The theme of the net $\mathscr{N}$ of quadrics containing $\mathscr{C}$ is resumed in § 21.4.

As the final part of this section we consider the projection of $\mathscr{C}$ from a point Q not on $\mathscr{C}$ onto a plane, and show that the various possibilities coincide with those for plane singular cubics in Table 11.5 of § 11.3.

Lemma 21.1.11: *The number n_Q of osculating planes through a point Q of $PG(3, q)\backslash\mathscr{C}$ is given by Table* 21.1, *where $q \equiv c \pmod 3$, the line l_Q is the unique chord of $\mathscr{C}$ through Q and N_l is the number of Q of that type on l_Q.*

Proof: First, let $q\equiv 0 \pmod 3$. Since the osculating planes to $\mathscr{C}$ form a pencil, a point Q either lies in them all or in just one.

Now, take $q\equiv\pm1 \pmod 3$. If Q lies on a real or imaginary chord $l(t, s)$ and in an osculating plane $\pi(r)$, then

$$Q=\mathbf{P}(t^3+\lambda s^3, t^2+\lambda s^2, t+\lambda s, 1+\lambda)$$

and $\pi(r)=\boldsymbol{\pi}(1, -3r, 3r^2, -r^3)$. Since $Q\in\pi(r)$,

$$(t-r)^3+\lambda(s-r)^3=0. \tag{21.9}$$

Table 1.2 in § 1.10 gives the result; for t and s real, equation (21.9) has 0 or 3 solutions in γ for r when $q\equiv 1 \pmod 3$ and has always one solution for $q\equiv -1 \pmod 3$; for t and s imaginary, the situation is reversed.

When Q is on the tangent $l(t)$, then $Q=\mathbf{P}(3t^2+\lambda t^3, 2t+\lambda t^2, 1+\lambda t, \lambda)$. If it also lies in $\pi(r)$, then

$$3(t-r)^2+\lambda(t-r)^3=0. \tag{21.10}$$

So, for $\lambda\neq 0$, equation (21.10) has the solutions $r=t$, $t+3/\lambda$; for $\lambda=0$, we have $r=t, \infty$. $\square$

Table 21.1

	l_Q real chord				l_Q tangent				l_Q imaginary chord			
c	0	−1	1	1	0	0	−1	1	0	−1	−1	1
n_Q	1	1	3	0	1	$q+1$	2	2	1	3	0	1
N_l	$q-1$	$q-1$	$(q-1)/3$	$2(q-1)/3$	$q-1$	1	q	q	$q+1$	$(q+1)/3$	$2(q+1)/3$	$q+1$

Corollary: *$\mathcal{M}_3$, $\mathcal{M}_4$, $\mathcal{M}_5$ are orbits of points as in Lemma* 21.1.3, *Corollary* 5.

(i) *For* $q \equiv 1 \pmod 3$,

$$\mathcal{M}_3 \cup \mathcal{M}_5 = \{\text{points on a real chord}\},$$
$$\mathcal{M}_4 = \{\text{points on an imaginary chord}\}.$$

(ii) *For* $q \equiv -1 \pmod 3$,

$$\mathcal{M}_3 \cup \mathcal{M}_5 = \{\text{points on an imaginary chord}\},$$
$$\mathcal{M}_4 = \{\text{points on a real chord}\}. \quad \square$$

For Q in $PG(3, q)\backslash\mathscr{C}$, let π be a plane not through Q. Let $\mathscr{C}' = \{P' = PQ \cap \pi \mid P \in \mathscr{C}\}$. By Theorem 21.1.9, there is a unique chord l_Q of $\mathscr{C}$ through Q.

Theorem 21.1.12: (i) *For each q, there are four, projectively distinct, irreducible, singular plane cubic curves.*

(ii) *$\mathscr{C}'$ is a singular cubic with a node, a cusp, or an isolated double point according as the chord l_Q is real, a tangent, or imaginary.*

(iii) *For l_Q a real or imaginary chord, $\mathscr{C}'$ has n_Q inflexions as in Table* 21.1; *for l_Q a tangent, $\mathscr{C}'$ has $n_Q - 1$ inflexions.*

Proof: (i) By Corollary 5 to Lemma 21.1.3, there are four orbits of points of $PG(3, q)\backslash\mathscr{C}$ under G_q. Each orbit gives a different $\mathscr{C}'$.

(ii) When l_Q is a real chord with points of contact P_1 and P_2, then the tangents at P_1 and P_2 project to the tangents of $\mathscr{C}'$ at the node $P_1' = P_2'$; similarly for the case that l_Q is an imaginary chord. When l_Q is a tangent, the osculating plane through l_Q projects to the cuspidal tangent of $\mathscr{C}'$.

(iii) The point of contact of each osculating plane through Q projects to an inflexion of $\mathscr{C}'$, apart from the case that it projects to the cusp. $\square$

21.2 Characterization of the twisted cubic for q odd

A k-arc $\mathscr{K}$ in $PG(3, q)$ is a set of k points, no four of which are coplanar. If $k > 3$, then no three points of $\mathscr{K}$ are collinear. If $k \leqslant 3$, then $\mathscr{K}$ lies in a plane and so we revert to the definition of a plane k-arc (see § 8.1); thus again, no three points of $\mathscr{K}$ are collinear.

As defined in § 3.3, let $m(3, q)$ be the maximum value of k. Our aim is to find $m(3, q)$ and to characterize k-arcs with $k = m(3, q)$. A k-arc $\mathscr{K}$ is *complete* if there exists no $(k+1)$-arc containing $\mathscr{K}$; that is, the planes through triads of points of $\mathscr{K}$ together contain all the points of $PG(3, q)$.

Lemma 21.2.1: *For* $q = 2$, 3, *and* 4,

(i) $m(3, q) = 5$;

(ii) *each* 5*-arc is projectively equivalent to* $\mathscr{K} = \{\mathbf{U}_0, \mathbf{U}_1, \mathbf{U}_2, \mathbf{U}_3, \mathbf{U}\}$;

(iii) *these are the only complete arcs in these spaces.*

Proof: $\mathscr{K}$ is a 5-arc. Any point P of $PG(3, q)$ with at least one coordinate zero lies in a face of the tetrahedron of reference. As $q \leqslant 4$, if P has no coordinate zero, then at least two of its coordinates are equal; so P lies in a plane spanned by $\mathbf{U}$ and two of the $\mathbf{U}_i$. So $\mathscr{K}$ is complete, every 5-arc is projectively equivalent to $\mathscr{K}$ and no subset of $\mathscr{K}$ is complete. □

To find $m(3, q)$ we first need to consider when a $(q+1)$-arc is a twisted cubic; the converse is true by Theorem 21.1.1(iv).

Lemma 21.2.2: *If $\mathscr{K}$ is a k-arc in $PG(3, q)$ with $k \geqslant 6$ such that every projection of $\mathscr{K}$ from one of its points onto a plane is contained in a conic, then $\mathscr{K}$ lies on a unique twisted cubic.*

Proof: Let P_0, P_1 be points of $\mathscr{K}$ and let π be a plane containing neither P_0 nor P_1. Let $\mathscr{K}_0$ and $\mathscr{K}_1$ be the projections of $\mathscr{K}$ onto π from P_0 and P_1 respectively; that is, $\mathscr{K}_i = \{P' \mid P' = PP_i \cap \pi, P \in \mathscr{K} \backslash \{P_i\}\}$, $i = 0, 1$. If three points P', Q', R' of $\mathscr{K}_i$ were collinear, then the corresponding points P, Q, R of $\mathscr{K}$ would be coplanar with P_i. So each $\mathscr{K}_i$ is a plane k'-arc. Since no three points of $\mathscr{K}$ are collinear, each projection of $\mathscr{K}$ is a bijection of $\mathscr{K} \backslash \{P_i\}$ to $\mathscr{K}_i$; so $k' = k - 1$. By hypothesis, each $\mathscr{K}_i$ is contained in a conic $\mathscr{C}_i$. As $k - 1 \geqslant 5$, each $\mathscr{C}_i$ is unique.

Now consider the two cones $P_0\mathscr{C}_0$ and $P_1\mathscr{C}_1$. The line P_0P_1 is a generator of both cones. Let the residual curve of intersection be $\mathscr{F}$. By construction, $\mathscr{K}$ lies on both $P_0\mathscr{C}_0$ and $P_1\mathscr{C}_1$; so $\mathscr{K} \backslash \{P_0, P_1\} \subset \mathscr{F}$. The cones cannot have a second generator in common. So, either they touch along P_0P_1 and meet residually in a conic or $\mathscr{F}$ is a twisted cubic. As $|\mathscr{K} \backslash \{P_0, P_1\}| \geqslant 4$, so $\mathscr{F}$ contains at least four points and no four of these are coplanar. Thus $\mathscr{F}$ cannot be a conic and must be a twisted cubic. However, by Lemma 21.1.7(ii), a twisted cubic lying on a cone contains the vertex. So $\mathscr{F}$ contains P_0 and P_1, and hence $\mathscr{K}$. Thus $\mathscr{K}$ lies on the twisted cubic $\mathscr{F}$. As $k \geqslant 6$, $\mathscr{F}$ is unique, by Theorem 21.1.1(v). □

From this lemma, we can deduce the three theorems of this section.

Theorem 21.2.3: *In $PG(3, q)$, q odd, a $(q+1)$-arc $\mathscr{K}$ is a twisted cubic.*

Proof: For $q = 3$ and $q = 5$, the result follows from Theorem 21.1.1(v). However $q = 5$ is not excluded from the general case.

Let $q \geqslant 5$ so that $q + 1 \geqslant 6$. By Theorem 8.6.10, a plane q-arc lies on a conic. So, by the previous lemma, $\mathscr{K}$ lies on a twisted cubic $\mathscr{C}$. As $|\mathscr{K}| = |\mathscr{C}| = q + 1$, so $\mathscr{K} = \mathscr{C}$. □

Theorem 21.2.4: *For q odd,*

$$m(3, q) = \max(5, q + 1).$$

Proof: By Lemma 21.2.1, $m(3, 3) = 5$. So we need to show that $m(3, q) = q + 1$ for $q \geqslant 5$.

Suppose there exists a $(q+2)$-arc $\mathscr{K}$. By Theorem 8.2.4, a plane

$(q+1)$-arc is a conic. So, again by Lemma 21.2.2, $\mathcal{K}$ is contained in a twisted cubic $\mathcal{C}$. As $|\mathcal{C}|=q+1$, this is a contradiction. So $\mathcal{K}$ cannot exist. However, a twisted cubic $\mathcal{C}$ is a $(q+1)$-arc, by Theorem 21.1.1(iv). So $m(3,q)=q+1$. □

Theorem 21.2.5: *In $PG(3,q)$, q odd and $q>49$, a k-arc $\mathcal{K}$ with $k>q-\sqrt{q}/4+11/4$ is contained in a unique twisted cubic.*

Proof: The condition $q>49$ is required to make the theorem say something, since, if $q\leq 49$, $k>q+1$.

By Theorem 10.4.4, a plane k-arc with $k>q-\sqrt{q}/4+7/4$ is contained in a conic. So Lemma 21.2.2 gives the result. □

21.3 $(q+1)$-arcs for q even

Throughout this section $q=2^h$. Let $\mathcal{K}$ be a k-arc in $PG(3,q)$. A unisecant l to $\mathcal{K}$ with point of contact P is *special* if every plane through l meets $\mathcal{K}$ in at most one point other than P. A plane π through a special unisecant meeting $\mathcal{K}$ at P is a *contact plane* at P if $\pi\cap\mathcal{K}=\{P\}$. For example, if $\mathcal{K}$ is a twisted cubic, then each tangent is a special unisecant and each osculating plane is a contact plane.

We now proceed to find $m(3,q)$, the maximum value of k for which a k-arc exists.

Lemma 21.3.1: *If $\mathcal{K}$ is a k-arc in $PG(3,q)$ and Q is a point off $\mathcal{K}$, then $\mathcal{K}\cup\{Q\}$ is a $(k+1)$-arc if and only if QP is a special unisecant to $\mathcal{K}$ for each point P in $\mathcal{K}$.*

Proof: Let QP be a special unisecant to $\mathcal{K}$ for each P in $\mathcal{K}$ and suppose that Q lies in the plane $P_1P_2P_3$, where $P_1, P_2, P_3\in\mathcal{K}$. Then this plane contains the special unisecant QP_1 as well as the points P_2, P_3 of $\mathcal{K}$; this contradicts the definition of a special unisecant. So $\mathcal{K}\cup\{Q\}$ is a $(k+1)$-arc.

Conversely, let $\mathcal{K}\cup\{Q\}$ be a $(k+1)$-arc. If, for some P in $\mathcal{K}$, the line QP is not a special unisecant to $\mathcal{K}$, then there exists a plane π through QP containing two other points of $\mathcal{K}$; that is, there are four coplanar points of $\mathcal{K}\cup\{Q\}$. □

Lemma 21.3.2: *If $\mathcal{K}$ is a k-arc in $PG(3,q)$, then $k\leq q+3$.*

Proof: Each of the $q+1$ planes through a bisecant l of $\mathcal{K}$ contains at most one point of $\mathcal{K}\backslash l$. So $k\leq 2+(q+1)=q+3$. □

Let $\mathcal{K}$ be a k-arc in $PG(3,q)$ and let $\mathcal{K}_{P,\pi}$ be the plane $(k-1)$-arc obtained by projecting $\mathcal{K}\backslash\{P\}$ from the point P of $\mathcal{K}$ onto the plane π, where P is not in π.

Lemma 21.3.3: (i) *If $Q \in \pi \backslash \mathcal{K}_{P,\pi}$, then PQ is a special unisecant to $\mathcal{K}$ if and only if $\mathcal{K}_1 = \mathcal{K}_{P,\pi} \cup \{Q\}$ is a plane k-arc.*

(ii) *$\mathcal{K}$ has $q-k+2$ contact planes through any special unisecant at P.*

(iii) *If $\mathcal{K}_{P,\pi}$ is complete, then so is $\mathcal{K}$.*

Proof: (i) Let $l = PQ$ be a special unisecant to $\mathcal{K}$ at P and suppose that Q lies on a bisecant l' of $\mathcal{K}_{P,\pi}$. Then $l' \cap \mathcal{K}_{P,\pi} = \{P_1', P_2'\}$, where P_1' and P_2' are the respective projections of the points P_1 and P_2 of $\mathcal{K}$. Therefore the plane ll' contains the points P, P_1, and P_2 of $\mathcal{K}$ contradicting that the unisecant l is special. So there is no bisecant of $\mathcal{K}_{P,\pi}$ through Q, whence $\mathcal{K}_1$ is a k-arc.

Conversely, if $\mathcal{K}_1$ is a k-arc, then $l' = P_0'Q$ is a unisecant of $\mathcal{K}_{P,\pi}$ for any point P_0' in $\mathcal{K}_{P,\pi}$. So ll' meets $\mathcal{K}$ in P and P_0 where P_0' is the projection of P_0. Therefore l is a special unisecant to $\mathcal{K}$.

(ii) Let l be a special unisecant to $\mathcal{K}$ at P. If $Q = l \cap \pi$, then $\mathcal{K}_1 = \mathcal{K}_{P,\pi} \cup \{Q\}$ is a k-arc, whence there are $q+2-k$ unisecants to $\mathcal{K}_1$ through Q in π, by Lemma 8.1.1. If l_0 is any one of them, then the plane ll_0 meets $\mathcal{K}$ in the point P only and is therefore a contact plane to $\mathcal{K}$ at P. If l_1 is a bisecant of $\mathcal{K}$ through Q, then ll_1 contains P and another point of $\mathcal{K}$, whence ll_1 is not a contact plane.

(iii) If $\mathcal{K}_{P,\pi}$ is complete, then there are no special unisecants to $\mathcal{K}$ at P, by part (i). Thus, by Lemma 21.3.1, $\mathcal{K}$ is complete. □

Lemma 21.3.4: *Let $\mathcal{K}$ be a k-arc in $PG(3, q)$ such that $k > \max(q - \sqrt{q} + 2, 3)$. Then*

(i) *the number of special unisecants at any point of $\mathcal{K}$ is*

$$\tau = q - k + 3,$$

(ii) *the number of contact planes at any point of $\mathcal{K}$ is*

$$\nu = \tau(\tau - 1)/2.$$

Proof: (i) If P is any point of $\mathcal{K}$, then $\mathcal{K}_{P,\pi}$ is a $(k-1)$-arc and $k-1 > q - \sqrt{q} + 1$. Therefore, by Theorem 10.3.3, Corollary 2, $\mathcal{K}_{P,\pi}$ can be completed uniquely to an oval $\mathcal{O}$. So, by Lemma 21.3.3(ii),

$$\tau = |\mathcal{O} \backslash K_{P,\pi}| = (q+2) - (k-1) = q - k + 3.$$

(ii) By Lemma 21.3.3(ii), there are $\tau - 1$ contact planes through each special unisecant. The only points Q of π for which $\mathcal{K}_{P,\pi} \cup \{Q\}$ is a plane k-arc are the points of $\mathcal{O} \backslash \mathcal{K}_{P,\pi}$. The join of any two points of $\mathcal{O} \backslash \mathcal{K}_{P,\pi}$ contains only these two points of $\mathcal{O}$ and no points of $\mathcal{K}_{P,\pi}$. So the contact planes at P are the planes through pairs of special unisecants at P, and no contact plane can contain three special unisecants. So $\nu = \mathbf{c}(\tau, 2)$. □

Corollary: *For $q > 2$, a k-arc $\mathcal{K}$ in $PG(3, q)$ has the following values of τ*

and ν:

k	τ	ν
$q+3$	0	0
$q+2$	1	0
$q+1$	2	1 . □

Lemma 21.3.5: *If $\mathscr{K}$ is a $(q+3)$-arc in $PG(3,q)$, then there is no plane meeting $\mathscr{K}$ in exactly two points.*

Proof: Let l be a bisecant of $\mathscr{K}$; then each of the $q+1$ planes through l contains exactly one point of $\mathscr{K}\backslash l$. □

Lemma 21.3.6: *If $\mathscr{K}$ is a $(q+3)$-arc in $PG(3,q)$, then there are points of the space on no bisecant of $\mathscr{K}$.*

Proof: The number N of points on no bisecant of $\mathscr{K}$ is

$$N=(q^2+1)(q+1)-(q+3)-c(q+3,2)(q-1)$$
$$=(q^2-1)(q-2)/2.$$

So, for $q>2$, we have $N>0$. □

In the language of § 16.1, this lemma says that, for $q>2$, $\mathscr{K}$ is not a complete cap.

Lemma 21.3.7: *If $\mathscr{K}$ is a $(q+2)$-arc in $PG(3,q)$, then it is incomplete.*

Proof: Let P be a point of $\mathscr{K}$ and let $\mathscr{K}'=\mathscr{K}\backslash\{P\}=\{P_i \mid i\in\mathbf{N}_{q+1}\}$. By the corollary to Lemma 21.3.4, $\mathscr{K}$ has one special unisecant at each of its points. Denote the special unisecants at P and P_i by l and l_i respectively. By Lemma 21.3.1, the line $PP_i=l_i'$ is a special unisecant to $\mathscr{K}'$ at P_i. So the contact plane at P_i to $\mathscr{K}'$ is l_il_i'. However, the plane ll_i' contains the special unisecant l to $\mathscr{K}$ at P and the point P_i of $\mathscr{K}$. So, ll_i' contains only the point P_i of $\mathscr{K}'$ and hence $ll_i'=l_il_i'$. Thus l meets l_i; that is, any two special unisecants of $\mathscr{K}$ intersect. So the special unisecants to $\mathscr{K}$ are either coplanar or concurrent. They cannot be coplanar, as then $\mathscr{K}$ would lie in a plane. So the special unisecants are concurrent at a point Q. Hence $\mathscr{K}\cup\{Q\}$ is a $(q+3)$-arc. □

Theorem 21.3.8: *For q even,*

$$m(3,q)=\max(5,q+1).$$

Proof: From Lemma 21.2.1, $m(3,q)=5$ for $q=2$ and $q=4$. It remains therefore to show that $m(3,q)=q+1$ for $q\geqslant 8$. Since a twisted cubic contains $q+1$ points, by Theorem 21.1.1, so $m(3,q)\geqslant q+1$. By Lemma

21.3.7, if a $(q+2)$-arc exists, then so does a $(q+3)$-arc. Therefore, let $\mathcal{K}$ be a $(q+3)$-arc; it suffices to show that $\mathcal{K}$ cannot exist.

By Lemma 21.3.6, there is a point Q on no bisecant of $\mathcal{K}$. By Lemma 21.3.5, every plane meeting $\mathcal{K}$ in two points meets it in three. So the number of planes through Q meeting $\mathcal{K}$ in three points is $\mathbf{c}(q+3,2)/3 = (q+3)(q+2)/6 = (2^h+3)(2^{h-1}+1)/3$, where $q=2^h$. Hence 3 divides $2^{h-1}+1$, which means that h is even.

Now let l be a unisecant to $\mathcal{K}$ with point of contact P_0. By Lemma 21.3.5, every plane through l meets $\mathcal{K}$ in P_0 only or in P_0 and two other points. So the number of planes through l meeting $\mathcal{K}$ in three points is $(q+2)/2$.

Let P_1 and P_2 be points of $\mathcal{K}$, and let Q_0 be another point of P_1P_2. For any point P of $\mathcal{K}\backslash\{P_1, P_2\}$, the line Q_0P is a unisecant. So the number of planes through Q_0P apart from PP_1P_2 meeting $\mathcal{K}$ in three points is $q/2$. Thus the number of planes through Q_0 meeting $\mathcal{K}$ in three points and not containing P_1P_2 is $(q+1)(q/2)/3 = 2^{h-1}(2^h+1)/3$. So 3 divides 2^h+1, which is impossible for h even. Hence $\mathcal{K}$ does not exist. □

It remains to characterize $(q+1)$-arcs and to determine in particular whether or not a $(q+1)$-arc is necessarily a twisted cubic.

By the corollary to Lemma 21.3.4, a $(q+1)$-arc has two special unisecants at each point, whose join is the contact plane at that point. The next two lemmas should be compared to Lemma 21.1.5 for the twisted cubic.

Lemma 21.3.9: *If $\mathcal{K}$ is a $(q+1)$-arc in $PG(3,q)$, $q>2$, then no three of its special unisecants are coplanar.*

Proof: If l_1, l_2, l_3 are special unisecants to $\mathcal{K}$ and $\mathcal{K}\cap l_i = \{P_i\}$, $i=1,2,3$, then $|\{P_1, P_2, P_3\}| = 2$ or 3. If, say $P_1 = P_2$, then l_1l_2 is the contact plane at P_1 and so cannot contain P_3; so l_1, l_2, l_3 are not coplanar. If $P_1 \neq P_2$, then l_1l_2 contains l_1 and P_2. So, by the definition of a special unisecant, l_1l_2 does not contain P_3. □

Theorem 21.3.10: *If $\mathcal{K}$ is a $(q+1)$-arc in $PG(3,q)$, $q>2$, then the special unisecants to $\mathcal{K}$ are the generators of a quadric $\mathcal{H}_3$.*

Proof: Let $\mathcal{K} = \{P_i \mid i \in \mathbf{N}_{q+1}\}$ and let l_i and l'_i be the special unisecants to $\mathcal{K}$ at P_i. Consider a bisecant $l = l_{ij} = P_iP_j$. Of the $q+1$ planes through l, exactly $q-1$ contain one point of $\mathcal{K}\backslash\{P_i, P_j\}$. The two planes through l containing no point of $\mathcal{K}\backslash\{P_i, P_j\}$ are ll_i and ll'_i or, equally well, ll_j and ll'_j. So $ll_i = ll'_j$, say, and $ll'_i = ll_j$; that is, l_i meets l'_j, and l'_i meets l_j. Therefore we may take it that $l_2, l_3, \ldots, l_{q+1}$ meet l'_1 and that $l'_2, l'_3, \ldots, l'_{q+1}$ meet l_1.

It must now be shown that the lines $l_2, l_3, \ldots, l_{q+1}$ are mutually skew:

we already have that these lines are skew to l. Suppose, therefore, that l_2 meets l_3. As no three special unisecants to $\mathcal{K}$ are coplanar (Lemma 21.3.9), the lines l_2, l_3, and l'_1 are concurrent at a point Q off $\mathcal{K}$. Then, by the result of the previous paragraph, l'_2 meets l'_3, and they are concurrent with l_1 at a point Q' off $\mathcal{K}$.

Now consider l_4. It meets both lines of one of the pairs $\{l'_2, l'_3\}$, $\{l'_2, l_3\}$, $\{l_2, l'_3\}$, $\{l_2, l_3\}$. If l_4 meets l'_2 and l'_3, then these three lines are concurrent at Q'; so l_4 meets l_1, contradicting that l_4 meets l'_1. If l_4 meets l'_2 and l_3, then l_4, l_3, l'_1 are concurrent at Q. As l_2 meets l_3 and l'_1, it also contains Q. So l_4 meets l_2 and l'_2, a contradiction. Similarly, l_4 cannot meet l_2 and l'_3. Therefore l_4 meets l_2 and l_3: hence they are concurrent. So $l'_1, l_2, l_3, \ldots, l_{q+1}$ are all concurrent at Q. So, by Lemma 21.3.1, $\mathcal{K} \cup \{Q\}$ is a $(q+2)$-arc, contradicting Theorem 21.3.8. Thus we conclude that no two of the lines l_i meet and, similarly, no two of the l'_i meet. Hence each l_i meets each l'_j; that is, the special unisecants to $\mathcal{K}$ are generators of a hyperbolic quadric $\mathcal{H}_3$. □

In Lemma 21.1.5 we saw that the special unisecants to a twisted cubic consisted of the regulus of tangents and the complementary regulus. So far, then, a $(q+1)$-arc looks very much like a twisted cubic.

Lemma 21.3.11: *Let $\mathcal{K}$ be a $(q+1)$-arc in $PG(3, q)$, $q>2$, and let $\mathcal{H}_3$ be the quadric containing its special unisecants. Then the restriction to $\mathcal{K}$ of the stereographic projection of $\mathcal{H}_3$ determines an oval $\mathcal{O}$, which is regular if and only if $\mathcal{K}$ is a twisted cubic.*

Proof: As in Theorem 21.3.10, $\mathcal{K}$ lies on $\mathcal{H}_3$ such that each generator contains exactly one point of $\mathcal{K}$. Let P be a point of $\mathcal{K}$ and l and l' the generators of $\mathcal{H}_3$ through P. Let π be a plane not containing P and let $A = l \cap \pi$, $A' = l' \cap \pi$. Then, since no three points of $\mathcal{K} \backslash \{P\}$ are coplanar with P, so $\mathcal{K}_{P,\pi}$ is a plane q-arc; see § 16.3,I. Also, $\mathcal{O} = \mathcal{K}_{P,\pi} \cup \{A, A'\}$ is a plane $(q+2)$-arc or oval.

If $\mathcal{K}$ is a twisted cubic and l the tangent to the cubic at P, then $\mathcal{K}_{P,\pi} \cup \{A\}$ is a conic, by Lemma 21.1.3, Corollary 2, and so $\mathcal{O}$ is a regular oval; see § 8.4.

Suppose now that $\mathcal{O}$ is regular, so that $\mathcal{O} = \mathcal{P}_2 \cup \{N\}$, where $\mathcal{P}_2$ is a conic and N its nucleus. If A and A' are both in $\mathcal{P}_2$, then the cone $P\mathcal{P}_2$ meets $\mathcal{H}_3$ in the lines l and l' and in a conic $\mathcal{P}'_2$ containing $\mathcal{K} \backslash \{P\}$, which is impossible. So, suppose that $A' = N$. Then $P\mathcal{P}_2$ meets $\mathcal{H}_3$ in l and a cubic curve $\mathcal{F}$ containing $\mathcal{K} \backslash \{P\}$. The only possibility is that $\mathcal{F}$ is a twisted cubic, which, as it lies on $P\mathcal{P}_2$, contains P, by Lemma 21.1.7(ii). Thus $\mathcal{K} = \mathcal{F}$. □

Corollary: *In $PG(3, q)$ with $q = 2$, 4 or 8, a $(q+1)$-arc $\mathcal{K}$ is a twisted cubic.*

Proof: For $q=2$, all 3-arcs are projectively equivalent and, for $q=4$, all 5-arcs are. So, in both cases, $\mathcal{K}$ is a twisted cubic.

For $q=8$, every oval is regular, by Theorems 8.4.4 and 9.2.3. So, by the lemma, $\mathcal{K}$ is a twisted cubic. □

Now, we use Theorem 21.3.10 to obtain an analytical form for a $(q+1)$-arc with $q=2^h$. As in § 1.3(v), $\mathcal{P}(q; x)$ denotes the set of permutation polynomials of $GF(q)$ in the indeterminate x.

Theorem 21.3.12: *A $(q+1)$-arc $\mathcal{K}$ in $PG(3, q)$, $q=2^h$ with $h>1$, can be written as*

$$\mathcal{C}(F)=\{P(tF(t), F(t), t, 1) \mid t\in\gamma^+\}$$

such that

(i) $F\in\mathcal{P}(q; x)$ *with* $F(0)=0$, $F(1)=1$;
(ii) $F(x)=a_2x^2+a_4x^4+\ldots+a_{q-2}x^{q-2}$.

Proof: Let the special unisecants of $\mathcal{K}$ be generators of $\mathcal{H}_3=\mathbf{V}(x_0x_3+x_1x_2)$ as in Theorem 21.3.10. Then each generator of $\mathcal{H}_3$ contains exactly one point of $\mathcal{K}$. Since, in each regulus, we may select three generators arbitrarily, choose $\mathbf{U}_0$, $\mathbf{U}_3$, $\mathbf{U}$ in $\mathcal{K}$. Since each generator contains exactly one point of $\mathcal{K}$, so

$$\mathcal{K}\backslash\{\mathbf{U}_0\}=\{\mathbf{P}(t_is_i, t_i, s_i, 1) \mid i\in\mathbf{N}_q\},$$

where $\{t_i \mid i\in\mathbf{N}_q\}=\{s_i \mid i\in\mathbf{N}_q\}=\gamma$. Hence $s_i\mapsto t_i$, $i\in\mathbf{N}_q$, defines a permutation of γ with corresponding permutation polynomial F. Thus

$$\mathcal{K}\backslash\{\mathbf{U}_0\}=\{\mathbf{P}(tF(t), F(t), t, 1) \mid t\in\gamma\}$$

and

$$\mathcal{K}=\{P(t)=\mathbf{P}(tF(t), F(t), t, 1) \mid t\in\gamma^+\},$$

where $P(\infty)=\mathbf{U}_0$. Since $\mathbf{U}_0, \mathbf{U}_3\in\mathcal{K}$, so $F(0)=0$, $F(1)=1$.

By Lemma 21.3.11, the projection of $\mathcal{K}$ from $\mathbf{U}_0$ onto the plane $\mathbf{u}_0$ gives the oval

$$\mathcal{K}'=\{\mathbf{P}(0, F(t), t, 1) \mid t\in\gamma^+\}\cup\{\mathbf{U}_2\}.$$

By Theorem 8.4.2, Corollary 1, the polynomial F has the required form. □

When $F(x)=x^r$, write $\mathcal{C}(F)=\mathcal{C}(r)$. Let $G(\mathcal{K})$ be the group of projectivities fixing the set $\mathcal{K}$.

Lemma 21.3.13: (i) $G(\mathcal{C}(2^n))>PGL(2, q)$;
(ii) $G(\mathcal{C}(2^n))$ *acts triply transitively on* $\mathcal{C}(2^n)$.

Proof: When $F(x)=x^r$ with $r=2^n$, the mapping $t\mapsto(at+b)/(ct+d)$

implies that

$$\begin{aligned}&\mathbf{P}(t^{r+1}, t^r, t, 1)\\ &\quad\to \mathbf{P}((at+b)^{r+1}, (at+b)^r(ct+d), (at+b)(ct+d)^r, (ct+d)^{r+1})\\ &\quad= \mathbf{P}((t^{r+1}, t^r, t, 1)T),\end{aligned}$$

where

$$T=\begin{bmatrix} a^{r+1} & a^r c & ac^r & c^{r+1}\\ a^r b & a^r d & bc^r & c^r d\\ ab^r & b^r c & ad^r & cd^r\\ b^{r+1} & b^r d & bd^r & d^{r+1}\end{bmatrix}.$$

So each element of $PGL(2, q)$ gives an element of $G(\mathscr{C}(2^n))$, whence (i) and (ii) follow. □

Lemma 21.3.14: *$\mathscr{C}(2^n)$ is a $(q+1)$-arc in $PG(3, q)$, $q=2^h$ with $h>1$, if and only if n and h are coprime.*

Proof: We must consider the condition that no four points of $\mathscr{C}(2^n)$ are coplanar. By the triple transitivity of $G(\mathscr{C}(2^n))$, we may take $\mathbf{U}_0$, $\mathbf{U}_3$, and $\mathbf{U}$ as three of the four. Then the condition is that, for all t in γ_{01},

$$\begin{vmatrix} 1 & 0 & 0 & 0\\ 0 & 0 & 0 & 1\\ 1 & 1 & 1 & 1\\ t^{r+1} & t^r & t & 1\end{vmatrix}\neq 0;$$

that is, $t^r+t\neq 0$. Since $t\neq 0$, so $t^{r-1}\neq 1$. However, $t^{r-1}=1$ has the unique solution $t=1$ if and only if $(r-1, q-1)=1$, by § 1.5. But, $(r-1, q-1)=(2^n-1, 2^h-1)=2^{(n,h)}-1$, which is 1 if and only if $(n, h)=1$. □

Theorem 21.3.15: *In $PG(3, q)$, $q=2^h$ and $h>1$, every $(q+1)$-arc $\mathscr{K}$ is projectively equivalent to $\mathscr{C}(2^n)$ for some n with $(n, h)=1$.*

Proof: By Theorem 21.3.12, take $\mathscr{K}$ as $\mathscr{C}(F)$, where

$$\mathscr{C}(F)=\{P(t)=\mathbf{P}(tF(t), F(t), t, 1)\mid t\in\gamma^+\}$$

with $F(0)=0$, $F(1)=1$ and $F(x)=a_2x^2+a_4x^4+\ldots+a_{q-2}x^{q-2}$. Recall that $\mathbf{U}_0$, $\mathbf{U}_3$, $\mathbf{U}$ are on $\mathscr{K}$. By exactly the same argument, $\mathscr{K}$ may also be written as

$$\mathscr{K}=\{\mathbf{P}(1, s, G(s), sG(s))\mid s\in\gamma^+\}$$

with $G(0)=0$, $G(1)=1$ and $G(x)=b_2x^2+b_4x^4+\ldots+b_{q-2}x^{q-2}$. So, for $s, t\neq 0, \infty$,

$$\frac{tF(t)}{1}=\frac{F(t)}{s}=\frac{t}{G(s)}=\frac{1}{sG(s)},$$

whence

$$st = 1 = F(t)G(s).$$

Thus, for all t in γ_0,

$$F(t)G(t^{-1}) = 1;$$

that is,

$$(a_2t^2 + a_4t^4 + \ldots + a_{q-2}t^{q-2})(b_2t^{-2} + b_4t^{-4} + \ldots + b_{q-2}t^{-(q-2)}) = 1$$

or

$$(a_2t^2 + a_4t^4 + \ldots + a_{q-2}t^{q-2})(b_{q-2}t + b_{q-4}t^3 + \ldots + b_2t^{q-3}) = 1 = t^{q-1}.$$

Therefore, with $c_i = b_{q-1-i}$, we have for all t in γ,

$$(a_2t^2 + a_4t^4 + \ldots + a_{q-2}t^{q-2})(c_1t + c_3t^3 + \ldots + c_{q-3}t^{q-3}) = t^{q-1}. \quad (21.11)$$

When the left-hand side of (21.11) is reduced modulo $t^q - t$, then (21.11) is an identity. Suppose (21.11) is, with the zero coefficients omitted,

$$(a_rt^r + \ldots + a_Rt^R)(c_mt^m + \ldots + c_Mt^M) = t^{q-1}. \quad (21.12)$$

Each term $a_ic_jt^{i+j}$ has $i+j$ odd with $3 \leqslant i+j \leqslant 2q-5$. But, if $i+j > q-1$, then $a_ic_jt^{i+j} = a_ic_jt^{i+j-(q-1)}$. The term of degree $R+M$ is $a_Rc_Mt^{R+M}$. As there are no terms of even degree on the right-hand side of (21.12), so $R+M \leqslant q-1$. The term of degree $r+m$ on the left-hand side of (21.12) is $a_rc_mt^{r+m}$. Therefore

$$r+m = R+M = q-1,$$

whence $r = R$ and $m = M$. Thus $F(x) = x^r$ and

$$\mathscr{K} = \mathscr{C}(r) = \{P(t) = \mathbf{P}(t^{r+1}, t^r, t, 1) \mid t \in \gamma^+\}.$$

The projectivity $\mathfrak{T}$ given by

$$\mathbf{P}(x_0, x_1, x_2, x_3)\mathfrak{T} = \mathbf{P}(x_3, x_2, x_1, x_0)$$

fixes $\mathscr{K}$ but interchanges $\mathbf{U}_0$ and $\mathbf{U}_3$, which are arbitrarily chosen points of $\mathscr{K}$. So $G(\mathscr{K})$ acts transitively on $\mathscr{K}$. The contact plane to $\mathscr{K}$ at $\mathbf{U}_3$ is $\mathbf{u}_0$ and meets $\mathscr{K}$ at points $P(t)$, where $t^{r+1} = 0$, that is $r+1$ times at $\mathbf{U}_3$. The contact plane at $P(s)$ is the tangent plane to H_3 at $P(s)$ and is

$$\boldsymbol{\pi}(1, s, s^r, s^{r+1}),$$

which meets $\mathscr{K}$ at points $P(t)$ where

$$t^{r+1} + t^rs + ts^r + s^{r+1} = 0;$$

that is,

$$(t+s)(t^r + s^r) = 0. \quad (21.13)$$

By the transitivity of $G(\mathscr{K})$, equation (21.13) has the solution $t=s$ of multiplicity $r+1$. So $t^r+s^r=(t+s)^r$, whence $r=2^n$. Lemma 21.3.14 implies that $(n,h)=1$. □

For $q=2$, we regard $\mathscr{C}(2)$ as the set $\{\mathbf{U}_0, \mathbf{U}_3, \mathbf{U}\}$.

Now we determine which $(q+1)$-arcs are projectively distinct. Given $\mathscr{K}=\mathscr{C}(F)$, let $\bar{\mathscr{K}}=\mathscr{C}(\bar{F})$ where $F(\bar{F}(x))=\bar{F}(F(x))=x$. If $\mathscr{K}$ and $\mathscr{K}'$ are projectively equivalent, write $\mathscr{K}\sim\mathscr{K}'$.

Lemma 21.3.16:

(i) $\bar{\mathscr{K}}\sim\mathscr{K}$;

(ii) $\bar{\mathscr{C}}(r)=\mathscr{C}(q/r)$;

(iii) $\bar{\mathscr{C}}(2^n)=\mathscr{C}(2^{h-n})$;

(iv) $\bar{\mathscr{C}}(2^n)=\mathscr{C}(2^n) \Leftrightarrow n=h/2$;

(v) *if* $(n,h)=1$, *then* $\bar{\mathscr{C}}(2^n)=\mathscr{C}(2^n) \Leftrightarrow h=2, n=1$;

(vi) *if* $\mathscr{C}(2^m)$ *and* $\mathscr{C}(2^n)$ *are* $(q+1)$*-arcs, then* $\mathscr{C}(2^m)\sim\mathscr{C}(2^n) \Leftrightarrow m=n$ *or* $m=h-n$.

Proof: (i)

$$\begin{aligned}\mathscr{K}&=\{\mathbf{P}(tF(t), F(t), t, 1)\}\\ &\sim\{\mathbf{P}(tF(t), t, F(t), 1)\}\\ &=\{\mathbf{P}(\bar{F}(F(t))F(t), \bar{F}(F(t)), F(t), 1)\}\\ &=\{\mathbf{P}(t\bar{F}(t), \bar{F}(t), t, 1)\}\\ &=\bar{\mathscr{K}}.\end{aligned}$$

(ii)–(v) These follow immediately.

(vi) Each $\mathscr{C}(r)$ contains $\mathbf{U}_0$, $\mathbf{U}_3$, and $\mathbf{U}$. Also each generator of $\mathscr{H}_3=\mathbf{V}(x_0x_3+x_1x_2)$ contains one point of a $(q+1)$-arc $\mathscr{C}(r)$. So a projectivity $\mathfrak{T}$ fixing $\mathscr{C}(r)$ fixes $\mathscr{H}_3$, and $\mathfrak{T}$ either fixes both reguli of $\mathscr{H}_3$ or interchanges them. If $\mathfrak{T}$ fixes both reguli and $\mathbf{U}_0$, $\mathbf{U}_3$, $\mathbf{U}$, it is the identity. If $\mathfrak{T}$ interchanges the reguli, then $\mathscr{C}(2^n)\mathfrak{T}=\bar{\mathscr{C}}(2^n)=\mathscr{C}(2^{h-n})$. Hence $m=n$ or $m=h-n$. □

Corollary: *Let* $\mathscr{K}$ *be a* $(q+1)$*-arc in* $PG(3,q)$, q *even.*

(i) *If* $q>4$, *then* $G(\mathscr{K})=PGL(2,q)$.

(ii) *If* $q=4$, *then* $G(\mathscr{K})=\mathbf{Z}_2PGL(2,4)\cong\mathbf{S}_5$.

Proof: By Lemma 21.3.13, $G(\mathscr{K})>PGL(2,q)$. The same argument as part (vi) of the lemma gives the result. □

The results of this section are summarized in the final theorem. Here $\phi(h)$ is the Euler totient function.

Theorem 21.3.17: *In* $PG(3,q)$ *with* $q=2^h$, *let* $\mathscr{K}$ *be a* $(q+1)$*-arc.*

(i) $\mathscr{K}\sim\mathscr{C}(2^n)$ *for some* n *with* $(n,h)=1$.

(ii) $\mathscr{C}(2^n)$ *is a twisted cubic if and only if* $n=1$ *or* $h-1$.

(iii) *For $q>4$, the group $G(\mathscr{K})$ is $PGL(2,q)$; for $q=4$, $G(\mathscr{K})$ is* $\mathbf{S}_5$.
(iv) *The number N_h of projectively distinct $(q+1)$-arcs satisfies*

$$N_h = 1 \quad \text{for} \quad h = 1, 2,$$
$$N_h = \tfrac{1}{2}\phi(h) \quad \text{for} \quad h \geqslant 3;$$

h	1	2	3	4	5	6	7	8	9	10	11	12	13	14	15	16
N_h	1	1	1	1	2	1	3	2	3	2	5	2	6	3	4	4

(v) *In $PG(3,q)$, q even, every $(q+1)$-arc is a twisted cubic if and only if $q=2$, 4, 8, 16, or 64.* □

21.4 Further properties of the twisted cubic

The fact that the chords of a twisted cubic $\mathscr{C}$ pass exactly once through each point off $\mathscr{C}$ enables us to construct examples of spreads and complete spans that complement the examples of Chapter 17. The construction can be mirrored for the other examples of $(q+1)$-arcs in § 21.3, but this is omitted.

Theorem 21.4.1: *In $PG(3,q)$ with $(q+1,3)=1$, the imaginary chords and the tangents to $\mathscr{C}$ form a complete (q^2+q+2)-span $\mathscr{P}$.*

Proof: By Theorem 21.1.9(i), $\mathscr{P}$ is a k-span. By Lemma 21.1.4, $\mathscr{P} = \mathcal{O}_2 \cup \mathcal{O}_3$ *and* $k=(q+1)+q(q-1)/2=(q^2+q+2)/2$.

If l is an external line of $\mathscr{C}$ other than a chord, then, by Theorem 21.1.9(ii), there is a unique chord of $\mathscr{C}$ through each of its points. If P_1P_2 and P_1P_3 are real chords of $\mathscr{C}$ meeting l with P_1, P_2, P_3 on $\mathscr{C}$, then P_2P_3 is also a real chord of $\mathscr{C}$ meeting l. As the plane P_1l can only meet $\mathscr{C}$ in three points (over any extension of γ), there is no other chord meeting l and containing P_1, P_2, or P_3. Thus the chords of $\mathscr{C}$ meeting l comprise

m triangles of real chords,
n_2 real chords each skew to the other real chords meeting l,
n_1 tangents,
n_0 imaginary chords.

Counts of the points on l and $\mathscr{C}$ on these chords give respectively

$$3m + n_2 + n_1 + n_0 = q+1,$$
$$3m + 2n_2 + n_1 \leqslant q+1.$$

If $n_0+n_1=0$, then $n_2=0$ and $3m=q+1$. As $(q+1,3)=1$, this is impossible. So $n_0+n_1 \geqslant 1$, which means that some line of $\mathscr{P}$ meets l. □

We now investigate the case that $(q+1,3)=3$. This is equivalent to saying that $q \equiv -1 \pmod 3$ and that x^2+x+1 is irreducible over γ. This in turn means that, for q odd, -3 is a non-square and, for q even, that q is an odd power of 2 and that 1 is in $\mathscr{C}_1$ (see § 1.5(xii)).

For q odd, $y^2+y+1=0$ and $y=x/(2b)-(c+1)/2$ imply that $F_{bc}=0$, where

$$F_{bc}=x^2-2bcx+b^2(c^2+3).$$

As b varies in γ_0 and c in γ, the quadratics F_{bc} vary through all $q(q-1)/2$ monic irreducible quadratics, § 1.6(vii), since $F_{bc}=F_{de}$ if and only if $(d, e)=(b, c)$ or $(-b, -c)$. Therefore, from (21.5), the set of imaginary chords of $\mathscr{C}$ is

$$\begin{aligned}\mathcal{O}_3=\{L(b, c)\\ =\mathbf{l}(b^4(c^2+3)^2, 2b^3c(c^2+3), 3b^2(c^2-1), b^2(c^2+3), -2bc, 1)\};\end{aligned} \tag{21.14}$$

from (21.7), the set of imaginary axes of Γ is

$$\begin{aligned}\mathcal{O}_3'=\{L'(b, c)\\ =\mathbf{l}(b^4(c^2+3)^2, 2b^3c(c^2+3), 3b^2(c^2+3), b^2(c^2-1), -2bc, 1)\}.\end{aligned} \tag{21.15}$$

For q even, $y^2+y+1=0$ and $y=x/R+\sigma$ imply that $G_{R\sigma}=0$, where

$$G_{R\sigma}=x^2+Rx+R^2\rho \quad \text{with} \quad \rho=\sigma^2+\sigma+1.$$

As R varies in γ_0 and σ in γ, the quadratics $G_{R\sigma}$ vary through all $q(q-1)/2$ monic irreducible quadratics, § 1.6(g), since $G_{R\sigma}=G_{S\mu}$ if and only if $(S, \mu)=(R, \sigma)$ or $(R, \sigma+1)$. So, from (21.5) and (21.7) respectively, the set of imaginary chords of $\mathscr{C}$ is

$$\mathcal{O}_3=\{L(R, \rho)=\mathbf{l}(R^4\rho^2, R^3\rho, R^2(\rho+1), R^2\rho, R, 1)\}, \tag{21.16}$$

and the set of imaginary axes of Γ is

$$\mathcal{O}_3'=\{L'(R, \rho)=\mathbf{l}(R^4\rho^2, R^3\rho, R^2\rho, R^2(\rho+1), R, 1)\}. \tag{21.17}$$

Theorem 21.4.2: *In $PG(3, q)$ with $(q+1, 3)=3$, the set $\mathscr{S}=\mathcal{O}_2\cup\mathcal{O}_3\cup\mathcal{O}_3'$, comprising the tangents to $\mathscr{C}$, the imaginary chords of $\mathscr{C}$ and the imaginary axes of Γ, is a spread. $\mathscr{S}$ contains no regulus if q is odd and exactly one regulus if q is even with $q>2$.*

Proof: From Lemma 21.1.4, $|\mathscr{S}|=q^2+1$. By Theorem 21.1.9 and its corollary, the only possibility of two lines of $\mathscr{S}$ intersecting is a pair (l, l') with l in $\mathcal{O}_3$ and l' in $\mathcal{O}_3'$.

For q odd, let $l=L(b, c)$ and $l'=L'(B, C)$ as in (21.14) and (21.15). Then

$$\varpi(l, l')=[b^2(c^2+3)+B^2(C^2+3)-2bcBC]^2+12b^2B^2.$$

If $\varpi(l, l')=0$, then $-12b^2B^2$ is a square, whence -3 is a square since $bB\neq 0$. This is impossible for $(q+1, 3)=3$.

For q even, let $l=L(R, \rho)$, $l'=L'(S, \mu)$ with $\rho=\sigma^2+\sigma+1$, $\mu=\nu^2+\nu+1$, as in (21.16) and (21.17). Then

$$\begin{aligned}\varpi(l, l') &= (R^2\rho+S^2\mu)^2+RS(R^2\rho+S^2\mu)+R^2S^2(\rho+\mu+1)\\ &= \alpha^2+T\alpha+T^2(\tau^2+\tau+1),\end{aligned}$$

where $\alpha=R^2\rho+S^2\mu$, $T=RS$, $\tau=\sigma+\nu$. If $\varpi(l, l')=0$, then $G_{T\tau}=x^2+Tx+T^2(\tau^2+\tau+1)$ is reducible, which is again impossible. Hence, for q both even and odd, $\mathcal{S}$ is a spread.

If q is even and $q>2$, then $q\geqslant 8$. As the tangents to $\mathscr{C}$ form a regulus (Theorem 21.1.2(iii)), any other regulus $\mathscr{R}$ in $\mathcal{S}$ contains at most two tangents and therefore at least four lines of $\mathcal{O}_3$ or four lines of $\mathcal{O}_3'$. If $\mathscr{R}$ contains four lines of $\mathcal{O}_3$, then the quadric $\mathscr{H}(\mathscr{R})$ contains $\mathscr{C}$. So, by Lemma 21.1.8, $\mathscr{R}$ contains at least $(q-1)/2$ real chords of $\mathscr{C}$, whence $\mathscr{R}$ is not contained in $\mathcal{S}$. If $\mathscr{R}$ contains four lines of $\mathcal{O}_3'$, then application of the null polarity $\mathfrak{A}$ induced by $\mathscr{C}$ (Theorem 21.1.2(iv)), gives the previous situation.

For q odd and $q\geqslant 11$, the same argument applies and $\mathcal{S}$ contains no regulus. For $q=5$, the only possibilities for a regulus $\mathscr{R}$ in $\mathcal{S}$ such that $\mathscr{C}$ does not lie on $\mathscr{H}(\mathscr{R})$ or $\mathscr{H}(\mathscr{R}\mathfrak{A})$ are that the lines of $\mathscr{R}$ are partitioned as follows:

$\mathcal{O}_2$	$\mathcal{O}_3$	$\mathcal{O}_3'$
2	2	2 .
0	3	3

In the first case, the two tangents may be taken as $l(\infty)$ and $l(0)$. Then an inspection of the list of the ten lines of $\mathcal{O}_3$ and the ten lines of $\mathcal{O}_3'$ shows that $\mathscr{R}$ cannot exist. For the second case, a longer computation is necessary, but again there is no regulus. □

We now resume from § 21.1 the consideration of the net $\mathcal{N}$ of quadrics containing $\mathscr{C}=\{P(t)=\mathbf{P}(t^3, t^2, t, 1)\mid t\in\gamma^+\}$. From Lemma 21.1.6, the cubic $\mathscr{C}$ lies on all $\mathscr{F}_{\lambda\mu\nu}$ in $\mathcal{N}$, where

$$\mathscr{F}_{\lambda\mu\nu}=\mathbf{V}(\lambda(x_0x_3-x_1x_2)+\mu(x_0x_2-x_1^2)+\nu(x_1x_3-x_2^2));$$

for $q\geqslant 7$, these are the only quadrics containing $\mathscr{C}$. We recall that $\mathscr{F}_{\lambda\mu\nu}$ is a cone for $\lambda^2=\mu\nu$ and a hyperbolic quadric otherwise.

Lemma 21.4.3: *For q odd, $\mathscr{F}_{\lambda\mu\nu}$ contains 2, 1, or 0 tangents to $\mathscr{C}$ according as $\lambda^2-\mu\nu$ is a non-zero square, zero, or a non-square. For q even, $\mathscr{F}_{\lambda\mu\nu}$ contains $q+1$ or 1 tangent to $\mathscr{C}$ according as $\mu=\nu=0$ or not.*

Proof: From (21.6), the tangent $l(t)$ to $\mathscr{C}$ at $P(t)$ is

$$\mathbf{l}(t^4, 2t^3, 3t^2, t^2, -2t, 1);$$

this is the join of $\mathbf{P}(3t^2, 2t, 1, 0)$ and $\mathbf{P}(-2t^3, -t^2, 0, 1)$. If these two points lie on $\mathscr{F}_{\lambda\mu\nu}$, then with $P(t)$ there are three points of $l(t)$ on $\mathscr{F}_{\lambda\mu\nu}$ and so the whole line lies on $\mathscr{F}_{\lambda\mu\nu}$. The conditions for the two points to lie on $\mathscr{F}_{\lambda\mu\nu}$ are

$$2t\lambda + t^2\mu + \nu = 0,$$
$$t^2(2t\lambda + t^2\mu + \nu) = 0.$$

So $l(t)$ lies on $\mathscr{F}_{\lambda\mu\nu}$ if and only if $\mu t^2 + 2\lambda t + \nu = 0$, whence the lemma follows, by § 1.4(i) for q odd, and immediately for q even. □

Theorem 21.4.4: *The mapping* $\mathfrak{T}: \mathcal{N} \to PG(2, q)$ *given by* $\mathscr{F}_{\lambda\mu\nu}\mathfrak{T} = \mathbf{P}(\lambda, \mu, \nu)$ *has the following properties.*

(i) *The cones in* $\mathcal{N}$ *map to points of the conic* $\mathscr{P}_2 = \mathbf{V}(x_0^2 - x_1x_2)$.

(ii) *For q odd,* $\mathbf{P}(\lambda, \mu, \nu)$ *is external to* $\mathscr{P}_2$, *on* $\mathscr{P}_2$, *or internal to* $\mathscr{P}_2$, *according as* $\mathscr{F}_{\lambda\mu\nu}$ *contains* 2, 1, *or* 0 *tangents to* $\mathscr{C}$.

(iii) *For q even,* $\mathbf{P}(\lambda, \mu, \nu)$ *is the nucleus of* $\mathscr{P}_2$, *on* $\mathscr{P}_2$, *or another point of the plane, according as* $\mathscr{F}_{\lambda\mu\nu}$ *contains all* $q+1$ *tangents of* $\mathscr{C}$, *is a cone, or is a hyperbolic quadric containing exactly one tangent of* $\mathscr{C}$.

(iv) *The chords of* $\mathscr{C}$ *map to lines of the plane; if* $\mathscr{F}_{\lambda\mu\nu} \cap \mathscr{F}_{\lambda'\mu'\nu'} = \mathscr{C} + l$, *then* $l\mathfrak{T} = \mathbf{P}(\lambda, \mu, \nu)\mathbf{P}(\lambda', \mu', \nu')$. *Also,* $l\mathfrak{T}$ *is a bisecant, a tangent, or an external line to* $\mathscr{P}_2$, *according as l is a real chord, a tangent, or an imaginary chord of* $\mathscr{C}$.

Proof: (i) $\mathscr{F}_{\lambda\mu\nu}$ is a cone if and only if $\lambda^2 - \mu\nu = 0$, by Lemma 21.1.6.

(ii) This follows from the previous lemma and from Theorem 8.3.2; see also § 8.2.

(iii) The nucleus of $\mathbf{V}(x_0^2 - x_1x_2)$ is $\mathbf{P}(1, 0, 0)$ and the regulus of tangents to $\mathscr{C}$ lies on the quadric $\mathbf{V}(x_0x_3 - x_1x_2) = \mathscr{F}_{100}$.

(iv) Any two quadrics of $\mathcal{N}$ meet residually in a chord of $\mathscr{C}$, by Lemma 21.1.6, Corollary 3. Conversely, through $\mathscr{C}$ and any of its chords, there passes a pencil of quadrics in $\mathcal{N}$, by Lemma 21.1.6, Corollary 2.

Explicitly, the chord $l(t_1, t_2)$ of (21.5) lies on $\mathscr{F}_{\lambda\mu\nu}$ if a point of it besides $P(t_1)$ and $P(t_2)$ does. So, consider

$$\mathbf{P}(t_1^2 + t_1t_2 + t_2^2, t_1 + t_2, 1, 0) = \mathbf{P}(\alpha_1^2 - \alpha_2, \alpha_1, 1, 0),$$

where $\alpha_1 = t_1 + t_2$, $\alpha_2 = t_1t_2$. This point lies on $\mathscr{F}_{\lambda\mu\nu}$ if

$$\lambda\alpha_1 + \mu\alpha_2 + \nu = 0.$$

Hence

$$l(t_1, t_2)\mathfrak{T} = \boldsymbol{\pi}(\alpha_1, \alpha_2, 1).$$

But the line $\boldsymbol{\pi}(\alpha_1, \alpha_2, 1)$ meets $\mathbf{V}(x_0^2 - x_1x_2)$ where

$$x_0^2 + \alpha_1 x_0 x_1 + \alpha_2 x_1^2 = 0.$$

So $\boldsymbol{\pi}(\alpha_1, \alpha_2, 1)$ is a bisecant, a tangent, or an external line of $\mathscr{P}_2$ according as $x^2 + \alpha_1 x + \alpha_2$ has 2, 1, or 0 roots in γ. However, by the definition in § 21.1, $l(t_1, t_2)$ is a real chord, a tangent or an imaginary tangent to $\mathscr{C}$ according as $x^2 - \alpha_1 x + \alpha_2$ has 2, 1, or 0 roots in γ. □

There is another representation of $\mathscr{N}$ in the plane with attractive properties. Let π be a plane meeting $\mathscr{C}$ in a trio of 3-conjugate complex points; that is, if $\pi = \boldsymbol{\pi}(a_0, a_1, a_2, a_3)$, then $F(x) = a_0x^3 + a_1x^2 + a_2x + a_3$ is irreducible over γ. For any q, there exists d in γ such that

$$f_0(x) = dx^3 + x + 1$$

is irreducible over γ; see § 1.10. For definiteness, we will take this polynomial in what follows. Thus $\pi = \boldsymbol{\pi}(d, 0, 1, 1)$.

Now, π contains no chords of $\mathscr{C}$, as otherwise f_0 would have at least one root in γ. So π meets each hyperbolic quadric in $\mathscr{N}$ in a conic. As π contains no real point of $\mathscr{C}$, it does not pass through the vertex of any cone in $\mathscr{N}$, by Lemma 21.1.7(ii); so π meets each cone in $\mathscr{N}$ in a conic.

Let $\mathscr{N}' = \{\pi \cap \mathscr{F}_{\lambda\mu\nu} \mid \mathscr{F}_{\lambda\mu\nu} \in \mathscr{N}\}$; that is, $\mathscr{N}'$ is the net of conics in π through the three complex points Q_1, Q_2, Q_3 in which π meets $\mathscr{C}$, where $Q_i = P(t_i)$ with t_1, t_2, t_3 the roots of f_0.

Theorem 21.4.5: *The map $\mathfrak{S}: \mathscr{N} \to \mathscr{N}'$ given by $\mathscr{F}_{\lambda\mu\nu}\mathfrak{S} = \pi \cap \mathscr{F}_{\lambda\mu\nu}$ has the following properties.*

(i) *If l is a chord of $\mathscr{C}$, then $l\mathfrak{S} = l \cap \pi$.*

(ii) *Every two conics in $\mathscr{N}'$ have exactly one real point P in common; if $\mathscr{F} \cap \mathscr{F}' = \mathscr{C} + l$, then $\mathscr{F}\mathfrak{S} \cap \mathscr{F}'\mathfrak{S} = \{Q_1, Q_2, Q_3, P = l \cap \pi\}$.*

(iii) *Every two conics in $\mathscr{N}'$ have a unique common tangent.*

Proof: We prove (iii) by considering the conics

$$\mathscr{C}_1 = \mathbf{V}(x_0x_2 - x_1^2) = \{P'(t) = \mathbf{P}(t^2, t, 1) \mid t \in \gamma^+\},$$
$$\mathscr{C}_2 = \mathbf{V}(dx_0x_1 + x_1x_2 + x_2^2).$$

They meet in $\mathbf{U}_0$ and the 3-conjugate complex points $P'(t_i)$, where t_i is a root of $f_0(x) = dx^3 + x + 1$. The tangent l' to $\mathscr{C}_1$ at $P'(t)$ is $\mathbf{V}(x_0 - 2tx_1 + t^2x_2)$. This meets $\mathscr{C}_2$ where

$$2dtx_1^2 + (1 - dt^2)x_1x_2 + x_2^2 = 0.$$

Hence, for q even, l' is tangent to $\mathscr{C}_2$ if and only if $t = 1/\sqrt{d}$. Alternatively, the nucleus P_1 of $\mathscr{C}_1$ is $\mathbf{U}_1$ and the nucleus P_2 of $\mathscr{C}_2$ is $\mathbf{P}(1, 0, d)$, by Lemma 7.2.3, Corollary 5. So the unique common tangent of $\mathscr{C}_1$ and $\mathscr{C}_2$ is P_1P_2. Similarly, all conics in $\mathscr{N}'$ have a different nucleus.

For q odd, l' is tangent to $\mathscr{C}_2$ if and only if $(1-d^2t^2)-4\cdot 2dt=0$; that is,

$$F(t)=d^2t^4-2dt^2-8dt+1=0.$$

It must be shown that F has exactly one root in γ.

Let $f(x)=a_0x^4+a_1x^3+a_2x^2+a_3x+a_4$; its resolvent cubic ϕ is given by

$$\phi(x)=x^3+2a_2x^2+(a_1a_3+a_2^2-4a_0a_4)x+(a_1a_2a_3-a_0a_3^2-a_1^2a_4),$$

by Lemma 1.11.3. Also $\Delta(\phi)=\Delta(f)$, by Lemma 1.11.4, Corollary 1. If ϕ is irreducible, then f has 0 or 1 root in γ, by Theorem 1.11.6. Further, if ϕ is irreducible, then $\Delta(\phi)$ is a square, by Theorem 1.8.8, Corollary 1(ii). Conversely, if $\Delta(f)$ is a non-zero square, then, by part (iii) of the same corollary, f has one root in γ or four roots in γ or is the product of two irreducible quadratics. So, if ϕ is irreducible, either f has exactly one root in γ or is the product of two irreducible quadratics.

The resolvent Φ of F is

$$\Phi(x)=x^3-4dx^2-64d^4.$$

Let

$$\Phi_1(x)=-\Phi(-4dx)/(64d^3)=x^3+x^2+d$$

and

$$\Phi_2(x)=x^3\Phi_1(1/x)=dx^3+x+1.$$

As Φ_2 is irreducible, so is Φ.

If we show that F is not the product of two irreducible quadratics, then F has exactly one root in γ. Suppose therefore that

$$d^2x^4-2dx^2-8dx+1=(d^2x^2+b_1x+c_1)(x^2+b_2x+c_2).$$

Then

$$b_1+b_2d^2=0, \qquad b_1b_2+c_1+c_2d^2=-2d,$$
$$b_1c_2+b_2c_1=-8d, \qquad c_1c_2=1.$$

The elimination of b_1, c_1, c_2 gives

$$b_2^6d^2-4b_2^4d-64=0.$$

So b_2^2 is a root of

$$\Phi_0(x)=d^2x^3-4dx^2-64.$$

However, $d^4\Phi_0(x/d^2)=\Phi(x)$, which is irreducible. So Φ_0 is irreducible, and F is not the product of two irreducible quadratics. □

21.5 Notes and references

§ 21.1. Theorem 21.1.1 for q odd was given by Segre (1955b), who also considered the twisted cubic for q even in Segre (1956). The remainder of the section, apart from the derivation of plane singular cubics is based on Hirschfeld (1971) and Bruen and Hirschfeld (1977). See also Hirschfeld (1967a), D'Orgeval (1967b), Lüneburg (1980, § 43).

§ 21.2. This comes from Segre (1955b) apart from Theorem 21.2.5, which is part of a more general theorem of Thas (1968a).

§ 21.3. As far as Theorem 21.3.10, this account follows Casse (1969). The result of Theorem 21.3.8, that $m(3, q) = \max(5, q+1)$, was also obtained by Gulati and Kounias (1970). For a survey of generalizations of this result, see Hirschfeld (1983b). Theorem 21.3.12 and Lemmas 21.3.13 and 21.3.14 come from Hirschfeld (1971), while the form of the polynomial F comes from Segre and Bartocci (1971); see also Theorem 8.4.2, Corollary 1. The main result, Theorem 21.3.15, follows Casse and Glynn (1982). Casse and Glynn (1984) have also shown that a q-arc in $PG(3, q)$, q even and $q \geq 8$, is contained in a unique $(q+1)$-arc.

§ 21.4. Theorems 21.4.1 and 21.4.2 follow Bruen and Hirschfeld (1977). The corresponding translation planes were found by Hering (1970), (1971) for q odd and by Ott (1975*) for q even. Bruen and Hirschfeld (1977*) extended Theorems 21.4.1 and 21.4.2 to the $(q+1)$-arcs $\mathscr{C}(2^n)$. The corresponding translation planes were found by H.-J. Schaeffer, Translationsebenen auf denen die Gruppe $SL(2, p^n)$ operiert, Diplomarbeit, Tübingen, 1975. For an account of the planes, see Lüneburg (1980, §§ 43–45). See also Liebler (1978).

k-arcs give rise to *maximum distance separable* or *optimal* codes; see MacWilliams and Sloane (1977, Chapter 11). For k-arcs and matroid theory, see Fenton and Vámos (1982).

APPENDIX III

Orders of and isomorphisms among the semi-linear groups

AIII.1 Definitions

(i) We define the groups $DX(n, q)$ for $n \geqslant 2$ and $q = p^h$, where

$$D = I, S, S', G, \Gamma, \Gamma S, P, PS, PS', PG, P\Gamma, P\Gamma S$$

and

$$X = L, O, O_+, O_-, U, Sp, Ps, Ps^*.$$

For $D = I$, S, S', G, Γ, ΓS, the group $DX(n, q)$ is a group of semi-linear transformations in $V(n, q)$. For $D = P$, PS, PS', PG, $P\Gamma$, $P\Gamma S$, the group $DX(n, q)$ is a group of collineations in $PG(n-1, q)$. The symbol X refers to the form (quadratic, Hermitian, bilinear) involved. In particular, the full linear and semi-linear groups of varieties and polarities in $V(n, q)$ are given by the respective rows $GX(n, q)$ and $\Gamma X(n, q)$ of Table AIII.1. The full projective and collineation groups of varieties or polarities in $PG(n-1, q)$ are given by the respective rows $PGX(n, q)$ and $P\Gamma X(n, q)$.

(ii) $IX(n, q) = X(n, q)$.

$X(n, q)$ = group of linear transformations leaving invariant the canonical form specified in the table.

$SX(n, q)$ = subgroup of X of linear transformations with determinant one.

$S'X(n, q) = SX(n, q)'$ = commutator subgroup of $SX(n, q)$.

$GX(n, q)$ = group of linear transformations leaving the canonical form invariant up to a scalar multiple.

$\Gamma X(n, q)$ = group of semi-linear transformations leaving the canonical form invariant up to an automorphism of $\gamma = GF(q)$ and a scalar multiple.

$\Gamma SX(n, q)$ = group of semi-linear transformations that are products of an element of $SX(n, q)$ and an automorphism of γ.

(iii) The groups $DX(n, q)$ beginning with the symbol 'P' are the corresponding projective groups.

The centre of $\Gamma L(n, q)$ is $Z = \{\lambda I \mid \lambda \in \gamma_0,\ I \text{ identity of } \Gamma L(n, q)\}$. Then, PX, PSX, $PS'X$, PGX, $P\Gamma X$, $P\Gamma SX$ are the respective canonical images of

X, SX, $S'X$, GX, ΓX, ΓSX under $\Gamma L(n, q) \to \Gamma L(n, q)/Z$. *Thus*

$$PX(n, q) = X(n, q)/[Z \cap X(n, q)], \qquad PSX(n, q) = SX(n, q)/[Z \cap SX(n, q)],$$

$$PS'X(n, q) = S'X(n, q)/[Z \cap S'X(n, q)] = PSX(n, q)',$$

$$PGX(n, q) = GX(n, q)/[Z \cap GX(n, q)],$$

$$P\Gamma X(n, q) = \Gamma X(n, q)/[Z \cap \Gamma X(n, q)],$$

$$P\Gamma SX(n, q) = \Gamma SX(n, q)/[Z \cap \Gamma SX(n, q)].$$

(iv) Commonly, one sees

$$\Gamma SX = \Sigma X, \qquad P\Gamma SX = P\Sigma X,$$

$$S'O = \Omega, \qquad PS'O = P\Omega,$$

$$S'O_+ = \Omega_+, \qquad PS'O_+ = P\Omega_+,$$

$$S'O_- = \Omega_-, \qquad PS'O_- = P\Omega_-,$$

with the lower indices often written as upper indices.

(v) For q square, it is sometimes opportune to consider only the involutory and the identity automorphisms of $\gamma = GF(q)$ instead of all automorphisms. Then we write γX instead of ΓX and $P\gamma X$ instead of $P\Gamma X$.

(vi) Commonly, but not here, when q is even, the orthogonal groups $SO_+(n, q)$ and $SO_-(n, q)$ are the kernels, not of the determinant map, but taken with respect to the Dickson invariant. In that case, $|SO_\pm(n, q)| = |O_\pm(n, q)|/2$.

AIII.2 Comparative orders and isomorphisms in the same column of Table AIII.1

For all X,

$$|GX|/(q-1) = |\Gamma X|/[h(q-1)] = |P\Gamma X|/h = |PGX|.$$

Now we consider the rows I, S, P, PS, PG.

(i) $X = L$

Orders:

$$|L| = |GL|, \qquad |PGL| = |SL| = |PL| = |L|/(q-1),$$

$$|PSL| = |SL|/(n, q-1).$$

Isomorphisms:

$$L = GL, \qquad PL = PGL, \qquad PGL \cong SL \cong PSL \iff (n, q-1) = 1.$$

(ii) $X = 0$

Orders:

$$|PGO| = |PSO| = |PO| = |SO| = |O|/(q-1, 2).$$

Table AIII.1 *Orders of the Semi-Linear Groups* $DX(n, q)$ $(q = p^h,\ n \geq 2)$. The $\lambda_\varepsilon(m, r) = r^{m(m-1)/2}\prod_{i=1}^{m}(r^i - \varepsilon^i)$; $l_{ij} = x_i y_j - x_j y_i$; $|n, q| =$ order of the group

X:	L	O	O_+	O_-
F:	$(q-1)^{-1}\times$	$q^{(n-1)/2}\times$	$q^{n-2}(q^{n/2}-1)\times$	$q^{n-2}(q^{n/2}+1)\times$
	$\lambda_1(n, q)\times$	$\lambda_1\left(\frac{n-1}{2}, q^2\right)\times$	$\lambda_1\left(\frac{n-2}{2}, q^2\right)\times$	$\lambda_1\left(\frac{n-2}{2}, q^2\right)\times$
D				
I	$q-1$	$(q-1, 2)$	2	2
S	1	1	$(q, 2)$	$(q, 2)$
S' (†)	1	$(q-1, 2)^{-1}$	$(q-1, 2)^{-1}$	$(q-1, 2)^{-1}$
(†) except	$\|2, 2\| = 3$	$\|3, 2\| = 3$	$\|4, 2\| = 18$	
	$\|2, 3\| = 8$	$\|5, 2\| = 360$		
G	$q-1$	$q-1$	$2(q-1)$	$2(q-1)$
Γ	$h(q-1)$	$h(q-1)$	$2h(q-1)$	$2h(q-1)$
ΓS	h	h	$h(q, 2)$	$h(q, 2)$
P	1	1	$(q, 2)$	$(q, 2)$
PS	$(n, q-1)^{-1}$	1	$2^{(-1)^q}$	$2^{(-1)^q}$
PS' (†)	$(n, q-1)^{-1}$	$(q-1, 2)^{-1}$	$(q^{n/2}-1, 4)^{-1}$	$(q^{n/2}+1, 4)^{-1}$
(†) except	$\|2, 2\| = 3$	$\|3, 2\| = 3$	$\|4, 2\| = 18$	
	$\|2, 3\| = 4$	$\|5, 2\| = 360$		
PG	1	1	2	2
$P\Gamma$	h	h	$2h$	$2h$
$P\Gamma S$	$h(n, q-1)^{-1}$	h	$2^{(-1)^q}h$	$2^{(-1)^q}h$
n	all	odd	even	even
q	all	all	all	all
Invariant in $PG(n-1, q)$	$PG(n-1, q)$	$\mathscr{P}_{n-1}$	$\mathscr{H}_{n-1}$	$\mathscr{E}_{n-1}$
Canonical form of invariant	–	$x_1^2 + x_2x_3 + \ldots + x_{n-1}x_n$	$x_1x_2 + x_3x_4 + \ldots + x_{n-1}x_n$	$f(x_1, x_2) + x_3x_4 + \ldots + x_{n-1}x_n$, f irreducible

orders are the table entries multiplied by the factors F given for each column. $DX(n, q)$; (n_1, n_2) = greatest common divisor of n_1 and n_2.

U	Sp	Ps	Ps^*	$:X$
$(\sqrt{q}+1)^{-1}\times$	$q^{n/2}\times$	$q^{(n-1)/2}\times$	$q^{(3n-4)/2}\times$	$:F$
$\lambda_{-1}(n, \sqrt{q})\times$	$\lambda_1\left(\frac{n}{2}, q^2\right)\times$	$\lambda_1\left(\frac{n-1}{2}, q^2\right)\times$	$\lambda_1\left(\frac{n-2}{2}, q^2\right)\times$	
				D
$\sqrt{q}+1$	1	1	1	I
1	1	1	1	S
1	1	1	1	S' (†)
$\lvert 2, 4\rvert = 3$	$\lvert 2, 2\rvert = 3$	$\lvert 3, 2\rvert = 3$	$\lvert 2, q\rvert = 1$	(†) except
$\lvert 2, 9\rvert = 8$	$\lvert 2, 3\rvert = 8$	$\lvert 5, 2\rvert = 360$	$\lvert 4, 2\rvert = 12$	
$\lvert 3, 4\rvert = 54$	$\lvert 4, 2\rvert = 360$		$\lvert 6, 2\rvert = 11\,520$	
$q-1$	$q-1$	$q-1$	$q-1$	G
$h(q-1)$	$h(q-1)$	$h(q-1)$	$h(q-1)$	Γ
h	h	h	h	ΓS
1	$(q-1, 2)^{-1}$	1	1	P
$(n, \sqrt{q}+1)^{-1}$	$(q-1, 2)^{-1}$	1	1	PS
$(n, \sqrt{q}+1)^{-1}$	$(q-1, 2)^{-1}$	1	1	PS' (†)
$\lvert 2, 4\rvert = 3$	$\lvert 2, 2\rvert = 3$	$\lvert 3, 2\rvert = 3$	$\lvert 2, q\rvert = 1$	(†) except
$\lvert 2, 9\rvert = 4$	$\lvert 2, 3\rvert = 4$	$\lvert 5, 2\rvert = 360$	$\lvert 4, 2\rvert = 12$	
$\lvert 3, 4\rvert = 9$	$\lvert 4, 2\rvert = 360$		$\lvert 6, 2\rvert = 11\,520$	
1	1	1	1	PG
h	h	h	h	$P\Gamma$
$h(n, \sqrt{q}+1)^{-1}$	$h(q-1, 2)^{-1}$	h	h	$P\Gamma S$
all	even	odd	even	n
square	all	even	even	q
$\mathcal{U}_{n-1}$	null polarity	pseudo polarity	pseudo polarity	Invariant in $PG(n-1, q)$
$x_1^{\sqrt{q}+1} + \ldots + x_n^{\sqrt{q}+1}$	$l_{12}+l_{34} + \ldots + l_{n-1,n}$	$x_1y_1+l_{23}+l_{45} + \ldots + l_{n-1,n}$	$x_1y_1+l_{12}+ l_{34}+\ldots+l_{n-1,n}$	Canonical form of invariant

Isomorphisms:

$$q \text{ even}, \quad PGO \cong PSO \cong PO \cong SO \cong O;$$

$$q \text{ odd}, \quad PGO \cong PSO \cong PO \cong SO.$$

(iii) $X = O_+, O_-$
Orders:

$$q \text{ even}, \quad |PGX| = |PSX| = |PX| = |SX| = |X|;$$

$$q \text{ odd}, \quad 4\,|PSX| = 2\,|SX| = 2\,|PX| = |X| = |PGX|.$$

Isomorphisms:

$$q \text{ even}, \quad PGX \cong PSX \cong PX \cong SX \cong X.$$

(iv) $X = U$
Orders:

$$|PGU| = |SU| = |PU| = |U|/(\sqrt{q}+1),$$

$$|PSU| = |SU|/(n, \sqrt{q}+1).$$

Isomorphisms:

$$PGU \cong PSU \cong SU \cong PU \;\Leftrightarrow\; (n, \sqrt{q}+1) = 1.$$

(v) $X = Sp$
Orders:

$$|PGSp| = |SSp| = |Sp|,$$

$$|PSSp| = |PSp| = |Sp|/(q-1, 2).$$

Isomorphisms:

$$q \text{ even}, \quad PGSp \cong PSSp \cong PSp \cong SSp \cong Sp;$$

$$q \text{ odd}, \quad SSp \cong Sp, \; PSSp \cong PSp.$$

(vi) $X = Ps, Ps^*$
Orders:

$$|PGX| = |PSX| = |PX| = |SX| = |X|.$$

Isomorphisms:

$$PGX \cong PSX \cong PX \cong SX \cong X.$$

In rows S' and PS' of Table AIII.1, some specific equations $|n, q| = N$ appear. This indicates that $|DX(n, q)| = N$ and that for this value of n and q, the order of $DX(n, q)$ is not given by the general formula appearing immediately above.

AIII.3 Simple groups

(i) $PSL(n, q)$ except $PSL(2, 2)$, $PSL(2, 3)$;
(ii) $PS'O(n, q)$ except $PS'O(5, 2)$, $PS'O(3, 2)$, $PS'O(3, 3)$;
(iii) $PS'O_+(n, q)$ except $PS'O_+(4, q)$, $PS'O_+(2, q)$ when $(q-1)/(q-1, 4)$ is not prime;
(iv) $PS'O_-(n, q)$ except $PS'O_-(2, q)$ when $(q+1)/(q+1, 4)$ is not prime;
(v) $PSU(n, q)$ except $PSU(3, 4)$, $PSU(2, 4)$, and $PSU(2, 9)$;
(vi) $PSp(n, q)$ except $PSp(4, 2)$, $PSp(2, 2)$, and $PSp(2, 3)$;
(vii) $PPs(n, q)$ except $PPs(5, 2)$, $PPs(3, 2)$;
(viii) $PPs^*(n, q)$ except $PPs^*(6, 2)$, $PPs^*(4, 2)$, and $PPs^*(2, q)$ for $q > 2$.

AIII.4 Isomorphisms between $DX(n, q)$ and $\mathbf{A}_m$ or $\mathbf{S}_m$

$\mathbf{S}_3 \cong PGL(2, 2) \cong PSL(2, 2)$	§ 6.4(i)
$\mathbf{A}_4 \cong PSL(2, 3)$	§ 6.4(ii)
$\mathbf{S}_4 \cong PGL(2, 3)$	§ 6.4(ii)
$\mathbf{A}_5 \cong PGL(2, 4) \cong PSL(2, 4)$	§ 6.4(iii)
$\mathbf{A}_5 \cong PSL(2, 5)$	§ 6.4(iv)
$\mathbf{S}_5 \cong P\Gamma L(2, 4)$	§ 6.4(iii)
$\mathbf{S}_5 \cong PGL(2, 5)$	§ 6.4(iv)
$\mathbf{A}_6 \cong PSL(2, 9)$	§ 6.4(vii)
$\mathbf{A}_6 \cong PS'O_-(4, 3)$	§ 6.4(vii), Theorem 15.3.17
$\mathbf{S}_6 \cong P\Gamma SL(2, 9)$	§ 6.4(vii)
$\mathbf{S}_6 \cong PO_-(4, 3)$	§ 6.4(vii), Theorem 15.3.17
$\mathbf{S}_6 \cong PGSp(4, 2)$	Theorems 17.5.5, 20.3.1
$\mathbf{S}_6 \cong PGO(5, 2)$	Theorem 17.5.5
$\mathbf{A}_8 \cong PGL(4, 2)$	Theorem 17.5.5
$\mathbf{A}_8 \cong PS'O_+(6, 2)$	Theorem 17.5.5
$\mathbf{S}_8 \cong PGO_+(6, 2)$	Theorem 17.5.5, Corollary 1
$\mathbf{S}_8 \cong$ group of projectivities and correlations of $PG(3, 2)$	Theorem 17.5.5, Corollary 1.

AIII.5 Isomorphisms among $DX(n, q)$ for different characteristic p

$PSL(2, 7) \cong PSL(3, 2)$;
$PSL(2, 5) \cong PSL(2, 4)$;
$PGL(2, 5) \cong P\Gamma L(2, 4)$;
$P\Gamma SL(2, 9) \cong PSp(4, 2)$;
$PSp(4, 3) \cong PS'O(5, 3) \cong PSU(4, 4)$;
$\cong$ group of 27 lines of a cubic surface fixing each half of the double-sixes (order 25 920);
$PGSp(4, 3) \cong PGO(5, 3) \cong P\Gamma U(4, 4) \cong PGO_-(6, 2)$
$\cong$ full group of 27 lines of a cubic surface (order 51 840).

AIII.6 Isomorphisms among $DX(n, q)$ for the same characteristic p

These isomorphisms all follow from the projective equivalence of a pair of geometrical configurations. In each case, the geometrical fact (A) is followed by the concomitant isomorphisms (B).

(i)(A) A linear complex in $PG(1, q)$ is $PG(1, q)$.

(B) (1) $P\Gamma Sp(2, q) \cong P\Gamma L(2, q)$;

(2) $PGSp(2, q) \cong PGL(2, q)$;

(3) $PSp(2, q) \cong PSL(2, q)$ (SIMPLE for $q > 3$).

(ii)(A) For q square, $\mathcal{U}_{1,q}$ is a subline $PG(1, \sqrt{q})$ on $PG(1, q)$, Lemma 6.2.1.

(B) For q square,

(1) $PGU(2, q) \cong PGL(2, \sqrt{q})$;

(2) $PSU(2, q) \cong PSL(2, \sqrt{q})$ (SIMPLE for $q > 9$).

(iii)(A) In $PG(2, q)$, there is a bijection between a conic $\mathcal{P}_2$ and a line Π_1, by Lemma 7.2.2.

(B) (1) $P\Gamma O(3, q) \cong P\Gamma L(2, q)$;

(2) $PGO(3, q) \cong PGL(2, q)$;

(3) $PS'O(3, q) \cong PSL(2, q)$ (SIMPLE for $q > 3$).

(iv)(A) For q square, there is a bijection from the set of sublines $PG(1, \sqrt{q})$ of $PG(1, q)$ to the set $PG(3, \sqrt{q}) \backslash \mathcal{E}_{3,\sqrt{q}}$, the points off an elliptic quadric $\mathcal{E}_{3,\sqrt{q}}$, Theorem 15.3.11.

(B) For q square,

(1) $P\Gamma L(2, q) \cong P\Gamma O_-(4, \sqrt{q})$;

(2) $P\gamma L(2, q) \cong PGO_-(4, \sqrt{q})$;

(3) $P\gamma SL(2, q) \cong PO_-(4, \sqrt{q})$;

(4) $PSL(2, q) \cong PS'O_-(4, \sqrt{q})$ (SIMPLE).

(v)(A) A quadric $\mathcal{H}_{3,q}$ has as generators two complementary reguli, § 15.3, III.

(B) (1) $PGL(2, q) \times PGL(2, q) \cong$ subgroup of index two of $PGO_+(4, q)$ fixing both reguli;

(2) $PSL(2, q) \times PSL(2, q) \cong PS'O_+(4, q)$ for $q > 2$.

(vi)(A) The Grassmann map $\mathfrak{G}: PG^{(1)}(3, q) \to \mathcal{H}_{5,q}$ is a bijection, § 15.4.

(B) (1) The group of collineations and reciprocities of $PG(3, q) \cong P\Gamma O_+(6, q)$;

(2) The group of projectivities and correlations of $PG(3, q) \cong PGO_+(6, q)$;

(3) $P\Gamma L(4, q) \cong$ subgroup of index two of $P\Gamma O_+(6, q)$ leaving both systems of generators of $\mathcal{H}_{5,q}$ fixed;

(4) $PGL(4, q) \cong$ subgroup of index two of $PGO_+(6, q)$ leaving both systems of generators of $\mathcal{H}_{5,q}$ fixed;

(5) $PSL(4, q) \cong PS'O_+(6, q)$ (SIMPLE).

(vii)(A) Under $\mathfrak{G}$, a linear complex maps to $\mathscr{P}_{4,q}$, § 15.4.

(B) (1) $P\Gamma Sp(4, q) \cong P\Gamma O(5, q)$;

(2) $PGSp(4, q) \cong PGO(5, q)$;

(3) $PSp(4, q) \cong PS'O(5, q)$ for $q > 2$ (SIMPLE).

(viii)(A) Under $\mathfrak{G}$, for q square, $\mathscr{U}^{(1)}_{3,q}$ maps to $\mathscr{E}_{5,\sqrt{q}}$ on $\mathscr{H}_{5,q}$, Theorem 19.2.2.

(B) For q square,

(1) $P\Gamma U(4, q) \cong P\Gamma O_-(6, \sqrt{q})$;

(2) $P\gamma U(4, q) \cong PGO_-(6, \sqrt{q})$;

(3) $PGU(4, q) \cong$ subgroup of index two of $PGO_-(6, \sqrt{q})$;

(4) $PSU(4, q) \cong PS'O_-(6, \sqrt{q})$ (SIMPLE).

(ix)(A) In $PG(2m, q)$ for q even, the quadric $\mathscr{P}_{2m,q}$ induces a null polarity on any subspace Π_{2m-1} not through the nucleus, Theorem 5.3.4.

(B) For q even,

(1) $P\Gamma O(2m+1, q) \cong P\Gamma Sp(2m, q)$;

(2) $PGO(2m+1, q) \cong PGSp(2m, q)$;

(3) $PSO(2m+1, q) \cong PSp(2m, q)$ (SIMPLE for $(m, q) \neq (1, 2)$, $(2, 2)$);

(4) in particular, $PGO(7, 2) \cong PGSp(6, 2) \cong$ group of 28 bitangents of a non-singular quartic curve, § 20.7.

(x)(A) In $PG(2m, q)$, q even, the self-conjugate lines of a pseudo-polarity form a linear complex in the Π_{2m-1} of self-conjugate points, Theorem 5.3.3.

(B) For q even,

(1) $P\Gamma Ps(2m+1, q) \cong P\Gamma Sp(2m, q)$;

(2) $PGPs(2m+1, q) \cong PGSp(2m, q)$;

(3) $Ps(2m+1, q) \cong Sp(2m, q)$ (SIMPLE for $(m, q) \neq (1, 2)$, $(2, 2)$).

AIII.7 Notes and references

§ AIII.1. The same definitions are used in Appendix I in *Projective geometries over finite fields*. The reason for using a lower index plus or minus for the orthogonal groups is that Dieudonné (1971) uses an upper index plus to indicate determinant one. The symbol $S'O$ has been used instead of Ω to indicate the origin of the definition, namely the commutator subgroup of SO. Although this is confusing for the experts, it was felt to be more helpful for easy reference. The symbol ΓS has been preferred to Σ for the same reason.

§ AIII.2. For remarks on this section, see Appendix I.

§ AIII.3. For proofs of (i)–(vi), see Dieudonné (1971), Huppert (1967),

Dickson (1901), Higman (1978). For a detailed analysis of (vii) and (viii), I am most grateful to G. E. Wall.

§§ AIII.4–6. Apart from the references for § AIII.3, see van der Waerden (1935), Dieudonné (1954), Dembowski (1968), Artin (1955a, 1955b), Biggs (1971), Blichfeldt (1917), Burnside (1911), Carmichael (1937), Jordan (1870).

APPENDIX IV

The Number of Points on an Algebraic Variety

AIV.1 The Weil conjectures

This appendix amplifies the discussion of the Hasse-Weil theorem of § 10.2 and corrects the definition of the number M of model points of a curve in $PG(2, q)$.

Let F be a form of degree m in $\gamma[x_0, x_1, \ldots, x_n]$ and let $\mathscr{F} = \mathbf{V}_{n,q}(F)$ be an absolutely irreducible, non-singular primal in $PG(n, q)$. Also, let $\mathscr{F}_i = \mathbf{V}_{n,q^i}(F)$ and let $N_i = |\mathscr{F}_i|$; that is, N_i is the number of points over $GF(q^i)$ of the variety defined by F, which has all its coefficients in $\gamma = GF(q)$. The *zeta function* of $\mathscr{F}$ is

$$\zeta(\mathscr{F}) = \zeta(F; x; q) = \exp\left(\sum N_i x^i/i\right).$$

The Hasse–Weil–Dwork–Deligne theorem says the following for primals.

Theorem AIV.1.1:

(i) $\zeta(\mathscr{F}) = f(x)^{(-1)^n}[(1-x)(1-qx)\ldots(1-q^{n-1}x)]^{-1}$.

(ii) $f(x) = (1-\alpha_1 x)\ldots(1-\alpha_r x) \in 1 + x\mathbf{Z}[x]$ *and has the properties*:

(a) $r = [(m-1)/m][(m-1)^n - (-1)^n]$;

(b) *if* α_j^{-1} *is a root of* f *in* $\mathbf{C}$, *then so is* α_j/q^{n-1};

(c) $|\alpha_j| = q^{(n-1)/2}$, $j = 1, \ldots, r$. □

Note that if $\alpha_j \in \mathbf{R}$, that is $\alpha_j^{-1} = \pm q^{(n-1)/2}$, then (b) does not imply the existence of another root of f.

Corollary 1:

$$N_i = 1 + q^i + q^{2i} + \ldots + q^{(n-1)i} + (-1)^{n+1}(\alpha_1^i + \ldots + \alpha_r^i).$$

Proof: Take logarithms of both sides in (i) of the theorem and expand formally. □

Corollary 2:

$$|N_i - (1 + q^i + \ldots + q^{(n-1)i})| \leq r q^{(n-1)i/2}.$$

Proof: By Corollary 1,

$$\begin{aligned} |N_i - (1 + q^i + \ldots + q^{(n-1)i})| &= |\alpha_1^i + \ldots + \alpha_r^i| \\ &\leq |\alpha_1|^i + \ldots + |\alpha_r|^i \\ &= r q^{(n-1)i/2}. \quad \square \end{aligned}$$

In particular, if $f(x) = 1 + c_1 x + c_2 x^2 + \ldots + c_r x^r$, we obtain the following.

Corollary 3:

$$N_1 = 1+q+q^2+\ldots+q^{n-1}+(-1)^{n+1}(\alpha_1+\ldots+\alpha_r)$$
$$= 1+q+q^2+\ldots+q^{n-1}+(-1)^n c_1$$

and

$$|N_1-(1+q+\ldots+q^{n-1})| \leqslant rq^{(n-1)/2}. \quad \square$$

Corollary 4: *For non-singular curves of order m in PG(2, q),*

$$|N_1-(1+q)| \leqslant (m-1)(m-2)\sqrt{q}. \quad \square$$

Corollary 5: *For non-singular surfaces of order m in PG(3, q),*

$$|N_1-(1+q+q^2)| \leqslant (m-1)(m^2-3m+3)q. \quad \square$$

Example 1: Let $\mathcal{F}$ be a non-singular quadric primal in $PG(n, q)$, so that $m=2$, whence $r=\frac{1}{2}[1-(-1)^n]$.

When n is even, $r=0$, $f(x)=1$ and $N_1=1+q+q^2+\ldots+q^{n-1}$; this is the case of the parabolic quadric, § 5.2.

When n is odd, $r=2$ and $f(x)=1+c_1x$. By (ii)(c) of the theorem, $c_1=\pm q^{(n-1)/2}$. So, Corollary 3 gives $N_1=1+q+\ldots+q^{n-1}\pm q^{(n-1)/2}$ corresponding to the cases that the quadric is hyperbolic or elliptic; see § 5.2.

Example 2: Let $\mathcal{F}$ be a cubic surface with 27 lines in $PG(3, q)$. By Theorem 20.2.3 and its corollary, $N_i = q^{2i}+7q^i+1$, whence

$$\zeta(\mathcal{F}) = [(1-x)(1-qx)^7(1-q^2x)]^{-1}.$$

Note that the upper bound in Corollary 5 is achieved in this case.

Example 3: Let $\mathcal{F}$ be a Hermitian variety $\mathcal{U}_n$ in $PG(n, q)$, q square. Then, from Theorem 5.1.8,

$$N_1 = [q^{(n+1)/2}+(-1)^n][q^{n/2}-(-1)^n]/(q-1).$$

With $m=\sqrt{q}+1$, we have $r=[\sqrt{q}/(\sqrt{q}+1)][(\sqrt{q})^n-(-1)^n]$ and

$$1+q+\ldots+q^{n-1}+rq^{(n-1)/2} = (q^n-1)/(q-1)+q^{n/2}[q^{n/2}-(-1)^n]/(\sqrt{q}+1)$$
$$= N_1.$$

So $\mathcal{U}_n$ gives an example of high order for which the upper bound in Corollary 3 is attained.

AIV.2 Curves

If $\mathcal{F}$ is a plane curve of order m with ordinary singularities over $\bar{\gamma}$ of multiplicities $r_1, \ldots, r_s$, then the *genus*

$$g = \tfrac{1}{2}(m-1)(m-2) - \sum_{i=1}^{s} r_i(r_i-1).$$

For any absolutely irreducible curve $\mathscr{F}$, the number M of *model points* is the number of points on a non-singular model of $\mathscr{F}$. In § 21.1, it is shown that a singular, plane cubic $\mathscr{F}$ is the projection of a twisted cubic, whether the singularity of $\mathscr{F}$ is a node, a cusp or an isolated double point; hence $M = q+1$. In these cases, the singularity of $\mathscr{F}$ was counted according to the number of tangents at the point lying over γ. This is not always true for more complicated singularities as is shown by analysing the tacnode at $\mathbf{U}_0$ of the curve

$$\mathscr{C} = \mathbf{V}[x_0^2(x_1+x_2)^2 + x_0(x_1+x_2)^3 + x_1^4 + x_2^4].$$

Two principles for estimating M may be enunciated:

(1) each simple point of $\mathscr{F}$ is counted once;

(2) if $\mathscr{F}$ has N simple points and s singularities of multiplicities $r_1, \ldots, r_s$, then

$$N \leq M \leq N + r_1 + \ldots + r_s.$$

Now, Theorem 10.2.1 and its corollaries are correct.

For curves, we restate Theorem AIV.1.1 in a slightly more general situation: $\mathscr{F}$ is not necessarily plane.

Theorem AIV.2.1: *If $\mathscr{F}$ is an absolutely irreducible non-singular curve of genus g, then*

$$\begin{aligned}\zeta(\mathscr{F}) &= \exp\left(\sum N_i x^i/i\right)\\ &= \frac{1 + c_1 x + \ldots + c_{2g-1}x^{2g-1} + q^g x^{2g}}{(1-x)(1-qx)}\\ &= \frac{(1-\beta_1 x + qx^2)\ldots(1-\beta_g x + qx^2)}{(1-x)(1-qx)},\end{aligned}$$

where each root α^{-1} of the numerator has $|\alpha| = \sqrt{q}$. □

Corollary: *With $\mathscr{F}$ as in the theorem,*

$$|N_1 - (q+1)| \leq 2g\sqrt{q}. \quad \square$$

An improvement is given by the next result, where $[y]$ denotes the integral part of y.

Theorem AIV.2.2:

$$|N_1 - (q+1)| \leq g[2\sqrt{q}]. \quad \square$$

For the case of elliptic curves, that is with $g = 1$, more can be said.

Theorem AIV.2.3: *For an elliptic curve $\mathscr{F}$,*

$$\zeta(\mathscr{F}) = \frac{1 - \beta x + qx^2}{(1-x)(1-qx)}.$$

Also

(i) $(\sqrt{q}-1)^2 \leq N_1 \leq (\sqrt{q}+1)^2$;

(ii) $N_1 = q+1-\beta$;

(iii) $N_2 = (q+1)^2 - \beta^2 = N_1[2(q+1)-N_1]$;

(iv) $N_3 = q^3+1+3q\beta-\beta^3$
$= N_1[3(q^2+q+1)-3(q+1)N_1+N_1^2]$. □

Example 4: Let $q=2$ and $F = x_0^3+x_1^3+x_2^3$. Since $y^3 = y$ over $GF(2)$, so $\mathscr{F}$ is a line. Hence $3 = 1+q-\beta = 3-\beta$ and $\beta = 0$. Thus

$$\zeta(\mathscr{F}) = (1+2x^2)/[(1-x)(1-2x)],$$
$$\log \zeta(\mathscr{F}) = \sum x^i/i + \sum (2x)^i/i + 2\sum (-1)^{j-1}(2x^2)^j/(2j).$$

Therefore,

$$\text{for} \quad h \equiv 1 \pmod 2,\ N_h = 1+2^h;$$
$$\text{for} \quad h \equiv 2 \pmod 4,\ N_h = 1+2^h+2\cdot 2^{h/2};$$
$$\text{for} \quad h \equiv 0 \pmod 4,\ N_h = 1+2^h-2\cdot 2^{h/2}.$$

The question posed in § 11.9 as to the possible number of points on a plane non-singular cubic has been completely answered.

Theorem AIV.2.4: *For every integer $N = q+1-t$ with $|t| \leq 2\sqrt{q}$, there exists a non-singular elliptic curve over $GF(q)$, $q = p^h$, with exactly N points, providing one of the following holds:*

	Condition on t	*Condition on p^h*
(i)	*p does not divide t*	
(ii)	$t=0$	h odd or $p \not\equiv 1 \pmod 4$
(iii)	$t = \pm\sqrt{q}$	h even and $p \not\equiv 1 \pmod 3$
(iv)	$t = \pm 2\sqrt{q}$	h even
(v)	$t = \pm\sqrt{(2q)}$	h odd and $p=2$
(vi)	$t = \pm\sqrt{(3q)}$	h odd and $p=3$. □

Corollary 1: *The number N assumes every integer value in the interval $q+1-[2\sqrt{q}] \leq N \leq q+1+[2\sqrt{q}]$ if and only if*

(i) $q = p$,

or

(ii) $q = p^2$ *with* $p=2$ *or* $p=3$ *or* $p \equiv 11 \pmod{12}$. □

Define $N_q(g) = \max N_1$ and $L_q(g) = \min N_1$, where N_1 is the number of points on $\mathscr{C}$ over $GF(q)$ and $\mathscr{C}$ runs over all non-singular curves of genus g defined over $GF(q)$. Then the Hasse–Weil theorem, that is the corollary to Theorem AIV.2.1, says that

$$N_q(g) \leq q+1+[2g\sqrt{q}], \tag{AIV.1}$$

whereas Theorem AIV.2.2 says that

$$N_q(g) \leqslant q+1+g[2\sqrt{q}]. \qquad \text{(AIV.2)}$$

The integer $q = p^h$ we will call *exceptional* if h is odd, $h \geqslant 3$ and p divides $[2\sqrt{q}]$. The only exceptional integers less than 10 000 are $2^7 = 128$, $2^{11} = 2048$ and $3^7 = 2187$. From Theorem AIV.2.4, the following is immediately deduced.

Corollary 2:

(i) $N_q(1) = \begin{cases} q+[2\sqrt{q}], & \text{if } q \text{ is exceptional} \\ q+1+[2\sqrt{q}], & \text{otherwise;} \end{cases}$

(ii) $L_q(1) = \begin{cases} q+2-[2\sqrt{q}], & \text{if } q \text{ is exceptional} \\ q+1-[2\sqrt{q}], & \text{otherwise.} \end{cases}$ □

Table AIV.1 gives the values of $N_q(1)$ and $L_q(1)$ for $q \leqslant 128$, as well as those values of N between $N_q(1)$ and $L_q(1)$ which are not the number of points on an elliptic curve.

The value of $N_q(2)$ has also been determined for all q. The number $q = p^h$ is *special* if h is odd and one of the following holds:

(a) p divides $[2\sqrt{q}]$;
(b) $q = n^2+1$ for some n in $\mathbf{Z}$;
(c) $q = n^2+n+1$ for some n in $\mathbf{Z}$;
(d) $q = n^2+n+2$ for some n in $\mathbf{Z}$.

Write $\{y\} = y - [y]$, the fractional part of y.

Theorem AIV.2.5: (i) *If q is special, then*

$$N_q(2) = \begin{cases} q+2[2\sqrt{q}], & \text{if } \{2\sqrt{q}\} > \tfrac{1}{2}(\sqrt{5}-1) \\ q-1+2[2\sqrt{q}], & \text{otherwise.} \end{cases}$$

(ii) *If q is not special, then*

$$N_q(2) = \begin{cases} 2q+2, & \text{if } q = 4 \text{ or } 9 \\ q+1+2[2\sqrt{q}], & \text{otherwise.} \end{cases} \quad □$$

For $q \leqslant 128$, the only special q with $\{2\sqrt{q}\} > \frac{1}{2}(\sqrt{5}-1)$ are $q = 2$, 8, 128; the other special q in this range are 3, 5, 7, 13, 17, 31, 32, 37, 43, 73, 101.

Finally, some results for larger g are given.

Lemma AIV.2.6: *For $g = \frac{1}{2}(q-\sqrt{q})$,*

$$N_q(g) = q+1+2g\sqrt{q}.$$

Proof: The Hermitian curve $\mathscr{U}_{2,q}$, § 7.3, is non-singular of degree $\sqrt{q}+1$ and hence of genus $g = \frac{1}{2}\sqrt{q}(\sqrt{q}-1) = \frac{1}{2}(q-\sqrt{q})$. It has $q\sqrt{q}+1$ points, attaining thus the upper bound of (AIV.1). □

Table AIV.1 *The Number of Points on an Elliptic Curve for $q \leqslant 128$*

q	$L_q(1)$	$N_q(1)$	Forbidden N
2	1	5	
3	1	7	
4	1	9	
5	2	10	
7	3	13	
8	4	14	7, 11
9	4	16	
11	6	18	
13	7	21	
16	9	25	11, 15, 19, 23
17	10	26	
19	12	28	
23	15	33	
25	16	36	26
27	18	38	22, 25, 31, 34
29	20	40	
31	21	43	
32	22	44	23, 27, 29, 31, 35, 37, 39, 43
37	26	50	
41	30	54	
43	31	57	
47	35	61	
49	36	64	43, 57
53	40	68	
59	45	75	
61	47	77	
64	49	81	51, 53, 55, 59, 61, 63, 67, 69, 71, 75, 77, 79
67	52	84	
71	56	88	
73	57	91	
79	63	97	
81	64	100	67, 70, 76, 79, 85, 88, 94, 97
83	66	102	
89	72	108	
97	79	117	
101	82	122	
103	84	124	
107	88	128	
109	90	130	
113	93	135	
121	100	144	
125	104	148	106, 111, 116, 121, 131, 136, 141, 146
127	106	150	
128	108	150	109, 111, 115, 117, 119, 121, 123, 125, 127, 131, 133, 135, 137, 139, 141, 143, 147, 149

Theorem AIV.2.7:

$$N_q(g) \leqslant q+1-\tfrac{1}{2}g+\sqrt{[2(q+\tfrac{1}{8})g^2+(q^2-q)g]}. \quad \square$$

This theorem gives a better result that (AIV.1) and (AIV.2) when $g > \frac{1}{2}(q-\sqrt{q})$.

To discover what happens for large g, let

$$A_q = \limsup N_q(g)/g.$$

Then (AIV.1) implies that $A_q \leqslant 2\sqrt{q}$ and Theorem AIV.2.7 that $A_q \leqslant \frac{1}{2}[\sqrt{(8q+1)}-1]$.

Theorem AIV.2.8:

$$A_q \leqslant \sqrt{q}-1,$$

and equality holds when q is a square. □

AIV.3 Notes and references

§ AIV.1. See § 10.5 for some history. Theorem AIV.1.1 was proved as far as (ii)(b) by Dwork (1960) and completely by Deligne (1974). For an exposition of Dwork's proof, see Koblitz (1984). See also Katz (1976*).

§ AIV.2. Theorem AIV.2.1 is Weil's theorem (1948a) and Theorem AIV.2.3 is Hasse's (1934). Theorem AIV.2.2 is due to Serre (1983b). Theorem AIV.2.4 is due to Waterhouse (1969*); for another proof, see Ughi (1983). On elliptic curves, see also Honda (1968*), Tate (1974*), Bedocchi (1980, 1981), De Groote and Hirschfeld (1980). The number of isomorphism classes of elliptic curves having a fixed number of points has been determined by R. J. Schoof; for a discussion of these results see Hirschfeld (1983d). In the latter paper there are several incorrect statements: the Counting Principle (corrected at the beginning of § AIV.2), Corollary 2 of Theorem 1 (this may be false for n odd), Corollary 2 of Theorem 5 (corrected in Corollary 2 of Theorem AIV.2). The author is grateful to J.-P. Serre for the corrections. Theorem AIV.2.5 is also due to Serre (1983a, 1983b), who gives values of $N_q(3)$ for some q and $N_2(g)$ for some g. Theorem AIV.2.7 is due to Ihara (1981).

Goppa (1981, 1983) derived from curves over $GF(q)$ some remarkable codes. This was responsible for the investigation of A_q; see Manin (1981). The inequality of Theorem AIV.2.8 is due to Drinfeld and Vladut (1983), while the equality for q square is shown both by Ihara (1981) and Tsfasman, Vladut, and Zink (1982); see also Serre (1983b). For other accounts of Goppa's codes see Tsfasman (1982), Beth (1982), Voloch (1983), Hirschfeld (1984).

APPENDIX V

Errata for *Projective geometries over finite fields*

Where there is no ambiguity, only the correct version is given. Below 28^3 means 'page 28, line 3' whereas 28_5 means 'page 28, line 5 from the bottom'.

2^{17}	$\mathscr{G}[x]$
4^4	Add '(1.4)'
6^{13}	$\lvert\mathscr{N}(m, q)\rvert$
11^7	$\Delta(F)$
11^{13}	Delete '1.8.1 and'
12^4	$\alpha_i^q = \alpha_{i+1}$, where $\alpha_{n+1} = \alpha_1$.
12_6	$F = F_1F_2 \ldots F_k$
13^1	$t \in \gamma_0$
13^3	$C: \gamma_0 \cup \{2\} \rightarrow \{0, 1\}$
17^8	in γ
17_3	$c^{2^{h-1}}$
18_3	$3(b_1\beta_1^2 + b_2\beta_2^2)$
18_2	$\dfrac{a_2 - 3a_0\beta_1^2}{a_2 - 3a_0\beta_2^2}$
21^1	$-A_0^3 b_1 b_2(\beta_1 - \beta_2)^6$
21^2	(1.8)
21^{19}	then α_1
23^5	Add '(1.10)'
24_8	$G = a_0 x_1^2$
27^4	$s \notin \gamma$
29_{20}	For 'are' put 'is'
30^2	Π_s,
31^6	simplifies
33_{11}	$\tilde{T}T^{*-1}$
34_{15}	$T = -T^*$
36	Ia(2) . . . $t_{ij} = -t_{ji}$
	II . . . $t_{ij} = \bar{t}_{ji}$
44^1	Delete 'of . . . set'
61^8	planes
65_{13}	$X = (x_0, x_1, \ldots, x_n)$
77_7	*regular*
79^5	difference
82^6	$\bar{\mathbf{N}}_l$
82^8	$\bar{\mathbf{N}}_l$
82_6	$\bar{\mathbf{N}}_l$
85_6	For 'faces' put 'layers'
91_{14}	$\mathfrak{S} = \mathfrak{A}\mathfrak{S}'\mathfrak{A}^{-1}$
92–96	running heads should read 'PARTITIONS'
94_6	$\rho_0(n, q)$
98_6	$x_0(t_{10}\tilde{x}_1 + \ldots + t_{n0}\tilde{x}_n)$
98_3	$i \geqslant 1$
98_2	$i \geqslant 1$
102^{12}	$Q = x_0x_1 + \sum_{i=2}^{n}{}' \, b_{ij}x_ix_j$
102_3	number $\mu(n, q)$ of points
105_5	$(x_1 - bx_0)(x_1 + bx_0)$
114^{11}	$+(x_{2s-1}y_{2s} + x_{2s}y_{2s-1})$
115_4	§ 5.3
125^6	$x^2 - a_1x - a_0$
133_{15}	$t_3 + t_4$
137^{14}	§ 2.8
147^3	theorem 5.1.8
158^6	, then
163_{12}	τ_1
165^{11}	$\mathscr{K} \cap \mathscr{K}'$
167_3	a
177_4	Appendix II
178^7	$\mathbf{Z}_2^h\mathbf{Z}_{q-1}$
178^{17}	$G_2 \ldots \cong \mathbf{Z}_2^h$
179^4	$\mathscr{D}(k_1)$
188^1	$\mathscr{C} =$

BIBLIOGRAPHY

This continues the Bibliography in *Projective geometries over finite fields*. It contains mainly works published since 1978. Some published before 1978 are included, and any reference in the text to these has been embellished with a star. Any work not so embellished and published before 1978 may be found in the Bibliography of *PGOFF*.

As in *PGOFF*, the Bibliography is as complete as possible regarding works dealing with properties of $PG(n, q)$. It also contains representative works of adjacent material. Papers in the course of publication have not been included.

The abbreviations for periodical titles follow the current style of *Mathematical Reviews*.

This Bibliography contains 579 items.

Abatangelo, L. M. (1980). Calotte complete in uno spazio di Galois $PG(3, q)$, q pari. *Atti Sem. Mat. Fis. Univ. Modena* 29, 215–21.

—— (1982). On the number of points of caps obtained from an elliptic quadric of $PG(3, q)$. *European J. Combin.* 3, 1–7.

—— (1983). Complete arcs in $PG(2, q)$, q even. *Ann. Discrete Math.* 18, 13–15.

—— (1984). Fibrazioni di $PG(3, q)$ dotate di particolari famiglie di regoli. *Rend. Sem. Mat. Brescia* 7, 3–11.

—— and Pertichino, M. (1982). Calotte complete in $PG(3, 8)$. *Note Mat.* 2, 131–43.

—— and Raguso, G. (1981a). Una caratterizzazione degli archi chiusi giacenti su una conica di un piano pascaliano di caratteristica due. *Rend. Mat.* 1, 39–45.

—— and —— (1981b). Una caratterizzazione grafica dei k-archi di un piano di Galois d'ordine pari che risultano essere ovali di un sottopiano proprio. *Rend. Mat.* 1, 371–9.

—— and —— (1981c). Una caratterizzazione grafica dei k-archi di un piano di Galois di ordine dispari che risultano essere ovali di un sottopiano proprio. *Note Mat.* 1, 1–17.

—— and —— (1982). On the $n(2n+1)$-set of class $[0, 1, n, 2n]$. *Ann. Discrete Math.* 14, 77–82.

—— and —— (1984). Finite projective planes with an affine partition. *Rend. Mat. Sem. Brescia* 7, 13–18.

—— Abatangelo, V. and Korchmáros, G. (1984). A translation plane of order 25. *J. Geom.* 22, 108–16.

Abatangelo, V. (1983). A class of complete $[(q+8)/3]$-arcs of $PG(2, q)$, with $q=2^h$ and h ($\geqslant 6$) even. *Ars Combin.* 16, 103–11.

—— (1984). Un nuovo procedimento per la construzione di calotte complete di $PG(3, q)$, q pari. *Rend. Sem. Mat. Brescia* 7, 19–25.

—— (1984). A translation plane of order 81 and its full collineation group. *Bull. Austral. Math. Soc.* 29, 19–34.

Aigner, M. and Jungnickel, D. (1981). *Geometries and groups* (Berlin, 1981). Lecture Notes in Math. 893. Springer, Berlin, 250pp.

Amici, O. and Casciaro, B. (1981). Sui q-archi in un piano proiettivo sopra un corpo finito. *Rend. Sem. Fac. Sci. Univ. Cagliari* 51, 65–9.

—— and —— (1983). Intorno ad un teorema di Buekenhout sulle ovali pascaliane. *Ann. Discrete Math.* 18, 17–27.

Arf, C. (1943). Untersuchungen über quadratische Formen in Körper der Charakteristik 2. II. *Rev. Fac. Sci. Univ. Istanbul* (*A*) 8, 297–327.

Arnoux, G. (1911). *Essai de géométrie analytique.* Gauthier-Villars, Paris, 159pp.

Baeza, R. (1978). *Quadratic forms over semilocal rings.* Lecture Notes in Math. 655. Springer, Berlin, 199pp.

—— (1981). Discriminants of polynomials and of quadratic forms. *J. Algebra* 72, 17–28.

Baker, R. D. (1976). Partitioning the planes of $AG_{2m}(2)$ into 2-designs. *Discrete Math.* 15, 205–11.

—— (1984). Orthogonal line packings of $PG_{2m-1}(2)$. *J. Combin. Theory Ser. A* 36, 245–8.

Baldisserri, N. (1983). Sui fasci di coniche e di quadriche in un campo di caratteristica 2. *Boll Un. Mat. Ital. B*2, 483–97.

Barlotti, A. (Ed.) (1982). *Combinatorial and geometric structures and their applications* (Trento, 1980). Ann. Discrete Math. 14. North Holland, 292pp.

——, Ceccherini, P. V., and Tallini, G. (Eds.) (1983). *Combinatorics* '81 (Rome, 1981). Ann. Discrete Math. 18. North Holland, Amsterdam, 823pp.

——, Marchi, M., and Tallini, G. (Eds.) (1984). *Geometria combinatoria e di incidenza: fondamenti e applicazioni* (La Mendola, 1982). Rend. Sem. Mat. Brescia 7, Vita e Pensiero, Milan, 656pp.

Barnabei, M. (1976). Alcuni risultati sui (k, n)-archi con tre caratteri. *Istit. Lombardo Accad. Sci. Lett. Rend. A* 110, 351–8.

—— (1979). On arcs with weighted points. *J. Statist. Plann. Inference* 3, 279–86.

Barnabei, M. and Zucchini, C. (1978). Archi astratti di tipo (m, n). *Boll. Un. Mat. Ital. A* 15, 585–91.

——, Searby, D., and Zucchini, C. (1978). On small $\{k; q\}$-arcs in planes of order q^2. *J. Combin. Theory Ser.* A. 24, 241–6.

Bartocci, U. and Faina, G. (1979). Archi chiusi di un piano proiettivo finito. *Rend. Mat.* 12, 297–311.

—— and Ughi, E. (1981). Terne di quadrati consecutivi in un campo di Galois. *Atti. Accad. Naz. Lincei Rend.* 71, 151–5.

Basile, A. and Brutti, P. (1975). Proprietà delle ovali di un piano desarguesiano finito. *Rend. Circ. Mat. Palermo* 24, 233–43.

—— and —— (1979). Planes and ovals. *J. Geom.* 13, 101–7.

Batten, L. M. and Buekenhout, F. (1981). Quadriques partielles d'indice deux. *J. Geom.* 16, 93–102.

Baumann, B. (1984). Symmetrische Singer-Zyklen über Körpern der Charakteristik 2. *Mitt. Math. Sem. Giessen* 163, 135–40.

Bedocchi, E. (1980). Cubiche ellittiche su F_p. *Boll. Un. Mat. Ital. B* 17, 269–77.

—— (1981). Classi di isomorfismo delle cubiche di F_q. *Rend. Circ. Mat. Palermo* 30, 397–415.

Benz, W. (1979). A functional equation in finite geometry. *Abh. Math. Sem. Univ. Hamburg* 48, 231–40.

Berardi, L. and Eugeni, F. (1984a). Limitazioni inferiori e superiori per le t-fibrazioni massimali in $PG(n, q)$ con fissato livello. *Rend. Sem. Mat. Brescia* 7, 67–82.

—— and —— (1984b). On the cardinality of blocking sets in $PG(2, q)$. *J. Geom.* 22, 5–14.

—— and —— (1984c). On blocking sets in affine planes. *J. Geom.* 22, 167–77.

Bernasconi, C. and Vincenti, R. (1981). Spreads induced by varieties V_2^3 of $PG(4, q)$ and Baer subplanes. *Boll. Un. Mat. Ital. B* 18, 821–30.
Beth, T. (1981). The rank of the $AG(2, n)$ incidence matrix over any prime field. *J. Geom.* 15, 89–92.
—— (1982). Some aspects of coding theory between probability, algebra, combinatorics and complexity theory. *Combinatorial theory*, Lecture Notes in Math. 969, Springer, Berlin, pp. 12–29.
—— and Jungnickel, D. (1981). Mathieu groups, Witt designs, and Golay codes. *Geometries and groups*, Lecture Notes in Math. 893, Springer, Berlin, pp. 157–79.
—— and —— (1982). Variations on seven points: an introduction to the scope and methods of coding theory and finite geometry. *Aequationes Math.* 25, 153–76.
——, ——, and Lenz, H. (1984). Design Theory. *Bibliographisches Institut*, Mannheim, 688pp.
Beutelspacher, A. (1978a). Embeddings of partial spreads in spreads. *Arch. Math.* 30, 317–24.
—— (1978b). Partitions of finite vector spaces: an application of the frobenius number in geometry. *Arch. Math.* 31, 202–8.
—— (1979). On t-covers in finite projective spaces. *J. Geom.* 12, 10–16.
—— (1980). Blocking sets and partial spreads in finite projective spaces. *Geom. Dedicata* 9, 425–49.
—— (1983). On Baer subspaces of finite projective spaces. *Math. Z.* 184, 301–19.
—— (1984). Embedding the complement of a Baer subplane or a unital in a finite projective plane. *Mitt. Math. Sem. Giessen* 163, 189–202.
Bichara, A. (1975). Sui k-archi di un bipiano. *Riv. Mat. Univ. Parma* 1, 247–64.
—— (1977). Sui k-insiemi di $S_{3,q}$ di tipo $((n-1)q+1, nq+1)_2$. *Atti Accad. Naz. Lincei Rend.* 62, 480–8.
—— (1978). Caratterizzazione dei sistemi rigati immersi in $A_{3,q}$. *Riv. Mat. Univ. Parma* 4, 277–90.
—— (1980). Sui k-insiemi di $PG(r, q)$ di classe $[0, 1, 2, n.]_2$. *Rend. Mat.* 13, 359–70.
—— (1981). On k-sets of class $[0, 1, 2, n]_2$ in $PG(r, q)$. *Finite geometries and designs*, London Math. Soc. Lecture Note Series 49, Cambridge University Press, Cambridge, pp. 31–9.
—— and Korchmáros, G. (1980). n^2-sets in a projective plane which determine exactly n^2+n lines. *J. Geom.* 15, 175–81.
—— and —— (1982). Note on $(q+2)$-sets in a Galois plane of order q. *Ann. Discrete Math.* 14, 117–22.
—— and Mazzocca, F. (1982a). On a characterization of Grassmann space representing the lines in an affine space. *Simon Stevin* 56, 129–41.
—— and —— (1982b). On the independence of the axioms defining the affine and projective Grassmann spaces. *Ann. Discrete Math.* 14, 123–8.
—— and —— (1983). On a characterization of the Grassmann spaces associated with an affine space. *Ann. Discrete Math.* 18, 95–112.
—— and Somma, C. (1984). A characterization of Schubert manifold associated with three-dimensional projective space. *Rend. Sem. Mat. Brescia* 7, 89–110.
—— and Tallini, G. (1982). On a characterization of the Grassmann manifold representing the planes in a projective space. *Ann. Discrete Math.* 14, 129–50.
—— and —— (1983). On a characterization of Grassmann space representing the

h-dimensional subspaces in a projective space. *Ann. Discrete Math.* 18, 113–32.
—— and Venezia, A. (1982). Sui $(q+1)^2$-insiemi rigati di classe $[0, 1, 2, q+1]_1$ congiunti da $PG(4, q)$. *Atti Sem. Mat. Fis. Univ. Modena* 31, 163–80.
——, Mazzocca, F., and Somma, C. (1979). On the classification of generalized quadrangles in a finite affine space $AG(3, 2^h)$. *Boll. Un. Mat. Ital. B* 16, 298–307.
Bierbrauer, J. (1980). On minimal blocking sets. *Arch. Math.* 35, 394–400.
—— (1981). Blocking sets of maximal type in finite projective planes. *Rend. Sem. Mat. Univ. Padova* 65, 85–101.
—— (1982). On the weight distribution in binary codes generated by projective planes. *Quart. J. Math. Oxford* 33, 275–9.
Biggs, N. L. (1981). T. P. Kirkman, mathematician. *Bull. London Math. Soc.* 13, 97–120.
—— and Hoare, M. J. (1983). The sextet construction for cubic graphs. *Combinatorica* 3, 153–65.
Bilo, J. (1980). A geometrical problem depending on the equation $ax - xa = 2$ in skew fields. *Simon Stevin* 54, 27–62.
Biscarini, P. (1976). Una proprietà delle bissestuple in $S_{3,4}$. *Riv. Mat. Univ. Parma* 2, 263–8.
—— (1984). Archi hermitiani di $PG(2, q^2)$ con un gruppo di collineazioni transitivo sull'insieme delle $(q+1)$-secanti. *Rend. Sem. Mat. Brescia* 7, 111–24.
—— and Conti, F. (1982). On $(q+2)$-sets in a non-desarguesian projective plane of order q. *Ann. Discrete Math.* 14, 159–68.
—— and Korchmáros, G. (1984). Ovali di un piano di Galois di ordine pari dotate di un gruppo di collineazioni transitivo sui loro punti. *Rend. Sem. Mat. Brescia* 7, 125–35.
Bollobas, B. (Ed.) (1979). *Surveys in Combinatorics.* London Math. Soc. Lecture Note Series 38, Cambridge University Press, Cambridge, 261pp.
Bombieri, E. (1976). Hilbert's 8th problem: an analogue. *Proc. Sympos. Pure Math.* 28, *Mathematical Developments Arising from Hilbert Problems* (1974), American Math. Soc., Providence, pp. 269–74.
Bose, R. C. (1961). On some connections between the design of experiments and information theory. *Bull. Inst. Internat. Statist.* 38, 257–71.
—— and Ray-Chaudhuri, D. K. (1960a). On a class of error correcting binary group codes. *Inform. and Control* 3, 68–79.
—— and —— (1960b). Further results on error correcting binary group codes. *Inform. and Control* 3, 279–90.
——, Freeman, J. W., and Glynn, D. G. (1980). On the intersection of two Baer subplanes in a finite projective plane. *Utilitas Math.* 17, 65–77.
Bramwell, D. (1979). A note on $k-3$ caps in three-dimensional Galois space. *Math. Proc. Cambridge Philos. Soc.* 86, 21–3.
—— (1981). On maximal cubic arcs of a finite projective plane of order 7. *Proc. Third Caribbean Conf. on Combinatorics and Computing*, Univ. West Indies, Bridgetown, pp. 56–66.
Bridges, W. G., Hall, M., and Hayden, J. L. (1981). Codes and designs. *J. Combin. Theory Ser. A* 31, 155–74.
Brouwer, A. E. (1980). A series of separable designs with application to pairwise orthogonal Latin squares. *European J. Combin.* 1, 39–41.
—— (1981). Some unitals on 28 points and their embeddings in projective planes

of order 9. *Geometries and groups.* Lecture Notes in Math. 893, Springer, Berlin, pp. 183–8.

—— and Schrijver, A. (1978). The blocking number of an affine space. *J. Combin. Theory Ser. A* 24, 251–3.

—— and Wilbrink, H. A. (1982). Blocking sets in translation planes. *J. Geom.* 19, 200.

—— and —— (1983). The structure of near polygons with quads. *Geom. Dedicata* 14, 145–76.

Brown, T. C. and Buhler, J. P. (1982). A density version of a geometric Ramsey theorem. *J. Combin. Theory Ser. A* 32, 20–34.

Bruen, A. A. (1980). Blocking sets and skew subspaces of projective space. *Canad. J. Math.* 32, 628–30.

—— (1982a). Intersection of Baer subgeometries. *Arch. Math.* 39, 285–8.

—— (1982b). Arcs in planes of even order. *European J. Combin.* 3, 17–18.

—— (1982c). Lower bounds for complete $\{k; n\}$-arcs. *J. Combin. Theory Ser. A* 33, 109–11.

—— (1983). Blocking sets and translation nets. *Finite Geometries.* Lecture Notes in Pure and Appl. Math. 82, Dekker, New York, pp. 77–92.

—— and de Resmini, M. J. (1983). Blocking sets in affine planes. *Ann. Discrete Math.* 18, 169–75.

—— and Freeman, J. W. (1982). Intersections of t-reguli, rational curves, and orthogonal Latin squares. *Linear Algebra Appl.* 46, 103–16.

—— and Hirschfeld, J. W. P. (1977). Applications of line geometry over finite fields, I: The twisted cubic. *Geom. Dedicata* 6, 495–509.

—— and —— (1978). Applications of line geometry over finite fields, II: The Hermitian surface. *Geom. Dedicata* 7, 333–53.

—— and Rothschild, B. L. (1979). Characterizing subspaces. *Proc. Tenth Southeastern Conf. on Combinatorics, Graph Theory and Computing,* Congr. Numer. 23–24, Utilitas Math., Winnipeg, pp. 199–210.

—— and Silverman, R. (1981). Arcs and blocking sets. *Finite geometries and designs,* London Math. Soc. Lecture Note Series 49, Cambridge University Press, Cambridge, pp. 52–60.

—— and —— (1983). On the non-existence of certain M.D.S. codes and projective planes. *Math. Z.* 183, 171–6.

—— and Thas, J. A. (1977). Blocking sets. *Geom. Dedicata* 6, 193–203.

—— and —— (1982). Hyperplane coverings and blocking sets. *Math. Z.* 181, 407–9.

—— and —— (1983). A combinatorial characterization of maximal $(k; n)$ arcs. *Simon Stevin* 57, 141–3.

——, Rothschild, B. L., and van Lint, J. H. (1980). On characterizing subspaces. *J. Combin. Theory Ser. A* 29, 257–60.

Brylawski, T. (1982). Finite prime field characteristic sets for planar configurations. *Linear Algebra Appl.* 46, 155–76.

——, Lo Re, P. M., Mazzocca, F., and Olanda, D (1980). Alcune applicazioni della teoria dell' intersezione alle geometrie di Galois. *Ricerche Mat.* 29, 65–84.

Bu, T. (1980). Partitions of a vector space. *Discrete Math.* 31, 79–83.

Buekenhout, F. (1975a). Foundations of one dimensional projective geometry based on perspectivities. *Abh. Math. Sem. Univ. Hamburg* 43, 21–9.

—— (1975b). On weakly perspective subsets of Desarguesian projective lines.

Abh. Math. Sem. Univ. Hamburg 43, 30–7.

—— (1979). A characterization of polar spaces. *Simon Stevin* 53, 3–7.

—— (1981). Les plans de Benz: une approche unifiée des plans de Moebius, Laguerre et Minkowski. *J. Geom.* 17, 61–8.

—— (1982). Geometries for the Mathieu group M_{12}. *Combinatorial Theory*, Lecture Notes in Math. 969. Springer, Berlin, pp. 74–85.

—— (1983). (g, d^*, d)-gons. *Finite geometries*, Lecture Notes in Pure and Appl. Math. 82, Dekker, New York, pp. 93–111.

—— and Lefèvre, C. (1976). Semi-quadratic sets in projective spaces. *J. Geom.* 7, 17–24.

—— and Sprague, A. (1982). Polar spaces having some line of cardinality two. *J. Combin. Theory Ser. A* 33, 223–8.

Burde, K. (1980). Pythagoräische Tripel und Reziprozität in Galoisfeldern. *J. Number Theory* 12, 278–82.

Burton, C. T. and Chakravarti, I. M. (1982). On the commutant algebras corresponding to the permutation representations of the full collineation groups of $PG(k, s)$ and $EG(k, s)$, $s = p^r$, $k \geq 2$. *J. Math. Anal. Appl.* 89, 489–514.

Calderbank, R. (1982). On uniformly packed $[n, n-k, 4]$ codes over $GF(q)$ and a class of caps in $PG(k-1, q)$. *J. London Math. Soc.* 26, 365–84.

Cameron, P. J. (Ed.) (1977). *Combinatorial surveys*. Academic Press, London, 226pp.

—— (1982). Dual polar spaces. *Geom. Dedicata* 12, 75–85.

—— and Kantor, W. M. (1978). Rank 3 groups and biplanes. *J. Combin. Theory Ser. A* 24, 1–23..

—— and —— (1979). 2-transitive and antiflag transitive collineation groups of finite projective spaces. *J. Algebra* 60, 384–422.

—— and Liebler, R. A. (1982). Tactical decompositions and orbits of projective groups. *Linear Algebra Appl.* 46, 91–102.

——, Delsarte, P., and Goethals, J. M. (1979). Hemisystems, orthogonal configurations, and dissipative conference matrices. *Philips J. Res.* 34, 147–62.

——, Hirschfeld, J. W. P., and Hughes, D. R. (Eds.) (1981). *Finite geometries and designs*. London Math. Soc. Lecture Note Series 49, Cambridge University Press, Cambridge, 371pp.

Camion, P. (1981). Factorisation des polynômes de $F_q[X]$. *Révue du CETHEDEC* 18, 5–21.

Capursi, M. (1981). Catene di cerchi ottenibili mediante punti pseudoregolari rispetto ad una conica di un piano di Galois. *Note Mat.* 1, 113–26.

Car, M. (1982). Factorisation dans $F_q[X]$. *C. R. Acad. Sci. Paris Sér. I Math.* 294, 147–50.

Carter, D. S. and Vogt, A. (1980). Collinearity-preserving functions between Desarguesian planes. *Mem. Amer. Math. Soc.* 27, no. 235.

Casse, L. R. A. and Glynn, F. G. (1982). The solution to Beniamino Segre's problem $I_{r,q}$, $r=3$, $q=2^h$. *Geom. Dedicata* 13, 157–64.

—— and —— (1984). On the uniqueness of $(q+1)_4$-arcs of $PG(4, q)$, $q=2^h$, $h \geq 3$. *Discrete Math.* 48, 173–86.

—— and Wild, P. R. (1983). k-sets of $(n-1)$-dimensional subspaces of $PG(3n-1, q)$. *Combinatorial Mathematics X*, Lecture Notes in Math. 1036, Springer, Berlin, pp. 410–16.

Cater, F. S. (1978). On Desarguesian projective planes. *Geom. Dedicata* 7, 433–41.

Dickson, L. E. (1908). On the canonical forms and automorphs of ternary cubic forms. *Amer. J. Math.* 30, 117–28.

Dienst, K. J. (1974a). Eine charakteristische Spiegelungseigenschaft hermitescher Quadriken. *J. Geom.* 5, 67–81.

—— (1974b). Über die Eigenschattengrenze hermitescher Quadriken. *Math. Z.* 138, 191–7.

—— (1979a). Über schwach-perspektive hermitesche Mengen. *J. Geom.* 13, 83–9.

—— (1979b). Hermitesche Mengen vom Index 2 und ihre Bedeutung für Minkowski-Ebenen. *Arch. Math.* 33, 193–203.

—— (1980a). Verallgemeinerte Vierecke in projektiven Raümen. *Arch. Math.* 35, 177–86.

—— (1980b). Verallgemeinerte Vierecke in pappusschen projektiven Raümen. *Geom. Dedicata* 9, 199–206.

Dieudonné, J. (1951). On the automorphisms of the classical groups. *Mem. Amer. Math. Soc.* 2.

D'Orgeval, B. (1967). Sur certains 2-systèmes et le problème de Kirkman. *Acad. Roy. Belg. Bull. Cl. Sci.* 53, 21–5.

Drake, D. A. and Freeman, J. W. (1979). Partial t-spreads and group constructible (s, r, μ)-nets. *J. Geom.* 13, 210–16.

Drinfeld, V. G. and Vladut, S. G. (1983). The number of points of an algebraic curve. *Functional Anal. Appl.* 17, 53–4.

Dye, R. H. (1977a). A geometric characterization of the special orthogonal groups and the Dickson invariant. *J. London Math. Soc.* 15, 472–6.

—— (1977b). Partitions and their stabilizers for line complexes and quadrics. *Ann. Mat. Pura Appl.* 114, 173–94.

—— (1978). On the Arf invariant. *J. Algebra* 53, 36–9.

—— (1979a). Symmetric groups as maximal subgroups of orthogonal and symplectic groups over the field of two elements. *J. London Math. Soc.* 20, 227–37.

—— (1979b). Interrelations of symplectic and orthogonal groups in characteristic two. *J. Algebra* 59, 202–21.

—— (1980a). Maximal subgroups of $GL_{2n}(K)$, $SL_{2n}(K)$, $PGL_{2n}(K)$ and $PSL_{2n}(K)$ associated with symplectic polarities. *J. Algebra* 66, 1–11.

—— (1980b). On the maximality of the orthogonal groups in the symplectic groups in characteristic two. *Math. Z.* 172, 203–12.

—— (1981). Alternating groups as maximal subgroups of the special orthogonal groups over the field of two elements. *J. Algebra* 71, 472–80.

—— (1983). A maximal subgroup of $PSp_6(2^m)$ related to a spread. *J. Algebra* 84, 128–35.

—— (1984). Maximal subgroups of symplectic groups stabilizing spreads. *J. Algebra* 87, 493–509.

Ebert, G. L. (1977). Disjoint circles: a classification. *Trans. Amer. Math. Soc.* 232, 83–109.

—— (1978a). Maximal strictly partial spreads. *Canad. J. Math.* 30, 483–9.

—— (1978b). Blocking sets in projective spaces. *Canad. J. Math.* 30, 856–62.

—— (1978c). Translation planes of order q^2: asymptotic estimates. *Trans. Amer. Math. Soc.* 238, 301–8.

—— (1979). A new lower bound for msp spreads. *Proc. Tenth Southeastern Conf. on Combinatorics, Graph Theory and Computing* Vol. 1, Congr. Numer. 23, Utilitas Math., Winnipeg, pp. 413–21.

—— (1983). Subregular 1-spreads of $PG(2n+1, 2)$. *Geom. Dedicata* 14, 343–53.

Edge, W. L. (1984). $PGL(2, 7)$ and $PSL(2, 7)$. *Mitt. Math. Sem. Giessen* 164, 137–50.

Esposito, R. (1977). Sui k-insiemi di tipo $(0, 1, m_d)_d$ di uno spazio di Galois $S_{r,q}$ $(2 \leq d < r, r \geq 3)$. *Riv. Mat. Univ. Parma* 3, 131–40.

—— (1981). Sui k-insiemi di classe $[0, 1, m]_1$ di uno spazio di Galois $S_{r,q}$ $(r \geq 3, q > 2)$. *Rend. Mat.* 1, 203–18.

Eugeni, F. and Ferri, O. (1984). Sulle fibrazioni mediante piani in $PG(n, q)$. *Boll. Un. Mat. Ital.* A 3, 221–8.

Faina, G. (1980). Una estensione agli ovali astratti del teorema di Buekenhout sugli ovali Pascaliani. *Boll. Un. Mat. Ital. Suppl.* 2, 355–64.

—— (1983). A characterization of the tangent lines to a hermitian curve. *Rend. Mat.* 3, 553–7.

—— (1984a). Un esempio di ovale astratto non proiettivo a tangenti pascaliane il cui gruppo degli automorfismi è risolubile e due volte transitivo. *Rend. Sem. Mat. Brescia* 7, 289–96.

—— (1984b). The B-ovals of order $q \leq 8$. *J. Combin. Theory Ser. A* 36, 307–14.

—— and Cecconi, G. (1981). Sull'ordine minimo degli ovali astratti (o di Buekenhout) non proiettivi e unicità dell'ovale astratto di ordine sette. *Note Mat.* 1, 93–111.

—— and —— (1982). A finite Buekenhout oval which is not projective. *Simon Stevin* 56, 121–7.

—— and Korchmáros, G. (1980a). Una caratterizzazione del gruppo lineare $PGL(2, K)$ e delle coniche astratte nel senso di Buekenhout. *Boll. Un. Mat. Ital. Suppl.* 2, 195–208.

—— and —— (1980b). Su una classe di ovali di un piano desarguesiano di ordine pari che danno luogo ad un piano inversivo. *Atti Sem. Mat. Fis. Univ. Modena* 29, 187–96.

—— and —— (1981). Risultati intorno ad una congettura relativi agli archi chiusi. *Rend. Mat.* 1, 55–61.

—— and —— (1982). Desargues configurations inscribed in an oval. *Ann. Discrete Math.* 14, 207–10.

—— and —— (1983). A graphic characterization of Hermitian curves. *Ann. Discrete Math.* 18, 335–42.

Farmer, K. B. (1981). Hermitian geometries in projective space. *Linear Algebra Appl.* 35, 37–50.

—— and Hale, M. P. (1980). Dual affine geometries and alternative bilinear forms. *Linear Algebra Appl.* 30, 183–99.

Fenton, N. E. and Vámos P. (1982). Matroid interpretation of maximal k-arcs in projective spaces. *Rend. Mat.* 2, 573–80.

Ferri, O. (1975). Sulle curve che contengono tutti i punti di un piano lineare finito. *Riv. Mat. Univ. Parma* 1, 213–25.

—— (1976). Su di una caratterizzazione grafica della superficie di Veronese di un $S_{5,q}$. *Atti Accad. Naz. Lincei Rend.* 61, 603–10.

—— (1980a). Le calotte a due caratteri rispetto ai piani in uno spazio di Galois $S_{3,q}$. *Riv. Mat. Univ. Parma* 6, 55–63.

—— (1980b). Su certi k-insiemi di uno spazio di Galois $S_{r,q}$ a tre caratteri rispetto agli S_4. *Rend. Mat.* 13, 165–77.

—— (1981). Una caratterizzazione grafica dello insieme dei punti esterni ad una ovale in un piano Π_q (q dispari). *Rend. Mat.* 1, 31–8.

—— (1982). On type $((q-3)/2, (q-1)/2, q-1)$ k-sets in an affine plane $A_{2,q}$. *Ann. Discrete Math.* 14, 211–18.

—— (1983). I k-insiemi di classe $[0, (q-1)/2, (q+1)/2, q]$ in un piano proiettivo di ordine q dispari. *Rend. Mat.* 3, 33–41.

—— and Tallini, G. (1984). A characterization of the family of secant lines of an elliptic quadric in $PG(3, q)$, q odd. *Rend. Mat. Sem. Brescia* 7, 297–305.

Fisher, J. C. and Thas, J. A. (1979). Flocks in $PG(3, q)$. *Math. Z.* 169, 1–11.

Freeman, J. W. (1980). Reguli and pseudo-reguli in $PG(3, s^2)$. *Geom. Dedicata* 9, 267–80.

Fuji-Hara, R. and Vanstone, S. A. (1982). Orthogonal resolutions of lines in $AG(n, q)$. *Discrete Math.* 41, 17–18.

—— and —— (1983). Affine geometries obtained from projective planes and skew resolutions on $AG(3, q)$. *Ann. Discrete Math.* 18, 355–75.

—— and —— (1984). On a line partitioning problem for $PG(2k, q)$. *Rend. Sem. Mat. Brescia* 7, 337–41.

Fulton, J. D. (1979). Gauss sums and solutions to simultaneous equations over $GF(2^y)$. *Acta Arith.* 35, 17–24.

Games, R. A. (1983). The packing problem for projective geometries over $GF(3)$ with dimension greater than five. *J. Combin. Theory Ser.* A 35, 126–44.

Georgiades, J. (1981). Cyclic $(q+1, k)$-codes of odd order q and even dimension k are not optimal. *Atti Sem. Mat. Fis. Univ. Modena* 30, 284–5.

Gilbert, E. N. (1952). A comparison of signalling alphabets. *Bell System Tech. J.* 31, 504–22.

Giudici, R. E. and Margaglio, C. (1980). A geometric characterization of the generators in a quadratic extension of a finite field. *Rend. Sem. Mat. Univ. Padova* 62, 103–14.

Glynn, D. G. (1982). A lower bound for maximal partial spreads in $PG(3, q)$. *Ars Combin.* 13, 39–40.

—— (1983a). On the characterization of certain sets of points in finite projective geometry of dimension three. *Bull. London Math. Soc.* 15, 31–4.

—— (1983b). Two new sequences of ovals in finite Desarguesian planes of even order. *Combinatorial Mathematics X*, Lecture Notes in Math. 1036, Springer, Berlin, pp. 217–29.

Goethals, J.-M. (1973). Some combinatorial aspects of coding theory. *A survey of combinatorial theory*, North-Holland, Amsterdam, pp. 189–208.

—— and Delsarte, P. (1968). On a class of majority-logic decodable cyclic codes. *IEEE Trans. Inform. Theory* IT-14, 182–8.

Goldberg, D. Y. (1980). A generalized weight for linear codes and a Witt-MacWilliams theorem. *J. Combin. Theory Ser. A* 29, 363–7.

Goppa, V. D. (1981). Codes on algebraic curves. *Soviet Math. Dokl.* 24, 170–2.

—— (1983). Algebraico-geometric codes. *Math. USSR-Izv.* 21, 75–91.

Goudarzi, M. G. (1967a). Sur le problème de Kirkman. *Acad. Roy. Belg. Bull. Cl. Sci.* 53, 183–99.

—— (1967b). Sur le problème de Kirkman généralisé. *Acad. Roy. Belg. Bull. Cl. Sci.* 53, 1379–84.

Gow, R. (1981). The number of equivalence classes of non-degenerate bilinear and sesquilinear forms over a finite field. *Linear Algebra Appl.* 41, 175–81.

Grundhöfer, T. (1981). Über Abbildungen mit eingeschränktem Differenzenprodukt auf einem endlichen Körper. *Arch. Math.* 37, 59–62.

—— (1982). Projektivitätengruppen von ovoidalen Möbius- und Laguerreebenen. *Geom. Dedicata* 13, 125–47.

Guerra, L. and Ughi, E. (1983). On a conjecture of S. Ilkka. *Ann. Discrete Math.* 18, 419–26.

Gulati, B. R. (1971b). On orthogonal arrays of strength five. *Trabajos Estadist.* 23, 51–77.

—— (1972). More about maximal (n, r)-sets. *Inform. and Control* 20, 188–91.

Gunji, H. and Arnon, D. (1981). On polynomial factorization over finite fields. *Math. Comp.* 36, 281–7.

Halder, H.-R. (1980). Zur Existenz von k-Kurven in endlichen Ebenen. *J. Geom.* 14, 71–4.

Hall, J. I. (1980). On identifying $PG(3, 2)$ and the complete 3-design on seven points. *Ann. Discrete Math.* 7, 131–41.

—— (1983). Identifying classical geometries. *Finite geometries*, Lecture Notes in Pure and Appl. Math. 82, Dekker, New York, pp. 175–95.

Hall, M. and Roth, R. (1984). On a conjecture of R. H. Bruck. *J. Combin. Theory Ser.* A 37, 22–31.

Hamada, N. and Tamari, F. (1976). Construction of maximal t-linearly independent sets. *Essays in Probability and Statistics*, Keibundo Matsumoto Printing Co., Tokyo, pp. 41–55.

—— and —— (1978). On a geometrical method of construction of maximal t-linearly independent sets. *J. Combin. Theory Ser.* A 25, 14–28.

—— and —— (1980). Construction of optimal codes and optimal fractional factorial designs using linear programming. *Ann Discrete Math.* 6, 175–88.

—— and —— (1982). Construction of optimal linear codes using flats and spreads in a finite projective geometry. *European J. Combin.* 3, 129–41.

Häring, M. and Heise, W. (1979). On B. Sergre's construction of an ovaloid. *Acta Arith.* 35, 187–8.

Hartshorne, R. (1977). *Algebraic geometry*. Springer, New York, 496pp.

Heise, W. (1978). Es gibt keinen optimalen $(n+2, 3)$-Code einer ungeraden ordnung n. *Math. Z.* 164, 67–8.

—— (1984). The full equivalence theorem for cyclic codes. *Rend. Sem. Mat. Brescia* 7, 355–6.

Hering, C. and Schaeffer, H.-J. (1982). On the new projective planes of R. Figueroa. *Combinatorial theory*, Lecture Notes in Math. 969, Springer, Berlin, pp. 187–90.

Herzer, A. and Lunardon, G. (1980). Regoli, pseudoregoli e varietà algebriche. *Boll. Un. Mat. Ital.* A 17, 323–9.

Higman, D. G. (appendix by Taylor, D. E) (1978). *Classical groups*. Department of Mathematics, Technological University, Eindhoven, 85pp.

Hill, R. (1978a). Caps and codes. *Discrete Math.* 22, 111–37.

—— (1978b). Packing problems in Galois geometries over $GF(3)$. *Geom. Dedicata* 7, 363–73.

—— (1978c). Some results concerning linear codes and $(k, 3)$-caps in three-dimensional Galois space. *Math. Proc. Cambridge Philos. Soc.* 84, 191–205.

—— (1983). On Pellegrino's 20-caps in $S_{4,3}$. Ann. Discrete Math. 18, 433–47.

—— (1984). Some problems concerning (k, n)-arcs in finite projective planes. *Rend. Sem. Mat. Brescia* 7, 367–83.

—— and Mason, J. R. M. (1981). On (k, n)-arcs and the falsity of the Lunelli-Sce conjecture. *Finite geometries and designs*, London Math. Soc. Lecture Note Series 49, Cambridge University Press, Cambridge, pp. 153–68.

Hirschfeld, J. W. P. (1979). *Projective geometries over finite fields*. Oxford University Press, Oxford, 474pp.

—— (1981). Cubic surfaces whose points all lie on their 27 lines. *Finite geometries and designs*, London Math. Soc. Lecture Note Series 49, Cambridge University Press, Cambridge, pp. 169–71.

—— (1982). Del Pezzo surfaces over finite fields. *Tensor* 37, 79–84.

—— (1983a). Caps in elliptic quadrics. *Ann. Discrete Math.* 18, 449–66.

—— (1983b). Maximum sets in finite projective spaces. *Surveys in combinatorics*, London Math. Soc. Lecture Note Series 82, Cambridge University Press, Cambridge, pp. 55–76.

—— (1983c). The characterization of Hermitian geometries over finite fields. *Austral. Math. Soc. Gaz.* 10, 32–6.

—— (1983d). The Weil conjectures in finite geometry. *Combinatorial mathematics X*, Lecture Notes in Math. 1036, Springer, Berlin, pp. 6–23.

—— (1984). Linear codes and algebraic curves. *Geometrical combinatorics*, Research notes in mathematics 114, Pitman, London, pp. 35–53.

—— and Hubaut, X. (1980). Sets of even type in $PG(3, 4)$, alias the binary $(85, 24)$ projective code. *J. Combin. Theory Ser. A* 29, 101–12.

—— and Sadeh, A. R. (1984). The projective plane over the field of eleven elements. *Mitt. Math. Sem. Giessen* 164, 245–57.

—— and Thas, J. A. (1980a). Sets of type $(1, n, q+1)$ in $PG(d, q)$. *Proc. London Math. Soc.* 41, 254–78.

—— and —— (1980b). The characterization of projections of quadrics over finite fields of even order. *J. London Math. Soc.* 22, 226–38.

——, Hubaut, X. and Thas, J. A. (1979). Sets of type $(1, n, q+1)$ in finite projective spaces of even order q. *C. R. Math. Rep. Acad. Sci. Canada* 1, 133–6.

Hoggar, S. G. (1983). A complex polytope as generalized quadrangle. *Proc. Roy. Soc. Edinburgh Sect. A* 95, 1–5.

Hölz, G. (1981). Construction of designs which contain a unital. *Arch. Math.* 37, 179–83.

Honda, T. (1968). Isogeny classes of abelian varieties over finite fields. *J. Math. Soc. Japan* 20, 83–95.

Hubaut, X. and Metz, R. (1983). A class of strongly regular graphs related to orthogonal groups. *Ann. Discrete Math.* 18, 469–72.

Hughes, D. R. and Piper, F. C. (1985). *Design Theory*. Cambridge University Press, Cambridge, 240pp.

Ihara, Y. (1981). Some remarks on the number of rational points of algebraic curves over finite fields. *J. Fac. Sci. Univ. Tokyo Sect. IA Math.* 28, 721–4.

Jamison, R. E. (1978). Dimensions of hyperplane spaces over finite fields. *Math. Z.* 162, 101–11.

Janko, Z. and Tran Van Trung, (1981). Determination of projective planes of order 9 with a nontrivial perspectivity. *Studia Sci. Math. Hungar.* 16, 119.

—— (1982). The classification of projective planes of order 9 which possess an involution. *J. Combin. Theory Ser. A* 33, 65–75.

Janusz, G. and Rotman, J. (1982). Outer automorphisms of S_6. *Amer. Math. Monthly* 89, 407–10.

Jha, V. (1982). On subgroups and factor groups of $GL(n, q)$ acting on spreads with the same characteristic. *Discrete Math.* 41, 43–51.

Johnson, N. L., Kallaher, M. J., and Long, C. T. (Eds.) (1983). *Finite geometries: proceedings of a conference in honor of T. G. Ostrom.* Lecture Notes in Pure and Applied Mathematics 82, Dekker, New York, 454 pp.

Joni, S. A. and Rota, G.-C. (1980). A vector space analog of permutations with restricted position. *J. Combin. Theory Ser. A* 29, 59–73.

Jungnickel, D. (1984a). Maximal partial spreads and nets of small deficiency. *J. Algebra* 90, 119–32.

—— (1984b). The number of designs with classical parameters grows exponentially. *Geom. Dedicata* 16, 167–78.

—— and Vedder, K. (1982). *Combinatorial theory* (Schloss Rauischholzhausen, 1982). Lecture Notes in Math. 969, Springer, Berlin, 326pp.

—— and —— (1984). On the geometry of planar difference sets. *European J. Combin.* 5, 143–8.

Kahn, J. (1980). Locally projective-planar lattices which satisfy the bundle theorem. *Math. Z.* 175, 219–47.

—— (1982). Finite inversive planes which satisfy the bundle theorem. *Geom. Dedicata* 12, 171–87.

Kantor, W. M. (1978). *Classical groups from a non-classical viewpoint.* Mathematical Institute, Oxford, 91pp.

—— (1979). Permutation representations of the finite classical groups of small degree or rank. *J. Algebra* 60, 158–68.

—— (1980a). Linear groups containing a Singer cycle. *J. Algebra* 62, 232–4.

—— (1980b). Generalized quadrangles associated with $G_2(q)$. *J. Combin. Theory Ser. A* 29, 212–19.

—— (1982a). Spreads, translation planes and Kerdock sets. I. *SIAM J. Algebraic Discrete Methods* 3, 151–65.

—— (1982b). Spreads, translation planes and Kerdock sets. II. *SIAM J. Algebraic Discrete Methods* 3, 308–18.

—— (1982c). Translation planes of order q^6 admitting $SL(2, q^2)$. *J. Combin. Theory Ser. A* 32, 299–302.

—— (1982d). Strongly regular graphs defined by spreads. *Israel J. Math.* 41, 298–312.

—— (1982e). Ovoids and translation planes. *Canad. J. Math.* 34, 1195–203.

—— (1983a). Expanded, sliced and spread spreads. *Finite geometries*, Lecture Notes in Pure and Appl. Math. 82. Dekker, New York, pp. 251–61.

—— (1983b). Non-Desarguesian planes, partial geometries, strongly regular graphs and codes arising from hyperbolic quadrics. *Ann. Discrete Math.* 18, 511–17.

—— and Liebler, R. A. (1982). The rank 3 permutation representations of the finite classical groups. *Trans. Amer. Math. Soc.* 271, 1–71.

Karteszi, F. (1977). Su una congettura di Seppo Ilkka. *Ann. Univ. Sci. Budapest Eötvös Sect. Math.* 20, 167–75.

Karzel, H. and Marchi, M. (1983). The projectivity groups of ovals and of quadratic sets. *Ann. Discrete Math.* 18, 519–33.

Katz, N. M. (1976). An overview of Deligne's proof of the Riemann Hypothesis for varieties over finite fields. *Mathematical Developments Arising from the Hilbert Problems* (1974), Proc. Sympos. Pure Math. 28, 275–305.

Keedwell, A. D. (1979). When is a (k, n)-arc of $PG(2, q)$ embeddable in a unique algebraic plane curve of order n? *Rend. Mat.* 12, 397–410.

Kelly, L. M. and Nwankpa, S. (1973). Affine embeddings of Sylvester–Gallai designs. *J. Combin. Theory Ser. A* 14, 422–38.

Kestenband, B. C. (1980). Projective geometries that are disjoint unions of caps. *Canad. J. Math.* 32, 1299–305.

—— (1981a). Hermitian configurations in odd-dimensional projective geometries. *Canad. J. Math.* 33, 500–12.

—— (1981b). Unital intersections in finite projective planes. *Geom. Dedicata* 11, 107–17.

—— (1982). Degenerate unital intersections in finite projective planes. *Geom. Dedicata* 13, 101–6.

—— (1983). Partitions of finite affine geometries into caps. *Linear and Multilinear Algebra* 14, 257–70.

Koblitz, N. (1982). Why study equations over finite fields? *Math. Mag.* 55, 144–9.

—— (1984). *p-adic numbers, p-adic analysis, and zeta-functions.* Second edition. Springer, New York, 150pp.

Korchmáros, G. (1976). Su una classificazione delle ovali dotate di automorfismi. *Rend. Accad. Naz. XL* 1/2, 77–86.

—— (1977/78). Sulle ovali di traslazione in un piano di Galois d'ordine pari. *Rend. Accad. Naz. XL* 3, 55–65.

—— (1978a). Una proprietà gruppale delle involuzioni planari che mutano in sé un'ovale di un piano proiettivo finito. *Ann. Mat. Pura Appl.* 116, 189–206.

—— (1978b). Gruppi di collineazioni transitivi sui punti di un ovale $[(q+2)$-arco$]$ di $S_{2,q}$, q pari. *Atti Sem. Mat. Fis. Univ. Modena* 27, 89–105.

—— (1980). On n^2-sets of type $(0, 1, n)$ in projective planes. *J. Geom.* 15, 170–4.

—— (1981a). Example of a chain of circles on an elliptic quadric of $PG(3, q)$, $q = 7, 11$. *J. Combin. Theory Ser. A* 31, 98–100.

—— (1981b). Una generalizzazione del teorema di F. Buekenhout sulle ovali pascaliane. *Boll. Un. Mat. Ital. B* 18, 673–87.

—— (1983). New examples of complete k-arcs in $PG(2, q)$. *European J. Combin.* 4, 329–34.

—— and Olanda, D. (1983). On egglike inversive planes. *J. Geom.* 21, 53–8.

Kung, J. P. S. (1981). The cycle structure of a linear transformation over a finite field. *Linear Algebra Appl.* 36, 141–55.

Kustaanheimo, P. and Qvist, B. (1952). On differentiation in Galois fields. *Ann. Acad. Sci. Fenn. Ser. A*, No. 137.

Lam, C. W. H., Thiel, L., Swiercz, S. and McKay, J. (1983). The nonexistence of ovals in a projective plane of order 10. *Discrete Math.* 45, 319–21.

Lander, E. S. (1983). *Symmetric designs: an algebraic approach.* London Math. Soc. Lecture Note Series 74, Cambridge University Press, Cambridge, 306pp.

Larato, B. (1980). Sull'esistenza delle ovali del tipo $D(x^k)$. *Atti Sem. Mat. Fis. Univ. Modena* 29, 345–8.

—— (1984). Strutture d'incidenza associate a curve hermitiane. *Boll. Un. Mat. Ital. A* 3, 87–95.

—— and Raguso, G. (1984). Il gruppo delle collineazioni di un piano di traslazione di ordine 13^2. *Rend. Sem. Mat. Brescia* 7, 453–70.

Lefèvre-Percsy, C. (1977a). Semi-quadriques en tant que sous-ensembles des espaces projectifs. *Bull. Soc. Math. Belg.* 29, 175–83.

—— (1977b). Sur les semi-quadriques en tant qu'espaces de Shult projectifs. *Acad. Roy. Belg. Bull. Cl. Sci.* 63, 160–4.

—— (1980). An extension of a theorem of G. Tallini. *J. Combin. Theory Ser. A* 29, 297–305.

—— (1981a). Polar spaces embedded in a projective space. *Finite geometries and designs*, London Math. Soc. Lecture Note Series 49, Cambridge University Press, Cambridge, pp. 216–20.

—— (1981b). Classification d'une famille d'ensembles de classe $(0, 1, n, q+1)$ de $PG(d, q)$. *J. Combin. Theory Ser. A* 31, 270–6.

—— (1981c). Espaces polaires degenerés des espaces projectifs. *Simon Stevin* 55, 237–46.

—— (1981d). Projectivités conservant un espace polaire faiblement plongé. *Acad. Roy. Belg. Bull. Cl. Sci.* 67, 45–50.

—— (1981e). Quadrilatères generalisés faiblement plongés dans $PG(3, q)$. *European J. Combin.* 2, 249–55.

—— (1981f). Ensembles de classe $(0, 1, q, q+1)$ de $PG(d, q)$. *J. Geom.* 15, 93–8.

—— (1981g). Espaces polaires faiblement plongés dans un espace projectif. *J. Geom.* 16, 126–37.

—— (1981h). Characterization of Buekenhout-Metz unitals. *Arch. Math.* 36, 565–8.

—— (1982a). Characterization of Hermitian curves. *Arch. Math.* 39, 476–80.

—— (1982b). Quadriques Hermitiennes et ensembles de classe $(0, 1, n, q+1)$ de $PG(d, q)$. *Geom. Dedicata* 12, 209–13.

—— (1982c). Geometries with dual affine planes and symplectic quadrics. *Linear Algebra Appl.* 42, 31–7.

—— (1983). Copolar spaces fully embedded in projective spaces. *Ann. Discrete Math.* 18, 553–66.

Lemay, F. (1983). Le dodécaèdre et la géométrie géométrie projective d'ordre 5. *Finite geometries*, Lecture Notes in Pure and Appl. Math. 82, Dekker, New York, pp. 279–306.

Lewandowski, H. and Makowiecka, H. (1979a). Some remarks on the Havliček-Tietze configuration. *Časopis Pěst. Mat.* 104, 180–4.

—— (1979b). A geometrical characterization of the projective plane of order 4. *Časopsis Pěst. Mat.* 104, 185–7.

Liang, J. J. (1978). On the solutions of trinomial equations over finite fields. *Bull. Calcutta Math. Soc.* 70, 379–82.

Lidl, R. and Niederreiter, H. (1983). *Finite fields.* Encyclopedia of Mathematics and its Applications 20. Addison-Wesley, Reading, Mass., 755pp.

Liebler, R. A. (1978). A note on the Hering–Ott planes. *J. Combin. Theory Ser. A* 25, 202–4.

—— (1981). On relations among the projective geometry codes. *Finite geometries and designs.* London Math. Soc. Lecture Note Series 49, Cambridge University Press, Cambridge, pp. 221–5.

Limbos, M. (1982). Plongements de $PG(n, q)$ et $AG(n, q)$ dans $PG(m, q')$, $m < n$. *C. R. Math. Rep. Acad. Sci. Canada* 4, 65–8.

Lloyd, E. K. (Ed.) (1983). *Surveys in combinatorics.* London Math. Soc. Lecture Note Series 82. Cambridge University Press, Cambridge, 256pp.

Lo Re, P. M. and Olanda, D. (1981). Grassmann spaces. *J. Geom.* 17, 50–60.

—— and —— (1983). On embeddable Minkowski planes. *J. Geom.* 21, 138–45.
——, Mazzocca, F. and Olanda, D. Esagoni generalizzati e spazi polari. *Rend. Mat.* 13, 401–8.
Lorimer, P. J. (1983). Some of the projective planes. *Math. Intelligencer* 5, 41–50.
Lozanov, C. (1981). Incomplete 6-arcs in a Desarguesian projective plane of order 9 (in Bulgarian). *Annuaire Univ. Sofia Fac. Math. Méc.* 70 (1975/76), 219–26.
Lunardon, G. (1976). Proposizioni configurazionali in una classe di fibrazioni. *Boll. Un. Mat. Ital. A* 13, 404–13.
—— (1977). Una classificazione dei piani di traslazione in relazione alle fibrazioni ad essi associate. I. *Atti Accad. Naz. Lincei Rend.* 63, 504–8.
—— (1978). Una classificazione dei piani di traslazione in relazione alle fibrazioni ad essi associate. II. *Atti Accad. Naz. Lincei Rend.* 64, 59–64.
Lüneburg, H. (1980). *Translation planes.* Springer, Berlin, 278pp.
—— (1981). Some new results on groups of projectivities. *Geometry—von Staudt's point of view*, Reidel, Dordrecht, pp. 231–48.
Macdonald, I. G. (1981). Numbers of conjugacy classes in some finite classical groups. *Bull. Austral. Math. Soc.* 23, 23–48.
MacDougall, J. A. (1981). Bivectors over a finite field. *Canad. Math. Bull.* 24, 489–90.
Madden, D. J. (1981). Polynomials and primitive roots in finite fields. *J. Number Theory* 13, 499–514.
Malykh, A. E. and Alyabeva, V. G. (1980). On the question of the origin of finite projective geometries (in Russian). *History and methodology of the natural sciences*, No. 25, 57–66.
Manin, J. I. (1981). What is the maximum number of points on a curve over F_2. *J. Fac. Sci. Univ. Tokyo Sect. IA Math.* 28, 715–20.
Martinov, N. I. and Lozanov, C. G. (1977). Complete 6-arcs in Desarguesian projective planes (in Russian). *Serdica* 3, 106–16.
Mason, G. (1983). Orthogonal geometries over $GF(2)$ and actions of extra-special 2-groups of translation planes. *European J. Combin.* 4, 347–57.
Mason, J. R. M. (1982). On the maximum sizes of certain (k, n)-arcs in finite projective geometries. *Math. Proc. Cambridge Philos. Soc.* 91, 153–70.
—— (1984). A class of $((p^n - p^m)(p^n - 1),\ p^n - p^m)$-arcs in $PG(2, p^n)$. *Geom. Dedicata* 15, 355–61.
Mathews, R. (1982). Orthogonal systems of polynomials over a finite field with coefficients in a subfield. *Contemp. Math.* 9, 295–302.
Matzeu, P. M. (1978). Algebra lineare su piani di Galois. *Rend. Sem. Fac. Sci. Univ. Cagliari* 48, 87–105.
Mäurer, H. (1981). Symmetries of quadrics. *Geometry–von Staudt's point of view*, Reidel, Dordrecht, pp. 197–229.
Mavron, V. C. (1981). A characterisation of some symmetric substructures of projective and affine geometries. *Arch. Math.* 36, 281–8.
Mazzocca, F. and Melone, N. (1984). Caps and Veronese varieties in projective Galois spaces. *Discrete Math.* 48, 243–52.
—— and Olanda, D. (1979). Alcune caratterizzazione dei sistemi rigati di prima specie. *Ricerche Mat.* 28, 101–8.
McDonough, T. P. (1980). Two group isomorphisms and a little projective geometry. *Math. Gaz.* 64, 245–54.
McEliece, R. J. (1972). Weight congruences for p-ary cyclic codes. *Discrete Math.*

3, 177–92.

Melone, N. (1983). Veronese spaces. *J. Geom.* 20, 169–80.

—— and Olanda, D. (1981). Spazi pseudoprodotto e varietà di C. Segre. *Rend. Mat.* 1, 381–97.

Mena, R. A. (1982). The characteristic polynomials of the plane of order 2. *J. Combin. Inform. System Sci.* 7, 91–3.

Menghini, M. (1981). Una classe di blocking-sets in $PG(2, q)$. *Atti Sem. Mat. Fis. Univ. Modena* 30, 239–42.

Metz, R. (1979). On a class of unitals. *Geom. Dedicata* 8, 125–6.

—— (1981). Der affine Raum verallgemeinerter Reguli. *Geom. Dedicata* 10, 337–67.

Mielants, W. and Leemans, H. (1983). Z_2-cohomology of projective spaces of odd order. *Ann. Discrete Math.* 18, 635–51.

Migliori, G. (1981). Calotte complete, non ovaloidi aventi intersezione massima con una quadrica ellittica in $PG(3, q)$. *Rend. Mat.* 1, 619–22.

Milne, S. C. (1982). Mappings of subspaces into sets. *J. Combin. Theory Ser. A* 33, 36–47.

Moreno, O. (1980). Counting traces of powers over $GF(2^m)$. *Proc. Eleventh Southeastern Conference on Combinatorics, Graph Theory and Computing*, Vol. II, Congr. Numer. 29, Utilitas Math., Winnipeg, pp. 673–9.

Morikawa, M. (1983). On a certain homology of finite projective spaces. *Nagoya Math. J.* 90, 57–62.

Mortimer, B. (1980). The modular permutation representations of the known doubly transitive groups. *Proc. London Math. Soc.* 41, 1–20.

Mukhopadhyay, A. C. (1978). Lower bounds on $m_t(r, s)$. *J. Combin. Theory Ser. A* 25, 1–13.

Mullen, G. L. (1982). Polynomials over finite fields which commute with linear permutations. *Proc. Amer. Math. Soc.* 84, 315–17.

Mwene, B. (1982). On some subgroups of $PSL(4, q)$, q odd. *Geom. Dedicata* 12, 189–99.

Neumaier, A. (1981). Distance matrices and n-dimensional designs. *European J. Combin.* 2, 165–72.

Nevanlinna, R. and Kustaanheimo, P. E. (1976). *Grundlagen der Geometrie.* Birkhäuser, Basle, 131pp.

Niederreiter, H. and Robinson, K. H. (1982). Complete mappings of finite fields. *J. Austral. Math. Soc. Ser. A* 33, 197–212.

Olanda, D. (1972). Quadragoni di Tits e sistemi rigati. *Rend. Acad. Sci. Fis. Mat. Napoli* 39, 81–7.

—— (1973). Sistemi rigati immersi in uno spazio proiettivo. *Relazione* 26, Istituto di Matematica, Università di Napoli.

—— (1977). Sistemi rigati immersi in uno spazio proiettivo. *Atti Accad. Naz. Lincei Rend.* 62, 489–99.

Orzech, G. and Orzech, M. (1981). *Plane algebraic curves.* Dekker, New York, 225pp.

Ott, U. (1975). Eine neue Klasse endlicher Translationsebenen. *Math. Z.* 143, 181–5.

Payne, S. E. (1977). Generalized quadrangles with symmetry. II. *Simon Stevin* 50, 209–45.

—— (1980). Generalized quadrangles as group coset geometries. *Proc. Eleventh Southeastern Conf. on Combinatorics, Graph Theory and Computing*, Vol. II, Congr. Numer. 29, Utilitas Math., Winnipeg, pp. 717–34.

—— (1983a). Collineations of finite generalized quadrangles. *Finite geometries*, Lecture Notes in Pure and Appl. Math. 82, Dekker, New York, pp. 361–90.

—— (1983b). On the structure of translation generalized quadrangles. *Ann. Discrete Math.* 18, 661–5.

—— and Conklin, J. E. (1978). An unusual generalized quadrangle of order sixteen. *J. Combin. Theory Ser. A* 24, 50–74.

—— and Maneri, C. C. (1982). A family of skew-translation generalized quadrangles of even order. *Proc. Thirteenth Southeastern Conf. on Combinatorics, Graph Theory and Computing*, Cong. Numer. 36, Utilitas Math., Winnipeg, pp. 127–35.

—— and Thas, J. A. (1981). Moufang conditions for finite generalized quadrangles. *Finite geometries and designs*, London Math. Soc. Lecture Note Series 49, Cambridge University Press, Cambridge, pp. 275–303.

—— and —— (1984). *Finite generalized quadrangles*, Pitman, London, 312pp.

Pellegrino, G. (1974). Costruzione di una classe di schemi di associazione e di P.B.I.B.-disigni. *Accad. Naz. Sci. Lett. Arti Modena Atti Mem.* 16, 5–17.

—— (1976a). Schemi di associazione negli spazi lineari finiti. *Rend. Mat.* 9, 645–56.

—— (1976b). Sulle calotte massime dello spazio $S_{4,3}$. *Atti Accad. Sci. Lett. Arti Palermo* 34, 297–328.

—— (1977). Un'osservazione sul problema dei k-archi completi in $S_{2,q}$, con $q \equiv 1$ (mod 4). *Atti Accad. Naz. Lincei Rend.* 63, 33–44.

—— (1978). t-designs associated with non degenerate conics in a Galois plane of odd order. *J. Statist. Plann. Inference* 2, 307–12.

—— (1979). Alcune elementari proposizioni aritmetiche e loro applicazioni alla teoria dei k-archi. *Boll. Un. Mat. Ital. A* 16, 322–30.

—— (1980). Sui campi di Galois, di ordine dispari, che ammettono terne di elementi quadrati (non quadrati) consecutivi. *Boll. Un. Mat. Ital. B* 17, 1482–95.

—— (1981). Archi completi di ordine $(q+3)/2$ nei piani di Galois, $S_{2,q}$, con $q \equiv 3$ (mod 4). *Rend. Circ. Mat. Palermo* 30, 311–20.

—— (1982a). Archi completi, di ordine $(q+3)/2$, nei piani di Galois, $S_{2,q}$, con $q \equiv 1$ (mod 4). *Rend. Mat.* 2, 59–66.

—— (1982b). Sulle sostituzioni lineari, sui campi finiti di ordine dispari, che conservano oppure scambiano il carattere quadratico degli elementi trasformati. *Boll. Un. Mat. Ital. B* 1, 211–23.

—— (1983). Sur les k-arcs complets des plans de Galois d'ordre impair. *Ann. Discrete Math.* 18, 667–94.

—— (1984). Sugli archi completi dei piani di Galois, di ordine dispari, contenenti $(q+3)/2$ punti di un'ovale. *Rend. Sem. Mat. Brescia* 7, 495–523.

—— and Korchmáros, G. (1982). Translation planes of order 11^2. *Ann. Discrete Math.* 14, 249–64.

Percsy, N. (1981). Embedding geometric lattices into a projective space. *Finite geometries and designs*, London Math. Soc. Lecture Note Series 49, Cambridge University Press, Cambridge, pp. 304–15.

—— (1983). The bundle axiom and egglike subsets of projective spaces. *Ann. Discrete Math.* 18, 695–8.

Peterson, W. W. and Weldon, E. J. (1972). *Error-correcting codes*. MIT Press, Cambridge, Mass., 560pp.

Pickert, G. (1982). Von der Desargues-Konfiguration zum 5-dimensionalen projektiven Raum mit 63 Punkten. *Math. Semesterber.* 29, 51–67.

Plaumann, P. and Strambach, K. (Eds.) (1981). *Geometry–von Staudt's point of view* (Bad Windsheim 1980). Reidel, Dordrecht, 430pp.

Prohaska, O. and Walker, M. (1977). A note on the Hering type of inversive planes of even order. *Arch. Math.* 28, 431–2.

Raguso, G. (1982). Example of chain of circles on an elliptic quadric of $PG(3, q)$, $q = 9$, 13. *J. Combin. Theory Ser.* A 33, 99–101.

Ralston, T. (1979). On the distribution of squares in a finite field. *Geom. Dedicata* 8, 207–12.

Rao, C. R. (1947). Factorial experiments derivable from combinatorial arrangements of arrays. *J. Roy. Statist. Soc., Suppl.* 9, 128–39.

Ray-Chaudhuri, D. K. and Sprague, A. P. (1979). A combinatorial characterization of attenuated spaces. *Utilitas Math.* 15, 3–29.

Roos, C. (1982). A generalization of the BCH bound for cyclic codes, including the Hartmann-Tzeng bound. *J. Combin. Theory Ser. A* 33, 229–32.

Rosati, L. A. (1983). Sulle ovali dei piani desarguesiani finiti d'ordine pari. *Ann. Discrete Math.* 18, 713–20.

Sachar, H. (1979). The F_p span of the incidence matrix of a finite projective plane. *Geom. Dedicata* 8, 407–15.

Sce, M. (1981). Geometria combinatoria e geometrie finite. *Rend. Sem. Mat. Fis. Milano* 51, 77–123.

Schellekens, G. J. (1962). On a hexagonic structure I; II. *Nederl. Akad. Wetensch. Proc. Ser. A* 65, 201–17; 218–34.

Scherk, P. (1977). On the intersection number of two plane curves. *J. Geom.* 10, 57–68.

Senato, D. (1982). Blocking sets di indice tre. *Rend. Accad. Sci. Fis. Mat. Napoli* 49, 89–95.

Serre, J.-P. (1983a). Nombres de points des courbes algébriques sur F_q. *Séminaire de Théorie des Nombres de Bordeaux*, exposé no. 22.

—— (1983b). Sur le nombre des points rationnels d'une courbe algébrique sur un corps fini. *C. R. Acad. Sci. Paris Sér. I* 296, 397–402.

Shank, H. (1979). Some parity results on binary vector spaces. *Ars Combin.* 8, 107–8.

Sherk, F. A. (1979). Indicator sets in an affine space of any dimension. *Canad. J. Math.* 31, 211–24.

Sherman, B. (1983). On sets with only odd secants in geometries over $GF(4)$. *J. London Math. Soc.* 27, 539–51.

Shull, R. (1984). Collineations of projective planes of order 9. *J. Combin. Theory Ser. A.* 37, 99–120.

Shult, E. E. (1983). Characterizations of the Lie incidence geometries. *Surveys in combinatorics*, London Math. Soc. Lecture Note Series 82, Cambridge University Press, Cambridge, pp. 157–86.

Small, C. (1982). On the equation $xyz = x + y + z = 1$. *Amer. Math. Monthly* 89, 736–49.

Smit-Ghinelli, D. (1973). Varietà hermitiane e strutture finite, II. *Rend. Mat.* 6, 13–17.

Snapper, E. (1979). Quadratic spaces over finite fields and codes. *C. R. Math. Rep. Sci. Canada* 1, 149–52.

Somma, C. (1982). Generalized quadrangles with parallelism. *Ann. Discrete Math.* 14, 265–82.

Stahnke, W. (1973). Primitive binary polynomials. *Math. Comp.* 27, 977–80.

Stangarone, R. and Terrusi, A. (1982). Blocking-sets contenuti nell'unione di tre rette formanti fascio. *Note Mat.* 2, 167–76.

Stanley, R. P. (1977). Some combinatorial aspects of the Schubert calculus. *Combinatoire et répresentation du groupe symétrique*, Lecture Notes in Math. 579, Springer, Berlin, pp. 217–51.

Suzuki, M. (1962). On a class of doubly transitive groups. *Ann. of Math.* 75, 105–45.

Sved, M. (1982). On configurations of Baer subplanes of the projective plane over a finite field of square order. *Combinatorial mathematics IX*, Lecture Notes in Math. 952, Springer, Berlin, pp. 423–43.

—— (1983). Baer subspaces in the n-dimensional projective space. *Combinatorial mathematics X*, Lecture Notes in Math. 1036, Springer, Berlin, pp. 375–91.

Swinnerton-Dyer, H. P. F. (1981). Universal equivalence for cubic surfaces over finite and local fields. *Symposia Mathematica* vol. 24, Academic Press, London, pp. 111–43.

Tallini, G. (1976b). I k-insiemi di classe $[0, 1, n, q+1]$ regolari di $S_{r,q}$. *Convegno di geometria combinatoria*, Consiglio Nationale delle Ricerche, Florence, pp. 101–10.

—— (1981a). On a theorem by W. Benz characterizing plane Lorentz transformations in Jaernefelt's world. *J. Geom.* 17, 171–3.

—— (1981b). On a characterization of the Grassmann manifold representing the lines in a projective space. *Finite geometries and designs*, London Math. Soc. Lecture Note Series 49, Cambridge University Press, Cambridge, pp. 354–8.

—— (1981c). Su una caratterizzazione della grassmanniana delle rette di uno spazio proiettivo. *Rend. Sem. Mat. Univ. Brescia* 6, 82–6.

—— (1981d). Commemorazione di Beniamino Segre. *Rend. Mat.* 1, 1–29.

—— (1982a). The geometry on Grassmann manifolds representing subspaces in a Galois space. *Ann. Discrete Math.* 14, 9–38.

—— (1982b). On line k-sets of type $(0, n)$ with respect to pencils of lines in $PG(d, q)$. *Ann. Discrete Math.* 14, 283–92.

Tallini Scafati, M. (1981). I k-insiemi di tipo (m, n) di uno spazio affine $A_{r,q}$. *Rend. Mat.* 1, 63–80.

—— (1982). On k-sets of kind (m, n) of a finite projective or affine space. *Ann. Discrete Math.* 14, 39–56.

—— (1983). Two characters k-sets with respect to a singular space in $PG(r, q)$. *Ann. Discrete Math.* 18, 731–44.

—— (1984). d-dimensional two-character k-sets in an affine space $AG(r, q)$. *J. Geom.* 22, 75–82.

Tamari, F. (1981). On an $\{f, m; t, s\}$-max · hyper and a $\{k, m; t, s\}$-min · hyper in a finite projective geometry $PG(t, s)$. *Bull. Fukuoka Univ. Ed. III* 31, 35–43.

Tate, J. T. (1974). The arithmetic of elliptic curves. *Invent. Math.* 23, 179–206.

Teirlinck, L. (1982). Factorization properties for isometries of matroids into projective spaces. *European J. Combin.* 3, 353–86.

Temperley, H. N. V. (Ed.) (1981). *Combinatorics.* London Math. Soc. Lecture Note Series 52. Cambridge University Press, Cambridge, 190pp.

Thas, J. A. (1977). Combinatorial characterizations of the classical generalized quadrangles. *Geom. Dedicata* 6, 339–51.

—— (1978a). Partial geometries in finite affine spaces. *Math. Z.* 158, 1–13.

—— (1978b). Combinatorial characterizations of generalized quadrangles with parameters $s=q$ and $t=q^2$. *Geom. Dedicata* 7, 223–32.

—— (1979a). Generalized quadrangles satisfying at least one of the Moufang conditions. *Simon Stevin* 53, 151–62.

—— (1979b). A restriction on the parameters of a suboctagon. *J. Combin. Theory Ser. A* 27, 385–7.

—— (1979c). Geometries in finite projective and affine spaces. *Surveys in combinatorics*, London Math. Soc. Lecture Note Series 38, Cambridge University Press, Cambridge, pp. 181–211.

—— (1980a). Partial three-spaces in finite projective spaces. *Discrete Math.* 32, 299–322.

—— (1980b). Construction of maximal arcs and dual ovals in translation planes. *European J. Combin.* 1, 189–92.

—— (1980c). Polar spaces, generalized hexagons and perfect codes. *J. Combin. Theory Ser. A* 29, 87–93.

—— (1981a). Ovoids and spreads of finite classical polar spaces. *Geom. Dedicata* 10, 135–44.

—— (1981b). Some results on quadrics and a new class of partial geometries. *Simon Stevin* 55, 129–39.

—— (1981c). New combinatorial characterizations of generalized quadrangles. *European J. Combin.* 2, 299–303.

—— (1982). Combinatorics of finite generalized quadrangles: a survey. *Ann. Discrete Math.* 14, 57–76.

—— (1983a). Elementary proofs of two fundamental theorems of B. Segre without using the Hasse-Weil theorem. *J. Combin. Theory Ser. A* 34, 381–4.

—— (1983b). Semi-partial geometries and spreads of classical polar spaces. *J. Combin. Theory Ser. A* 35, 58–66.

—— (1983c). Geometries in finite projective spaces: recent results. *Combinatorial mathematics X*, Lecture Notes in Math. 1036, Springer, Berlin, pp. 96–110.

—— and De Clerck, F. (1977). Partial geometries satisfying the axiom of Pasch. *Simon Stevin* 51, 123–37.

—— and De Winne, P. (1977). Generalized quadrangles in finite projective spaces. *J. Geom.* 10, 126–37.

—— and Payne, S. E. (1981). Generalized quadrangles and the Higman-Sims technique. *European J. Combin.* 2, 79–89.

—— and —— (1983). Moufang conditions for finite generalized quadrangles. *Ann. Discrete Math.* 18, 745–52.

Thomas, A. D. (1977). *Zeta-functions: an introduction to algebraic geometry.* Research notes in mathematics 12. Pitman, London, 230pp.

Timmermann, H. (1977). Descrizioni geometriche sintetiche di geometrie proiettive con caratteristica $p>0$. *Ann. Mat. Pura Appl.* 114, 121–39.

Tran van Trung (1982). The existence of symmetric block designs with parameters (41, 16, 6) and (66, 26, 10). *J. Combin. Theory Ser. A* 33, 201–4.

Tsfasman, M. A. (1982). Goppa codes that are better than the Varshamov-Gilbert bound. *Problems Inform. Transmission* 18, 163–6.

——, Vladut, S. G., and Zink, T. (1982). Modular curves, Shimura curves and Goppa codes, better than Varshamov-Gilbert bound. *Math. Nachr.* 109, 21–8.

Tsuzuku, T. (1982). *Finite groups and finite geometries.* Cambridge University Press, Cambridge, 328pp.

Uhlig, F. (1979). A rational canonical pair form for a pair of symmetric matrices over an arbitrary field F with char $F \neq 2$ and applications to finest simultaneous block diagonalizations. *Linear and Multilinear Algebra* 8, 41–67.

—— (1981). Inertia and eigenvalue relations between symmetrized and symmetrizing matrices for the real and general field case. *Linear Algebra Appl.* 35, 203–26.

Ughi, E. (1983). On the number of points of elliptic curves over a finite field and a problem of B. Segre. *European J. Combin.* 4, 263–70.

van Leijenhorst, D. C. (1981). Orbits on the projective line. *J. Combin. Theory Ser. A* 31, 146–54.

Varshamov, R. R. (1957). Estimate of the number of signals in error correcting codes. *Dokl. Akad. Nauk SSSR*, 117, 739–41.

Vedder, K. (1981a). Affine subplanes of projective planes. *Finite geometries and designs*, London Math. Soc. Lecture Note Series 49, Cambridge University Press, Cambridge, pp. 359–64.

—— (1981b). A note on the intersection of two Baer subplanes. *Arch. Math.* 37, 287–8.

Venezia, A. (1983). Sui k-insiemi di punti della Grassmanniana delle rette di $PG(3, q)$ di tipo $(0, 1, 2, q+1)_1$ rispetto alle rette e ad un sol carattere rispetto ai piani. *Rend. Mat.* 3, 9–16.

—— (1984). On a characterization of the set of lines which either belong to or are tangent to a non-singular quadric in $PG(3, q)$, q odd. *Rend. Sem. Mat. Brescia* 7, 617–23.

Vincenti, R. (1980). Alcuni tipi di varieta V_2^3 di $S_{4,q}$ e sottopiani di Baer. *Boll. Un. Mat. Ital. Suppl.* 2, 32–44.

—— (1983). A survey on varieties of $PG(4, q)$ and Baer subplanes of translation planes. *Ann. Discrete Math.* 18, 775–9.

—— (1984). Cryptoreguli and derivable translation planes. *Rend. Sem. Mat. Brescia* 7, 625–33.

Voloch, J. F. (1983). Codes and curves. *Eureka* 43, 53–61.

Wagner, A. (1977). An observation on the degrees of projective representations of the symmetric and alternating group over an arbitrary field. *Arch. Math.* 29, 583–9.

—— (1978a). The minimal number of involutions generating some finite three-dimensional groups. *Boll. Un. Mat. Ital. A* 15, 431–9.

—— (1978b). Collineation groups generated by homologies of order greater than 2. *Geom. Dedicata* 7, 387–98.

—— (1980). Determination of the finite primitive reflection groups over an arbitrary field of characteristic not two. Part I. *Geom. Dedicata* 9, 239–53.

—— (1981a). Determination of the finite primitive reflection groups over an arbitrary field of characteristic not two. Part II. *Geom. Dedicata* 10, 191–203.

—— (1981b). Determination of the finite primitive reflection groups over an arbitrary field of characteristic not two. Part III. *Geom. Dedicata* 10, 475–523.

—— (1983). Some reflections on the history of the study of finite projective planes. *Finite geometries.* Lecture Notes in Pure and Appl. Math. 82. Dekker, New York, pp. 419–26.

Walker, M. (1976). The collineation groups of derived translation planes. *Geom. Dedicata* 5, 87–95.

—— (1977). On the structure of finite collineation groups containing symmetries of generalized quadrangles. *Invent. Math.* 40, 245–65.

—— (1979). A characterization of some translation planes. *Abh. Math. Sem. Univ. Hamburg* 49, 216–33.

Wall, G. E. (1980). Conjugacy classes in projective and special linear groups. *Bull. Austral. Math. Soc.* 22, 339–64.

Wallis, W. D. (1973). Configurations arising from maximal arcs. *J. Combin. Theory* 15, 115–19.

Walton, P. N. and Welsh, D. J. A. (1980). On the chromatic number of binary matroids. *Mathematika* 27, 1–9.

—— and —— (1982). Tangential 1-blocks over *GF*(3). *Discrete Math.* 40, 319–20.

Waterhouse, W. G. (1969). Abelian varieties over finite fields. *Ann. Sci. École. Norm. Sup.* 2, 521–60.

Wells, A. L. (1983). Universal projective embeddings of the Grassmannian, half spinor, and dual orthogonal geometries. *Quart. J. Math. Oxford* 34, 375–86.

Welsh, D. J. A. (1980). Colourings, flows and projective geometry. *Nieuw Arch. Wisk.* 28, 159–76.

Wesselkamper, T. C. (1979). The algebraic representation of partial functions. *Discrete Appl. Math.* 1, 137–42.

Whitesides, S. H. (1979). Collineations of projective planes of order 10, Part I; II. *J. Combin. Theory Ser. A* 26, 249–68; 269–77.

Wilbrink, H. (1983). A characterization of classical unitals. *Finite geometries.* Lecture Notes in Pure and Appl. Math. 82, Dekker, New York, pp. 445–54.

Willems, M. L. H. (1983). Restricted Mi-spaces, restricted Li-spaces, optimal codes and n-arcs. *Ann. Discrete Math.* 18, 789–802.

—— and Thas, J. A. (1983). A note on the existence of special Laguerre i-structures and optimal codes. *European J. Combin.* 4, 93–6.

Williamson, J. (1935). The equivalence of non-singular pencils of Hermitian matrices in an arbitrary field. *Amer. J. Math.* 57, 475–90.

Wilson, B. J. (1982). Incompleteness of $(nq+n-q-2, n)$-arcs in finite projective planes of even order. *Math. Proc. Cambridge Philos. Soc.* 91, 1–8.

Witt, E. (1954). Über eine invariante quadratische Formen mod 2. *J. Reine Angew. Math.* 193, 119–20.

Wlodarski, L. (1977). On different notions of the interior of a conic. *Bull. Acad. Polon. Sci.* 25, 965–7.

Yanushka, A. (1976). Generalized hexagons of order (t, t). *Israel J. Math.* 23, 309–24.

Zalesskii, A. E. (1981). Linear groups. *Russian Math. Surveys* 36, 63–128.

Zeitler, H. (1973). Moderne Mathematik am Tetraeder. *Didaktik Math.* 1, 32–45.

—— (1980). Quadratische Mengen in endlichen projektiven Räumen. *Bayreuth. Math. Schr.* No. 6, 33–207.

—— (1981). Regelflächen in endlichen projektiven Räumen. *Math. Semesterber.* 28, 113–38.

Zucchini, C. (1976). Alcuni risultati sulle (k, n)-calotte con tre caratteri. *Istit. Lombardo Accad. Sci. Lett. Rend. A* 110, 359–70.

Zvereva, J. N. (1976a). Arcs in a Desarguesian projective plane of order 8. (in Russian). *Perm. Gos. Univ. Ucen. Zap.* No. 152, 28–39.

—— (1976b). Arcs in a Desarguesian projective plane of order 9 (in Russian). *Perm. Gos. Univ. Ucen. Zap.* No. 152, 40–3.

INDEX OF NOTATION

Operations on sets and groups

$\|X\|$	number of elements in the set
$X \cap Y$	the intersection of X and Y
$X \cup Y$	the union of X and Y
$X \backslash Y$	the set of elements of X not in Y
$X \subset Y$	X is a subset of Y
$\varnothing$	the empty set
$x \in X$	x is an element of X
$x \notin X$	x is not an element of X
$G \cong H$	the groups G and H are isomorphic
$G < H$	G is a subgroup of H
$G \lhd H$	G is a normal subgroup of H
$G \times H$	the direct product of the groups G and H
GH	a semi-direct product of the groups G and H

Miscellaneous sets and numbers

$\mathbf{N}$	the natural numbers $\{1, 2, 3, \ldots\}$
$\mathbf{N}_r$	$\{1, 2, \ldots, r\}$
$\bar{\mathbf{N}}_r$	$\{0, 1, \ldots, r\}$
$\theta(n)$	$(q^{n+1}-1)/(q-1)$
$\boldsymbol{\phi}(n)$	$\|\{m \in N \mid 1 \leqslant m \leqslant n, (m, n) = 1\}\|$
$\mathbf{c}(n, r)$	$n(n-1)\ldots(n-r+1)/r!$
$\mathbf{h}(n, r)$	$n(n+1)\ldots(n+r-1)/r!$
$(m_1, \ldots, m_r)$	greatest common divisor of $m_1, \ldots, m_r$

Fields: subsets, functions, polynomials

K	an arbitrary field
K_0	$K \backslash \{0\}$
$\mathbf{C}$	the complex numbers
$\mathbf{R}$	the real numbers
$\gamma, GF(q)$	the Galois field of $q = p^h$ elements
γ_0	$\gamma \backslash \{0\}$
γ_{01}	$\gamma \backslash \{0, 1\}$
γ^+	$\gamma \cup \{\infty\}$
q	order of γ
p	characteristic of γ

h	order of the automorphism group of γ
$\bar{\gamma}$	algebraic closure of γ
γ'	$GF(q^2)$
γ''	$GF(q^3)$
$K[X], K[x_0, x_1, \ldots, x_n]$	polynomial ring in $x_0, \ldots, x_n$ over K
$\gamma[X], \gamma[x_0, x_1, \ldots, x_n]$	polynomial ring in $x_0, \ldots, x_n$ over γ
$\mathcal{P}(q; x)$	permutation polynomials in x over γ
$D(t), D_2(t)$	$t+t^2+t^4+\ldots+t^{2^{h-1}}, q=2^h$
$\mathcal{C}_0$	$\{t \in \gamma \mid D(t)=0\}$
$\mathcal{C}_1$	$\{t \in \gamma \mid D(t)=1\}$
$C(t)$	$D_2(t)$ when $p=2$
$C(t)$	$(1-t^{(q-1)/2})/2$ when $p>2$
$\sum f(i_1, \ldots, i_r)$	sum of all terms $f(i_1, \ldots, i_r)$
$\sum' f(i_1, \ldots, i_r)$	sum of all terms $f(i_1, \ldots, i_r)$ with $i_1 \leq i_2 \leq \ldots \leq i_r$
$\sum'' f(i_1, \ldots, i_r)$	sum of all terms $f(i_1, \ldots, i_r)$ with $i_1 < i_2 < \ldots < i_r$

Spaces, subspaces, numbers

$V(n, K)$	n-dimensional vector space over the field K
$V(n, q)$	$V(n, K)$ when $K=\gamma$
$PG(n, K)$	n-dimensional projective space over the field K
$PG(n, q)$	$PG(n, K)$ when $K=\gamma$
$AG(n, K)$	n-dimensional affine space over the field K
$AG(n, q)$	$AG(n, K)$ when $K=\gamma$
$PG^{(r)}(n, q)$	set of r-spaces of $PG(n, q)$
Π	$PG^{(0)}(3, q)$
$\mathcal{L}$	$PG^{(1)}(3, q)$
Φ	$PG^{(2)}(3, q)$
$\mathbf{P}(X), \mathbf{P}(x_0, x_1, \ldots, x_n)$	point of $PG(n, K)$ with vector $X=(x_0, x_1, \ldots, x_n)$
$\boldsymbol{\pi}(U), \boldsymbol{\pi}(u_0, u_1, \ldots, u_n)$	prime of $PG(n, K)$ with vector $U=(u_0, u_1, \ldots, u_n)$
$\mathbf{U}_i$	$\mathbf{P}(0, \ldots, 0, 1, 0, \ldots, 0)$ with 1 in the $(i+1)$-th place
$\mathbf{U}$	$\mathbf{P}(1, 1, \ldots, 1)$
$\mathbf{u}_i$	$\boldsymbol{\pi}(0, \ldots, 0, 1, 0, \ldots, 0)$ with 1 in the $(i+1)$-th place
$\mathbf{u}$	$\boldsymbol{\pi}(1, 1, \ldots, 1)$
$\hat{A}$	$(a_5, a_4, a_3, a_2, a_1, a_0)$, where $A=(a_0, a_1, a_2, a_3, a_4, a_5)$
Π_r	an r-dimensional subspace of $PG(n, K)$, specific or generic
$\Pi_r\Pi_s$	the join of Π_r and Π_s

$\Pi_r \cap \Pi_s$	the meet of Π_r and Π_s
Π_{-1}	the empty subspace
$k_{n,d,q}$	k-set of type $(1, n, q+1)$ in $PG(d, q)$
$\phi(r; n, q)$	$\lvert PG^{(r)}(n, q)\rvert$
$\phi(0; n, q)$, $\theta(n, q)$	$\lvert PG(n, q)\rvert$
$\chi(s, r; n, q)$	$\lvert\{\Pi_r$ through a fixed Π_s in $PG(n, q)\}\rvert$
$\psi(t, s, r; n, q)$	$\lvert\{(\Pi_r, \Pi_s) \mid \Pi_r \cap \Pi_s = \text{some } \Pi_t \text{ in } PG(n, q)\}\rvert$

Lines

l_{ij}	$x_i y_j - x_j y_i$
$\mathbf{l}(L)$	line of $PG(3, q)$ with vector $L = (l_{01}, l_{02}, l_{03}, l_{12}, l_{31}, l_{23})$
$\varpi(l)$	$l_{01}l_{23} + l_{02}l_{31} + l_{03}l_{12}$
$\varpi(l, l')$	LL'^*, the mutual invariant of l and l'
$\mathscr{A}(l)$	$a_{01}l_{01} + a_{02}l_{02} + a_{03}l_{03} + a_{12}l_{12} + a_{31}l_{31} + a_{23}l_{23}$
$\varpi(A)$	$a_{01}a_{23} + a_{02}a_{31} + a_{03}a_{12}$
$\boldsymbol{\lambda}(A)$, $\boldsymbol{\lambda}(AL^*)$	$\{\mathbf{l}(L) \mid AL^* = 0\}$
$\mathscr{R}(l_1, l_2, l_3)$	the regulus containing the lines l_1, l_2, l_3
λ_i	the number of lines of $\mathscr{L}/\mathscr{S}$ meeting exactly i lines of the k-span $\mathscr{S}$
$\Pi(\mathscr{S})$	the subset of $PG^{(0)}(3, q)$ on no line of the k-span $\mathscr{S}$
$\Phi(\mathscr{S})$	the subset of $PG^{(2)}(3, q)$ containing no line of the k-span $\mathscr{S}$

Varieties

$\mathbf{V}(F_1, \ldots, F_r)$, $\mathbf{V}_n(F_1, \ldots, F_r)$, $\mathbf{V}_{n,K}(F_1, \ldots, F_r)$	$\{\mathbf{P}(X) \in PG(n, K) \mid F_1(X) = \ldots = F_r(X) = 0$, F_i forms in $K[X]\}$
$\mathbf{V}(F_1, \ldots, F_r)$, $\mathbf{V}_n(F_1, \ldots, F_r)$, $\mathbf{V}_{n,q}(F_1, \ldots, F_r)$	$\mathbf{V}_{n,K}(F_1, \ldots, F_r)$ when $K = \gamma$
$\mathscr{V}^{(1)}$	the set of lines on the variety $\mathscr{V}$
$m_P(\mathscr{F})$	multiplicity of P on the variety $\mathscr{F}$
$\pi_P(\mathscr{F})$	tangent plane to the surface $\mathscr{F}$ at P
$\mathscr{Q}_n$	a non-singular quadric in $PG(n, q)$
$\mathscr{U}_n$, $\mathscr{U}_{n,q}$	a non-singular Hermitian variety in $PG(n, q)$
$\mathscr{U}_{n,q}^{(1)}$	the set of lines on $\mathscr{U}_{n,q}$
$\mathscr{P}_{2s}$, $\mathscr{P}_{2s,q}$	a non-singular (parabolic) quadric in $PG(2s, q)$
$\mathscr{H}_{2s-1}$, $\mathscr{H}_{2s-1,q}$	a non-singular hyperbolic quadric in $PG(2s-1, q)$
$\mathscr{E}_{2s-1}$, $\mathscr{E}_{2s-1,q}$	a non-singular elliptic quadric in $PG(2s-1, q)$

$\mu(n, q)$	$\|\mathcal{U}_n\|$
ψ, $\psi(2s, q)$	$\|\mathcal{P}_{2s}\|$
ψ_+, $\psi_+(2s-1, q)$	$\|\mathcal{H}_{2s-1}\|$
ψ_-, $\psi_-(2s-1, q)$	$\|\mathcal{E}_{2s-1}\|$
$\mathcal{G}_{r,n}$	Grassmannian of Π_r's in Π_n
$\Pi_r\mathcal{S}$	a cone with vertex Π_r and base $\mathcal{S}$
$\mathcal{F}_1 \sim \mathcal{F}_2$	the varieties $\mathcal{F}_1$ and $\mathcal{F}_2$ are projectively equivalent

Surfaces in $PG(3, q)$

$\mathcal{U}_3$, $\mathcal{U}_{3,q}$	a non-singular Hermitian surface
$\Pi_0\mathcal{U}_{2,q}$	a cone with vertex Π_0 and base a curve $\mathcal{U}_{2,q}$
$\Pi_1\mathcal{U}_{1,q}$	a cone with vertex Π_1 and base $\mathcal{U}_{1,q}$, forming $\sqrt{q}+1$ collinear planes
$\mathcal{U}'_3$, $\mathcal{U}'_{3,q}$	a sub-Hermitian surface
$\mathcal{H}_3$, $\mathcal{H}_{3,q}$	a hyperbolic quadric
$\mathcal{E}_3$, $\mathcal{E}_{3,q}$	an elliptic quadric
$\Pi_0\mathcal{P}_2$	a quadric cone with vertex Π_0 and base the conic $\mathcal{P}_2$
$\Pi_1\mathcal{H}_1$	a plane pair
$\Pi_1\mathcal{E}_1$	a line
$\Pi_2\mathcal{P}_0$	a repeated plane
$\mathcal{R}_3$	a projection of the quadric $\mathcal{P}_4$
$\mathcal{H}(\mathcal{R})$	the $\mathcal{H}_3$ containing a regulus $\mathcal{R}$
$\mathcal{H}(\mathcal{R}, \mathcal{R}')$	the $\mathcal{H}_3$ containing the reguli $\mathcal{R}$, $\mathcal{R}'$

Transformations

Mappings are written on the right of the elements being mapped.

T^*	transpose of the matrix T
$\mathbf{M}(T)$	projectivity defined by the matrix T
$\mathfrak{U}$, $\mathfrak{U}_n$	polarity induced by $\mathcal{U}_n$
$\mathfrak{P}$, $\mathfrak{P}_{2s}$	polarity induced by $\mathcal{P}_{2s}$
$\mathfrak{H}$, $\mathfrak{H}_{2s-1}$	polarity induced by $\mathcal{H}_{2s-1}$
$\mathfrak{E}$, $\mathfrak{E}_{2s-1}$	polarity induced by $\mathcal{E}_{2s-1}$
$\mathfrak{G}$	the Grassman map from $\mathcal{L}$ to $\mathcal{H}_5$
$\mathcal{S}_1 \sim \mathcal{S}_2$	the subsets $\mathcal{S}_1$ and $\mathcal{S}_2$ of $PG(n, q)$ are projectively equivalent
$p(n, q)$	$\|PGL(n, q)\|$
I, I_n	identity in $GL(n, q)$
$\mathfrak{I}$	identity in $PGL(n, q)$
diag $(a_1, a_2, \ldots, a_n)$	the matrix with diagonal elements $a_1, a_2, \ldots, a_n$ and zeros elsewhere

Groups

$\mathbf{S}_n$	the symmetric group of degree n
$\mathbf{A}_n$	the alternating group of degree n
$\mathbf{D}_n$	the dihedral group of order $2n$
$\mathbf{Z}_n$	the cyclic group of order n
$\mathbf{Z}$	the integers
$GL(n, q)$	non-singular linear transformations of $V(n, q)$
$\Gamma L(n, q)$	non-singular semi-linear transformations of $V(n, q)$
$SL(n, q)$	elements of $GL(n, q)$ of determinant one
$PGL(n, q)$	projectivities of $PG(n-1, q)$
$P\Gamma L(n, q)$	collineations of $PG(n-1, q)$
$PSL(n, q)$	projectivities of $PG(n-1, q)$ of determinant one
$P\Gamma SL(n, q)$	collineations of $PG(n-1, q)$ of determinant one
$PGU(n, q)$	subgroup of $PGL(n, q)$ fixing $\mathscr{U}_{n-1}$
$PGO(2s+1, q)$	subgroup of $PGL(2s+1, q)$ fixing $\mathscr{P}_{2s}$
$PGO_+(2s, q)$	subgroup of $PGL(2s, q)$ fixing $\mathscr{H}_{2s-1}$
$PGO_-(2s, q)$	subgroup of $PGL(2s, q)$ fixing $\mathscr{E}_{2s-1}$
$PGSp(2s, q)$	subgroup of $PGL(2s, q)$ fixing a null polarity in $PG(2s-1, q)$
$PGPs(2s+1, q)$	subgroup of $PGL(2s+1, q)$ fixing a pseudo polarity in $PG(2s, q)$
$PGPs^*(2s, q)$	subgroup of $PGL(2s, q)$ fixing a pseudo polarity in $PG(2s-1, q)$
$G(\mathscr{F})$	subgroup of $PGL(n+1, q)$ fixing a subset $\mathscr{F}$ of $PG(n, q)$

See also Appendix III

Constants for a k-cap $\mathscr{K}$ in $PG(3, q)$

$m_2(3, q)$	maximum value of k
t	number of tangents through P in $\mathscr{K}$
τ_i	number of i-secants to $\mathscr{K}$
$\sigma_i, \sigma_i(Q)$	number of i-secants through Q in $PG(3, q)\backslash\mathscr{K}$
r_i	number of points of $PG(3, q)\backslash\mathscr{K}$ through which i bisecants pass
κ_i	number of planes meeting $\mathscr{K}$ in an i-arc
c_i, c_i^P	number of planes through P in $\mathscr{K}$ meeting $\mathscr{K}$ in an i-arc

k-arcs in $PG(3, q)$

$m(3, q)$	maximum value of k
$\mathscr{C}(F)$	$\{\mathbf{P}(F(t), F(t), t, 1) \mid t \in \gamma^+\}$
$\mathscr{C}(r)$	$\mathscr{C}(F)$ when $F(x) = x^r$

Constants for a set $\mathscr{K}$ in $PG(3, q)$

τ_i	number of i-secants to $\mathscr{K}$
ρ_i, $\rho_i(P)$	number of i-secants to $\mathscr{K}$ through P in $\mathscr{K}$
σ_i, $\sigma_i(Q)$	number of i-secants to $\mathscr{K}$ through Q not in $\mathscr{K}$

Other symbols

$\mathscr{A}\mathscr{B}\mathscr{C}\mathscr{D}\mathscr{E}\mathscr{F}\mathscr{G}\mathscr{H}\mathscr{I}\mathscr{J}\mathscr{K}\mathscr{L}\mathscr{M}\mathscr{N}\mathscr{O}\mathscr{P}\mathscr{Q}\mathscr{R}\mathscr{S}\mathscr{T}\mathscr{U}\mathscr{V}\mathscr{W}\mathscr{X}\mathscr{Y}\mathscr{Z}$ script alphabet

$\mathfrak{A}\mathfrak{B}\mathfrak{C}\mathfrak{D}\mathfrak{E}\mathfrak{F}\mathfrak{G}\mathfrak{H}\mathfrak{I}\mathfrak{J}\mathfrak{K}\mathfrak{L}\mathfrak{M}\mathfrak{N}\mathfrak{O}\mathfrak{P}\mathfrak{Q}\mathfrak{R}\mathfrak{S}\mathfrak{T}\mathfrak{U}\mathfrak{V}\mathfrak{W}\mathfrak{X}\mathfrak{Y}\mathfrak{Z}$ German alphabet

$\square$ end of the proof of theorem or lemma

AUTHOR INDEX

This index contains all authors cited in the last section of the chapters and whose work is given there or in the Bibliography. Other authors, mostly classical, are named in the General index.

Abatangelo, L. M. 111
Abatangelo, V. 111
Al-Dhahir, M. W. 32
Amici, O. 32
André, J. 90
Artin, E. 268
Artzy, R. 32

Baker, H. F. 228
Barlotti, A. 51, 111, 180
Bartocci, U. 259
Beth, T. 180, 275
Beutelspacher, A. 91, 92, 181
Biggs, N. L. 91, 268
Biscarini, P. 228
Blichfeldt, H. J. 268
Bose, R. C. 51, 91, 178, 228
Bruck, R. H. 91
Bruen, A. A. 91, 92, 178, 179, 181, 259
Buekenhout, F. 32, 51, 92, 178, 180, 181
Burnside, W. 268

Cameron, P. J. 92, 179
Campbell, A. D. 228
Carlitz, L. 228
Carmichael, R. D. 268
Casciaro, B. 32
Casse, L. R. A. 259
Chakravarti, I. M. 178
Coble, A. B. 228
Cole, F. N. 91
Conti, G. 32
Conwell, G. M. 91
Coxeter, H. S. M. 32, 228

de Finis, M. 228
De Groote, R. 275
Deligne, P. 275
Delsarte, P. 179, 180
De Meur, G. 181
Denniston, R. H. F. 91
de Resmini, M. J. 31, 181, 228

Dickson, L. E. 228, 268
Di Comite, C. 111
Dienst, K. J. 181
Dieudonné, J. A. 178, 267, 268
D'Orgeval, B. 259
Drake, D. A. 92
Drinfeld, V. G. 275
Dwork, B. 275

Ebert, G. L. 92
Edge, W. L. 91, 92, 179, 228

Faina, G. 92
Farmer, K. B. 181
Fellegara, G. 51
Fenton, N. E. 259
Fisher, J. C. 91
Frame, J. S. 228
Freeman, J. W. 92
Fuji-hara, R. 91

Gewirtz, A. 179
Glynn, D. G. 91, 92, 180, 259
Goethals, J.-M. 179, 180
Goppa, V. D. 275
Goudarzi, M. G. 91
Gulati, B. R. 259

Hall, M. 179
Hamada, N. 180
Häring, M. 51
Hasse, H. 275
Heise, W. 51
Hering, C. 259
Herzer, A. 92
Higman, D. G. 268
Hill, R. 179
Honda, T. 275
Hubaut, X. 179, 180, 228
Huppert, B. 267

Ihara, Y. 275

Jónsson, W. 179
Jordan, C. 268
Jungnickel, D. 92

Kantor, W. M. 92
Karzel, H. 32
Katz, N. M. 275
Koblitz, N. 275
Korchmáros, G. 32, 92
Kounias, E. G. 259

Lane, R. 179
Leemans, H. 31
Lefèvre (-Percsy), C. 51, 92, 178, 180
Liebler, R. A. 92, 180, 259
Lunardon, G. 92
Lüneburg, H. 51, 91, 259

MacWilliams, F. J. 180, 259
Manin, J. I. 228, 275
Mäurer, H. 181
Mazzocca, F. 180
McDonough, T. P. 91
McEliece, R. J. 180, 181
Mesner, D. M. 91
Metz, R. 92
Mielants, W. 31
Migliori, G. 111
Montague, S. 179
Morikawa, H. 31

Olanda, D. 51, 178
Orr, W. J. 91
Ostrom, T. G. 91
Ott, U. 259

Panella, G. 51, 92
Payne, S. E. 51, 179, 180
Pertichino, M. 111
Peterson, W. W. 180
Pickert, G. 91, 92
Primrose, E. J. F. 32

Qvist, B. 51

Ray-Chaudhuri, D. K. 32
Rigby, J. F. 32
Rosati, L. A. 228
Russo, A. 180

Schaeffer, H.-J. 259
Schoof, R. J. 275
Schröder, E. M. 51
Segre, B. 32, 51, 91, 110, 111, 178, 179, 181, 228, 259
Seidel, J. J. 179
Serre, J.-P. 275
Sherk, F. A. 92
Sherman, B. 180
Shult, E. E. 51, 180
Silverman, R. 92
Sloane, N. J. A. 259
Small, C. 228
Smit Ghinelli, D. 181
Smith, K. J. C. 180
Sörenson, K. 32
Suzuki, M. 51
Swinnerton-Dyer, H. P. F. 228

Tallini, G. 51, 180, 181, 228
Tallini Scafati, M. 180
Tate, J. T. 275
Thas, J. A. 51, 52, 91, 92, 179, 180, 181
Tits, J. 51, 52, 180
Todd, J. A. 31
Tsfasman, M. A. 275

Ughi, E. 275

Vámos, P. 259
van der Waerden, B. L. 268
Vanstone, S. A. 91
Vladut, S. G. 275
Voloch, J. F. 275

Wales, D. 179
Walker, M. 92
Wall, G. E. 181, 268
Wan, Z.-X. 181
Waterhouse, W. G. 275
Weil, A. 275
Weldon, E. J. 180
Wilbrink, H. 92

Yang, B.-F. 181

Zeitler, H. 32
Zink, T. 275

GENERAL INDEX

$\mathbf{A}_8$, isomorphism with $PGL(4, 2)$ 72–3
algebraic variety, number of points on 269–75
apolar linear complexes 6
  , image in $PG(5, q)$ 30
k-arc in $PG(3, q)$ 229–59
  , characterization of $(q+1)$-arc 242–53
  , contact plane of 244–5
  , maximum value of k 242–7
  , special unisecant of 244–9
aregular spread 64–6
  , from ovaloid 65–6
  , from λ-quadric 64–6
  , from twisted cubic 254–5
Aronhold set 223–4, 227
axis
  of osculating developable 232, 240
  of special linear complex 5

binode 222–3
bitangents of a quartic curve 215
  , Aronhold set 223–4, 227
  , Hesse arrangement 224, 227
  , representation in $PG(5, 2)$ 223–8
  , Steiner set 223, 226–7
blocking set, minimum 181
Buekenhout's theorem on Pascalian ovals 32

k-cap 93–111
  , bisecant of 33
  , complete 96–111
  , complete in $PG(3, 2)$ 96
  , complete in $PG(3, 3)$ 97
  , complete in $PG(3, 4)$ 97
  , complete in $PG(3, 8)$ 111
  , complete in $PG(5, 3)$ 135–6, 179–80
  , diophantine equations for 93–4
  , external line of 33
  in ovaloid 99–111
  in $PG(n, q)$ 52
  , maximum size 34
  , maximum size on a quadric 94–5
  , minimum size 95–6
  not in ovaloid 96–111
  , tangent of 33
carriers of a flock 60–2
chord of a twisted cubic 231, 235
class of a cubic surface 222–3
cohomology 32
complementary regulus 23, 62
  , image in $PG(5, q)$ 31
complete
  k-cap 96–111
  k-span 76, 79–86, 92
complex
  conjugate points 231, 257–8
  conjugate planes 231
  plane 231
  point 231, 257–8
  singular cubic surfaces 221–3
cone 38, 138, 140
(quadric) cone 13–17
  , characterization 37–40
  , external point 15
  , external vertex tangent 15
  , generator 13
  , image in $PG(5, q)$ 29
  , incidence tables 16–19
  , internal point 15
  , internal vertex tangent 15
  , nuclear line 14
conic, characterization of 154–5
conic node 222–3
conics
  , conjugate 60
  , flock of 60–2
λ-conjugate 43
conjugate conics 60
contact plane 244–5
coordinate vector
  , dual 4
  of a line 4
cubic curve
  , number of points 271–4
  , plane singular 241–2
  , section of cubic surface 191, 198–200
  , twisted 229–59
cubic surface 182–228
  , binode 222–3
  , class 222–3
  , complex singular 221–3
  , conic node 222–3
  , lines on a non-singular 228
  , node 222–3
  , tangent plane 191

cubic surface with 27 lines
  , diagonal surface 201–4, 208–10, 213
  , double-six 31, 182–91, 194–6, 216–19
  , Eckardt point 192–4, 197–214
  , equation 196
  , equianharmonic surface 200–6, 213
  , involutions on a line 196–7
  , lines 182, 191–7
  , mapping to the plane 213–15
  , number of points 192, 270
  over $GF(4)$ 202–5, 212–13
  over $GF(7)$ 205–6, 212–13
  over $GF(8)$ 206–8, 212–13
  over $GF(9)$ 208–13
  over $GF(11)$ 212–13
  over $GF(13)$ 212–13
  over $GF(16)$ 212–13
  , parabolic points 220
  , Schur quadrics 220
  , triad of trihedral pairs 195
  , trihedral pair 194–5
  , tritangent plane 192
  , vertex 194–5
curve
  , cubic 271–4
  , genus 270–5
  , plane singular cubic 241–2
  , quartic 215, 223–8
  , rational normal 229–30
  , sextic 215
  , twisted cubic 229–59

degenerate linear congruence 7–8
Deligne's theorem 269–75
dependence of lines, linear 7
Desargues configuration 199
developable, osculating 231, 234–5
diagonal surface 201–4, 208–10, 213
  , group of 201
disflection 170–8
double-six 182–91, 194–6
  , condition for existence 187–8
  , existence of self-polar 188–91
  , image in $PG(5,q)$ 31
  , image in $PG(5,2)$ 216–19
dual coordinate vector of a line 4
Dwork's theorem 269–75

Eckardt point 192–4, 197–214
elliptic (linear) congruence 7–8, 53–62, 80–1
  , image in $PG(5,q)$ 30
  , partition of 57–62
elliptic curve 271–5
elliptic polarity 9–10, 19–20, 31
elliptic quadric $\mathscr{E}_{3,q}$ 14, 17–22, 35, 88
  , bijection with $PG(1,q^2)$ 21–2
  , conjugate conics on 60
  , containing k-cap 95, 103–11
  , group 18, 27–8
  , incidence tables 20
  , partition 57–62
  , stereographic projection 40–1
elliptic quadric $\mathscr{E}_{n,q}$ 270
elliptic quadric $\mathscr{E}_{5,2}$ 215–21, 225–7
  , incidence table 220–1
envelope, cubic 231
equianharmonic surface 200–6, 213
  , group 201
even type, set of 168
exceptional integer 273
external vertex tangent 15

fifteen schoolgirls problem 68–75
flock of conics 60–2
  , carriers 60–2

general linear complex 5–6
  and Hermitian surface 122
  and λ-polarity 41–51
  and spreads 49–50
  and twisted cubic 232–3
  , bijection with $PG(3,q)$ 41–3
  , canonical form 6
  , characterization 12–13
  , group 6
  , image in $PG(5,q)$ 30
  , number 6
generalized quadrangle 51, 178–80
generator
  of cone 13
  of hyperbolic quadric 23
  of osculating developable 232
genus 270–5
Grassmannian 28
groups
  , isomorphisms 260–8
  , orders 260–8
  see also isomorphism

Hasse–Weil theorem 269–75
hemisystem 127
  , group 135
  on $\mathscr{U}_{3,9}$ 128–38
heptad of points off $\mathscr{H}_{5,2}$ 71–4
Hermitian arc 86–90, 92, 139–40
  in $PG(2,9)$ 90
  in $PG(2,16)$ 90
Hermitian curve 86–90
  , characterization 92
  , hyperbolic 86–7
  , parabolic 86–9
Hermitian polarity 7–11, 113–16, 121–3, 165–7

, commuting with collineation 121–2
, commuting with polarity 122–3
, line equation 11
Hermitian surface 112–81
, group 115–16
, incidence table 117
, lines 118–38
, not partitioned by lines 127
over $GF(4)$ 167, 169–74, 204, 215
, singular 112–3
, types 112
Hermitian variety 270
Hesse arrangement 224, 227
hexagon, skew 182–4, 188, 198
hyperbolic (linear) congruence 7–8
, axes 7
, image in $PG(5, q)$ 29–30
hyperbolic polarity 9–10, 24–6, 31
hyperbolic quadric $\mathscr{H}_{3,q}$ 13–14, 23–8
, characterization 37–40
, group 24, 27–8
, image in $PG(5, q)$ 31
, incidence tables 25
, permutable 124–6, 136–8
, stereographic projection 40
hyperbolic quadric $\mathscr{H}_{5,q}$ 28–31
hyperbolic quadric $\mathscr{H}_{n,q}$ 270
hyperbolic quadric $\mathscr{H}_{5,2}$ 68–77, 225–7
, incidence table 76–7

imaginary axis 232
imaginary chord 232
incidence matrix of points and lines of $PG(d, 4)$ 177–8
independence of lines, linear 7
index of subregular sequence 54
integer
, exceptional 273
, special 273
internal vertex tangent 15
involutions on a line of a cubic surface 196–7
isomorphism of
$\mathbf{A}_8$ and $PGL(4, 2)$ 72–3
$P\Gamma L(2, q^2)$ and $P\Gamma O_-(4, q)$ 27
$P\gamma L(2, q^2)$ and $PGO_-(4, q)$ 27
$P\gamma SL(2, q^2)$ and $PO_-(4, q)$ 27
$PGSp(6, 2)$ and group of bitangents 225–8
$P\gamma U(4, q)$ and $PGO_-(6, \sqrt{q})$ 121
$PSL(2, q^2)$ and $PS'O_-(4, q)$ 27
$\mathbf{S}_6$ and $PGSp(4, 2)$ 72–3

k-arc, *see* arc
k-cap, *see* cap
Kirkman's schoolgirls problem 68–75
k-span, *see* span
Kummer's configuration 195

λ-conjugate 43
λ-polar 43
λ-polarity 41–51
λ-pole 43
λ-quadric, *see* quadric
length of subregular sequence 54
line
coordinates 4–5
, partition into sublines 57–62
linear complex, *see* general, special
linear complexes, apolar 6, 30
linear congruence 6–8
, degenerate 7–8
, elliptic 7–8, 30, 53–62, 80–1
, hyperbolic 7–8, 29–30
, parabolic 7–8, 30
linear dependence of lines 7, 28
linear set of reguli 62
lines 1–32
, condition that five lines have a transversal 185–7
, Grassmannian 28
, linear dependence of 7–8
, representation in $PG(5, q)$ 29-31, 118–21

maximal arc 140
maximal partial spread 77
maximum number of points
no four in a plane 242–7
no three on a line 34
on an algebraic variety 269–75
model point 271

node 222–3
normal rational curve 229–30
normal set of type $(1, n, q+1)$ 164–7
nuclear line 14
number of points on an
algebraic curve 270–5
algebraic surface 270
algebraic variety 269–75
elliptic curve 271–4
null polarity in $PG(3, q)$ 9–10, 182–4, 188–91
commuting with Hermitian polarity 122–3
determined by linear complex 6
, image in $PG(5, q)$ 30–1
, line equation 10
null polarity in $PG(5, q)$ 224–5
null system 50

odd type
, set of 167–78
, standardized set of 175–8
orbits of
$PGO_+(4, q)$ 24
$PGO_-(4, q)$ 19
$PGU(4, q)$ 116
ordinary (=elliptic or hyperbolic) polarity 7–10
for double-six 188–91
for skew hexagon 182–4
, image in $PG(5, q)$ 31
, line equation 10
orthogonal group 19, 24
, isomorphism with special linear group 27
, transitivity 28
osculating developable 231, 235
, axis 232
, generator 232
, imaginary axis 232, 254–5
, real axis 232
osculating plane 231, 235–6
ovaloid 33–6
, aregular spread from 65–6
, caps in 99–111
, caps not in 96–111
, classification for q odd 35–6
, classification for $q = 4$ 35–6
, classification for $q = 8$ 51
, connection with elliptic quadric 35–6
, partition of 57–8
, polarity determined by 36
, spread from 49–50
, stereographic projection 41
, tangent plane 34
, tangents form linear complex 36
which is not elliptic quadric 45–6
ovoid 52

packing 66–8
, cyclic 91
, image in $PG(5, q)$
of $PG(3, 2)$ 68–76, 91
of $PG(3, 3)$ 91
of $PG(3, 4)$ 91
of $PG(3, 8)$ 91
parabolic (linear) congruence 7–8
, image in $PG(5, q)$ 30
partial spread 77
, maximal 77
partition of
elliptic congruence 57–62
elliptic quadric 57–62
line into sublines 57–62
ovaloid 57–8
plane into conics 57–62
regular spread 57–62
Pascal's theorem 32
pencil 4
pencil of conic sections of quadric
, elliptic 60
, hyperbolic 60
, parabolic 60
pencil of quadrics 53, 105–6, 151–2, 237
, spreads formed from 53–5
permutable quadric 124–6, 136–8
$PG(3, 2)$ 68–76, 79, 91, 96, 203
$PG(3, 3)$ 82–3, 91, 97, 104
$PG(3, 4)$ 81, 85, 91, 97, 160–78, 202–5, 212–13
$PG(3, 5)$ 104–5
$PG(3, 7)$ 205–6, 212–13
$PG(3, 8)$ 91, 111, 206–8, 212–13
$PG(3, 9)$ 128–38, 208–13, 215
$PG(3, 11)$ 212–13
$PG(3, 13)$ 212–13
$PG(3, 16)$ 212–13
$PG(5, 2)$ 68–76, 215–28
$PG(5, 3)$ 135–6
$PGL(4, 2)$, isomorphism with $\mathbf{A}_8$ 72–3
$PGO_+(4, q)$
, orbits 24
, transitivity 28
$PGO_-(4, q)$
, orbits 19
, transitivity 28
$PGO_-(6, \sqrt{q})$, isomorphism with $P\gamma U(4, q)$ 121
$PGSp(4, 2)$, isomorphism with $\mathbf{S}_6$ 72–3
$PGSp(6, 2)$
, isomorphism with group of bitangents 225–8
, subgroups 226
$P\gamma U(4, q)$, isomorphism with $PGO_-(6, \sqrt{q})$ 121
plane
, complex 231
, partition into conics 57–62
point, complex 231
λ-polar 43
polarity 7–11
defined by ovaloid 36
, elliptic 9–10, 19–20, 31
, Hermitian 7–11, 113–16, 121–3, 165–7
, hyperbolic 9–10, 24–6, 31
, line equation 10–11
, null 6, 9–10, 30–1, 122–3, 182–4, 188–91, 224–5
, ordinary 7–10, 31, 182–4, 188–91
, pseudo 9–11, 182–4, 188–91
λ-polarity 41–51
polarity of tactical configuration 41, 50
λ-pole 43

principle of duality 29
projection
of elliptic quadric $\mathscr{E}_3$ 40–1
of hyperbolic quadric $\mathscr{H}_3$ 40
of ovaloid 41
of quadric $\mathscr{P}_4$ 150–6
, stereographic 40–1
projectivity, line equation of 8
pseudo polarity 9–11
for double-six 188–91
for skew hexagon 182–4
, image in $PG(5, q)$ 31
, line equation 10
, self-conjugate line 11
, self-polar line 11

quadratic set 51
λ-quadric 45–51
, aregular spread from 65–6
, group 46–9
, identification with ovaloid 45
, plane section 52
, tangent space 45
quadrics 13–28
, canonical form
, number 26
through a twisted cubic 237–9, 255–8
see also elliptic, hyperbolic
quadrilaterals on $\mathscr{U}_{3,9}$ 128–38
, associated hexad of 131–8
, opposite 128–35
quartic curve 215

rational curve 229–30
rational normal curve 229–30
real axis 231
real chord 232
regular set of type $(1, n, q+1)$ 156–67
regular spread 53–62
, image in $PG(5, q)$ 30
in a packing 66–8
, partition 57–62
regular system of lines 123–38
regulus 4, 8, 23
, complementary 23, 31, 62
, image in $PG(5, q)$ 31
, linear set of reguli 62
, reversing 62
residual of a set containing a plane 146–9
reversed regulus 62
ruled set 37

$\mathbf{S}_6$
, isomorphism with $PGSp(4, 2)$ 72–3
, twelve subgroups $\mathbf{S}_5$ 203–4
Schur quadric 220
section
of type I–VI 139–40
of type VII 142
self-conjugate line 8
of pseudo polarity 11
self-polar line 8
of pseudo polarity 11
set of even type 167–78
set(s) of odd type 167–78
, classification in $PG(2, 4)$ 168–9
, classification in $PG(3, 4)$ 172–4
form a vector space 167–8, 177–8
, standaridized 175–7
set of type $(1, n, q+1)$ 138–81
containing a plane 143–56
, normal 164–7
, plane 139–40
, regular 156–67
, singular 138–40
, singular space of 139–40
set of type
$(0, n)$ 141–2
$(1, 2, q+1)$ 141
$(1, q, q+1)$ 141
$(0, 1, 2, q+1)$ 37–40
$(1, 3, 5)$ in $PG(3, 4)$ 160–78
sextic curve 215
singular cubic surfaces 221–3
singular plane cubic 241–2
skew quadrilateral 38–9
skew quadrilaterals on $\mathscr{U}_{3,9}$ 128–38
, associated hexad of 131–8
, opposite 128–35
k-span 76–86
, bounds 79, 85–6, 92
, complete 76–86
, deficiency 77–9
, diophantine equations 78
from twisted cubic 253–4
, planes residual to 77–80
, points residual to 77–80
special integer 273
special linear complex 5–6, 12–13
, axis 5
, image in $PG(5, q)$ 30
, number 6
spread 53
, aregular 64–6, 254–5
containing reguli but not subregular 91
from pencil of quadrics 53–5
from ovaloid 49–50
, image in $PG(5, q)$ 30
in a linear complex 49–50
in $PG(3, 2)$ 55
in $PG(3, 3)$ 55–7
, maximal partial 77
, partial 77
, regular 30, 53–62, 66–8

spread (*continued*)
, subregular 54–7, 62–4, 68
, subregular sequence 54
with one regulus 254–5
standardized set of odd type 175–7
star 4
Steiner set 223, 226–7
stereographic projection 40–1
subgroups of $GF(q)$ 154
sub-Hermitian surface 118
sublines of $PG(1, q^2)$ 21–2, 57–8, 62
subplane 139–40
subregular sequence 54
subregular spread 54–7, 62–4
in a packing 68
surface, *see* algebraic variety, cubic, Hermitian
system of generators of $\mathscr{H}_3$ 23
system of lines, regular 123–38
$Sz(q)$ 46–9
, transitivity 46–7
, simplicity 49

tangent line to twisted cubic 232
tangent plane to cubic surface 192
tetrahedral system of conics 106–11
triad of trihedral pairs 195
trihedral pair 194–5
tritangent plane 192
twisted cubic 230–59
, arcs contained in 244
, characterization for q odd 242–4
, characterization for $q = 2, 4, 8$ 248
, chord 231, 235, 240
, group 233–4
, imaginary chord 232, 235, 253–5
, linear complex containing tangents 232–3
, locus of tangents 240–1
, null polarity 232–3
, number 234
, orbits of group 234–6
, projection 234, 241–3
, quadrics containing 237–9, 255–8
, real chord 231, 235, 253
, regulus of tangents 232–3, 236
, tangent 232, 235, 240–1, 253–5
, unisecant 236–9

unital 92
see also Hermitian arc

Wedderburn's theorem 23, 32
week 73–5
Weil's theorem 269–75
Woolhouse's solutions to the schoolgirls problem 91

zeta function 269–72